AF745369

Springer Atmospheric Sciences

More information about this series at http://www.springer.com/series/10176

Xiaofan Li • Shouting Gao

Cloud-Resolving Modeling of Convective Processes

Second Edition

Springer

Xiaofan Li
Zhejiang University
Hangzhou, China

Shouting Gao
Chinese Academy of Sciences
Beijing, China

ISSN 2194-5217 ISSN 2194-5225 (electronic)
Springer Atmospheric Sciences
ISBN 978-3-319-26358-8 ISBN 978-3-319-26360-1 (eBook)
DOI 10.1007/978-3-319-26360-1

Library of Congress Control Number: 2015959927

© Springer International Publishing Switzerland 2016
This work is subject to copyright. All rights are reserved by the Publisher, whether the whole or part of the material is concerned, specifically the rights of translation, reprinting, reuse of illustrations, recitation, broadcasting, reproduction on microfilms or in any other physical way, and transmission or information storage and retrieval, electronic adaptation, computer software, or by similar or dissimilar methodology now known or hereafter developed.
The use of general descriptive names, registered names, trademarks, service marks, etc. in this publication does not imply, even in the absence of a specific statement, that such names are exempt from the relevant protective laws and regulations and therefore free for general use.
The publisher, the authors and the editors are safe to assume that the advice and information in this book are believed to be true and accurate at the date of publication. Neither the publisher nor the authors or the editors give a warranty, express or implied, with respect to the material contained herein or for any errors or omissions that may have been made.

Printed on acid-free paper

This Springer imprint is published by Springer Nature
The registered company is Springer International Publishing AG Switzerland

Preface

Clouds and cloud systems and their interactions with larger scales of motion, radiation, and the Earth's surface are extremely important parts of weather and climate systems. Their treatment in weather forecast and climate models is a significant source of errors and uncertainty. As computer power increases, it is beginning to be possible to explicitly resolve cloud and precipitation processes in these models, presenting opportunities for improving precipitation forecasts and larger-scale phenomena such as tropical cyclones which depend critically on cloud and precipitation physics.

This book by Professor Shouting Gao of the Institute of Atmospheric Physics in Beijing and Dr. Xiaofan Li of NOAA's National Environmental Satellite, Data, and Information Service (NESDIS) presents an update and review of results of high-resolution, mostly two-dimensional models of clouds and precipitation and their interactions with larger scales of motion and the Earth's surface. It provides a thorough description of cloud and precipitation physics, including basic governing equations and related physics, such as phase changes of water, radiation, and mixing. Model results are compared with observations from the 1992–1993 Tropical Ocean Global Atmosphere Coupled Ocean–Atmosphere Response Experiment (TOGA COARE) experiment. The importance of the ocean to tropical convective systems is clearly shown here in the numerical results of simulations with their air–sea coupled modeling system. While the focus is on tropical convection, the methodology and applicability can be extended to cloud and precipitation processes elsewhere.

The results described in this well-written book form a solid foundation for future high-resolution model weather forecasts and climate simulations that resolve clouds explicitly in three dimensions – a future that I believe has great promise for the understanding and prediction of weather and climate for the great benefit of society.

President, University Corporation for Atmospheric Research Richard Anthes
Boulder, CO, USA
June 2007

Contents

Introduction

The material in this book is based on our recent research work in the latest 10 years. It is the first book that focuses on cloud-resolving modeling of convective processes. Clouds play an important role in linking atmospheric and hydrological processes and have profound impacts on regional and global climate. The better description of clouds and associated cloud-radiation interaction is a key for successful simulations of cloud processes, which require the physical presence of cloud hydrometeors, prognostic cloud equations, and interactive radiative schemes in models. The cloud-resolving models have been developing for four decades toward providing the better understanding of cloud-scale processes associated with the convective development. With the explosive increase of computational powers, the cloud-resolving models, which were once used to develop cloud schemes for general circulation models, have been directly applied to a global domain with a high horizontal resolution (grid mesh is less than 5 km), whose preliminary results are promising.

This book starts with basic equations and physical packages used in cloud-resolving models and coupled ocean–cloud-resolving atmosphere model. The cloud-resolving model discussed in this book is the two-dimensional version of the Goddard Cumulus Ensemble Model. The model simulations are evaluated with available observations during Tropical Ocean Global Atmosphere Coupled Ocean–Atmosphere Response Experiment (TOGA COARE). The book covers many research aspects related to convective development, cloud, and precipitation. The material in this book has been used in a part of a graduate course at the Graduate School of the Chinese Academy of Sciences, Beijing, China. Therefore, this book can be used as a reference and textbook for graduate students and researchers whose research interests are mesoscale, cloud, and precipitation modeling.

This book is comprised of thirteen chapters. Chapter 1 presents governing equations, parameterization schemes of radiation, cloud microphysics, and subgrid-scale turbulence. Two model frameworks imposed by different large-scale forcing are intensively discussed. Chapter 2 describes thermal and vapor budgets, surface rainfall equation, energetics equation, and partitioning of convective and stratiform rainfall, which are frequently applied to the analysis of cloud-resolving model simulation data. The cloud-resolving model simulation data are evaluated with the high-quality observational data from TOGA COARE in terms of thermodynamic states, apparent heat sink and moisture source, surface radiative and latent heat fluxes, and surface rain rate in Chap. 3. Since most of the research work is from two-dimensional cloud-resolving modeling, the similarities and differences between two- and three-dimensional cloud-resolving modeling are discussed.

The surface rainfall equation is introduced to examine the contributions of water vapor and cloud hydrometeors in surface rainfall processes in Chap. 4. The intensive discussions of surface rainfall processes are conducted in raining stratiform, convective, non-raining stratiform, and clear-sky regions, respectively. In Chap. 5, kinematics and propagation of tropical cloud clusters are discussed. Chapter 6 addresses cloud microphysics and radiation. In particular, the depositional growth of snow from cloud ice as an important sink of cloud ice and precipitation–radiation interaction are examined. The vorticity vectors associated with tropical convection are discussed in Chap. 7. The dominant physical processes that are responsible for the diurnal variations of tropical convection including nocturnal and afternoon rainfall peaks and tropical convective and stratiform rainfall are quantitatively identified with analysis of surface rainfall equation in Chap. 8. The precipitation efficiencies and statistical equivalence of efficiencies defined with water vapor and cloud microphysics budgets are addressed in Chap. 9. The coupled ocean–cloud-resolving atmosphere model is developed to study the small-scale effects of precipitation in ocean mixing processes in Chap. 10. Effects of SST, diurnal variation, and cloud radiation on equilibrium states are discussed in Chap. 11. The microwave radiative transfer model with cloud-resolving model simulation data is applied to radiance simulations in Chap. 12. Finally, the future perspective of cloud-resolving modeling, including simplification of prognostic cloud schemes, and its application to the general circulation model and to the global domain are discussed in Chap. 13.

We would like to thank Dr. Richard A. Anthes, president of the University Corporation for Atmospheric Research, who read the book draft and wrote the preface for this book. Our sincere thanks also go to Dr. Wei-Kuo Tao and Dr. David Adamec at NASA/Goddard Space Flight Center (GSFC), Professor Ming-Dah Chou at National Taiwan University, and Professor Minghua Zhang at the State University of New York, Stony Brook, for providing the two-dimensional Goddard Cumulus Ensemble (GCE) Model, ocean mixed-layer model, radiative transfer code used in GCE model, and TOGA COARE forcing data, respectively. We also thank Dr. Hsiao-Ming Hsu at the National Center for Atmospheric Research for his comments; Drs. Fan Ping, Xiaopeng Cui, Yushu Zhou, and Lingkun Ran at the Institute of Atmospheric Physics, Chinese Academy of Sciences, for efficient and productive research collaborations; and Miss Di Li at the University of Maryland, College Park, for editing this book. Xiaofan Li would like to thank Dr. William K.-M. Lau, chief of the Laboratory for Atmospheres at NASA/GSFC, and Professor Chung-Hsiung Sui at the National Central University for their support, encouragement, and academic guidance when he worked at GSFC as a contract research scientist during 1994–2001, Drs. Fuzhong Weng and Quanhua Liu at NOAA/NESDIS/Center for Satellite Applications and Research for providing microwave radiative transfer model, and

Dr. Jian-Jian Wang at the Goddard Center for Earth Science and Technology, University of Maryland, Baltimore County, for research collaboration.

We are also indebted to Dr. Robert Doe and Ms. Nina Bennink at Springer for their editorial efforts. This work was supported by the national Key Basic Research and Development Projects of China under grants No.2004CB418301 and G1998040907 and Olympic Meteorological Service Projects under grants No. 2001BA904B09 and KACX1-02.

Beijing, China
Camp Springs, MD, USA
Fall 2007

Shouting Gao
Xiaofan Li

Introduction for the Second Edition

This book is written based on our research work in the last 17 years. In addition to the thirteen chapters from the first edition, the second edition has five more chapters which discuss the structures of precipitation systems, thermal impacts of doubled carbon dioxide on precipitation, precipitation predictability, and modeling of the depositional growth of ice crystal. The material in this book has been used in a part of a graduate course at the School of Earth Sciences, Zhejiang University, and Graduate School, Chinese Academy of Sciences, Beijing, China.

The second edition of this book is comprised of eighteen chapters. Chapter 1 discusses the equations, physical packages, and basic parameters of the cloud-resolving model. Chapter 2 describes methodologies for analyzing cloud-resolving model simulation data. Chapter 3 presents validation for cloud-resolving modeling of TOGA COARE using available observational data. Chapter 4 introduces surface precipitation budget and discusses the important contribution from cloud hydrometeors to precipitation. Chapters 5 and 6 examine structures of precipitation systems through the analyses of cloud content and budgets. New convective–stratiform rainfall separation schemes are developed based on the cloud content and cloud and precipitation budgets, respectively. The new scheme based on surface precipitation budget can be used to define maximum rainfall as water vapor, hydrometeor convergence, and atmospheric drying or, alternatively, as a 100 % precipitation efficiency.

Chapter 7 analyzes kinematics, propagation, and the merging processes of tropical cloud clusters. Chapter 8 discusses cloud radiative and microphysical processes associated with the production of precipitation. Chapter 9 examines the thermal aspects of the impacts of doubled carbon dioxide on precipitation based on a tropical rainfall event during the winter solstice and a pre-summer torrential event during the summer solstice. In particular, the effects of doubled carbon dioxide on rainfall responses to radiative processes are investigated. Chapter 10 introduces convective, moist, and dynamic vorticity vectors as important diagnostic tools for studying convective development and their linkages to thermal, water vapor, and cloud microphysics. Chapter 11 discusses the diurnal variations of tropical rainfall including nocturnal and afternoon rainfall peaks. Chapter 12 defines precipitation efficiency in the budgets of surface precipitation and cloud and rain microphysics and analyzes the relationship between these precipitation efficiencies and their connections to physical factors and processes.

Chapter 13 examines the small-scale effects of precipitation on the ocean mixing layer using simulation data from a coupled ocean–cloud-resolving atmosphere model. Chapter 14 discusses the effects of sea surface temperature, diurnal variation, and cloud radiative processes on equilibrium states. Chapter 15 addresses the application of cloud-resolving model simulation data to remote sensing with the aid of the microwave radiative transfer model. Chapter 16 studies precipitation predictability and its dependence on temporal and spatial-scale and large-scale forcing. Chapter 17 discusses improving the modeling of depositional growth of ice crystal using a modified parameterization scheme. Finally, Chap. 18 briefly discusses the simplification of prognostic microphysical schemes, coupled large-scale model with cloud-resolving model, and global cloud-resolving model.

We thank Drs. Yushu Zhou and Lingkun Ran at the Institute of Atmospheric Physics, Chinese Academy of Sciences, and Prof. Xinyong Shen at Nanjing University of Information Science and Technology for efficient and productive research collaborations, and Miss Di Li for editing this book. Xiaofan Li also thanks Zhejiang University for its support and his graduate students Wei Huang, Tingting Li, Lingyun Lou, Bin Wang, Jin Xin, Huiyan Xu, and Haoran Zhu for their assistance.

We are also indebted to Ms. Juliana Pitanguy and Ms. Rajeswari Balachandran at Springer for their editorial efforts. This work was supported by the National Natural Science Foundation of China (41475039), the Key Research Program of the Chinese Academy of Sciences (KZZD-EW-05-01), and National Key Basic Research and Development Project of China (2012CB417201 and 2015CB953601).

Hangzhou, China Xiaofan Li
Beijing, China Shouting Gao
Spring 2016

Chapter 1
Model and Physics

Abstract In this chapter, the governing equations for cloud-resolving model are derived. The model is imposed by large-scale forcing due to a small model domain. The differences in large-scale forcing between vertical velocity and advection are compared. The large-scale forcing includes vertical velocity and zonal wind in this book. The model has prognostic equations for potential temperature, specific humidity, mixing ratios of five cloud species (cloud water, raindrop, cloud ice, snow, and graupel), and perturbation zonal wind and vertical velocity. The model physical package including cloud microphysical and radiation parameterization schemes and turbulence closure is presented. The model boundary conditions and basic parameters for model setup are discussed. The governing equations for coupled ocean-cloud resolving atmosphere model are presented. The ocean model includes mixed-layer model and ocean circulation model.

Keywords Cloud-resolving model • Governing equations • Cloud microphysics • Radiation • Turbulence closure • Parameterization • Coupled ocean-cloud resolving atmosphere model

Cloud-resolving models differ from general circulation and mesoscale models in two ways. First, cloud-resolving models cannot simulate large-scale circulations due to small model domains whereas general circulation and mesoscale models can simulate large-scale circulations. A large-scale forcing is imposed in the cloud-resolving model. Second, cloud-resolving models with fine spatial resolutions use prognostic cloud microphysical parameterization to simulate cloud and precipitation processes. In contrast, general circulation and mesoscale models use diagnostic cumulus parameterization and/or prognostic cloud microphysical parameterization due to coarse spatial resolutions. Many cloud-resolving models have been developed to study convective responses to the large-scale forcing (Table 1.1). In this chapter, the cloud-resolving model and coupled ocean-cloud resolving atmosphere model will be described in a two-dimensional (2D) framework in terms of governing equations, large-scale forcing, parameterization schemes of cloud microphysics, radiation, subgrid-scale turbulence closure, ocean mixing closure, and boundary conditions.

© Springer International Publishing Switzerland 2016

X. Li, S. Gao, *Cloud-Resolving Modeling of Convective Processes*,
Springer Atmospheric Sciences, DOI 10.1007/978-3-319-26360-1_1

Table 1.1 A summary of cloud-resolving models

Model	Dynamic core	Cloud scheme	Turbulence closures	Radiation scheme
Clark (1977, 1979), Clark and Hall (1991), and Clark et al. (1996)	Anelastic and hydrostatic approximations	Water cloud (Kessler 1969), ice cloud (Koenig and Murray 1976)	Smagorinsky (1963), Lilly (1962)	Kiehl et al. (1994)
Soong and Ogura (1980) and Tao and Simpson (1993)	Anelastic approximation and non-hydrostatic form	Water and ice schemes (Lin et al. 1983; Rutledge and Hobbs 1983, 1984; Tao et al. 1989; Krueger et al. 1995)	Klemp and Wilhelmson (1978)	Solar (Chou et al. 1998) and IR infrared (Chou et al. 1991; Chou and Suarez 1994) schemes
Lipps and Hemler (1986, 1988, 1991)	Anelastic and hydrostatic approximations	Water cloud (Kessler 1969)	Klemp and Wilhelmson (1978)	Newtonian damping
Redelsperger and Sommeria (1986)	Deardorff (1972); Sommeria (1976); Redelsperger and Sommeria (1981a, b, 1982)	Water cloud (Kessler 1969)	Subgrid-scale turbulence closure (Redelsperger and Sommeria (1981a, b, 1982)	–
Hu and He (1988), Hu et al. (1998), and Zou (1991)	Nonhydrostatic primitive dynamic core (Zou 1991)	Prognostic equations for hydrometeors and number concentrations (Hu and He 1988)	Second-moment turbulence closure	–
Kruegger (1988)	Anelastic and hydrostatic approximations	Three-phase bulk scheme (Lin et al. 1983; Lord et al. 1984; Krueger et al. 1995)	Third-moment turbulence closure (Krueger 1988)	Harshvardhan et al. (1987)
Model	Dynamic core	Cloud scheme	Turbulence closures	Radiation scheme
Nakajima and Matsuno (1988)	Yamasaki (1975) and Ogura and Phillps (1962)	Water cloud (Kessler 1969)	Velocity deformation and static stability dependent	Horizontal homogeneous cooling
Tripoli (1992)	Nonhydrostatic core with enstrophy conservation	Flatau et al. (1989)	A turbulent kinetic closure for eddy diffusion	Chen and Cotton (1983)
Xu (1992) and Xu and Huang (1994)	Nonhydrostatic primitive dynamic core	Xu and Duan (1999)	Second-moment turbulence closure	–
Ferrier (1994) and Ferrier et al. (1995)	Anelastic approximation and non-hydrostatic form	Double-moment multiple-phase four-class ice scheme (Ferrier 1994)	Klemp and Wilhelmson (1978)	Chou (1984, 1986) and Chou and Kouvaris (1991)
Tompkins and Craig (1998)	Anelastic quasi-Boussineq approximation	Three-phase ice scheme (Swann 1994; Brown and Swann 1997); the water vapor and cloud water are combined into the total-water mixing ratio	First-order subgrid-scale turbulence closure (Shutts and Gray 1994)	Two-stream plane-parallel approximation (Edwards and Slingo 1996; Petch 1998)
Satoh et al. (2005)	Non-hydrostatic formulations in the icosahedral grids covering the global domain	Grabowski (1998)	Mellor Yamada (1974) level-2 closure	Nakajima et al. (2000)

1.1 Governing Equations

The cloud-resolving model was originally developed by Soong and Ogura (1980) and Soong and Tao (1980). This model was significantly improved by Tao and Simpson (1993) at the National Aeronautics and Space Administration (NASA) Goddard Space Flight Center (GSFC) and was modified by Sui et al. (1994, 1998). The model was named the Goddard cumulus ensemble (GCE) model. The cloud-resolving model used in this book is the 2D modified version of GCE model. The non-hydrostatic governing equations with anelastic approximation can be expressed by

$$\nabla \cdot \mathbf{V} + \frac{1}{\overline{\rho}}\frac{\partial}{\partial z}\overline{\rho}w = 0, \tag{1.1a}$$

$$\frac{\partial A}{\partial t} = -\nabla \cdot \mathbf{V}A - \frac{1}{\overline{\rho}}\frac{\partial}{\partial z}\overline{\rho}wA + S_A + D_A, \tag{1.1b}$$

$$\frac{\partial B}{\partial t} = -\nabla \cdot \mathbf{V}B - \frac{1}{\overline{\rho}}\frac{\partial}{\partial z}\overline{\rho}\left(w - w_{TB}\right)B + S_B + D_B. \tag{1.1c}$$

Here, **V** and w are horizontal wind vector and vertical velocity, respectively; $\overline{\rho}$ is a mean density, which is a function of height only; $A = (\theta, q_v, \mathbf{V}, w)$; $B = (q_c, q_r, q_i, q_s, q_g)$, θ and q_v are potential temperature and specific humidity, respectively; q_c, q_r, q_i, q_s, q_g are the mixing ratios of cloud water (small cloud droplets), raindrops, cloud ice (small ice crystals), snow (density 0.1 g cm^{-3}), and graupel (density 0.4 g cm^{-3}), respectively; w_{TB} is a terminal velocity that is zero for cloud water and ice; S_A is a source and sink in momentum, temperature, and moisture equations such as pressure gradient force, buoyancy force, condensational heating, and radiative heating; the radiation parameterization schemes will be addressed in Sect. 1.3; S_B is a cloud source and sink that is determined by microphysical processes, which will be discussed in Sect. 1.2; D_A and D_B are dissipation terms related to sub-grid scale turbulence closure, which will be elucidated in Sect. 1.4.

For model calculations by Li et al. (1999), it is convenient to partition (A, **V**) into area means ($\bar{A}$, $\overline{\mathbf{V}}$) and deviations (A', $\mathbf{V}'$), i.e.,

$$A = \overline{A} + A', \tag{1.2a}$$

$$\mathbf{V} = \overline{\mathbf{V}} + \mathbf{V}'. \tag{1.2b}$$

Applying (1.2) to (1.1b) leads to

$$\frac{\partial A}{\partial t} = -\nabla \cdot \left(\mathbf{V}'A' + \mathbf{V}'\overline{A} + \overline{\mathbf{V}}A'\right) - \frac{1}{\overline{\rho}}\frac{\partial}{\partial z}\overline{\rho}\left(w'A' + w'\overline{A} + \overline{w}A'\right)$$
$$+ S_A + D_A - \overline{\mathbf{V}} \cdot \nabla\overline{A} - \overline{w}\frac{\partial}{\partial z}\overline{A}. \tag{1.3}$$

Here, the area-mean continuity equation (1.1a) is used in the derivation of (1.3).

Taking an area mean over (1.3), we get the equation for $\overline{A}$,

$$\frac{\partial \overline{A}}{\partial t} = -\nabla \cdot \overline{\mathbf{V}'A'} - \frac{1}{\overline{\rho}}\frac{\partial}{\partial z}\overline{\rho}\overline{w'A'} + \overline{S}_A + \overline{D}_A - \overline{\mathbf{V}} \cdot \nabla \overline{A} - \overline{w}\frac{\partial}{\partial z}\overline{A}. \tag{1.4a}$$

Perturbation equation for A' is obtained by subtracting (1.4a) from (1.3):

$$\begin{aligned}\frac{\partial A'}{\partial t} =& -\nabla \cdot \left(\mathbf{V}'\overline{A} + \overline{\mathbf{V}}A'\right) - \frac{1}{\overline{\rho}}\frac{\partial}{\partial z}\overline{\rho}\left(w'\overline{A} + \overline{w}A'\right) - \nabla \cdot \left(\mathbf{V}'A' - \overline{\mathbf{V}'A'}\right)\\ &- \frac{1}{\overline{\rho}}\frac{\partial}{\partial z}\overline{\rho}\left(w'A' - \overline{w'A'}\right) + S_A - \overline{S}_A + D_A - \overline{D}_A.\end{aligned} \tag{1.4b}$$

Environment has an important impact on convective development. When convection develops, associated momentum, heat, and moisture transport upward through convective activity, which in turn modify environment significantly. Environment and convection interact in a nonlinear way (e.g., Chao 1962). Due to a small domain in the cloud-resolving model (e.g., 768 km in a 2D framework), the large-scale circulation cannot be simulated. Thus, the large-scale forcing needs to be imposed in the cloud-resolving model. Soong and Ogura (1980) were the first to develop ways to impose the observed large-scale variables into the cloud-resolving model to examine the convective response to the imposed large-scale forcing. The major forcing is vertical velocity and associated vertical advections. Thus, there are two ways to impose the large-scale forcing into the cloud model. The horizontally uniform and vertically varying vertical velocity can be imposed, as first introduced by Soong and Ogura (1980), or the horizontally uniform total advection of the heat and moisture can be imposed (e.g., Wu et al. 1998). Li et al. (1999) discussed the two model setups intensively.

For the model with the imposed vertical velocity ($\overline{w}^o$), horizontal wind ($\overline{\mathbf{V}}^o$) and the horizontal advection ($-\overline{\mathbf{V}}^o \cdot \nabla\overline{A}^o$) are also imposed. These forcing data denoted by subscript "o" are calculated from the observational data [e.g., Tropical Ocean Global Atmosphere Coupled Ocean–atmosphere Response Experiment (TOGA COARE) in Li et al. (1999) and Global Atmosphere Research Program (GARP) Atlantic Tropical Experiment (GATE) in Grabowski et al. (1996)]. With the assumption that $-\mathbf{V}' \cdot \nabla\overline{A}^o = 0$, the model equations for potential temperature and specific humidity can be expressed by

$$\begin{aligned}\frac{\partial A}{\partial t} =& -\nabla \cdot \mathbf{V}'A' - \overline{\mathbf{V}}^o \cdot \nabla A' - \frac{1}{\overline{\rho}}\frac{\partial}{\partial z}\overline{\rho}w'A' - \overline{w}^o\frac{\partial}{\partial z}A' - w'\frac{\partial}{\partial z}\overline{A}\\ &+ S_A + D_A - \overline{\mathbf{V}}^o \cdot \nabla\overline{A}^o - \overline{w}^o\frac{\partial}{\partial z}\overline{A}.\end{aligned} \tag{1.5}$$

The model consists of (1.1a) and (1.1c), perturbation momentum equation (1.4b), and equations for potential temperature and specific humidity (1.5).

For the model with imposed horizontally uniform total advection of the heat and moisture, horizontal wind is also imposed. With the assumption that

$-\nabla \cdot \overline{\mathbf{V}}^o A' - \partial \overline{\rho} A' \overline{w}^o / \overline{\rho} \partial z = 0$, the model equations for potential temperature and specific humidity can be written by

$$\frac{\partial A}{\partial t} = -\nabla \cdot \mathbf{V}' A - \frac{1}{\overline{\rho}} \frac{\partial}{\partial z} \overline{\rho} w' A + S_A + D_A - \overline{\mathbf{V}}^o \cdot \nabla \overline{A}^o - \overline{w}^o \frac{\partial}{\partial z} \overline{A}^o. \tag{1.6}$$

The model is comprised of (1.1a) and (1.1c), perturbation momentum equation (1.4b), and equations for potential temperature and specific humidity (1.6). Li et al. (1999) found that the terms omitted in (1.5) and (1.6) do not have any impact on the model simulations. The comparison between simulations by the two model setups will be discussed with the TOGA COARE data in chapter three.

The governing equations in the 2D cloud-resolving model can be expressed as follows:

$$\frac{\partial u'}{\partial x} + \frac{1}{\overline{\rho}} \frac{\partial (\overline{\rho} w')}{\partial z} = 0, \tag{1.7a}$$

$$\begin{aligned} \frac{\partial u'}{\partial t} = & -\frac{\partial}{\partial x}\left(2u'\overline{u}^o + u'u'\right) - \frac{1}{\overline{\rho}} \frac{\partial}{\partial z} \overline{\rho}\left(w'\overline{u}^o + \overline{w}^o u' + w'u' - \overline{w'u'}\right) \\ & -c_p \frac{\partial\left(\overline{\theta}\pi'\right)}{\partial x} + D_u - \overline{D}_u, \end{aligned} \tag{1.7b}$$

$$\begin{aligned} \frac{\partial w'}{\partial t} = & -\frac{\partial}{\partial x}\left(u'\overline{w}^o + \overline{u}^o w' + u'w'\right) - \frac{1}{\overline{\rho}} \frac{\partial}{\partial z} \overline{\rho}\left(2w'\overline{w}^o + w'w' - \overline{w'w'}\right) \\ & -c_p \frac{\partial\left(\overline{\theta}\pi'\right)}{\partial z} + g\left(\frac{\theta'}{\theta_o} + 0.61 q_v' - q_l'\right) + D_w - \overline{D}_w, \end{aligned} \tag{1.7c}$$

$$\begin{aligned} \frac{\partial \theta}{\partial t} = & -\frac{\partial (u'\theta')}{\partial x} - \overline{u}^o \frac{\partial \theta'}{\partial x} - \frac{1}{\overline{\rho}} \frac{\partial}{\partial z}\left(\overline{\rho} w'\theta'\right) - \overline{w}^o \frac{\partial \theta'}{\partial z} - w' \frac{\partial \overline{\theta}}{\partial z} \\ & + \frac{Q_{cn}}{\pi c_p} + \frac{Q_R}{\pi c_p} - \overline{u}^o \frac{\partial \overline{\theta}^o}{\partial x} - \overline{w}^o \frac{\partial \overline{\theta}}{\partial z} + D_\theta, \end{aligned} \tag{1.7d}$$

$$\begin{aligned} \frac{\partial q_v}{\partial t} = & -\frac{\partial (u'q_v')}{\partial x} - \overline{u}^o \frac{\partial q_v'}{\partial x} - \overline{w}^o \frac{\partial q_v'}{\partial z} - w' \frac{\partial \overline{q}_v}{\partial z} - \frac{1}{\overline{\rho}} \frac{\partial}{\partial z} \overline{\rho} w' q_v' \\ & - S_{qv} - \overline{u}^o \frac{\partial \overline{q}_v{}^o}{\partial x} - \overline{w}^o \frac{\partial \overline{q}_v}{\partial z} + D_{qv}, \end{aligned} \tag{1.7e}$$

$$\frac{\partial q_c}{\partial t} = -\frac{\partial (u q_c)}{\partial x} - \frac{1}{\overline{\rho}} \frac{\partial (\overline{\rho} w q_c)}{\partial z} + S_{qc} + D_{qc}, \tag{1.7f}$$

$$\frac{\partial q_r}{\partial t} = -\frac{\partial (u q_r)}{\partial x} - \frac{1}{\overline{\rho}} \frac{\partial}{\partial z} \overline{\rho}\left(w - w_{Tr}\right) q_r + S_{qr} + D_{qr}, \tag{1.7g}$$

$$\frac{\partial q_i}{\partial t} = -\frac{\partial (u q_i)}{\partial x} - \frac{1}{\overline{\rho}} \frac{\partial (\overline{\rho} w q_i)}{\partial z} + S_{qi} + D_{qi}, \tag{1.7h}$$

$$\frac{\partial q_s}{\partial t} = -\frac{\partial (uq_s)}{\partial x} - \frac{1}{\overline{\rho}}\frac{\partial}{\partial z}\overline{\rho}(w - w_{Ts})\,q_s + S_{qs} + D_{qs}, \tag{1.7i}$$

$$\frac{\partial q_g}{\partial t} = -\frac{\partial \left(uq_g\right)}{\partial x} - \frac{1}{\overline{\rho}}\frac{\partial}{\partial z}\overline{\rho}\left(w - w_{Tg}\right)q_g + S_{qg} + D_{qg}, \tag{1.7j}$$

where

$$\begin{aligned} Q_{cn} = {} & L_v\left(P_{CND} - P_{REVP}\right) + L_s\Big\{P_{DEP} + (1-\delta_1)\,P_{SDEP}\,(T < T_o) \\ & + (1-\delta_1)\,P_{GDEP}\,(T < T_o) - P_{MLTS}\,(T > T_o) - P_{MLTG}\,(T > T_o)\Big\} \\ & + L_f\Big\{P_{SACW}\,(T < T_o) + P_{SFW}\,(T < T_o) + P_{GACW}\,(T < T_o) \\ & + P_{IACR}\,(T < T_o) + P_{GACR}\,(T < T_o) + P_{SACR}\,(T < T_o) \\ & + P_{GFR}\,(T < T_o) - P_{RACS}\,(T > T_o) - P_{SMLT}\,(T > T_o) \\ & - P_{GMLT}\,(T > T_o) + P_{IHOM}\,(T < T_{oo}) - P_{IMLT}\,(T > T_o) \\ & + P_{IDW}\,(T_{oo} < T < T_o)\Big\}, \end{aligned} \tag{1.8a}$$

$$\begin{aligned} S_{qv} = {} & P_{CND} - P_{REVP} + P_{DEP} + (1-\delta_1)\,P_{SDEP}\,(T < T_o) \\ & + (1-\delta_1)\,P_{GDEP}\,(T < T_o) - P_{MLTS}\,(T > T_o) \\ & - P_{MLTG}\,(T > T_o)\,, \end{aligned} \tag{1.8b}$$

$$\begin{aligned} S_{qc} = {} & -P_{SACW} - P_{RAUT} - P_{RACW} - P_{SFW}\,(T < T_o) - P_{GACW} \\ & + P_{CND} - P_{IHOM}\,(T < T_{oo}) + P_{IMLT}\,(T > T_o) \\ & - P_{IDW}\,(T_{oo} < T < T_o)\,, \end{aligned} \tag{1.8c}$$

$$\begin{aligned} S_{qr} = {} & P_{SACW}\,(T > T_o) + P_{RAUT} + P_{RACW} + P_{GACW}\,(T > T_o) \\ & - P_{REVP} + P_{RACS}\,(T > T_o) - P_{IACR}\,(T < T_o) - P_{GACR}\,(T < T_o) \\ & - P_{SACR}\,(T < T_o) - P_{GFR}\,(T < T_o) + P_{SMLT}\,(T > T_o) \\ & + P_{GMLT}\,(T > T_o)\,, \end{aligned} \tag{1.8d}$$

$$\begin{aligned} S_{qi} = {} & -P_{SAUT}\,(T < T_o) - P_{SACI}\,(T < T_o) - P_{RACI}\,(T < T_o) \\ & - P_{SFI}\,(T < T_o) - P_{GACI}\,(T < T_o) + P_{IHOM}\,(T < T_{oo}) \\ & - P_{IMLT}\,(T > T_o) + P_{IDW}\,(T_{oo} < T < T_o) + P_{DEP}, \end{aligned} \tag{1.8e}$$

$$\begin{aligned} S_{qs} = {} & P_{SAUT}\,(T < T_o) + P_{SACI}\,(T < T_o) + \delta_4 P_{SACW}\,(T < T_o) \\ & + P_{SFW}\,(T < T_o) + P_{SFI}\,(T < T_o) + \delta_3 P_{RACI}\,(T < T_o) \\ & - P_{RACS}\,(T > T_o) - P_{GACS} - P_{SMLT}\,(T > T_o) \\ & - (1-\delta_2)\,P_{RACS}\,(T < T_o) + \delta_2 P_{SACR}\,(T < T_o) \\ & + (1-\delta_1)\,P_{SDEP}\,(T < T_o) - P_{MLTS}\,(T > T_o) \\ & + \delta_3 P_{IACR}\,(T < T_o) - (1-\delta_4)\,P_{WACS}\,(T < T_o)\,, \end{aligned} \tag{1.8f}$$

$$
\begin{aligned}
S_{qg} = &+ (1 - \delta_3) P_{RACI} (T < T_o) + P_{GACI} (T < T_o) \\
&+ P_{GACW} (T < T_o) + (1 - \delta_4) P_{SACW} (T < T_o) + P_{GACS} \\
&+ (1 - \delta_3) P_{IACR} (T < T_o) + P_{GACR} (T < T_o) \\
&+ (1 - \delta_2) P_{RACS} (T < T_o) + P_{GFR} (T < T_o) \\
&+ (1 - \delta_4) P_{WACS} (T < T_o) - P_{GMLT} (T > T_o) \\
&+ (1 - \delta_1) P_{GDEP} (T < T_o) - P_{MLTG} (T > T_o) \\
&+ (1 - \delta_2) P_{SACR} (T < T_o),
\end{aligned}
\tag{1.8g}
$$

and

$$\delta_1 = 1, \text{ only if } q_c + q_i > 10^{-8}\text{gg}^{-1}, T < T_o, \tag{1.8h}$$

$$\delta_2 = 1, \text{ only if } q_s + q_r < 10^{-4}\text{gg}^{-1}, T < T_o, \tag{1.8i}$$

$$\delta_3 = 1, \text{ only if } q_r > 10^{-4}\text{gg}^{-1}, T < T_o, \tag{1.8j}$$

$$\delta_4 = 1, \text{ only if } q_s \le 10^{-4}\text{gg}^{-1}, q_c > 5 \times 10^{-4}\text{gg}^{-1}, T < T_o. \tag{1.8k}$$

Here, u is zonal wind; θ_o is initial potential temperature; $q_l = q_c + q_r + q_i + q_s + q_g$; T is air temperature, and $T_o = 0$ °C, $T_{oo} = -35$ °C; $\pi = (p/p_o)^{\kappa}$, $\kappa = R/c_p$; R is the gas constant; c_p is the specific heat of dry air at constant pressure p, and $p_o = 1000$ mb; L_v, L_s, and L_f are latent heat of vaporization, sublimation, and fusion at 0 °C, respectively, and $L_s = L_v + L_f$; Q_R is the radiative heating rate due to convergence of the net flux of solar and infrared radiative fluxes, which will be discussed in Sect. 1.3. The cloud microphysical terms in (1.8b, 1.8c, 1.8d, 1.8e, 1.8f, and 1.8g) are defined in Table 1.2, which will be discussed in Sect. 1.2. The cloud microphysical processes are also summarized in Fig. 1.1.

When the model is integrated over the ocean, a time-invariant or temporally varied horizontally uniform sea surface temperature (*SST*) is imposed in both model setups. Li et al. (2000) developed a coupled ocean-cloud resolving atmosphere model to study the impacts of precipitation and associated salinity stratification in ocean mixed-layer temperature and salinity at small spatial scales. An embedded mixed layer-ocean circulation model was originally developed by Adamec et al. (1981). The mixed-layer equations in the 2D framework are

$$\frac{\partial h_m}{\partial t} = -\frac{\partial u_m h_m}{\partial x} + W_e, \tag{1.9a}$$

$$\frac{\partial u_m}{\partial t} = -\frac{W_e}{h_m} H(W_e)(u_m - u_e) - \frac{\tau_o}{\rho_r h_m}, \tag{1.9b}$$

$$\frac{\partial T_m}{\partial t} = -\frac{W_e}{h_m} H(W_e)(T_m - T_e) - \frac{Q_o + I(0) - I(h_m)}{\rho_r c_w h_m}, \tag{1.9c}$$

Table 1.2 List of microphysical processes and their parameterization schemes

Notation	Description	Scheme
P_{MLTG}	Growth of vapor by evaporation of liquid from graupel surface	RH84
P_{MLTS}	Growth of vapor by evaporation of melting snow	RH83
P_{REVP}	Growth of vapor by evaporation of raindrops	RH83
P_{IMLT}	Growth of cloud water by melting of cloud ice	RH83
P_{CND}	Growth of cloud water by condensation of supersaturated vapor	TSM
P_{GMLT}	Growth of raindrops by melting of graupel	RH84
P_{SMLT}	Growth of raindrops by melting of snow	RH83
P_{RACI}	Growth of raindrops by the accretion of cloud ice	RH84
P_{RACW}	Growth of raindrops by the collection of cloud water	RH83
P_{RACS}	Growth of raindrops by the accretion of snow	RH84
P_{RAUT}	Growth of raindrops by the autoconversion of cloud water	LFO
P_{IDW}	Depositional growth of cloud ice from cloud water	KFLC
P_{IACR}	Growth of cloud ice by the accretion of rain	RH84
P_{IHOM}	Growth of cloud ice by the homogeneous freezing of cloud water	
P_{DEP}	Growth of cloud ice by the deposition of supersaturated vapor	TSM
P_{SAUT}	Growth of snow by the conversion of cloud ice	RH83
P_{SACI}	Growth of snow by the collection of cloud ice	RH83
P_{SACW}	Growth of snow by the accretion of cloud water	RH83
P_{SFW}	Growth of snow by the deposition of cloud water	KFLC
P_{SFI}	Depositional growth of snow from cloud ice	KFLC
P_{SACR}	Growth of snow by the accretion of raindrops	LFO
P_{SDEP}	Growth of snow by the deposition of vapor	RH83
P_{GACI}	Growth of graupel by the collection of cloud ice	RH84
P_{GACR}	Growth of graupel by the accretion of raindrops	RH84
P_{GACS}	Growth of graupel by the accretion of snow	RH84
P_{GACW}	Growth of graupel by the accretion of cloud water	RH84
P_{WACS}	Growth of graupel by the riming of snow	RH84
P_{GDEP}	Growth of graupel by the deposition of vapor	RH84
P_{GFR}	Growth of graupel by the freezing of raindrops	LFO

The schemes are Rutledge and Hobbs (1983; 1984; RH83, RH84), Lin et al. (1983, LFO), Tao et al. (1989, TSM), and Krueger et al. (1995, KFLC)

$$\frac{\partial S_m}{\partial t} = -\frac{W_e}{h_m} H\left(W_e\right)\left(S_m - S_e\right) - \frac{S_m\left(P_s - E_s\right)}{\rho_r c_w h_m}. \tag{1.9d}$$

Here, T_m, S_m, u_m, and h_m are ocean mixed-layer temperature, salinity, zonal current, and depth respectively; T_e and S_e are temperature and salinity of the level just beneath the mixed layer, respectively; H is the heavyside step function in which $H = 1$ as $W_e > 0$ while $H = 0$ as $W_e < 0$; ρ_r is a constant reference sea water density; c_w is the heat capacity of water; $I = I_o\left[re^{-\gamma_1 z} + (1 - r)\, e^{-\gamma_2 z}\right]$, I_o is solar radiation at the ocean surface, γ_1, γ_2 are attenuation parameters for solar radiation penetration,

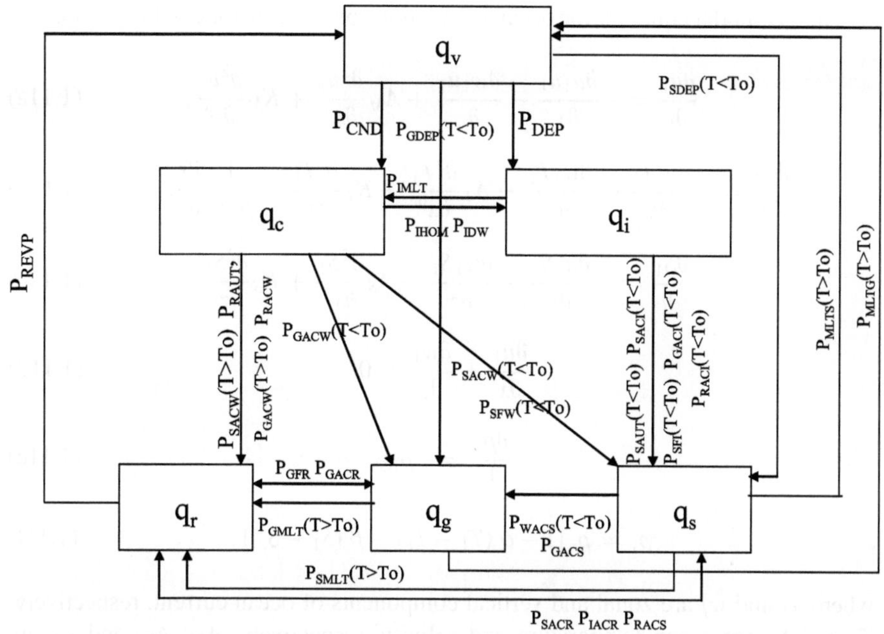

Fig. 1.1 A flow chart of the cloud microphysical schemes

and z is positive downward with $z = 0$ being the ocean surface; Q_o is the sum of longwave radiation, sensible and latent heat at the ocean surface; P_s and E_s denote the rates of precipitation and evaporation at the ocean surface, respectively; W_e is the entrainment velocity at the mixed-layer base, which can be obtained by calculating Kraus-Turner's equation that was originally derived by Niller and Kraus (1977), modified by Sui et al. (1997), and is similar to Garpar (1988),

$$\begin{aligned} &W_e H(W_e)\, h_m g \left[\alpha (T_m - T_e) - \beta (S_m - S_e)\right] \\ &= 2 m_s u_*{}^3 - \frac{h_m}{2}\left[(1 + m_b) B_o + (1 - m_b)\,|B_o|\right]. \end{aligned} \tag{1.10}$$

$$\begin{aligned} B_o = \frac{g\alpha}{\rho_r c_w} &\{Q_o + [1 + r e^{-\gamma_1 h_m} + (1 - r)\, e^{-\gamma_2 h_m} - \frac{2r}{\gamma_1 h_m}\left(1 - e^{-\gamma_1 h_m}\right) \\ &- \frac{2(1 - r)}{\gamma_2 h_m}\left(1 - e^{-\gamma_2 h_m}\right)] I_o\} + \frac{g}{\rho_r} \beta S_m (P_s - E_s), \end{aligned} \tag{1.10a}$$

where u_* is a surface friction velocity; α and β describe the logarithmic expansion of ocean water density ρ_r as a function of temperature and salinity, respectively; g is the gravitational acceleration; m_s and m_b are turbulent mixing factors due to wind stirring and convection, respectively.

The 2D model equation for ocean circulations on the equator can be written as

$$\frac{\partial u_1}{\partial t} = -\frac{\partial u_1 u_1}{\partial x} - \frac{\partial w_1 u_1}{\partial z} + A_M \frac{\partial^2 u_1}{\partial x^2} + K_M \frac{\partial^2 u_1}{\partial z^2}, \tag{1.11a}$$

$$\frac{\partial T_1}{\partial t} = -\frac{\partial u_1 T_1}{\partial x} - \frac{\partial w_1 T_1}{\partial z} + A_T \frac{\partial^2 T_1}{\partial x^2} + K_T \frac{\partial^2 T_1}{\partial z^2} - \frac{1}{\rho_r c_w} \frac{\partial I}{\partial z}, \tag{1.11b}$$

$$\frac{\partial S_1}{\partial t} = -\frac{\partial u_1 S_1}{\partial x} - \frac{\partial w_1 S_1}{\partial z} + A_S \frac{\partial^2 S_1}{\partial x^2} + K_S \frac{\partial^2 S_1}{\partial z^2}, \tag{1.11c}$$

$$\frac{\partial u_1}{\partial x} + \frac{\partial w_1}{\partial z} = 0, \tag{1.11d}$$

$$\frac{\partial p_1}{\partial z} = -\rho_1 g, \tag{1.11e}$$

$$\rho_1 = \rho_r \left[1 - \alpha (T_1 - T_r) - \beta (S_1 - S_r)\right], \tag{1.11f}$$

where u_1 and w_1 are zonal and vertical components of ocean current, respectively; T_1 and S_1 are ocean temperature and salinity, respectively; A_M, A_T, and A_S are horizontal momentum, heat and salinity diffusivity coefficients, respectively; K_M, K_T, and K_S are vertical momentum, heat and salinity diffusivity coefficients, respectively; T_r and S_r are the reference temperature and salinity, respectively; the mixed-layer model and the ocean circulation model communicate with each other through the embedding technique proposed by Adamec et al. (1981). The model also includes a convective adjustment scheme that ensures the static stability of the upper ocean.

1.2 Cloud Microphysical Parameterization Schemes

The formulations of cloud microphysical parameterization schemes are documented in this section. Table 1.2 shows the list of microphysical processes and their parameterization schemes. The schemes are Rutledge and Hobbs (1983, 1984), Lin et al. (1983), Tao et al. (1989), and Krueger et al. (1995).

$$P_{MLTG} = \frac{2\pi N_{OG} (S - 1)}{\rho (A' + B')} \left[\frac{0.78}{\lambda_G{}^2} + 0.31 \left(\frac{\overline{a}\rho}{\mu} \right)^{\frac{1}{2}} \left(\frac{\rho_o}{\rho} \right)^{\frac{1}{4}} \frac{\Gamma\left(\frac{\overline{b}+5}{2}\right)}{\lambda_G{}^{\frac{\overline{b}+5}{2}}} \right], \tag{1.12}$$

where N_{OG} ($=4 \times 10^6$ m^{-4}) is the intercept value in graupel size distribution; S ($=q_w/q_{ws}$), where q_{ws} is the saturated mixing ratio with respect to water; $\overline{a} \left(= 19.3 \text{ m}^{1-\overline{b}}\text{s}^{-1}\right)$ is the constant in fall-speed relation for graupel; $\overline{b}$ ($= 0.37$)

is the fall-speed exponent for graupel; $A' = L_v/K_aT\ (L_vM_w/RT - 1)$; $B' = RT/\chi M_w e_{ws}$ (Pruppacher and Klett 1978); K_a (=2.43 × 10^{-2} J m^{-1} s^{-1} K^{-1}) is the thermal conductivity coefficient of air; M_w (=18.016) is the molecular weight of water; $\chi\,(= 2.26 \times 10^{-5}\text{m}^2\text{s}^{-1})$ is the diffusivity coefficient of water vapor in air; $R\,(= 8.314 \times 10^3\text{Jkmol}^{-1}\text{K}^{-1})$ is the universal gas constant; e_{ws} is the saturation vapor pressure for water; $\lambda_G\left[= (\pi\rho_G N_{0G}/\rho q_g)^{\frac{1}{4}}\right]$ is the slope of graupel size distribution; $\rho_G\,(= 400\ \text{kgm}^{-3})$ is the density of graupel; $\mu\,(= 1.718 \times 10^{-5}\text{kgm}^{-1}\text{s}^{-1})$ is the dynamic viscosity of air; Γ is the Gamma function.

$$P_{MLTS} = \frac{4N_{0S}(S-1)}{\rho(A'+B')}\left[\frac{0.65}{{\lambda_S}^2} + 0.44\left(\frac{a''\rho}{\mu}\right)^2\left(\frac{\rho_o}{\rho}\right)^{\frac{1}{4}}\frac{\Gamma\left(\frac{b+5}{2}\right)}{{\lambda_S}^{\frac{b+5}{2}}}\right], \tag{1.13}$$

where $N_{0S}\,(= 4 \times 10^6\text{m}^{-4})$ is the intercept value in snowflake size distribution; $a''\,(= 1.139\ \text{m}^{1-b}\text{s}^{-1})$ is the constant in fall-speed relation for snow; $b\,(= 0.11)$ is the fall-speed exponent for snow; $\lambda_S\left[= (\pi\rho_S N_{0S}/\rho q_s)^{\frac{1}{4}}\right]$ is the slope of snow size distribution; $\rho_S\,(= 100\ \text{kgm}^{-3})$ is the density of snow.

$$P_{REVP} = \frac{2\pi N_{0R}(S-1)}{\rho(A'+B')}\left[\frac{0.78}{{\lambda_R}^2} + 0.31\left(\frac{a'\rho}{\mu}\right)^{\frac{1}{2}}\left(\frac{\rho_o}{\rho}\right)^{\frac{1}{4}}\frac{\Gamma(3)}{{\lambda_R}^3}\right], \tag{1.14}$$

where $N_{0R}\,(= 8 \times 10^6\text{m}^{-4})$ is the intercept value in raindrop size distribution; $a'\,(= 3 \times 10^3\text{s}^{-1})$ is the constant in linear fall-speed relation for raindrops; $\lambda_R\left[= (\pi\rho_L N_{0R}/\rho q_r)^{\frac{1}{4}}\right]$ is the slope of raindrop size distribution; $\rho_L\,(= 10^3\ \text{kgm}^{-3})$ is the density of raindrops.

$$P_{IMLT} = \frac{q_i}{\Delta t}, \tag{1.15}$$

where Δt is the time step.

$$P_{CND} = \frac{1}{\Delta t}\frac{T - T_{00}}{T_0 - T_{00}}\frac{q_v - (q_{qws} + q_{is})}{1 + \left(\frac{A_1 q_c q_{ws} + A_2 q_i q_{is}}{q_c + q_i}\right)\left(\frac{L_v(T-T_{00}) + L_s(T_0 - T)}{c_p(T_0 - T_{00})}\right)}, \tag{1.16}$$

where q_{is} is the saturation mixing ratio with respect to ice; $A_1 = 237.3B_1/(T\text{-}35.86)^2$; $A_2 = 265.5B_2/(T\text{-}7.66)^2$; $B_1 = 17.2693882$; $B_2 = 21.8745584$.

$$P_{GMLT} = -\frac{2\pi}{\rho L_f}K_a(T - T_o)N_{0G}\left[\frac{0.78}{{\lambda_G}^2} + 0.31\left(\frac{\bar{a}\rho}{\mu}\right)^{\frac{1}{2}}\left(\frac{\rho_o}{\rho}\right)^{\frac{1}{4}}\frac{\Gamma\left(\frac{\bar{b}+5}{2}\right)}{\lambda_G^{\frac{\bar{b}+5}{2}}}\right], \tag{1.17}$$

$$P_{SMLT} = -\frac{2\pi}{\rho L_f} K_a (T - T_o) N_{OS} \left[\frac{0.65}{\lambda_S{}^2} + 0.44 \left(\frac{a''\rho}{\mu} \right)^{\frac{1}{2}} \left(\frac{\rho_o}{\rho} \right)^{\frac{1}{4}} \frac{\Gamma\left(\frac{b+5}{2}\right)}{\lambda_S{}^{\frac{b+5}{2}}} \right], \quad (1.18)$$

$$P_{RACI} = \frac{\pi}{4} q_i E_{RI} N_{OR} \left(\frac{\rho_o}{\rho} \right)^{\frac{1}{2}} \left[\frac{a_o \Gamma(3)}{\lambda_R{}^3} + \frac{a_1 \Gamma(4)}{\lambda_R{}^4} + \frac{a_2 \Gamma(5)}{\lambda_R{}^5} + \frac{a_3 \Gamma(6)}{\lambda_R{}^6} \right], \quad (1.19)$$

where E_{RI} (=1) is the rain/cloud ice collection efficiency coefficient; $a_0 = 0.267\ \mathrm{m\,s^{-1}}, a_1 = 5.15 \times 10^3\ \mathrm{s^{-1}}, a_2 = -1.0225 \times 10^6\ \mathrm{m^{-1}\,s^{-1}}, a_3 = 7.55 \times 10^7\ \mathrm{m^{-2}}$ $\mathrm{s^{-1}}$, which are the coefficients in polynomial fall-speed relation for raindrops.

$$P_{RACW} = \frac{\pi}{4} q_c E_{RC} N_{OR} \left(\frac{\rho_o}{\rho} \right)^{\frac{1}{2}} \left[\frac{a_o \Gamma(3)}{\lambda_R{}^3} + \frac{a_1 \Gamma(4)}{\lambda_R{}^4} + \frac{a_2 \Gamma(5)}{\lambda_R{}^5} + \frac{a_3 \Gamma(6)}{\lambda_R{}^6} \right], \quad (1.20)$$

where E_{RC} (=1) is the rain/cloud water collection efficiency coefficient.

$$P_{RACS} = E_{SR} \pi^2 \frac{\rho_s}{\rho} \left| \overline{V}_R - \overline{V}_S \right| N_{OR} N_{OS} \left(\frac{\rho_o}{\rho} \right)^{\frac{1}{2}} \left[\frac{5}{\lambda_S{}^6 \lambda_R} + \frac{2}{\lambda_S{}^5 \lambda_R{}^2} + \frac{1}{2\lambda_S{}^4 \lambda_R{}^3} \right], \quad (1.21)$$

where E_{SR} (=1) is the snow/rain collection efficiency coefficient;

$$\overline{V}_R = \left(-0.267 + \frac{206}{\lambda_R} - \frac{2.045 \times 10^3}{\lambda_R{}^2} + \frac{9.06 \times 10^3}{\lambda_R{}^3} \right) \left(\frac{\rho_o}{\rho} \right)^{\frac{1}{2}}, \quad (1.21a)$$

$$\overline{V}_S = a'' \frac{\Gamma(4+b)}{6\lambda_S{}^b} \left(\frac{\rho_o}{\rho} \right)^{\frac{1}{2}}, \quad (1.21b)$$

where $\overline{V}_R$ and $\overline{V}_S$ are the mass-weighted fall-speed for rain and snow, respectively.

$$P_{RAUT} = \alpha (q_c - q_o), \quad (1.22)$$

where α (=$10^{-3}\ \mathrm{s^{-1}}$) is the rate coefficient for auto-conversion; q_o(= $1.25 \times 10^{-3} \mathrm{gg^{-1}}$) is the mixing ratio threshold.

$$P_{IDW} = \frac{n_0 e^{\frac{1}{2}|T-T_o|}}{10^3 \rho} b_1 \left(\frac{\rho q_i}{n_0 e^{\frac{1}{2}|T-T_o|}} \right)^{b_2}, \quad (1.23)$$

where $n_0 = 10^{-8} \mathrm{m^{-3}}$; b_1 and b_2 are the positive temperature-dependent coefficients tabulated by Koenig (1971).

$$P_{IACR} = n_{ci}E_{RI}\frac{\pi^2}{24}\frac{\rho_L}{\rho}N_{OR}\left(\frac{\rho_o}{\rho}\right)^{\frac{1}{2}}\left[\frac{a_o\Gamma(6)}{\lambda_R{}^6} + \frac{a_1\Gamma(7)}{\lambda_R{}^7} + \frac{a_2\Gamma(8)}{\lambda_R{}^8} + \frac{a_3\Gamma(9)}{\lambda_R{}^9}\right], \tag{1.24}$$

where $n_{ci}\left(= \rho q_i/\overline{M}_i\right)$ is the number concentration of cloud ice crystals; $\overline{M}_i\left(= 6 \times 10^{-12}\text{kg}\right)$ is the average mass of a cloud ice particle.

$$P_{IHOM} = \frac{q_c}{\Delta t}, \tag{1.25}$$

$$P_{DEP} = \frac{1}{\Delta t}\frac{T_0 - T}{T_0 - T_{00}}\frac{q_v - \left(q_{qws} + q_{is}\right)}{1 + \left(\frac{A_1 q_c q_{ws} + A_2 q_i q_{is}}{q_c + q_i}\right)\left(\frac{L_v(T - T_{00}) + L_s(T_0 - T)}{c_p(T_0 - T_{00})}\right)}, \tag{1.26}$$

$$P_{SAUT} = \frac{\rho q_i - M_{\max} n_0 e^{0.6(T - T_o)}}{\rho\Delta t}, \tag{1.27}$$

where $M_{\max}\left(= 9.4 \times 10^{-10}\text{kg}\right)$ is the maximum allowed crystal mass.

$$P_{SACI} = \frac{\pi a'' q_i E_{SI} N_{0S}}{4}\left(\frac{\rho_o}{\rho}\right)^{\frac{1}{2}}\frac{\Gamma\left(b + 3\right)}{\lambda_S{}^{b+3}}, \tag{1.28}$$

where E_{SI} (=0.1) is the snow/cloud ice collection efficiency coefficient.

$$P_{SACW} = \frac{\pi a'' q_c E_{SC} N_{0S}}{4}\left(\frac{\rho_o}{\rho}\right)^{\frac{1}{2}}\frac{\Gamma\left(b + 3\right)}{\lambda_S{}^{b+3}}, \tag{1.29}$$

where E_{SC} (=1) is the snow/cloud water collection efficiency coefficient.

$$P_{SFW} = \frac{q_i\Delta t}{m_{ir}\Delta t_1}\left(b_1 m_{ir}{}^{b_2} + \pi\rho q_c r^2 U_{ir}\right), \tag{1.30}$$

where $\Delta t_1\left[= \left(m_{ir}{}^{1-b_2} - m_{iro}{}^{1-b_2}\right)/b_1\left(1 - b_2\right)\right]$ is the timescale needed for a crystal to grow from radius r_o to radius r; $m_{ir}\left(= 3.84 \times 10^{-9}\text{kg}\right)$ and U_{ir} (=1 m s^{-1}) are the mass and terminal velocity of an ice crystal r (=10^2 μm); $m_{ir}\left(= 2.46 \times 10^{-10}\text{kg}\right)$ is the mass of an ice crystal r_o (=40 μm).

$$P_{SFI} = \frac{q_i}{\Delta t_1}, \tag{1.31}$$

$$P_{SACR} = E_{SR}\pi^2\frac{\rho_L}{\rho}\left|\overline{V}_S - \overline{V}_R\right| N_{OR}N_{OS}\left(\frac{\rho_o}{\rho}\right)^{\frac{1}{2}}\left[\frac{5}{\lambda_R{}^6\lambda_S} + \frac{2}{\lambda_R{}^5\lambda_S{}^2} + \frac{1}{2\lambda_R{}^4\lambda_S{}^3}\right], \tag{1.32}$$

$$P_{SDEP} = \frac{4N_{OS}(S_i - 1)}{\rho(A'' + B'')}\left[\frac{0.65}{\lambda_S{}^2} + 0.44\left(\frac{a''\rho}{\mu}\right)^{\frac{1}{2}}\left(\frac{\rho_o}{\rho}\right)^{\frac{1}{4}}\frac{\Gamma\left(\frac{b+5}{2}\right)}{\lambda_S{}^{\frac{b+5}{2}}}\right], \tag{1.33}$$

where $S_i\,(= q_v/q_{is})$; $A'' = L_v/K_aT\,(L_sM_w/RT - 1)$; $B'' = RT/\chi M_w e_{is}$; e_{is} is the saturation vapor pressure for ice.

$$P_{GACI} = \frac{\pi\bar{a}q_iE_{GI}N_{OG}}{4}\left(\frac{\rho_o}{\rho}\right)^{\frac{1}{2}}\frac{\Gamma\left(\bar{b}+3\right)}{\lambda_G{}^{\bar{b}+3}}, \tag{1.34}$$

where E_{GI} (=0.1) is the graupel/cloud ice collection efficiency coefficient.

$$P_{GACR} = E_{GR}\pi^2\frac{\rho_L}{\rho}\left|\overline{V}_G - \overline{V}_R\right|N_{OR}N_{OG}\left(\frac{\rho_o}{\rho}\right)^{\frac{1}{2}}\left[\frac{5}{\lambda_R{}^6\lambda_G} + \frac{2}{\lambda_R{}^5\lambda_G{}^2} + \frac{1}{2\lambda_R{}^4\lambda_G{}^3}\right], \tag{1.35}$$

where E_{GR} (=1) is the graupel/rain collection efficiency coefficient;

$$\overline{V}_G = \bar{a}\frac{\Gamma\left(4+\bar{b}\right)}{6\lambda_G{}^{\bar{b}}}\left(\frac{\rho_o}{\rho}\right)^{\frac{1}{2}}, \tag{1.35a}$$

where $\overline{V}_G$ is the mass-weighted fall-speed for graupel.

$$P_{GACS} = E_{GS}\pi^2\frac{\rho_s}{\rho}\left|\overline{V}_G - \overline{V}_S\right|N_{OS}N_{OG}\left(\frac{\rho_o}{\rho}\right)^{\frac{1}{2}}\left[\frac{5}{\lambda_S{}^6\lambda_G} + \frac{2}{\lambda_S{}^5\lambda_G{}^2} + \frac{1}{2\lambda_S{}^4\lambda_G{}^3}\right], \tag{1.36}$$

where E_{GS} (=0.1) is the graupel/snow collection efficiency coefficient.

$$P_{GACW} = \frac{\pi\bar{a}q_cE_{GC}N_{OG}}{4}\left(\frac{\rho_o}{\rho}\right)^{\frac{1}{2}}\frac{\Gamma\left(\bar{b}+3\right)}{\lambda_G{}^{\bar{b}+3}}, \tag{1.37}$$

where E_{GC} (=1) is the graupel/cloud water collection efficiency coefficient.

$$P_{WACS} = \bar{n}_cE_{SC}a''\frac{\pi^2}{24}\frac{\rho_s}{\rho}\left(\frac{\rho_o}{\rho}\right)^{\frac{1}{2}}\frac{\Gamma\left(b+6\right)}{\lambda_S{}^{b+6}}, \tag{1.38}$$

where $\bar{n}_c\left(= \rho q_c/\overline{M}_c\right)$ is the number concentration of cloud water droplets; $\overline{M}_c\left(= 4\times10^{-12}\right)$ is the average mass of cloud water droplets.

$$P_{GDEP} = \frac{2\pi N_{OG}\left(\overline{S}_i - 1\right)}{\rho\left(A'' + B''\right)}\left[\frac{0.78}{\lambda_S{}^2} + 0.31\left(\frac{\overline{a}\rho}{\mu}\right)^{\frac{1}{2}}\left(\frac{\rho_o}{\rho}\right)^{\frac{1}{4}}\frac{\Gamma\left(\frac{\overline{b}+5}{2}\right)}{\lambda_S{}^{\frac{\overline{b}+5}{2}}}\right], \quad (1.39)$$

$$P_{GFR} = 20\pi^2 B_3 N_{OR}\frac{\rho_L}{\rho}\frac{e^{A_3(T_o-T)-1}}{\lambda_R{}^7}, \quad (1.40)$$

where A_3 (=0.66 K^{-1}) is the constant in Bigg freezing; B_3 (=10^2 m^{-3} s^{-1}) is the constant in the raindrop freezing equation.

1.3 Radiation Parameterization Schemes

The parameterization of solar and infrared (IR) radiation is based on the model developed by Chou (1992), Chou et al. (1991), and Chou and Suarez (1994). The effective radii (μm) for cloud water (r_{ec}), raindrops (r_{er}), cloud ice and snow (r_{eis}), and graupel (r_{eg}) can be calculated by

$$r_{ec} = 15, \quad (1.41a)$$

$$r_{er} = \frac{3}{\lambda_R}, \quad (1.41b)$$

$$r_{eis} = 125 + 5 \times (T - 243.16) \quad \text{for } 243.16 \le T \le 223.16, \quad (1.41c)$$

$$r_{eg} = \frac{3}{\lambda_G}. \quad (1.41d)$$

The optical thicknesses for cloud water (τ_c), raindrops (τ_r), cloud ice and snow (τ_{isVIS} in the visible region and τ_{isIR} in the IR window), and graupel (τ_g) can be expressed as

$$\tau_c = \frac{\rho q_c \Delta z}{r_{ec}}, \quad (1.42a)$$

$$\tau_r = \frac{\rho q_r \Delta z}{r_{er}}, \quad (1.42b)$$

$$\tau_{isVIS} = \rho\left(q_i + q_s\right)\Delta z\left(-0.006656 + \frac{3.686}{r_{eis}}\right), \quad (1.42c)$$

$$\tau_{isIR} = \rho\left(q_i + q_s\right)\Delta z\left(-0.0115 + \frac{4.11}{r_{eis}} + \frac{17.3}{{r_{eis}}^2}\right), \tag{1.42d}$$

$$\tau_g = \frac{\rho q_g \Delta z}{r_{eg}}. \tag{1.42e}$$

Here, Δz is the thickness of a vertical layer.

The optical thicknesses for the calculations of solar (τ_{SW}) and IR (τ_{IR}) radiation parameterization schemes are

$$\tau_{SW} = \frac{3}{2}\left(\tau_c + \tau_r\right) + \tau_{isVIS} + \tau_g, \tag{1.43a}$$

$$\tau_{IR} = \frac{3}{4}\left(\tau_c + \tau_r\right) + \tau_{isIR} + \tau_g. \tag{1.43b}$$

Chou and Suarez (1994) broke down the spectrum into eight bands to compute IR fluxes due to water vapor, carbon dioxide, and ozone (see their Table 1.1). Water vapor line absorption covers the entire IR spectrum (0–3000 cm^{-1}), whereas water vapor continuum absorption is included in the spectral region of 540–1380 cm^{-1}. The absorptions due to carbon dioxide and ozone are included in the spectral ranges of 540–800 and 980–1100 cm^{-1}, respectively. All calculations of IR fluxes are based on k-distribution method with linear pressure scaling, table look-up with temperature and pressure scaling, and one-parameter temperature scaling.

Chou et al. (1998) divided the solar spectrum into an ultraviolet (UV) and visible region (wavelengths are smaller than 0.7 μm) and an IR region (wavelengths are between 0.7 and 10 μm) in the calculations of the solar radiation parameterization. The absorptions due to ozone and aerosols and scattering due to gases, clouds, and aerosols are included in the UV and visible region. The UV and visible region is further divided into eight spectral intervals. Thus, the fluxes in the photosynthetically active radiation (wavelengths are between 0.4 and 0.7 μm), UV-A (wavelengths are between 0.328 and 0.4 μm), UV-B (wavelengths are between 0.28 and 0.328 μm), and UV-C (wavelengths are between 0.18 and 0.28 μm) can be separately computed. The absorptions due to water vapor, oxygen, carbon dioxide, clouds, and aerosols are included in the IR region. The schemes developed by Chou and Lee (1996) were used to calculate the absorptions due to water vapor and ozone whereas those developed by Chou (1990) were used to compute the absorptions due to oxygen and carbon dioxide. Chou et al. (1998) calculated the solar radiation fluxes using varied cloud single-scattering properties including the extinction coefficient, single scattering albedo, and asymmetric factor.

1.4 Sub-grid Scale Turbulence Closure

The sub-grid scale turbulence closure developed by Klemp and Wilhelmson (1978) is used in the cloud-resolving model.

$$D_\eta = \frac{\partial}{\partial x}\left(K_m \frac{\partial \eta}{\partial x}\right) + \frac{\partial}{\partial z}\left(K_m \frac{\partial \eta}{\partial z}\right), \tag{1.44a}$$

$$D_\chi = \frac{\partial}{\partial x}\left(K_h \frac{\partial \chi}{\partial x}\right) + \frac{\partial}{\partial z}\left(K_h \frac{\partial \chi}{\partial z}\right), \tag{1.44b}$$

where $\eta = u, w$; $\chi = \theta,\ q_v, q_c,\ q_r,\ q_i,\ q_s,\ q_g$; K_m is the momentum eddy mixing coefficient, which can be determined by

$$K_m = c_m E^{\frac{1}{2}} l. \tag{1.45}$$

Here $c_m = 0.2$; $l = \Delta x \Delta z$, Δx and Δz are the zonal and vertical grid intervals, l is the appropriate length scale (Lilly 1967); E is the subgrid-scale kinetic energy, which can be calculated using the following prognostic equation

$$\begin{aligned}\frac{dE}{dt} = {} & g\overline{\overline{w''\left(\frac{\theta''}{\overline{\theta}} + 0.61 q_v'' - q_l''\right)}} - \overline{\overline{u''u''}}\frac{\partial u}{\partial x} - \overline{\overline{u''w''}}\left(\frac{\partial u}{\partial z} + \frac{\partial w}{\partial x}\right) \\ & - \overline{\overline{w''w''}}\frac{\partial w}{\partial z} + \frac{\partial}{\partial x}\left(K_m \frac{\partial E}{\partial x}\right) + \frac{\partial}{\partial z}\left(K_m \frac{\partial E}{\partial z}\right) - \left(\frac{c_e}{l}\right) E^{\frac{3}{2}},\end{aligned} \tag{1.46}$$

where the double overbar denotes an average over each grid point in the cloud model and the double primed variables are the deviation from the grid values; $c_e = 0.7$; and

$$\overline{\overline{u''u''}} = -2K_m \frac{\partial u}{\partial x} + \frac{2}{3}E, \tag{1.46a}$$

$$\overline{\overline{u''w''}} = -K_m \left(\frac{\partial u}{\partial z} + \frac{\partial w}{\partial x}\right), \tag{1.46b}$$

$$\overline{\overline{w''w''}} = -2K_m \frac{\partial w}{\partial z} + \frac{2}{3}E. \tag{1.46c}$$

(1.45) is similar to the equation used by Deardorff (1975), Mellor and Yamada (1974), and Schemm and Lipps (1976).

In a saturated area,

$$\overline{\overline{w''\left(\frac{\theta''}{\overline{\theta}} + 0.61 q_v'' - q_l''\right)}} = K_h \left(-\frac{1}{\overline{\theta}} \frac{1 + \frac{1.61\varepsilon L q_v}{R_d T}}{\frac{\varepsilon L^2 q_v}{c_p R_d T^2}} \frac{\partial \theta_e}{\partial z} + \frac{\partial q_l}{\partial z}\right) \tag{1.47a}$$

and $\varepsilon = 0.622$.

In an unsaturated area,

$$\overline{\overline{w''\left(\frac{\theta''}{\overline{\theta}} + 0.61q_v'' - q_l''\right)}} = -K_h\left(-\frac{1}{\overline{\theta}}\frac{\partial\theta}{\partial z} + 0.61\frac{\partial q_v}{\partial z}\right). \quad (1.47b)$$

Here, $K_h = 3K_m$ (Deardorff 1975).

1.5 Boundary Conditions and Basic Parameters

Lateral boundaries are cyclic. A free-slip condition is applied to the upper boundary, which allows no advection or diffusion of any quantity. The sensible heat flux from ocean surface is expressed by

$$\left(\overline{\overline{w''\theta''}}\right)_{z=0} = -\frac{C_D u_s (T_m - T_s)}{\pi}. \quad (1.48)$$

Here u_s and T_s are atmospheric zonal wind and temperature at ocean surface; C_D is the drag coefficient, which is given by Roll (1965) as

$$C_D = (1.1 + 0.04u_s) \times 10^{-3}, \quad (1.48a)$$

where u_s is larger than 4 m s^{-1}. The moisture flux from ocean surface is written as

$$\left(\overline{\overline{w''q_v''}}\right)_{z=0} = -C_D u_s (q_{ms} - q_s), \quad (1.49)$$

where q_{ms} is the saturation mixing ratio at *SST* or mixed-layer temperature; q_s is the atmospheric mixing ratio at ocean surface.

A Rayleigh relaxation (*RL*) term is imposed above 15 km with the form

$$RL = \frac{z - 15}{1000}. \quad (1.50)$$

Here z is in units of km.

Following Anthes (1970) and Wilhelmson and Chen (1982), a stretched grid is used in vertical coordinate. The grid is obtained by mapping the physical coordinate (z) onto the map coordinate (z'):

$$z = (c_1 + c_2 z')\, z', \quad (1.51)$$

where c_1 and c_2 are 10 and 0.01 cm^{-1}. A constant grid interval $\Delta z'$ (34 cm) is used in $z_I' = I\Delta z'$, I is the number of vertical level. The mapping factor M is defined by

$$M \equiv \frac{\partial z'}{\partial z} = \frac{1}{c_1 + 2c_2 z'}. \quad (1.52)$$

The vertical derivative of any function F in z coordinate ($\partial F/\partial z$) can be transformed into z' coordinate by

$$\frac{\partial F}{\partial z} = M\frac{\partial F}{\partial z'}. \tag{1.53}$$

The vertical grid resolution ranges from about 200 m near the surface to about 1 km near 100 mb. The top model level is 42 hPa. The time step is 12 s. The typical horizontal domain is 768 km with a horizontal grid resolution of 1.5 km.

References

Adamec D, Elsberry RL, Garwood RW, Haney RL (1981) An embedded mixed-layer-ocean circulation model. Dyn Atmos Oceans 6(2):69–96

Anthes RA (1970) Numerical experiments with a two-dimensional horizontal variable grid. Mon Weather Rev 98:810–822

Brown P, Swann H (1997) Evaluation of key microphysical parameters in three-dimensional cloud model simulations using aircraft and multiparameter radar data. Quart J Roy Meteor Soc 123:2245–2275

Chao J (1962) On the nonlinear impacts of stratification and wind on development of small-scale disturbances. Acta Meteor Sinica 32:164–176

Chen C, Cotton WR (1983) A one-dimensional simulation of the stratocumulus-capped mixed layer. Bound Layer Meteor 25:289–321

Chou MD (1984) Broadband water vapor transmission functions for atmospheric IR flux computation. J Atmos Sci 41:1775–1778

Chou MD (1986) Atmospheric solar heating rate in the water vapor bands. J Clim Appl Meteorol 25:1532–1542

Chou MD (1990) Parameterization for the absorption of solar radiation by O_2 and CO_2 with application to climate studies. J Climate 3:209–217

Chou MD (1992) A solar radiation model for use in climate studies. J Atmos Sci 49:762–772

Chou MD, Kouvaris L (1991) Calculations of transmission functions in the IR CO_2 and O_3 band. J Geophys Res 96:9003–9012

Chou MD, Lee KT (1996) Parameterization for the absorption of solar radiation by water vapor and ozone. J Atmos Sci 53:1203–1208

Chou MD, Suarez MJ (1994) An efficient thermal infrared radiation parameterization for use in General Circulation Model. NASA Tech Memo 104606:3

Chou MD, Kratz DP, Ridgway W (1991) IR radiation parameterization in numerical climate studies. J Climate 4:424–437

Chou MD, Suarez MJ, Ho CH, Yan MMH, Lee KT (1998) Parameterizations for cloud overlapping and shortwave single-scattering properties for use in general circulation and cloud ensemble models. J Climate 11:202–214

Clark TL (1977) A small-scale dynamic model using a terrain following coordinate transformation. J Comput Phys 24:186–215

Clark TL (1979) Numerical simulations with a three-dimension cloud model: lateral boundary condition experiments and multicellular severe storm simulations. J Atmos Sci 36:2191–2215

Clark TL, Hall WD (1991) Multi-domain simulations of the time dependent Navier–Stokes equations: benchmark error analysis of some nesting procedures. J Comput Phys 92:456–481

Clark TL, Hall WD, Coen JL (1996) Source code documentation for the Clark-Hall cloud-scale model: code version G3CH01. NCAR Tech. Note NCAR/TN-426 + STR, 137 pp [Avaiable from NCAR Information Service, P. O. Box 3000, Boulder, CO 80307.]

Deardorff JW (1972) Numerical investigation of neutral and unstable planetary boundary layers. J Atmos Sci 29:91–115

Deardorff JW (1975) The development of boundary-layer turbulence models for use in studying the severe storm environment. In: Proceedings of the SESAME opening meeting, Boulder, NOAA-ERL, pp 251–264

Edwards JM, Slingo A (1996) Studies with a flexible new radiation code. I: choosing a configuration for a large-scale model. Quart J Roy Meteor Soc 122:689–719

Ferrier BS (1994) A double-moment multiple-phase four-class bulk ice scheme. Part I: description. J Atmos Sci 51:249–280

Ferrier BS, Tao WK, Simpson J (1995) A double-moment multiple-phase four-class bulk ice scheme. Part II: simulations of convective storms in different large-scale environments and comparisons with other bulk parameterizations. J Atmos Sci 52:1001–1033

Flatau P, Tripoli GJ, Verlinde J, Cotton WR (1989) The CSU RAMS cloud microphysics module: general theory and code documentation, technical report 451, Colorado State University, USA

Garper P (1988) Modeling the seasonal cycle of the upper ocean. J Phys Oceanogr 18:161–180

Grabowski WW (1998) Toward cloud resolving modeling of large-scale tropical circulations: a simple cloud microphysics parameterization. J Atmos Sci 55:3283–3298

Grabowski WW, Wu X, Moncrieff MW (1996) Cloud-resolving model of tropical cloud systems during phase III of GATE. Part I: two-dimensional experiments. J Atmos Sci 53:3684–3709

Harshvardhan DR, Randall DA, Corsetti TG (1987) A fast radiation parameterization for general circulation models. J Geophys Res 92:1009–1016

Hu Z, He G (1988) Numerical simulation of microphysical processes in cumulonimbus. Part I: microphysical model. Acta Meteor Sinica 2:471–488

Hu Z, Lou X, Bao S, Wang X (1998) A simplified explicit scheme of mixed phase cloud and precipitation. Quart J Appl Meteor 9:257–264

Kessler E (1969) On the distribution and continuity of water substance in atmospheric circulations. Meteor monograph 32. American Meteorological Society, Boston

Kiehl JT, Hack JJ, Briegleb BP (1994) The simulated earth radiation budget of the National Center for Atmospheric Research community climate model CCM2 and comparisons with the Earth Radiation Budget Experiment (ERBE). J Geophys Res 99:20815–20827

Klemp JB, Wilhelmson RB (1978) The simulation of three-dimensional convective storm dynamics. J Atmos Sci 35:1070–1093

Koenig LR (1971) Numerical modeling of ice deposition. J Atmos Sci 28:226–237

Koenig LR, Murray FW (1976) Ice-bearing cumulus cloud evolution: numerical simulation and general comparison against observations. J Appl Meteorol 15:747–762

Krueger SK (1988) Numerical simulation of tropical cumulus clouds and their interaction with the subcloud layer. J Atmos Sci 45:2221–2250

Krueger SK, Fu Q, Liou KN, Chin HNS (1995) Improvement of an ice-phase microphysics parameterization for use in numerical simulations of tropical convection. J Appl Meteorol 34:281–287

Li X, Sui CH, Lau KM, Chou MD (1999) Large-scale forcing and cloud-radiation interaction in the tropical deep convective regime. J Atmos Sci 56:3028–3042, (c) American Meteorological Society. Reprinted with permission

Li X, Sui CH, Lau KM, Adamec D (2000) Effects of precipitation on ocean mixed-layer temperature and salinity as simulated in a 2-D coupled ocean-cloud resolving atmosphere model. J Meteor Soc Japan 78:647–659

Lilly DK (1962) On the numerical simulation of buoyant convection. Tellus 14:148–172

Lilly DK (1967) The representation of small-scale turbulence in numerical simulation experiment. In: Proceedings of the IBM scientific computing symposium on environmental sciences, White Plains

Lin YL, Farley RD, Orville HD (1983) Bulk parameterization of the snow field in a cloud model. J Clim Appl Meteorol 22:1065–1092

Lipps FB, Hemler RS (1986) Numerical simulation of deep tropical convection associated with large-scale convergence. J Atmos Sci 43:1796–1816

Lipps FB, Hemler RS (1988) Numerical modeling of a line of towering cumulus on day 226 of GATE. J Atmos Sci 45:2428–2444

Lipps FB, Hemler RS (1991) Numerical modeling of a midlatitude squall line: features of the convection and vertical momentum fluxes. J Atmos Sci 48:1909–1929

Lord SJ, Willoughby HE, Piotrowicz JM (1984) Role of a parameterized ice phase microphysics in an axisymmetric, nonhydrostatic tropical cyclone model. J Atmos Sci 41:2836–2848

Mellor GL, Yamada T (1974) A hierarchy of turbulence closure models for planetary boundary layers. J Atmos Sci 31:1791–1806

Nakajima K, Matsuno T (1988) Numerical experiments concerning the orgin of cloud clusters in the tropical atmosphere. J Meteor Soc Japan 66:309–329

Nakajima T, Tsukamoto M, Tsushima Y, Numaguti A, Kimura T (2000) Modeling of the radiative process in an atmospheric general circulation model. Appl Optics 39:4869–4878

Niiler PP, Kraus EB (1977) One-dimensional models. In: Kraus EB (ed) Modeling and prediction of the upper layers of the ocean. Pergamon, New York

Ogura Y, Philips NA (1962) Scale analysis of deep and shallow convection in the atmosphere. J Atmos Sci 19:173–179

Petch JC (1998) Improved radiative transfer calculations from information provided by bulk microphysical schemes. J Atmos Sci 55:1846–1858

Pruppacher HR, Klett JO (1978) Microphysics of clouds and precipitation. Reidel, Dordrecht, The Netherlands

Redelsperger JL, Sommeria G (1981a) Methode de representation de la turbulence d'echelle inferieure a la maille pour un medele tri-dimensional de convection nuageuse. Bound Layer Meteor 21:509–530

Redelsperger JL, Sommeria G (1981b) Three-dimensional simulations of a convective storm: sensitivity studies on subgrid parameterization and spatial resolution. J Atmos Sci 43:2619–2635

Redelsperger JL, Sommeria G (1982) Methode de representation de la turbulence associee aux precipitations dans un modele tri-dimensional de convection nuageuse. Bound Layer Meteor 24:231–252

Redelsperger JL, Sommeria G (1986) Three-dimensional simulation of a convective strom: sensitivity studies on subgrid parameterization and spatial resolution. J Atmos Sci 43:2619–2635

Roll HU (1965) Physics of the marine atmosphere. Academic Press, New York

Rutledge SA, Hobbs RV (1983) The mesoscale and microscale structure and organization of clouds and precipitation in midlatitude cyclones. Part VIII: a model for the "seeder-feeder" process in warm-frontal rainbands. J Atmos Sci 40:1185–1206

Rutledge SA, Hobbs RV (1984) The mesoscale and microscale structure and organization of clouds and precipitation in midlatitude clcones. Part XII: a diagnostic modeling study of precipitation development in narrow cold-frontal rainbands. J Atmos Sci 41:2949–2972

Satoh M, Tomita H, Miura H, Iga S, Nasumo T (2005) Development of a global resolving model – a multi-scale structure of tropical convections. J Earth Sim 3:1–9

Schemm CE, Lipps FB (1976) Some results from a simplified three-dimensional numerical model of atmospheric turbulence. J Atmos Sci 33:1021–1041

Shutts GJ, Gray MEB (1994) A numerical modeling study of the geostrophic adjustment following deep convection. Quart J Roy Meteor Soc 120:1145–1178

Smagorinsky J (1963) General circulation experiments with the primitive equations: I. The basic experiment. Mon Weather Rev 91:99–164

Sommeria G (1976) Three-dimensional simulation of turbulent process in an undisturbed trade wind boundary layer. J Atmos Sci 33:216–241

Soong ST, Ogura Y (1980) Response of tradewind cumuli to large-scale processes. J Atmos Sci 37:2035–2050

Soong ST, Tao WK (1980) Response of deep tropical cumulus clouds to mesoscale processes. J Atmos Sci 37:2016–2034

Sui CH, Lau KM, Tao WK, Simpson J (1994) The tropical water and energy cycles in a cumulus ensemble model. Part I: equilibrium climate. J Atmos Sci 51:711–728
Sui CH, Li X, Lau KM, Adamec D (1997) Multi-scale air-sea interactions during TOGA COARE. Mon Weather Rev 125:448–462
Sui CH, Li X, Lau KM (1998) Radiative-convective processes in simulated diurnal variations of tropical oceanic convection. J Atmos Sci 55:2345–2359
Swann H (1994) Cloud microphysical processes-a description of the parameterization used in the large Eddy model. UK meteorological office, JCMM internal report 10
Tao WK, Simpson J (1993) The goddard cumulus ensemble model. Part I: model description. Terr Atmos Oceanic Sci 4:35–72
Tao WK, Simpson J, McCumber M (1989) An ice-water saturation adjustment. Mon Weather Rev 117:231–235
Tompkins AM, Craig GC (1998) Radiative-convective equilibrium in a three-dimensional cloud-ensemble model. Quart J Roy Meteor Soc 124:2073–2097
Tripoli GJ (1992) A nonhydrostatic mesoscale model designed to simulate scale interaction. Mon Weather Rev 120:1342–1359
Wilhelmson RB, Chen CS (1982) A simulation of the development of successive cells along a cold outflow boundary. J Atmos Sci 39:1466–1483
Wu X, Grabowski WW, Moncrieff MW (1998) Long-term evolution of cloud systems in TOGA COARE and their interactions with radiative and surface processes. Part I: two-dimensional cloud-resolving model. J Atmos Sci 55:2693–2714
Xu H (1992) A meso-β modeling study. Comput Phys 9:731–734
Xu H, Duan Y (1999) Some issues associated with growth of cloud concentration numbers. Acta Meteor Sinica 57:450–460
Xu H, Huang S (1994) A three-dimensional meso-β modeling study: a numerical simulation of mesoscale structures associated with frontogenesis over the warm sector of a cyclone. Acta Meteor Sinica 52:165–171
Yamasaki M (1975) A numerical experiment of the interaction between cumulus convection and large-scale motion. Papers Meteor Geophys 26:63–91
Zou G (1991) Three-dimensional convective cloud model. Chinese Academy of Meteorological Sciences, Beijing

Chapter 2
Analysis Methodology

Abstract In this chapter, analysis methodologies used in this book are presented. Heat and water vapor budgets are used to examine thermal balance and water vapor source associated with convective development. Ocean mixed-layer thermal and saline budgets are used to study atmospheric effects on ocean mixed layer. Water vapor budget, cloud microphysical budget, and surface rainfall are linked through cloud microphysical processes mainly including net condensation denoted by condensation and depositions minus the evaporation of precipitation ice. Since it is directly responsible for the production of rainfall, rain budget is also discussed. The net condensation and rainfall are associated with the development of the secondary circulations that are denoted by perturbation kinetic energy. Energy budgets including mean and perturbation available potential energy and perturbation kinetic energy are presented in an Eulerian framework and convective available potential energy (CAPE) is introduced in a Lagrangian framework. Finally, rainfall is partitioned into convective and stratiform components, and their separation criteria are discussed.

Keywords Heat budget • Water vapor budget • Cloud budget • Rain budget • Surface rainfall budget • Energy budget • Ocean thermal and saline budget • Rainfall separation

To examine dominant physical processes associated with the development of convective systems, the simulation data can be analyzed in terms of heat, vapor, and cloud budgets, surface rainfall budget, and energy budget. Thus, these budgets in the 2D framework are summarized in this chapter. Convective and stratiform clouds could have different thermodynamic, microphysical, radiative, and rainfall properties. Convective–stratiform cloud partitioning analysis is an important part of the modeling study toward better understanding of clouds and associated microphysics and thermodynamics and their impacts on tropical hydrological and energy cycles, which is discussed in this chapter.

© Springer International Publishing Switzerland 2016

X. Li, S. Gao, *Cloud-Resolving Modeling of Convective Processes*,
Springer Atmospheric Sciences, DOI 10.1007/978-3-319-26360-1_2

2.1 Heat and Vapor Budgets

Li et al. (1999) derived domain-mean heat and vapor budgets and domain and mass-weighted mean heat budget and domain-mean precipitable water (*PW*) budget. For a domain-mean heat budget, multiplying (1.7d) by π and averaging the resulting equation over domain yield

$$\frac{\partial \overline{T}}{\partial t} = \frac{\overline{Q}_{cn}}{c_p} + \frac{\overline{Q}_R}{c_p} - \frac{\pi}{\overline{\rho}} \frac{\partial \left(\overline{\rho} \overline{w'\theta'} \right)}{\partial z} - \pi \overline{w}^o \frac{\partial \overline{\theta}}{\partial z} - \overline{u}^o \frac{\partial \overline{T}^o}{\partial x}. \quad (2.1a)$$

Thus, the local change of domain-mean temperature is contributed to by condensational heating, radiative heating, convergence of vertical heat flux, vertical temperature advection, and imposed horizontal temperature advection. A domain-mean mass-weighted heat budget can be derived by multiplying (2.1a) by $\overline{\rho}$ and integrating the resulting equation vertically and dividing it by the mass of the air column,

$$\frac{\partial < \overline{T} >}{\partial t} = \frac{< \overline{Q}_{cn} >}{c_p} + \frac{< \overline{Q}_R >}{c_p} + \overline{H}_s - < \pi \overline{w}^o \frac{\partial \overline{\theta}}{\partial z} > - < \overline{u}^o \frac{\partial \overline{T}^o}{\partial x} >. \quad (2.1b)$$

Here, $<F> = \int_0^{z_t} \overline{\rho}\, F dz / \int_0^{z_t} \overline{\rho}\, dz$, F is a variable, and z_t is the model top; $\overline{H}_s$ is the surface sensible heat flux. In the domain-mean mass-weighted heat budget, the local temperature change is determined by condensational heating, radiative heating, surface sensible heat flux, vertical temperature advection, and imposed horizontal temperature advection.

For a domain-mean vapor budget, averaging (1.7e) over domain leads to

$$\frac{\partial \overline{q}_v}{\partial t} = -\overline{S}_{qv} - \frac{1}{\overline{\rho}} \frac{\partial \left(\overline{\rho} \overline{w'q'_v} \right)}{\partial z} - \overline{w}^o \frac{\partial \overline{q}_v}{\partial z} - \overline{u}^o \frac{\partial \overline{q}_v^o}{\partial x}. \quad (2.2a)$$

Thus, the local change of domain-mean water vapor is contributed to by net condensation, convergence of vertical moisture flux, vertical moisture advection, and imposed horizontal moisture advection. A domain-mean *PW* budget is derived by multiplying (2.2a) by $\overline{\rho}$ and integrating the resulting equation vertically,

$$\frac{\partial \left[\overline{q}_v \right]}{\partial t} = - \left[\overline{S}_{qv} \right] - \overline{E}_s - \left[\overline{w}^o \frac{\partial \overline{q}_v}{\partial z} \right] - \left[\overline{u}^o \frac{\partial \overline{q}_v^o}{\partial x} \right]. \quad (2.2b)$$

Here, [F] $\left(= \int_0^{z_t} \overline{\rho} F\, dz \right)$ is a mass integration; $\bar{E}_s$ is the surface evaporation flux; and $[\overline{q}_v]$ is *PW*. In the domain-mean *PW* budget, the local PW change is determined by net condensation, surface evaporation flux, vertical moisture advection, and imposed horizontal moisture advection.

2.2 Surface Rainfall, Cloud, and Rain Microphysical Budgets

Gao et al. (2005) and Cui and Li (2006) derived the surface rainfall equation to study the role of water vapor and cloud source/sink in the surface rainfall processes. The equations of cloud hydrometeors (1.7f), (1.7g), (1.7h), (1.7i) and (1.7j) are added to yield

$$\frac{\partial q_l}{\partial t} = -\frac{\partial (u q_l)}{\partial x} - \frac{1}{\overline{\rho}}\frac{\partial}{\partial z}\overline{\rho} w q_l + \frac{1}{\overline{\rho}}\frac{\partial}{\partial z}\overline{\rho}\left(w_{Tr}q_r + w_{Ts}q_s + w_{Tg}q_g\right) + S_{qv}, \tag{2.3}$$

where $q_l = q_c + q_r + q_i + q_s + q_g$. The equation of cloud hydrometeors (2.3) and the equation of water vapor (1.7e) are added to eliminate S_{qv} and then are mass integrated to derive a surface rainfall equation (P_s),

$$P_s = Q_{WVT} + Q_{WVF} + Q_{WVE} + Q_{CM}, \tag{2.4}$$

where

$$P_s = \overline{\rho}\left(w_{Tr}q_r + w_{Ts}q_s + w_{Tg}q_g\right), \tag{2.4a}$$

$$Q_{WVT} = -\frac{\partial [q_v]}{\partial t}, \tag{2.4b}$$

$$\begin{aligned} Q_{WVF} = & -\left[\overline{u}^o \frac{\partial \overline{q}_v{}^o}{\partial x}\right] - \left[\overline{w}^o \frac{\partial \overline{q}_v}{\partial z}\right] - \left[\frac{\partial (u' q_v')}{\partial x}\right] \\ & -\left[\overline{u}^o \frac{\partial q_v'}{\partial x}\right] - \left[\overline{w}^o \frac{\partial q_v'}{\partial z}\right] - \left[w' \frac{\partial \overline{q}_v}{\partial z}\right], \end{aligned} \tag{2.4c}$$

$$Q_{WVE} = E_s, \tag{2.4d}$$

$$Q_{CM} = -\frac{\partial [q_l]}{\partial t} - \left[u \frac{\partial q_l}{\partial x}\right] - \left[w \frac{\partial q_l}{\partial z}\right]. \tag{2.4e}$$

In the tropics, $P_s = \overline{\rho}\,(w_{Tr}q_r)$. (2.4) indicates that the surface rain rate can be calculated if the wind, specific humidity, and cloud mixing ratios are known. Surface rain rate is contributed to by the local vapor change (Q_{WVT}), vapor convergence (Q_{WVF}), surface evaporation (Q_{WVE}), and cloud source/sink (Q_{CM}). Positive values of Q_{WVT}, Q_{WVF}, and Q_{CM} denote local vapor loss (atmospheric drying), vapor convergence, and local hydrometeor loss/hydrometeor convergence, respectively, whereas negative Q_{WVT}, Q_{WVF}, and Q_{CM} denote local vapor gain (atmospheric moistening), vapor divergence, and local hydrometeor gain/hydrometeor divergence, respectively.

In domain-mean surface rainfall equation,

$$\overline{P}_s = \overline{Q}_{WVT} + \overline{Q}_{WVF} + \overline{Q}_{WVE} + \overline{Q}_{CM}, \tag{2.5}$$

$$\overline{Q}_{WVT} = -\frac{\partial\left[\overline{q}_v\right]}{\partial t}, \tag{2.5a}$$

$$\overline{Q}_{WVF} = -\left[\overline{u}^o \frac{\partial \overline{q}_v{}^o}{\partial x}\right] - \left[\overline{w}^o \frac{\partial \overline{q}_v}{\partial z}\right], \tag{2.5b}$$

$$\overline{Q}_{WVE} = \overline{E}_s, \tag{2.5c}$$

$$\overline{Q}_{CM} = -\frac{\partial\left[\overline{q}_l\right]}{\partial t}. \tag{2.5d}$$

Note that in (2.5d) the convergence of cloud hydrometeors is zero due to cyclic lateral boundary conditions. Thus, positive value of domain-mean $\overline{Q}_{CM}$ denotes local hydrometeor loss whereas negative value is local hydrometeor gain.

The mass integration of (2.3) and (1.8b) yields mass-integrated cloud microphysical budget,

$$P_S - Q_{CM} = \sum_{I=1}^{7} P_I, \tag{2.6}$$

where

$$P_I = \left(\left[P_{CND}\right], \left[P_{DEP}\right], \left[P_{SDEP}\right], \left[P_{GDEP}\right], -\left[P_{REVP}\right], -\left[P_{MLTG}\right], -\left[P_{MLTS}\right]\right), \tag{2.6a}$$

$$Q_{CM} = Q_{CMC} + Q_{CMR} + Q_{CMI} + Q_{CMS} + Q_{CMG}, \tag{2.6b}$$

$$Q_{CMC} = -\frac{\partial\left[q_c\right]}{\partial t} = -Sqc, \tag{2.6c}$$

$$Q_{CMR} = -\frac{\partial\left[q_r\right]}{\partial t} = -Sqr + P_S, \tag{2.6d}$$

$$Q_{CMI} = -\frac{\partial\left[q_i\right]}{\partial t} = -Sqi, \tag{2.6e}$$

$$Q_{CMS} = -\frac{\partial\left[q_s\right]}{\partial t} = -Sqs, \tag{2.6f}$$

$$Q_{CMG} = -\frac{\partial\left[q_g\right]}{\partial t} = -Sqg. \tag{2.6g}$$

Here, Sqc, Sqr, Sqi, Sqs, and Sqg can be found in (1.8c), (1.8d), (1.8e), (1.8f), and (1.8g).

The mass integration of (1.7g) and (1.8d) leads to mass-integrated rain microphysical budget,

$$P_S - Q_{RM} = \sum_{I=1}^{12} RP_I, \tag{2.7}$$

where

$$\begin{aligned} RP_I = \Big(& [P_{SACW}(T > T_o)], [P_{RAUT}], [P_{RACW}], [P_{GACW}(T > T_o)], \\ & -[P_{REVP}], [P_{RACS}(T > T_o)], -[P_{IACR}(T < T_o)], -[P_{GACR}(T < T_o)], \\ & -[P_{SACR}(T < T_o)], -[P_{GFR}(T < T_o)], [P_{SMLT}(T > T_o)], \\ & [P_{GMLT}(T > T_o)] \Big), \end{aligned} \tag{2.7a}$$

$$Q_{RM} = -\frac{\partial [q_r]}{\partial t} - \left[u\frac{\partial q_r}{\partial x}\right] - \left[w\frac{\partial q_r}{\partial z}\right]. \tag{2.7b}$$

2.3 Energetics Equations in Moist Atmosphere and Convective Available Potential Energy

Lorenz (1955) first introduced the concept of available potential energy for a dry atmosphere that represents the portion of the potential energy that can be transferred into kinetic energy. He defined the available potential energy for a dry atmosphere as the difference between the actual total enthalpy and the minimum total enthalpy that could be achieved by rearranging the mass under an adiabatic flow. The dry enthalpy per unit mass is defined as the product of the temperature and the specific heat at constant pressure. In the absence of energy sources and sinks, the total kinetic energy and total enthalpy are conserved during adiabatic expansion. In a moist atmosphere, latent heat energy should be included in the energy conservation. The latent heat energy per unit mass is defined as the product of the specific humidity and the latent heat of vaporization at 0 °C. In the absence of energy sources and sinks, the total kinetic energy, enthalpy, and latent heat energy are conserved during dry and subsequent saturated adiabatic expansion. Therefore, the moist available potential energy is defined as the difference between the actual moist potential energy (sum of the enthalpy and latent heat energy) and the minimum moist potential energy that could be achieved by rearranging the mass under moist adiabatic processes. Li et al. (2002a) derived a set of equations for conversions between the moist available potential energy and kinetic energy in a Eulerian framework.

Domain-mean and perturbation moist available potential energy ($\overline{P}$, P') and perturbation kinetic energy (K') are, respectively, defined by

$$\overline{P} = \left[\frac{\Lambda}{2c_p}\left(\overline{h}^2 - {h_b}^2\right)\right], \tag{2.8a}$$

$$P' = \left[\overline{\frac{\Lambda}{2c_p}(h')^2}\right], \tag{2.8b}$$

$$K' = \left[\overline{\frac{(u')^2 + (w')^2}{2}}\right], \tag{2.8c}$$

where

$$h = c_p T + L_v q_v, \tag{2.8d}$$

$$\Lambda = -\frac{R\theta}{c_p p T} \frac{1}{\frac{\partial \theta_b}{\partial p} + \frac{L_v}{c_p \pi}\frac{\partial q_{vb}}{\partial p}}, \tag{2.8e}$$

h_b is a constant reference state and is calculated from the initial observed sounding.

The equation of h is derived by multiplying (1.7d) by $c_p\pi$ and (1.7e) by L_v and adding the resulting equations,

$$\begin{aligned}\frac{\partial h}{\partial t} =& -\frac{\partial (u'h')}{\partial x} - \overline{u}^o \frac{\partial h'}{\partial x} - \frac{c_p \pi}{\overline{\rho}} \frac{\partial}{\partial z} \overline{\rho} w' \theta' - c_p \pi \overline{w}^o \frac{\partial \theta'}{\partial z} - c_p \pi w' \frac{\partial \overline{\theta}}{\partial z} \\ & - \frac{L_v}{\overline{\rho}} \frac{\partial}{\partial z} \overline{\rho} w' q_v' - L_v \overline{w}^o \frac{\partial q_v'}{\partial z} - L_v w' \frac{\partial \overline{q}_v}{\partial z} + L_f P_{18} + Q_R - \overline{u}^o \frac{\partial \overline{h}^o}{\partial x} \\ & - c_p \pi \overline{w}^o \frac{\partial \overline{\theta}}{\partial z} - L_v \overline{w}^o \frac{\partial \overline{q}_v}{\partial z},\end{aligned} \tag{2.9}$$

where

$$\begin{aligned}P_{18} =& P_{DEP} + (1-\delta_1) P_{SDEP} (T < T_o) + (1-\delta_1) P_{GDEP} (T < T_o) \\ & - P_{MLTS} (T > T_o) - P_{MLTG} (T > T_o) + P_{SACW} (T < T_o) \\ & + P_{SFW} (T < T_o) + P_{GACW} (T < T_o) + P_{IACR} (T < T_o) \\ & + P_{GACR} (T < T_o) + P_{SACR} (T < T_o) + P_{GFR} (T < T_o) \\ & - P_{RACS} (T > T_o) - P_{SMLT} (T > T_o) - P_{GMLT} (T > T_o) \\ & + P_{IHOM} (T < T_{oo}) - P_{IMLT} (T > T_o) \\ & + P_{IDW} (T_{oo} < T < T_o).\end{aligned} \tag{2.9a}$$

The equations for $\overline{P}$ and P' can be derived by multiplying (2.9) by $\Lambda\left(\overline{h} - h_b\right)/c_p$ and $\Lambda h'/c_p$ and applying the domain mean and vertical integration on the resulting equations.

$$\frac{\partial \overline{P}}{\partial t} = C\left(P', \overline{P}\right) + G_R\left(\overline{P}\right) + G_{cn}\left(\overline{P}\right) + C_h\left(\overline{K}, \overline{P}\right) + C_v\left(\overline{K}, \overline{P}\right), \tag{2.10}$$

where

$$C\left(P',\overline{P}\right)=-\left[\frac{\Lambda}{c_p}\left(\overline{h}-h_b\right)\left(\frac{c_p\pi}{\overline{\rho}}\frac{\partial}{\partial z}\overline{\rho}\overline{w'\theta'}+\frac{L_v}{\overline{\rho}}\frac{\partial}{\partial z}\overline{\rho}\overline{w'q_v'}\right)\right],\tag{2.10a}$$

$$G_R\left(\overline{P}\right)=\left[\frac{\Lambda}{c_p}\overline{Q}_R\left(\overline{h}-h_b\right)\right],\tag{2.10b}$$

$$G_{cn}\left(\overline{P}\right)=\left[\frac{\Lambda L_f}{c_p}\overline{P}_{18}\left(\overline{h}-h_b\right)\right],\tag{2.10c}$$

$$C_h\left(\overline{K},\overline{P}\right)=-\left[\frac{\Lambda}{c_p}\left(\overline{h}-h_b\right)\overline{u}^o\left(c_p\pi\frac{\partial\overline{\theta}^o}{\partial x}+L_v\frac{\partial\overline{q}_v{}^o}{\partial x}\right)\right],\tag{2.10d}$$

$$C_v\left(\overline{K},\overline{P}\right)=-\left[\frac{\Lambda}{c_p}\left(\overline{h}-h_b\right)\overline{w}^o\left(c_p\pi\frac{\partial\overline{\theta}}{\partial z}+L_v\frac{\partial\overline{q}_v}{\partial z}\right)\right].\tag{2.10e}$$

Here, $C\left(P',\overline{P}\right)$ is the conversion between P' and $\overline{P}$ through covariance between $\overline{h}-h_b$ and convergence of vertical flux of potential temperature and moisture. $G_R\left(\overline{P}\right)$ and $G_{cn}\left(\overline{P}\right)$ are the generation terms of $\overline{P}$ through covariances between $\overline{h}-h_b$ and horizontal-mean radiative heating and between $\overline{h}-h_b$ and horizontal-mean heating due to phase change of the cloud contents, respectively. $C_h\left(\overline{K},\overline{P}\right)$ and $C_v\left(\overline{K},\overline{P}\right)$ are the conversion between $\overline{K}$ and $\overline{P}$ through covariances between $\overline{h}-h_b$ and imposed horizontal temperature and moisture advections and between $\overline{h}-h_b$ and the horizontal-mean vertical temperature and moisture advections by imposed vertical velocity, respectively.

$$\frac{\partial P'}{\partial t}=-C\left(P',\overline{P}\right)-C\left(P',K'\right)+G_R\left(P'\right)+G_{cn}\left(P'\right)+G\left(P'\right),\tag{2.11}$$

where

$$C\left(P',K'\right)=\left[g\frac{\overline{w'T'}}{T_b}\right],\tag{2.11a}$$

$$G_R\left(P'\right)=\left[\frac{\Lambda}{c_p}\overline{Q_Rh'}\right],\tag{2.11b}$$

$$G_{cn}\left(P'\right)=\left[\frac{\Lambda}{c_p}\overline{P_{18}h'}\right],\tag{2.11c}$$

$$G\left(P'\right) = -\left[\frac{gL_v}{c_pT_b}\overline{w'q_v'}\right] - \left[\frac{\Lambda}{c_p\overline{\rho}}\frac{\partial}{\partial z}\left(\overline{h} - h_b\right)\overline{\rho}\overline{h'w'}\right] - \left[\frac{g\Lambda}{c_pT_b}\left(\frac{\overline{T}}{T_b} - 1\right)\overline{h'w'}\right]$$
$$- \left[\frac{g\Lambda}{c_pT_b}\left(\overline{h} - h_b\right)\overline{w'T'}\right] - \left[\frac{\Lambda}{2c_p\overline{\rho}}\frac{\partial}{\partial z}\overline{\overline{\rho}(h')^2\left(\overline{w}^o + w'\right)}\right]$$
$$- \left[\frac{g\Lambda}{c_pT_b}\overline{h'T'\left(\overline{w}^o + w'\right)}\right]. \tag{2.11d}$$

Here, $C(P', K')$ is the conversion between P' and K' through covariance between perturbation vertical velocity and temperature. $G_R(P')$ and $G_{cn}(P')$ are the generation terms of P' through covariances between h' and perturbation radiative heating and between h' and perturbation heating due to phase changes of the cloud contents, respectively. $G(P')$ is the generation term of P'. Note that $C\left(\overline{P}, P'\right) + C\left(K', P'\right) + G\left(P'\right)$ causes changes of P' due to the vertical advection processes.

$$\frac{\partial K'}{\partial t} = C\left(\overline{K}, K'\right) + C\left(P', K'\right) + G_{qv}\left(K'\right) + G_{ql}\left(P'\right), \tag{2.12}$$

where

$$C\left(\overline{K}, K'\right) = -\left[\overline{u'w'}\frac{\partial \overline{u}^o}{\partial z}\right] - \left[\overline{w'w'}\frac{\partial \overline{w}^o}{\partial z}\right], \tag{2.12a}$$

$$G_{qv}\left(K'\right) = \left[0.61g\overline{w'q_v'}\right], \tag{2.12b}$$

$$G_{ql}\left(K'\right) = -\left[g\overline{w'q_l'}\right]. \tag{2.12c}$$

Here, $C\left(\overline{K}, K'\right)$ is the conversion between $\overline{K}$ and K' through covariance between perturbation zonal wind and vertical velocity under vertical shear of imposed horizontal-mean zonal wind and between perturbation vertical velocities under vertical shear of imposed horizontal-mean vertical velocity. $G_{qv}(K')$ and $G_{ql}(K')$ are the generation terms of K' through covariance between perturbation vertical velocity and specific humidity and between perturbation vertical velocity and cloud mixing ratio, respectively.

Li et al. (2002a) calculated the convective available potential energy (*CAPE*) using

$$CAPE = g\int_{LFC}^{z_c} \frac{\theta_{pcl}(z) - \theta_{env}(z)}{\theta_{env}(z)}dz. \tag{2.13}$$

Here, θ_{pcl} is the potential temperature of an air parcel lifted from the bottom of the atmosphere to the top of the atmosphere while not mixing with its environment

(θ_{env}). The air parcel is lifted dry adiabatically until it becomes saturated and then is lifted moist adiabatically thereafter. The level of free convection (*LFC*) is the height where $\theta_{pcl} > \theta_{env}$ and z_c is the level where $\theta_{pcl} = \theta_{env}$.

The *CAPE* can be calculated for a pseudoadiabatic process and a reversible moist adiabatic process, respectively. In the pseudoadiabatic process, an air parcel is lifted adiabatically while all condensed water drops out from the parcel. In the reversible moist adiabatic process, an air parcel is lifted adiabatically while all condensed water is kept in the parcel. Following Xu and Emanuel (1989), the virtual temperatures (T_{vpa}) for the pseudoadiabatic process and (T_{vre}) for the reversible moist adiabatic process are, respectively, expressed by

$$T_{vpa} = T_p \frac{1 + \frac{q_{vs}(T_p)}{0.622}}{1 + q_v} \tag{2.14a}$$

and

$$T_{vre} = T_p \frac{1 + \frac{q_{vs}(T_p)}{0.622}}{1 + q_{vs}(T_p)}, \tag{2.14b}$$

where T_p is the temperature of a pseudoadiabatically displaced air parcel; q_{vs} is the saturation specific humidity. The *CAPE* for the pseudoadiabatic process ($CAPE_{pa}$) and for the reversible moist adiabatic process ($CAPE_{re}$) is calculated by using (2.14a) and (2.14b), respectively. Li et al. (2002a) showed that both $CAPE_{pa}$ and $CAPE_{re}$ have the same evolution but different magnitudes.

2.4 Ocean Mixed-Layer Thermal and Saline Budgets

The domain-mean ocean mixed-layer thermal and saline budgets can be, respectively, expressed as

$$\frac{\partial \overline{T}_m}{\partial t} = -\overline{u_m \frac{\partial T_m}{\partial x}} - \overline{\frac{W_e}{h_m} \mathrm{H}(W_e)(T_m - T_e)} + \overline{\frac{Q_o + I(0) - I(h_m)}{\rho_r c_w h_m}}, \tag{2.15a}$$

$$\frac{\partial \overline{S}_m}{\partial t} = -\overline{u_m \frac{\partial S_m}{\partial x}} - \overline{\frac{W_e}{h_m} \mathrm{H}(W_e)(S_m - S_e)} + \overline{\frac{S_m(P_s - E_s)}{\rho_r c_w h_m}}. \tag{2.15b}$$

Equation (2.15a) shows that local tendency of ocean mixed-layer temperature is determined by horizontal thermal advection, thermal entrainment, and thermal forcing. Equation (2.15b) states that local tendency of ocean mixed-layer salinity is determined by horizontal saline advection, saline entrainment, and saline forcing.

2.5 Partition of Convective and Stratiform Clouds

Convective precipitation is associated with high rain rate, strong horizontal reflectivity gradients, and large vertical velocity. Stratiform precipitation that usually falls from the anvil clouds is associated with light to moderate rain rates, weak horizontal reflectivity gradients, the "bright band" near the melting level in radar echo, and weak vertical velocity. The primary microphysical process responsible for the growth of convective clouds and precipitation is a collection of cloud water by rain particles in the strong updraft cores whereas the primary microphysical process responsible for the growth of stratiform clouds and precipitation is vapor deposition on ice particles (Houghton 1968).

Convective–stratiform rainfall partition has been applied to radar data in most of the observational studies. Based on the assumption that convective cells have peak rainfall rates at least twice as high as the surrounding background rainfall rate, Churchill and Houze (1984) used a similar technique developed by Houze (1973) to classify the convective and stratiform precipitation. In their scheme, the cores of convective cells are first assigned to those data points in the radar reflectivity field that have rain rates twice as high as the average taken over the surrounding 400 km^2. These convective cores and the surrounding 150 km^2 of area are identified as convective precipitation. In addition, any radar echo 40 dBZ or more in intensity is considered convective precipitation. Stratiform precipitation is identified as any precipitation not designated as convective by either of the above criteria. Steiner and Houze (1993) developed a similar partition method, and Steiner et al. (1995) further refined this technique. They categorized the peaks of rain rate as convective when the peaks satisfy specified criteria regarding background rain rate. The remainder of the precipitation is categorized as stratiform after all the peaks and surrounding areas have been located and identified as convective. Steiner et al. (1995) argued that their method is fundamentally different from a method employed by Rosenfeld et al. (1995), which uses the bright band to separate the precipitation. Yuter and Houze (1997) employed an algorithm first developed for application to 4 km $\times$ 4 km grids (Churchill and Houze 1984) and later refined for application to 2 km $\times$ 2 km grids (Steiner et al. 1995) to study raindrop size distribution associated with convective and stratiform rainfall over the Pacific warm pool during TOGA COARE.

Alder and Negri (1988) developed a convective–stratiform technique for analysis of satellite infrared data that locates all local minima in the brightness temperature field. After an empirical screening to eliminate cirrus, these points are assumed to be convective centers. They used a brightness temperature threshold based on the mode temperature of thunderstorm anvils to determine the stratiform rain.

Convective–stratiform cloud partitioning analysis has been applied to the modeling studies. Tao and Simpson (1989) and Tao et al. (2000) developed a partition method that is similar to the method of Churchill and Houze (1984) and added more criteria in which the point is made convective if the cloud water and ice or the updraft exceeds certain threshold values. Xu (1995) developed the partition method using information of vertical motion. In addition, liquid water path is used to identify the stratiform rainfall, and cloud water path and rain water path are used for shallow convection. Lang et al. (2003) developed a new partition method based on the premise that the fall speed of precipitation particles is large relative to the vertical velocity in regions of stratiform precipitation. In this method, the model point is considered stratiform if the ratio of fall velocity to vertical velocity exceeds a threshold value (3.16 in their paper). Lang et al. (2003) compared six different partition methods including Churchill and House (1984), Tao and Simpson (1989), Caniaux et al. (1994), Xu (1995), Steiner et al. (1995), and Lang et al. (2003) in terms of surface rain rate, mass fluxes, apparent heating and moistening, hydrometeor contents, reflectivity and vertical velocity CFAD (contoured frequency by altitude diagram), microphysics, and latent heat retrieval. They showed that the method based on surface rain rate was consistently the most stratiform, whereas the method based on radar information below the melting level and the new method of Lang et al. (2003) was consistently the most convective.

Tao et al. (1990, 1991, and 1993) developed the convective–stratiform cloud partitioning method based on Churchill and Houze (1984) and Adler and Negri (1988), and Sui et al. (1994) modified this partitioning scheme. In Sui et al. (1994), each vertical column containing clouds is partitioned into convective or stratiform based on the following criterion. Model grid points in the surface rain field that have a rain rate twice as large as the average taken over the surrounding four grid points (two grid points at the left and two grid points at the right in the 2D framework) are identified as the cores of convective cells. For each core grid point, the one grid point on either side (in the 2D framework) is also considered as convective. In addition, any grid point with a rain rate of 20 mm h^{-1} or more is designated as convective regardless of the above criteria. All nonconvective cloudy points are regarded as stratiform. Since the above separation criterion is strictly based on surface precipitation, the stratiform region may actually include areas with tilted convective updrafts aloft. If may also include light- or nonprecipitating convective cells that are initiated ahead of the organized convective system. Therefore, grid points in the stratiform regions are further checked and classified as convective if (1) in the precipitating stratiform regions, cloud water below the melting level is greater than 0.5 g kg^{-1} or the maximum updraft above 600 mb exceeds 5 m s^{-1} or (2) in the nonprecipitating stratiform regions, cloud water exists (cloud water is greater than 0.025 g kg^{-1}) or the maximum updraft exceeds 5 m s^{-1} below the melting level. The main difference between Sui's method and Churchill and Houze's method is that in Sui's method the average is taken over 4 grids (6 km here), and only 3 grids surrounding rainfall peaks are considered as convective, whereas in Churchill and Houze's method, the average is taken over 400 km^2 (equivalent to 20 km in 2D or 13 grids), and 150 km^2 (equivalent to 12 km in 2D or 8 grids) surrounding rainfall peaks are considered as convective. The fractional cover of convective precipitation (Fig. 10b in Li et al. 2002b) and convective fraction of rain volume (Li et al. 1999) calculated using Sui's method are in the ranges of calculations using the methods of Churchill and Houze (1984) and Steiner et al.

(1995). The convective–stratiform cloud partitioning method developed by Tao et al. (1993) and modified by Sui et al. (1994) will be used in the following analysis throughout the book.

References

Adler RF, Negri AJ (1988) A satellite infrared technique to estimate tropical convective and stratiform rainfall. J Appl Meteorol 27:30–51

Caniaux G, Redelsperger JL, Lafore JP (1994) A numerical study of the stratiform region of a fast-moving squall line. Part I: general description and water and heat budgets. J Atmos Sci 51:2046–2074

Churchill DD, Houze RA Jr (1984) Development and structure of winter monsoon cloud clusters on 10 December 1978. J Atmos Sci 41:933–960

Cui X, Li X (2006) Role of surface evaporation in surface rainfall processes. J Geophys Res 111. doi:10.1029/2005JD006876

Gao S, Cui X, Zhou Y, Li X (2005) Surface rainfall processes as simulated in a cloud resolving model. J Geophys Res 110. doi:10.1029/2004JD005467

Houghton HG (1968) On precipitation mechanisms and their artificial modification. J Appl Meteorol 7:851–859

Houze RA Jr (1973) A climatological study of vertical transports by cumulus-scale convection. J Atmos Sci 30:1112–1123

Lang S, Tao WK, Simpson J, Ferrier B (2003) Modeling of convective-stratiform precipitation processes: sensitivity to partition methods. J Appl Meteorol 42:505–527

Li X, Sui CH, Lau KM, Chou MD (1999) Large-scale forcing and cloud-radiation interaction in the tropical deep convective regime. J Atmos Sci 56:3028–3042

Li X, Sui CH, Lau KM (2002a) Interactions between tropical convection and its environment: an energetics analysis of a 2-D cloud resolving simulation. J Atmos Sci 59:1712–1722

Li X, Sui CH, Lau KM (2002b) Dominant cloud microphysical processes in a tropical oceanic convective system: a 2-D cloud resolving modeling study. Mon Wea Rev 130:2481–2491

Lorenz EN (1955) Available potential energy and the maintenance of the general circulation. Tellus 7:157–167

Rosenfeld D, Amitai E, Wolff DB (1995) Classification of rain regimes by the three-dimensional properties of reflectivity fields. J Appl Meteorol 34:198–211

Steiner M, Houze RA Jr (1993) Three-dimensional validation at TRMM ground truth sites: some early results from Darwin, Australia. 26th international conference on radar meteorology, Norman, OK

Steiner M, Houze RA Jr, Yuter SE (1995) Climatological characterization of three-dimensional storm structure from operational radar and rain gauge data. J Appl Meteorol 34:1978–2007

Sui CH, Lau KM, Tao WK, Simpson J (1994) The tropical water and energy cycles in a cumulus ensemble model. Part I: equilibrium climate. J Atmos Sci 51:711–728

Tao WK, Simpson J (1989) Modeling study of a tropical squall-type convective line. J Atmos Sci 46:177–202

Tao WK, Simpson J, Lang S, McCumber M, Adler R, Penc R (1990) An algorithm to estimate the heating budget from vertical hydrometeor profiles. J Appl Meteorol 29:1232–1244

Tao WK, Simpson J, Soong ST (1991) Numerical simulation of a subtropical squall line over Taiwan Strait. Mon Weather Rev 119:2699–2723

Tao WK, Simpson J, Sui CH, Ferrier B, Lang S, Scala J, Chou MD, Pickering K (1993) Heating, moisture, and water budgets of tropical and midlatitude squall lines: comparisons and sensitivity to longwave radiation. J Atmos Sci 50:673–690

Tao WK, Lang S, Simpson J, Olson WS, Johnson D, Ferrier B, Kummerow C, Adler R (2000) Vertical profiles of latent heat release and their retrieval for TOGA COARE convective systems using a cloud resolving model, SSM/I, and ship-borne radar data. J Meteorol Soc Jpn 78:333–355

Xu KM (1995) Partitioning mass, heat, and moisture budgets of explicit simulated cumulus ensembles into convective and stratiform components. J Atmos Sci 52:1–23

Xu KM, Emanuel KA (1989) Is the tropical atmosphere conditionally unstable? Mon Weather Rev 117:1471–1479

Yuter SE, Houze RA Jr (1997) Measurements of raindrop size distribution over the Pacific warm pool and implications for Z-R relations. J Appl Meteorol 36:847–867

Chapter 3
Comparison Between Simulations and Observations

Abstract In this chapter, cloud-resolving model simulations are compared with available observations in terms of surface rain rate, vertical profiles of temperature and specific humidity, surface solar radiative and latent heat fluxes, apparent heat source and apparent water vapor source, and ocean surface temperature and salinity. Sensitivity of model simulations to uncertainties in initial water vapor conditions is evaluated, and the processes that are responsible for the uncertainty of model simulations are discussed. Finally, the similarities and differences between two- and three-dimensional cloud-resolving model simulations are briefly discussed.

Keywords Comparison • Observation • Initial moisture perturbation • Model responses • Two- and three-dimensional model simulations

The validation of cloud-resolving model simulations with observations highlights how well the models reproduce observed vertical structures of temperature and moisture, surface fluxes, and precipitation. The evaluation of simulations with observations lays down the foundation for further process studies with cloud-resolving models, aiming toward understanding dynamic, thermodynamic, cloud microphysical processes associated with the development of convection. In this chapter, the cloud-resolving model simulation and coupled ocean–cloud-resolving atmosphere model simulation are intensively compared with available observations based on Li et al. (1999, 2000).

The cloud-resolving model simulations are sensitive to initial conditions. The uncertainties of initial conditions could affect the simulations of clouds and precipitation. The sensitivity of cloud and precipitation simulations to initial water vapor conditions and associated physical processes are discussed based on Li et al. (2006) in this chapter.

Due to limitations in computational power, the cloud-resolving model simulations have been mainly carried out in the 2D (x–z) framework while the three-dimensional (3D) simulations have been conducted in small horizontal domains. Thus, the comparison between the 2D and 3D model simulations is discussed in this chapter.

© Springer International Publishing Switzerland 2016

X. Li, S. Gao, *Cloud-Resolving Modeling of Convective Processes*,
Springer Atmospheric Sciences, DOI 10.1007/978-3-319-26360-1_3

3.1 Comparison Between Simulations and Observations

Grabowski et al. (1996) integrated the 2D cloud model with the forcing of heat and vapor sources and sinks from Phase III of GATE for 7 days, conducted a comparison study between simulations and observations, and found that the differences of temperatures could be as large as 2–3 °C whereas the differences of specific humidity could be 1.5 g kg^{-1}. The simulated mid and upper troposphere is moister than what is observed, which may result from the lack of forcing data in upper-tropospheric ice and cyclic lateral boundary conditions. Xu and Randall (1996) conducted an 18-day 2D cloud model with the forcing of large-scale vertical velocity from Phase III of GATE and showed that the temperature differences between simulations and observations are as large as 3 °C and the differences in water vapor mixing ratio are as large as 2–3 g kg^{-1}. Wu et al. (1998) carried out a 39-day integration of cloud model with the forcing of heat and vapor sources and sinks from TOGA COARE and found that simulations agree well with observations when the forcing is strong whereas simulations show notable differences with observations (6 °C in temperature difference and 4 g kg^{-1} in specific-humidity difference) when the forcing is weak. The further sensitivity test shows that the big difference is from the vapor difference. Moncrieff et al. (1997) conducted an intercomparison study for the simulations of four cloud-resolving models in terms of temperature at 500 hPa, *PW*, upwelling IR flux at the top of atmosphere, cloud ice water path, and cloud mass flux at 500 hPa and found notable differences between model simulations such as a cold bias of 2 °C at 500 hPa.

Li et al. (1999) carried out the two experiments with different model setups to examine convective responses to the large-scale forcing data during TOGA COARE and conducted a comparison study between simulations and observations. Figure 3.1 shows the imposed large-scale vertical velocity in pressure coordinate and total temperature and water vapor advections (the sums of horizontal and vertical advections) for a selected 6-day period during the TOGA COARE Intensive Observation Period (IOP), as calculated by Sui et al. (1997). Strong upward motion centers occur on the 20, 23, and 25 of December, signaling quasi-2-day oscillation (Takayabu et al. 1996) embedded in the active phase of intraseasonal oscillation during COARE. Two moderate ascending motion centers appear on the 19 and 21 of December, forming diurnal variations with nocturnal rainfall peaks (Sui et al. 1997). Large-scale advective cooling and moistening are largely associated with these ascending motion centers. Two model setups are used. The model in experiment COARE1 uses vertical velocity, zonal wind, horizontal thermal and vapor advection, and *SST* as the forcing data. The large-scale vertical velocity is the major forcing during the integration in COARE1. The model in experiment COARE2 uses total temperature and vapor advections, zonal wind, and *SST*. The total advections are the major forcing during the integration in COARE2.

The simulated temperatures in COARE1 are 1–2 °C colder than the observed temperatures (Fig. 3.2a) whereas the simulated temperatures in COARE2 are 2–4 °C colder than the observed temperatures (Fig. 3.2b), indicating that the model

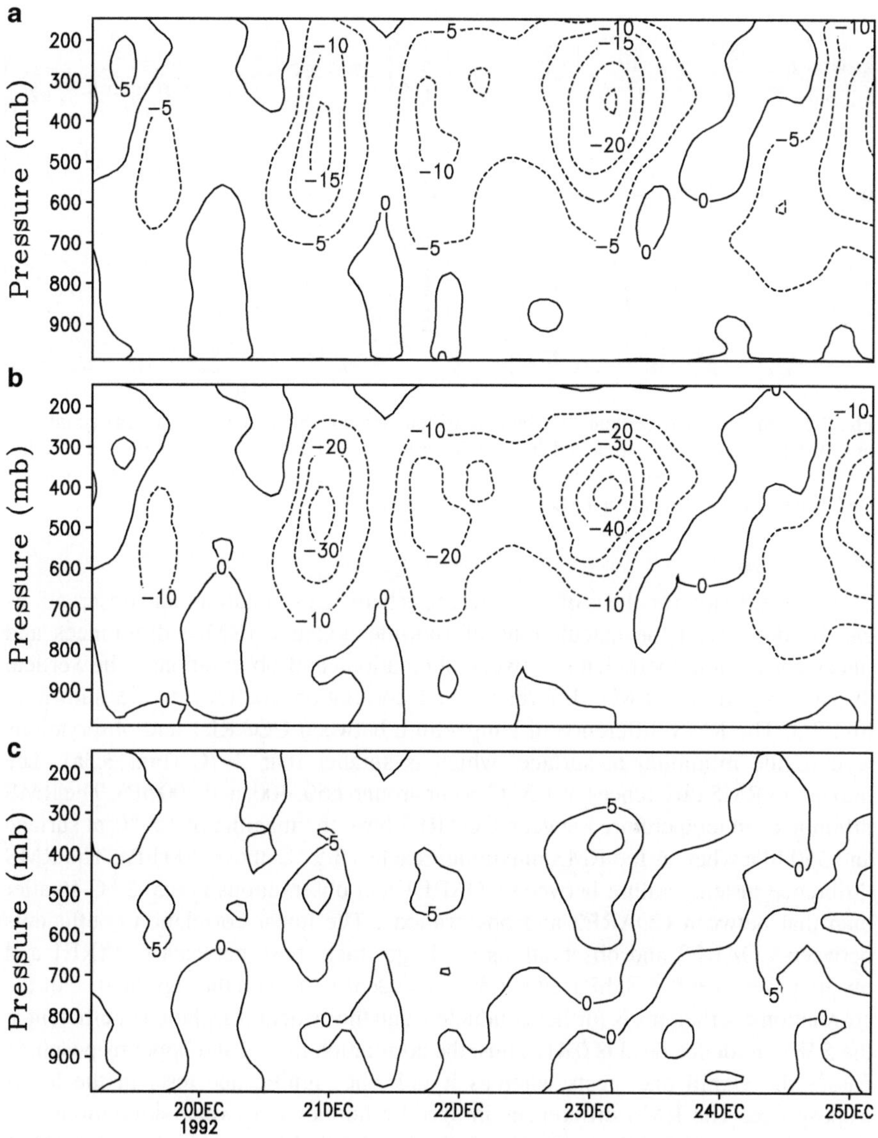

Fig. 3.1 Temporal and vertical distributions of vertical velocity in pressure coordinate (mb h^{-1}) in (**a**) and total advection of potential temperature (°C d^{-1}) in (**b**) and water vapor (g kg^{-1} d^{-1}) in (**c**) calculated from the TOGA COARE data for a selected 6-day period (After Li et al. 1999)

with imposed total advections produces a larger cooling bias than the model with imposed vertical velocity does. The differences in specific humidity between COARE1 and observations and between COARE 2 and observations are similar (Fig. 3.2c, d), but the latter is slightly larger than the former.

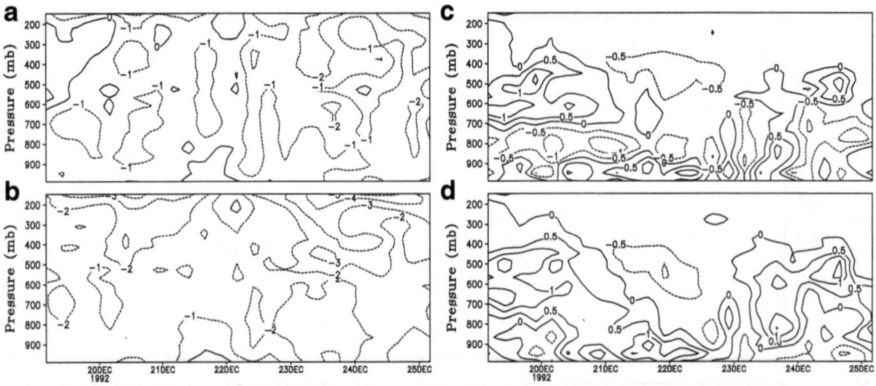

Fig. 3.2 Temporal and vertical distributions of temperature differences (°C) of (**a**) simulation in COARE1 minus observation and (**b**) simulation in COARE2 minus observation and specific-humidity differences (g kg^{-1}) of (**c**) simulation in COARE1 minus observation and (**d**) simulation in COARE2 minus observation (After Li et al. 1999)

The quantitative analysis of the differences between simulations and observations can be done with the calculations of root-mean-square (RMS) differences and linear correlation coefficients between simulations and observations. The vertical distributions of the RMS differences and correlation coefficients are shown in Fig. 3.3. The RMS difference in temperature between COARE1 and observations reaches the minimum at surface, which is smaller than 1 °C (Fig. 3.3a). The maximum RMS differences of 1.5 °C occur around 650, 400, and 200 hPa. The RMS differences in temperature between COARE2 have the maxima of 1.5 °C at surface and 550 hPa whereas the RMS maximum reaches 3.5 °C above 200 hPa. The RMS difference in temperature between COARE1 and observations is 0.5–2 °C smaller than that between COARE2 and observations. The linear correlation coefficients between COARE2 and observations are larger than those between COARE1 and observations in 300–800 hPa (Fig. 3.3c). A student's t-test on the significance of the correlation coefficients is further conducted, and the critical correlation coefficient at the 5 % confidence level is 0.41. Thus, the correlation in mid and upper troposphere passes the significance tests whereas it fails the significance tests in the lower troposphere. The RMS differences in specific humidity between simulations and observations are 0.5–0.7 g kg^{-1} (Fig. 3.3b). The RMS difference between COARE1 and observations is smaller than that between COARE2 and observations in mid-troposphere (500–700 hPa). The linear correlation coefficient of specific humidity between COARE1 and observations is much larger than that between COARE2 and observations around 400–800 hPa although the former is slightly smaller than the latter in the lower troposphere (below 800 hPa) (Fig. 3.3d).

Following Yanai et al. (1973), apparent heat source (Q_1) and apparent vapor source (Q_2) are calculated in Fig. 3.4. Simulated Q_1 and Q_2 basically follow observed Q_1 and Q_2, which is due to the imposed large-scale vertical velocity

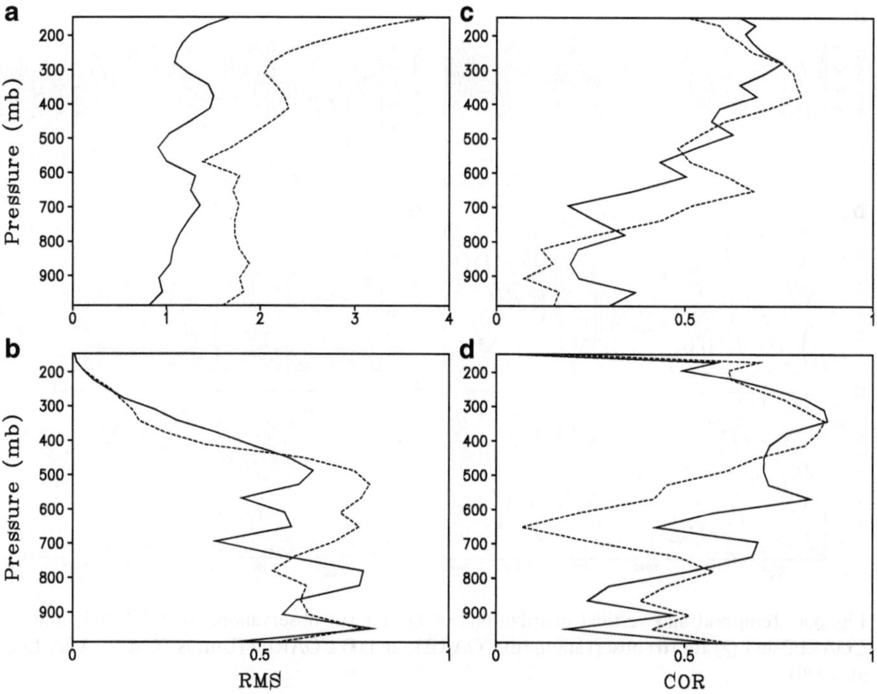

Fig. 3.3 Vertical distributions of root-mean-square (RMS) differences of (**a**) temperatures (°C) and (**b**) specific humidity (g kg^{-1}) and linear correlation coefficients (COR) of (**c**) temperatures and (**d**) specific humidity between simulations and observations. *Solid* and *dashed lines* denote COARE1 and COARE2, respectively

in COARE1 and large-scale advective cooling and moistening in COARE2. The differences of magnitudes between simulations and observations could be 5–10 °C d^{-1}. More positive centers of Q_1 occur in COARE1 than in COARE2 and observation, as a result of high-frequency convective response (Li et al. 1999).

Although the simulated solar fluxes follow the observed flux, the simulated solar fluxes (e.g., time mean in COARE1 is 205 W m^{-2}) are much larger than the observed flux (time mean is 148 W m^{-2}) (Fig. 3.5). The possible causes for the differences include that parameterized cloud–radiation interaction processes in the model may produce more convective cells and less stratiform cells (Sui et al. 1998) and that observational data quality is problematic as indicated by unclosed budgets. The simulated latent heat fluxes (e.g., time mean in COARE1 is 164 W m^{-2}) are also larger than the observed flux (time mean is 145 W m^{-2}), which may be caused by the treatment of surface gusty winds associated with strong convective events. The simulated surface rain rates have larger fluctuations than the observed rain rate does although the simulated rain rates basically follow the observation (Fig. 3.6). Since the surface rain rate is largely determined by the imposed large-scale forcing (Gao et al. 2005), the mismatch between vapor budget and surface rain rate in

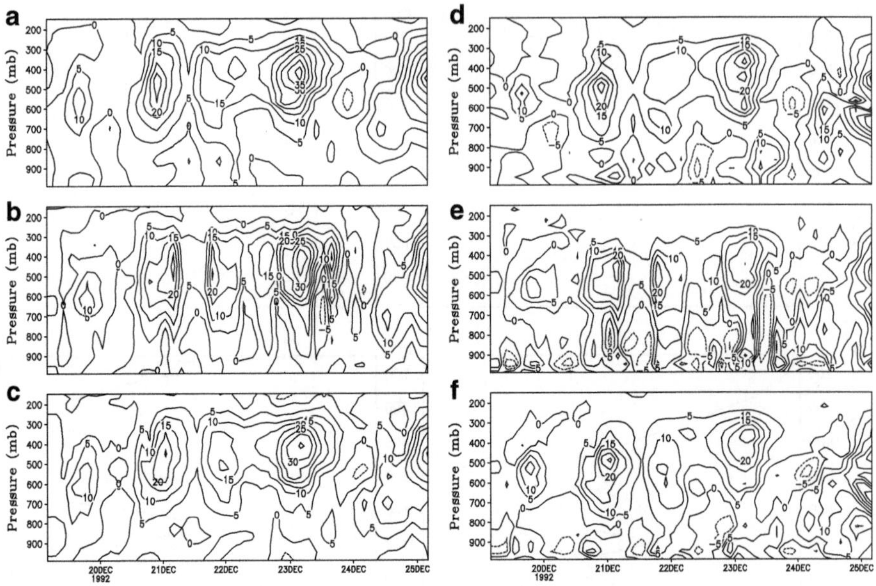

Fig. 3.4 Temporal and vertical distributions of Q_1 for (**a**) observation, (**b**) COARE1, and (**c**) COARE2 and Q_2 for (**d**) observation, (**e**) COARE1, and (**f**) COARE2 (Unit is °C d^{-1}. After Li et al. 1999)

observations may yield the difference. For example, a strong ascending motion and associated advective cooling and moistening occur late evening of the 22 and early morning of the 23 of December 1992 (Fig. 3.1), whereas the observed surface rain rate is small (Fig. 3.6). Li et al. (1999) also compared convective and stratiform rainfalls between COARE1 and COARE2 and found that COARE1 produces more convective rainfall (67 %) than COARE2 does (57 %).

Li et al. (2000) conducted a 2D coupled ocean–cloud-resolving atmosphere simulation with the forcing from TOGA COARE and showed remarkable similarity of surface fluxes between simulations and observations in both phase and amplitude (Fig. 3.7). However, the model simulations differ from the observations in three ways. First, simulated surface wind stresses are much smaller than observed wind stresses on 21–23 December 1992. Second, simulated surface net heat fluxes are larger than observed net heat fluxes on 20–22 December. The smaller magnitudes of observed surface net heat flux are caused by the smaller magnitudes of observed surface solar radiative flux. The larger magnitudes of solar radiative flux in the model simulation may be due to the fact that the model produces more convective cells and less stratiform cells. The third difference is that simulated freshwater fluxes are much larger than observed freshwater fluxes on 21–23 December. The larger amplitudes of simulated freshwater fluxes are accounted for by the larger simulated surface rain rate associated with stronger ascending motion imposed in the cloud-resolving model. The observed surface rain rate is not consistent with

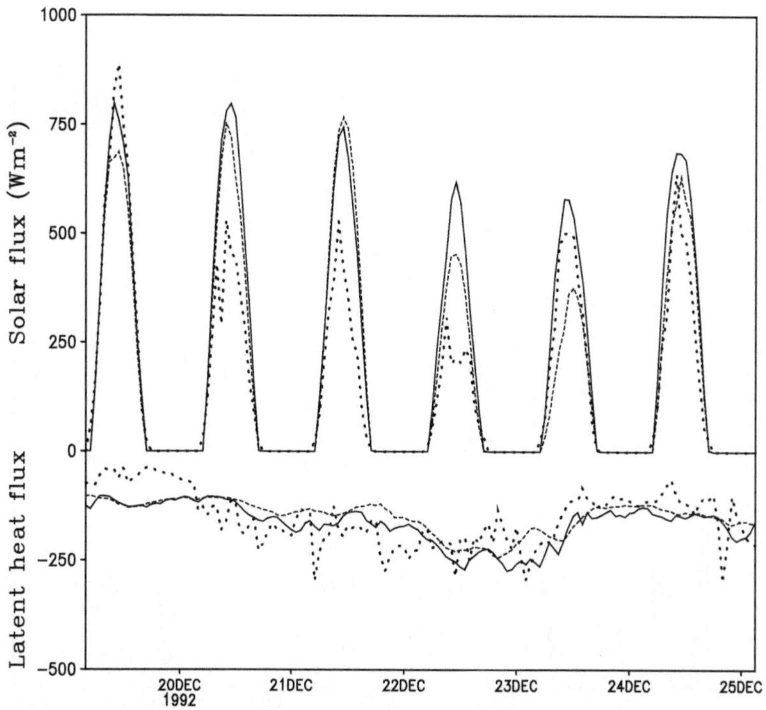

Fig. 3.5 Time series of surface solar radiative fluxes and latent heat fluxes simulated in COARE1 (*solid*) and COARE2 (*dashed*). *Dotted lines* denote observed fluxes (Unit is W m^{-2}. After Li et al. 1999)

observed ascending motion imposed in the model. The differences include the small simulated zonal wind stress, large simulated surface heat fluxes, and large simulated freshwater flux. The coupled model produces mixed-layer temperature and salinity reasonably well (Fig. 3.8), although the simulated diurnal temperature amplitudes are larger than the observed diurnal amplitudes. The RMS differences in mixed-layer temperature and 3-m salinity between simulations and observations are 0.28 °C and 0.09PSU, respectively.

3.2 Model Responses to Initial Moisture Perturbations

Accurate precipitation forecast may rely on accurate initial conditions. Initial conditions include temperature, moisture, and cloud hydrometeors. As a unique source for condensation, deposition, and surface rainfall, water vapor plays a crucial role in determining tropical convective development. The comparison between the National Oceanic and Atmospheric Administration (NOAA)/National Environmental Satellite, Data, and Information Service (NESDIS)/Microwave Surface and Precipitation

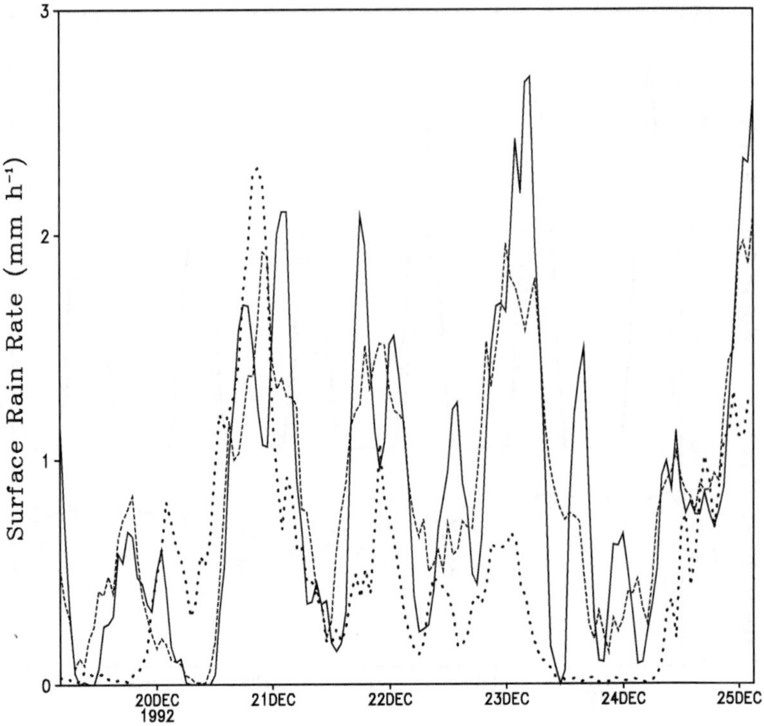

Fig. 3.6 Time series of surface rain rates simulated in COARE1 (*solid*) and COARE2 (*dashed*). *Dotted lines* denote observed rain rate. Unit is mm h^{-1} (After Li et al. 1999)

Products System (MSPPS) data and National Centers for Environmental Prediction (NCEP)/Global Data Assimilation System (GDAS) data showed that the RMS difference in ice water path (*IWP*) is 0.12 mm, which is larger than the standard deviation of *IWP* for GDAS (0.06 mm). The RMS difference in PW over cloudy regions is 4.5 mm, which is smaller than the standard deviation of *PW* for GDAS (6.0 mm). The area-mean *IWP* and *PW* for GDAS are 0.04 mm and 52.1 mm, respectively. Thus, the statistical error that is defined as the ratio of the RMS difference to the area mean is 300 % for *IWP* while it only is 8.6 % for *PW*. Does a small initial *PW* error lead to a large *IWP* error? Li et al. (2006) used 2D cloud-resolving model with imposed forcing (zonally uniform vertical velocity, zonal wind, along with thermal and moisture advection) based on 6-hourly NCEP/GDAS data over tropical area (averaged over 150–160°E, EQ) to study model responses to initial moisture perturbations. Daily-mean *SST* data retrieved from NASA/Tropical Rainfall Measuring Mission (TRMM) Microwave Imager (TMI) radiometer with a 10.7 GHz channel (Wentz et al. 2000) are also imposed in the model. Figure 3.9 shows temporal and vertical distribution of the large-scale vertical velocity and zonal wind from 1100 local standard time (LST) 18 to 1700 LST 26 April 2003 (8.25 days total), which are imposed in the model during the integrations. The

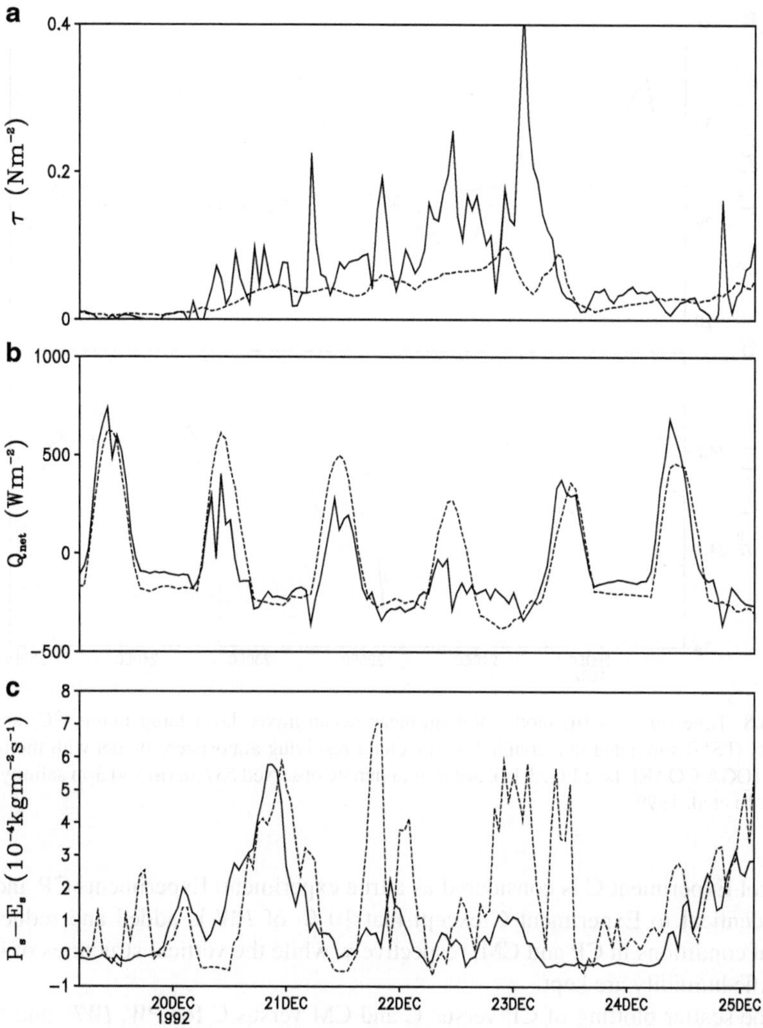

Fig. 3.7 Time series of (**a**) model domain-mean zonal wind stress (N m^{-2}), (**b**) net surface heat fluxes (W m^{-2}), and (**c**) freshwater flux (10^{-4} kg m^{-2} s^{-1}) observed during TOGA COARE (*solid*) and simulated in the 2D coupled ocean–cloud-resolving atmosphere model (*dashed*) with the forcing from TOGA COARE (see Fig. 3.1a) (After Li et al. 2000)

ascending motion with maximum of –4 hPa h^{-1} occurs around 300 hPa on 18 April 2003. Moderate upward motions of −2 mm h^{-1} appear daily in mid and lower troposphere from 20 to 22 April when westerly winds confine in the lower troposphere while easterly winds weaken gradually. Two strong ascending motion centers dominate the troposphere on 24 and 25 April when the westerly winds switch into the intensified easterly winds. Three experiments are designed. The

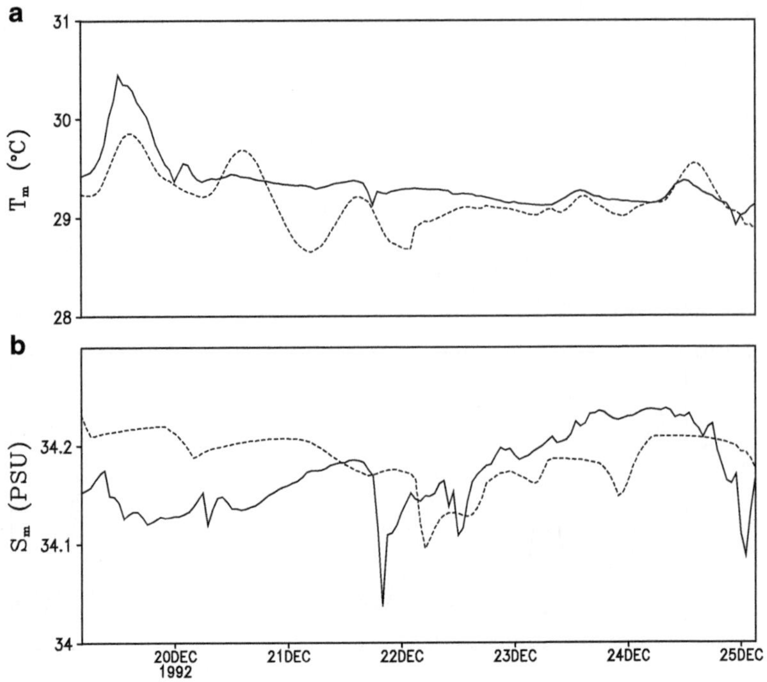

Fig. 3.8 Time series of (**a**) model domain-mean ocean mixed-layer temperature (°C) and 3-m salinity (PSU) simulated in a coupled ocean–cloud-resolving atmosphere model with the forcing from TOGA COARE (see Fig. 3.1a). *Solid lines* denote observed *SST* in (**a**) and 3-m salinity in (**b**) (After Li et al. 1999)

control Experiment C is considered as a true experiment. Experiments CP and CM are identical to Experiment C except that 10 % of *PW* is added and reduced in initial conditions in CP and CM, respectively, while the vertical structures of initial specific humidity are kept.

The scatter plotting of CP versus C and CM versus C for *PW*, *IWP*, and liquid water path (*LWP*) is shown in Fig. 3.10. The *PW* simulated in CP and C is along the diagonal line of the diagram, indicating a small RMS difference (1.1 mm), although the initial difference in PW is 5.1 mm. The *PW* simulated in CM and C is below the diagonal line of the diagram, suggesting a large RMS difference (3.3 mm). The RMS difference between CP and C is much smaller than the standard deviation of CP (3.3 mm) whereas that between CM and C is marginally smaller than the standard deviation of CM (3.8 mm).

Cloud hydrometeors (*LWP* and *IWP*) simulated in CP and CM versus those simulated in C show large scattering patterns that are away from the diagonal lines (Fig. 3.10c–f). The RMS difference in *IWP* between CP and C is 0.104 mm whereas that between CM and C is 0.107 mm. Both are larger than the standard deviations of CP (0.094 mm) and CM (0.093 mm). The small initial difference in *PW* produces

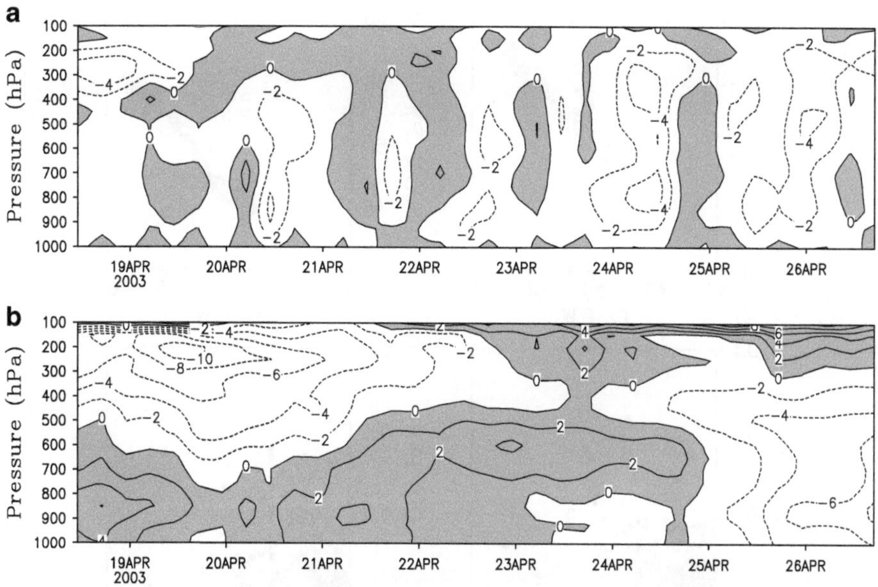

Fig. 3.9 Time–pressure cross sections of (**a**) *vertical* motion (mb h^{-1}) and (**b**) zonal wind (m s^{-1}) obtained from GDAS during the selected 8-day period. *Downward* motion in (*a*) and westerly wind in (*b*) are shaded (After Li et al. 2006)

the large difference in *IWP*. This demonstrates that the large RMS difference in *IWP* between MSPPS and GDAS data may be caused by the small RMS difference in *PW*. The large scattering in *IWP* between CP/CM and C associated with the small scattering in PW implies uncertainties in cloud microphysical parameterization schemes that are nonlinear functions of temperature and water vapor. The RMS difference in *LWP* between CP and C is 0.085 mm whereas that between CM and C is 0.096 mm. Both are smaller than the standard deviations of CP (0.098 mm) and CM (0.103 mm).

Surface rain rates simulated in CP and CM versus those simulated in C also show large scattering patterns (Fig. 3.11a, b). The RMS difference in P_s between CP and C is 0.28 mm h^{-1} whereas that between CM and C is 0.33 mm h^{-1}. Both are similar to the standard deviations of CP (0.29 mm) and CM (0.32 mm). Although the imposed large-scale ascending motion is identical in three experiments, the small differences in PW still produce the large differences in the surface rain rate.

Q_{WVT} and Q_{CM} display large scattering patterns between CP/CM and C whereas Q_{WVF} and Q_{WVE} show good relationships between CP and C and CM and C (Fig. 3.11). The RMS differences in Q_{WVF} (0.01 mm h^{-1}) and Q_{WVE} (0.02 mm h^{-1}) are significantly smaller than the standard deviations of CP (0.24 mm h^{-1}) and CM (0.04 mm h^{-1}), respectively. The small RMS difference in Q_{WVF} reflects the dominance of imposed vertical velocity during the integrations. The RMS differences in Q_{WVT} between CP and C (0.37 mm h^{-1}) and CM and C (0.4 mm h^{-1}) are larger than the standard deviations of CP (0.32 mm h^{-1}) and CM (0.3 mm h^{-1}),

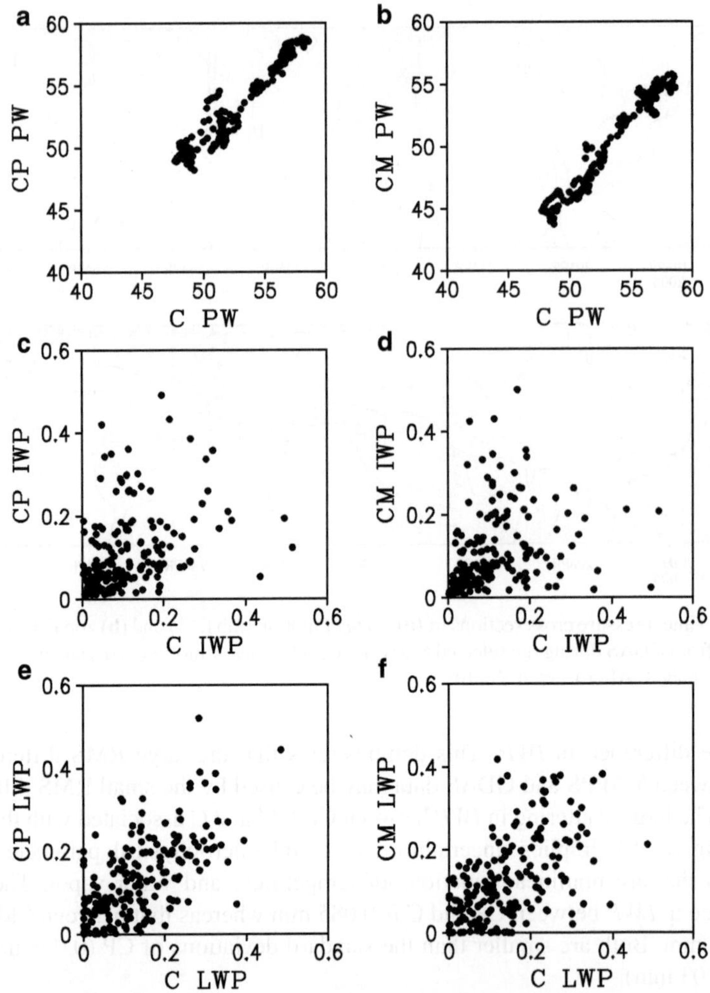

Fig. 3.10 Scatter plots of CP versus C for model domain averaged (**a**) *PW*, (**c**) *IWP*, and (**e**) *LWP* and CM versus C for (**b**) *PW*, (**d**) *IWP*, (**f**) *LWP* (Unit is in mm. After Li et al. 2006)

respectively, whereas the RMS differences in Q_{CM} between CP and C (0.27 mm h^{-1}) and CM and C (0.29 mm h^{-1}) are larger than the standard deviations of CP (0.19 mm h^{-1}) and CM (0.19 mm h^{-1}), respectively. This indicates that the RMS differences in Q_{WVT} and Q_{CM} largely contribute to the RMS differences in P_s.

To explain the large RMS differences in Q_{WVT} and Q_{CM}, the *PW* and total cloud budgets will be separately analyzed. The *PW* budget (2.2b) can be expressed as

$$Q_{WVT} + Q_{WVF} + Q_{WVE} = P_{CND} + P_{DEP} + P_{SDEP} + P_{GDEP} - P_{REVP} - P_{MLTS} - P_{MLTG}, \tag{3.1a}$$

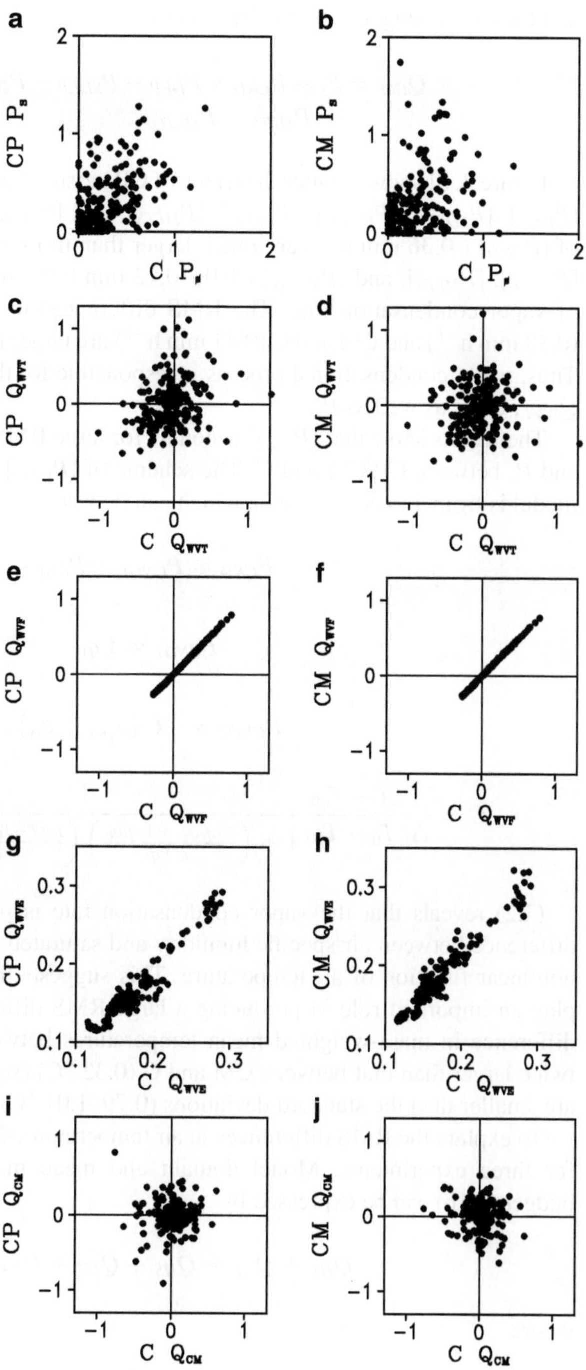

Fig. 3.11 Scatter plots of CP versus C for (**a**) P_s, (**c**) Q_{WVT}, (**e**) Q_{WVF}, (**g**) Q_{WVE}, (**i**) and Q_{CM} and CM versus C for (**b**) P_s, (**d**) Q_{WVT}, (**f**) Q_{WVF}, (**h**) Q_{WVE}, (**j**) Q_{CM}. (Units are in mm h^{-1}. After Li et al. 2006)

and the cloud budget (2.6) can be written as

$$Q_{CM} = P_s - P_{CND} - P_{DEP} - P_{SDEP} - P_{GDEP} + P_{REVP} + P_{MLTS} + P_{MLTG}. \tag{3.1b}$$

Figure 3.12 shows scatter diagrams of CP versus C and CM versus C for $[P_{CND}]$, $[P_{DEP}]$, $[P_{SDEP}]$, $[P_{GDEP}]$, $[P_{REVP}]$, $[P_{MLTS}]$, and $[P_{MLTG}]$. The standard deviations of $[P_{CND}]$ (0.36 mm h^{-1}) are much larger than those of $[P_{DEP}]$, $[P_{SDEP}]$, $[P_{GDEP}]$, $[P_{REVP}]$, $[P_{MLTS}]$, and $[P_{MLTG}]$ (0.06–0.12 mm h^{-1}), indicating a large fluctuation of vapor condensation rate. The RMS differences in $[P_{CND}]$ between CP and C (0.38 mm h^{-1}) and CM and C (0.41 mm h^{-1}) are larger than the standard deviations. Thus, vapor condensational process is responsible for the large RMS differences in Q_{WVT}, Q_{CM}, as well as P_s.

The results show that $[P_{CND}]$ accounts for large RMS differences in Q_{WVT}, Q_{CM}, and P_s between CP/CM and C. The scheme of $[P_{CND}]$ used in the cloud-resolving model is from Tao et al. (1989), which can be written as

$$P_{CND} = P_{CND1} + P_{CND2}, \tag{3.2}$$

$$P_{CND1} = Cq_v, \tag{3.2a}$$

$$P_{CND2} = -C\left(q_{qws} + q_{is}\right), \tag{3.2b}$$

$$C = \frac{1}{\Delta t}\frac{T - T_{00}}{T_0 - T_{00}}\frac{1}{1 + \left(\frac{A_1 q_c q_{ws} + A_2 q_i q_{is}}{q_c + q_i}\right)\left(\frac{L_v(T - T_{00}) + L_s(T_0 - T)}{c_p(T_0 - T_{00})}\right)}. \tag{3.2c}$$

(3.2) reveals that the vapor condensation rate is primarily determined by the difference between air specific humidity and saturated specific humidity that is the nonlinear function of air temperature. This suggests that the air temperature may play an important role in producing a large RMS difference in $[P_{CND}]$. The RMS difference in mass-weighted mean temperatures between CP and C (0.63 °C) is twice larger than that between CM and C (0.32 °C) (see Fig. 3.13a), although both are smaller than the standard deviations (0.79–1.01 °C).

To explain the RMS differences in air temperature, the heat budgets are analyzed for three experiments. Model domain and mean mass-weighted mean thermal budget (2.1b) can be expressed by

$$Q_{HT} + Q_{HF} + Q_{HS} + Q_{LH} + Q_{RAD} = 0, \tag{3.3}$$

where

$$Q_{HT} = -\frac{\partial < T >}{\partial t}, \tag{3.3a}$$

Fig. 3.12 Scatter plots of CP versus C for (**a**) –[P_{CND}], (**c**) –[P_{DEP}]-[P_{SDEP}], (**e**) –[P_{GDEP}], (**g**) [P_{REVP}], and (**i**) [P_{MLTS}] + [P_{MLTG}] and CM versus C for (**b**) –[P_{CND}], (**d**) –[P_{DEP}]-[P_{SDEP}], (**f**) –[P_{GDEP}], (**h**) [P_{REVP}], and (**j**) [P_{MLTS}] + [P_{MLTG}] (Units are in mm h^{-1}. After Li et al. 2006)

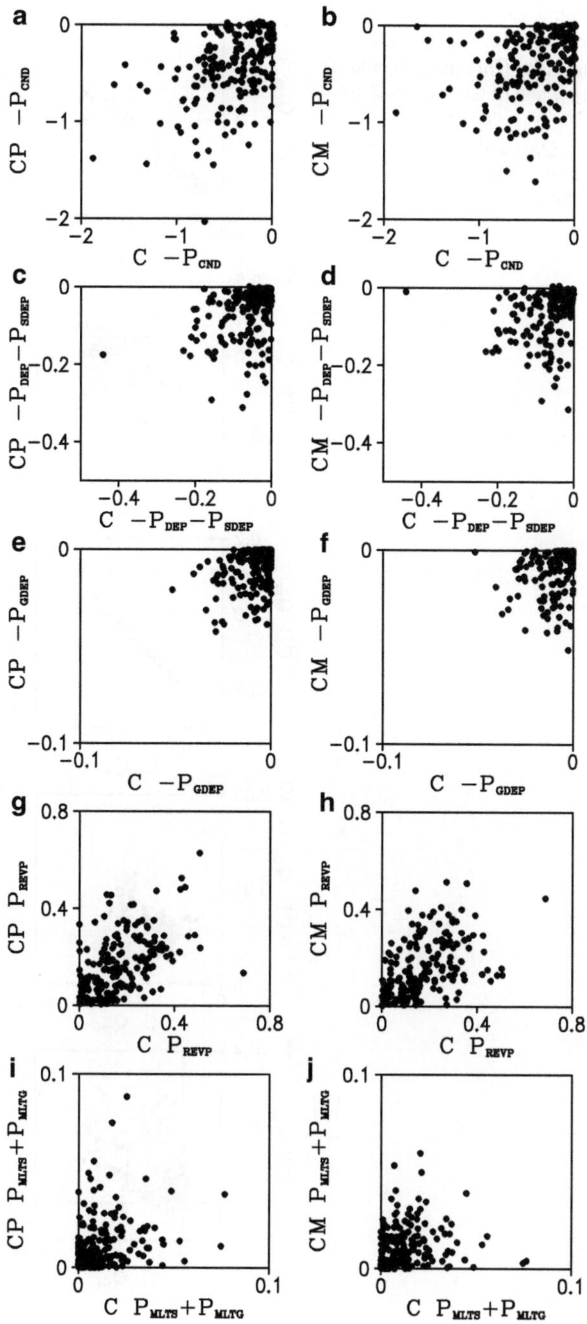

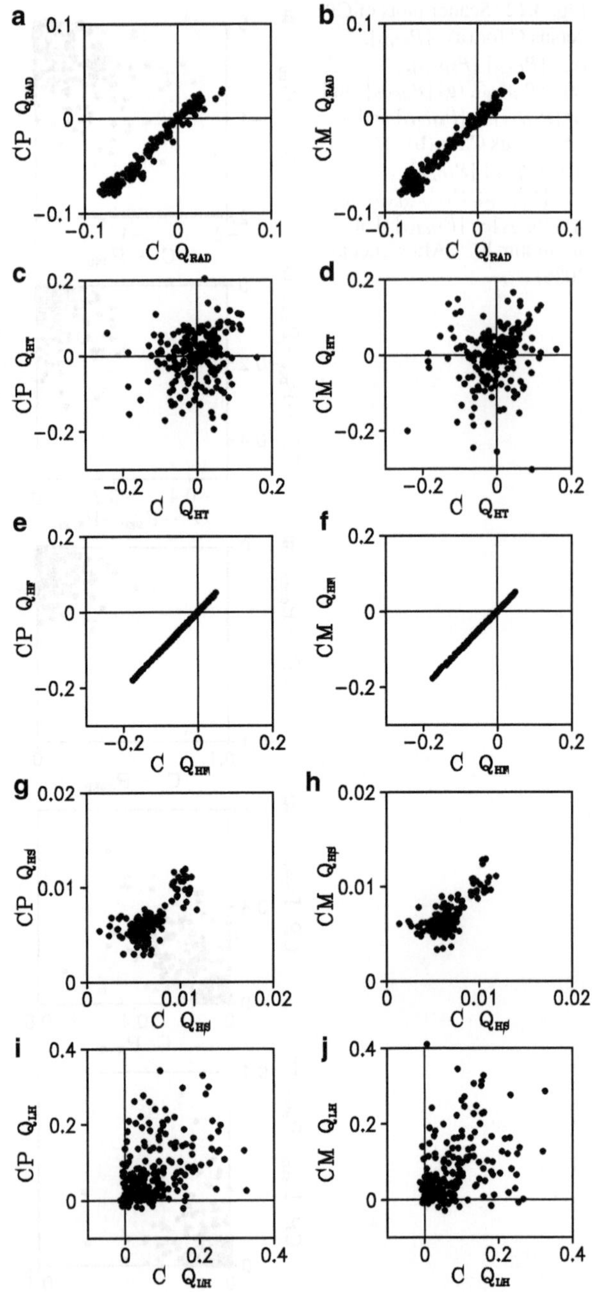

Fig. 3.13 Scatter plots of CP versus C for (**a**) Q_{RAD}, (**c**) Q_{HT}, (**e**) Q_{HF}, (**g**) Q_{HS}, and (**i**) Q_{LH} and of CM versus C for (**b**) Q_{RAD}, (**d**) Q_{HT}, (**f**) Q_{HF}, (**h**) Q_{HS}, and (**j**) Q_{LH} (Units are in °C h^{-1}. After Li et al. 2006)

$$Q_{HF} = - < \overline{u}^o \frac{\partial \overline{T}^o}{\partial x} > - < \pi \overline{w}^o \frac{\partial \overline{\theta}}{\partial z} >, \tag{3.3b}$$

$$Q_{HS} = H_s, \tag{3.3c}$$

$$Q_{LH} = \frac{1}{c_p} < Q_{cn} >, \tag{3.3d}$$

$$Q_{RAD} = \frac{1}{c_p} < Q_R > . \tag{3.3e}$$

In (3.3), Q_{HT} is local thermal change; Q_{HF} is thermal convergence, which is mainly contributed by vertical advective cooling/warming since imposed horizontal thermal advection is much smaller than the vertical advection; H_s is surface sensible heat flux; Q_{LH} denotes the net latent heat release through phase changes among different cloud species; and Q_{RAD} is the radiative heating rate due to convergence of net flux of solar and infrared radiative fluxes. The RMS differences between CP/CM and C (also see Fig. 3.13) are 0.01 °C for Q_{RAD}, 0.09 °C for Q_{HT}, 0.01 °C for Q_{FT}, 0.01 °C for Q_{HS}, and 0.09 °C for Q_{LH}, indicating that the RMS differences in condensational heating account for those in local thermal changes. The RMS differences in Q_{HT} and Q_{LH} are larger than the standard deviations (0.07 °C for Q_{HT} and 0.08 °C for Q_{LH}).

$[P_{CND}]$ can be broken down into $[P_{CND1}]$ and $[P_{CND2}]$ as indicated by (3.2). The variation of $[P_{CND1}]$ is determined by specific humidity, whereas the variation of $[P_{CND2}]$ is determined by saturated specific humidity that is a function of temperature. $[P_{CND1}]$ and $[P_{CND2}]$ are negatively correlated with similar magnitudes of about 3×10^3 mm h^{-1} (not shown) whereas $[P_{CND}]$ has the magnitudes of about 2 mm h^{-1} (see Fig. 3.12). Thus, vapor condensation rate is the small residual between the two large terms related to specific humidity and temperature-dependent saturated specific humidity. Furthermore, the variance in $[P_{CND}]$ between CP and C can be calculated by

$$\begin{aligned} Var\,(P_{CND,CP}, P_{CND,C}) &= Var\,(P_{CND1,CP}, P_{CND1,C}) \\ &+ 2CoVar\,(P_{CND1,CP} - P_{CND1,C}, P_{CND2,CP} - P_{CND2,C}) \\ &+ Var\,(P_{CND2,CP}, P_{CND2,C})\,, \end{aligned} \tag{3.4}$$

$$Var\,(P_{F,CP}, P_{F,C}) = \frac{1}{N}\sum_{I=1}^{N} [P_{F,CP}(I) - P_{F,C}(I)]^2, \tag{3.4a}$$

$$\begin{aligned} &CoVar\,(P_{CND1,CP} - P_{CND1,C}, P_{CND2,CP} - P_{CND2,C}) \\ &= \frac{1}{N}\sum_{I=1}^{N} [P_{CND1,CP}(I) - P_{CND1,C}(I)]\,[P_{CND2,CP}(I) - P_{CND2,C}(I)]\,, \end{aligned} \tag{3.4b}$$

where $N = 198$ and $F = CND, CND1, CND2$. (3.4) can be also applied to calculate the variance in P_{CND} between CM and C. $Var(P_{CND,CP}, P_{CND,C})$, $Var(P_{CND1,CP}, P_{CND1,C})$, $CoVar(P_{CND1,CP}\text{-}P_{CND1,C}, P_{CND2,CP}\text{-}P_{CND2,C})$, and $Var(P_{CND2,CP}, P_{CND2,C})$ are 0.142, 3151.803, −6304.007, and 3152.346 mm^2 h^{-2}, respectively. $Var(P_{CND,CM}, P_{CND,C})$, $Var(P_{CD1,CM}, P_{CND1,C})$, $CoVar(P_{CND1,CM}\text{-}P_{CND1,C}, P_{CND2,CM}\text{-}P_{CND2,C})$, and $Var(P_{CND2,CM}, P_{CND2,C})$ are 0.167, 26395.184, −52792.895, and 26397.871 mm^2 h^{-2}, respectively. The variances in $[P_{CND}]$ between CP/CM and C are four to five orders of magnitudes smaller than variances in $[P_{CND1}]$ and $[P_{CND2}]$ as well as covariances between CP/CM and C, implying that small perturbations in specific humidity and saturated specific humidity could cause large differences in $[P_{CND}]$.

The above analysis suggests that the improvement of cloud simulations may rely on the reduced error in initial conditions or improvement of accuracy of calculation of vapor condensation rate. Experiment CP shows that the small difference in PW during the integration can produce large difference in cloud simulation. Thus, the improvement of calculation of vapor condensation rate may be the only way to produce accurate cloud simulations.

3.3 Comparison Between 2D and 3D Simulations

Due to limitations in computational power, 3D cloud-resolving simulations have been carried out less frequently than 2D simulations. With increasing computer power in recent decades, 3D cloud-resolving simulations have been conducted to study deep convection (e.g., Wilhelmson 1974; Miller and Pearce 1974; Pastushkov 1975; Klemp and Wilhelmson 1978; Cotton and Tripoli 1978; Clark 1979; Tao and Soong 1986; Redelsperger and Sommeria 1986; Redelsperger and Lafore 1988; Grabowski et al. 1998; Tompkins and Craig 1998).

It should be noted that some cloud structures and mass circulations may not be well represented by a 2D model. For example, Moncrieff and Miller (1976) showed that the 3D crossover flow pattern associated with propagating tropical squall lines can only be simulated in the 3D framework. In contrast, Rotunno et al. (1988) found that the 2D framework captures well the basic dynamics associated with long-lived squall lines in strong low-level shear. Tao and Soong (1986) and Tao et al. (1987) conducted a comparison study between 2D and 3D cloud-resolving model simulations in terms of the collective thermodynamic feedback effects and vertical transports of mass, sensible heat, and moisture and found profound similarities since both 2D and 3D models simulate the convective line structures well. Grabowski et al. (1998) compared 2D and 3D cloud-resolving model simulations with the GATE forcing in terms of their thermodynamic fields, surface heat fluxes, and surface precipitation and showed similarity. Sui et al. (2005) showed the statistical equivalence between large-scale precipitation efficiency and cloud microphysics precipitation efficiency in the calculations with grid data from both 2D cloud-resolving model simulation of tropical squall lines with the TOGA COARE forcing and 3D cloud-resolving model simulation of typhoons without imposed forcing.

Wang et al. (2007) combined a 2D cloud-resolving model simulation with dual-Doppler and polarimetric radar analysis to study the evolution, dynamic structure, cloud microphysics, and rainfall process of a monsoon convection observed during the South China Sea (SCS) summer monsoon onset and found a good agreement between the model simulations and the radar observations. Therefore, we will use a 2D cloud-resolving model with the large-scale forcing of vertical velocity from the TOGA COARE in the following discussions unless otherwise clearly indicated.

References

Clark TL (1979) Numerical simulations with a three-dimension cloud model: lateral boundary condition experiments and multicellular severe storm simulations. J Atmos Sci 36:2191–2215

Cotton WR, Tripoli GJ (1978) Cumulus convection in shear flow: three-dimensional numerical experiments. J Atmos Sci 35:1503–1521

Gao S, Cui X, Zhou Y, Li X (2005) Surface rainfall processes as simulated in a cloud resolving model. J Geophys Res. doi:10.1029/2004JD005467

Grabowski WW, Wu X, Moncrieff MW (1996) Cloud-resolving model of tropical cloud systems during phase III of GATE. Part I: two-dimensional experiments. J Atmos Sci 53:3684–3709

Grabowski WW, Wu X, Moncrieff MW, Hall WD (1998) Cloud-resolving model of tropical cloud systems during phase III of GATE. Part II: effects of resolution and the third spatial dimension. J Atmos Sci 55:3264–3282

Klemp JB, Wilhelmson RB (1978) The simulation of three-dimensional convective storm dynamics. J Atmos Sci 35:1070–1093

Li X, Sui CH, Lau KM, Chou MD (1999) Large-scale forcing and cloud-radiation interaction in the tropical deep convective regime. J Atmos Sci 56:3028–3042, (c) American Meteorological Society. Reprinted with permission

Li X, Sui CH, Lau KM, Adamec D (2000) Effects of precipitation on ocean mixed-layer temperature and salinity as simulated in a 2-D coupled ocean-cloud resolving atmosphere model. J Meteorol Soc Jpn 78:647–659, (c) Meteorological Society of Japan. Reprinted with permission

Li X, Zhang S, Zhang DL (2006) Thermodynamic, cloud microphysics and rainfall responses to initial moisture perturbations in the tropical deep convective regime. J Geophys Res doi:10.1029/2005JD006968, (c) American Geophysical Union. Reprinted with permission

Miller MJ, Pearce RP (1974) A three-dimensional primitive equation model of cumulonimbus convection. Q J Roy Meteorol Soc 100:133–154

Moncrieff MW, Miller MJ (1976) The dynamics and simulation of tropical cumulonimbus and squall line. Q J Roy Meteorol Soc 102:373–394

Moncrieff MW, Krueger SK, Gregory D, Redelsperger JL, Tao WK (1997) GEWEX cloud system study (GCSS) working group 4: precipitating convective cloud systems. Bull Am Meteorol Soc 78:831–845

Pastushkov RS (1975) The effects of vertical wind shear on the evolution of convective clouds. Q J Roy Meteorol Soc 101:281–291

Redelsperger JL, Lafore JP (1988) A three-dimensional simulation of a tropical squall line: convective organization and thermodynamic vertical transport. J Atmos Sci 45:1334–1356

Redelsperger JL, Sommeria G (1986) Three-dimensional simulation of a convective storm: sensitivity studies on subgrid parameterization and spatial resolution. J Atmos Sci 43:2619–2635

Rotunno R, Klemp JB, Weisman ML (1988) A theory for strong, long-lived squall lines. J Atmos Sci 45:463–485

Sui CH, Lau KM, Takayabu Y, Short D (1997) Diurnal variations in tropical oceanic cumulus ensemble during TOGA COARE. J Atmos Sci 54:639–655

Sui CH, Li X, Lau KM (1998) Radiative-convective processes in simulated diurnal variations of tropical oceanic convection. J Atmos Sci 55:2345–2359

Sui CH, Li X, Yang MJ, Huang HL (2005) Estimation of oceanic precipitation efficiency in cloud models. J Atmos Sci 62:4358–4370

Takayabu YN, Lau KM, Sui CH (1996) Observation of a quasi-2-day wave during TOGA COARE. Mon Weather Rev 124:1892–1913

Tao WK, Soong ST (1986) The study of the response of deep tropical clouds to mesoscale processes: three-dimensional numerical experiments. J Atmos Sci 43:2653–2676

Tao WK, Simpson J, Soong ST (1987) Statistical properties of a cloud ensemble: a numerical study. J Atmos Sci 44:3175–3187

Tao WK, Simpson J, McCumber M (1989) An ice-water saturation adjustment. Mon Weather Rev 117:231–235

Tompkins AM, Craig GC (1998) Radiative-convective equilibrium in a three-dimensional cloud-ensemble model. Q J Roy Meteorol Soc 124:2073–2097

Wang JJ, Li X, Carey L (2007) Evolution, structure, cloud microphysical and surface rainfall processes of a monsoon convection during the South China Sea monsoon experiment. J Atmos Sci 64:360–379

Wentz FJ, Gentemann C, Smith D, Chelton D (2000) Satellite measurements of sea surface temperature through clouds. Science 288:847–850

Wilhelmson RB (1974) The life cycle of a thunderstorm in three dimensions. J Atmos Sci 31:1629–1651

Wu X, Grabowski WW, Moncrieff MW (1998) Long-term evolution of cloud systems in TOGA COARE and their interactions with radiative and surface processes. Part I: two-dimensional cloud-resolving model. J Atmos Sci 55:2693–2714

Xu KM, Randall DA (1996) Explicit simulation of cumulus ensembles with the GATE phase III data: comparison with observations. J Atmos Sci 53:3710–3736

Yanai M, Esbensen S, Chu JH (1973) Determination of bulk properties of tropical cloud clusters from large-scale heat and moisture budgets. J Atmos Sci 30:611–627

Chapter 4
Surface Rainfall Processes

Abstract In this chapter, surface rainfall budget is analyzed using model domain-mean cloud-resolving simulation data of tropical rainfall event during TOGA COARE. The rain rate is mainly associated with water vapor convergence, while it is significantly modified by local changes of water vapor and cloud hydrometeor at a short time scale. The analysis of time- and domain-mean data shows that the rain rate corresponds mainly to surface evaporation flux and water vapor convergence, while local changes of water vapor and cloud hydrometeor have minor contributions to the rain rate. The mean rain rate comes mainly from convective rainfall regions, where cloud hydrometeor is transported into stratiform rainfall regions. Stratiform rain rate is mainly related to atmospheric drying. The surface rainfall budget is also applied to lag correlation analysis and the analysis of convective development at different stages.

Keywords Surface rainfall budget • Water vapor convergence • Local change • Water vapor • Surface evaporation flux • Cloud hydrometeors

Surface rain rate is one of the most important parameters in meteorology and hydrology, and its accurate measurement and quantitative estimate and forecast have significant economic and social impacts in rainfall-rich countries. However, it is very difficult to accurately measure and predict the surface rain rate since surface rainfall processes are associated with multi-scale dynamic, thermodynamic, cloud microphysical, and radiative processes and their interactions. The roles of clouds in moist air mass conservation have been given attention in the meteorological research community, in which cloud source/sink is included in the governing equation of water vapor with prognostic variables for cloud hydrometeors (e.g., Ooyama 1990, 2001; Bannon 2002). Following (2.4), the surface rain rate can be expressed by

$$P_s = Q_{WV} + Q_{CM}. \tag{4.1}$$

© Springer International Publishing Switzerland 2016

X. Li, S. Gao, *Cloud-Resolving Modeling of Convective Processes*,
Springer Atmospheric Sciences, DOI 10.1007/978-3-319-26360-1_4

Kuo (1965, 1974) calculated the surface rain rate and local vapor increase by the large (~95 %) and small (~5 %) parts of the vapor sink (vapor convergence plus surface evaporation) in his cumulus parameterization scheme, respectively, i.e.,

$$P_s = 0.95\left(Q_{WVF} + Q_{WVE}\right), \tag{4.2a}$$

$$-Q_{WVT} = 0.05\left(Q_{WVF} + Q_{WVE}\right). \tag{4.2b}$$

The addition of (4.2a) and (4.2b) leads to

$$P_s = Q_{WVT} + Q_{WVF} + Q_{WVE} = Q_{WV}. \tag{4.3}$$

Thus, the comparison between (4.1) and (4.3) indicates that Q_{CM} is not included in the calculation of precipitation in Kuo's scheme. The impacts of clouds on surface rainfall processes as an important link between environment and precipitation are missing in Kuo's scheme. What roles do water vapor and cloud hydrometeors play in surface rainfall processes? How different are the surface rainfall processes in convective and stratiform clouds and in different stages of convective development? These questions will be addressed based on Gao et al. (2005), Cui and Li (2006), and Zhou et al. (2006) in this chapter.

4.1 Time Series of Model Domain-Mean Surface Rain Rate

The experiment COARE uses the model with the imposed vertical velocity, zonal wind, horizontal thermal and vapor advection, and sea surface temperature. Figure 4.1 shows the time evolution of vertical distribution of the large-scale atmospheric vertical velocity, zonal wind, and *SST* during 19 December 1992–8 January 1993 that are imposed during the 21-day integration in experiment COARE. During this period, a westerly wind burst occurred along with a strong upward motion, causing significant cooling over the ocean surface, which was associated with the intraseasonal variability (Webster and Lukas 1992; Sui et al. 1997b). From 19 to 25 December 1992, the upward motion was dominant, indicating strong convection. From 26 December 1992 to 3 January 1993, the downward motion became dominant, along with occasional upward motion, suggesting a dry phase. In the last few days, the moderate upward motion occurred. Diurnal and 2-day signals are also detected in Fig. 4.1a as indicated by Sui et al. (1997a) and Takayabu et al. (1996), respectively. The large-scale westerly winds increase significantly in the lower- and mid-troposphere and reach their maximum of 20 m s^{-1} at 600 mb around 3 January 1993 (Fig. 4.1b). Except for the first and last 4 days, the *SST* had only a weak diurnal variation with a slowly decreasing trend (Fig. 4.1c). COARE also includes the imposed horizontal temperature and moisture advections derived by Sui et al. (1997a) (not shown), which have much smaller amplitudes than the vertical advections, respectively.

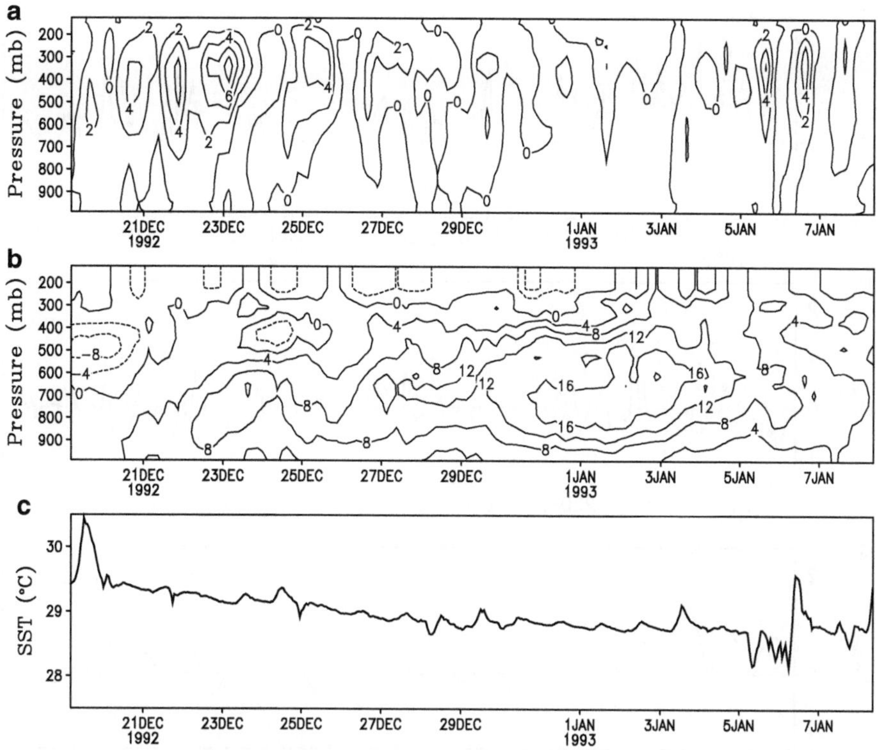

Fig. 4.1 Temporal and vertical distributions of (**a**) vertical velocity (cm s^{-1}), (**b**) zonal wind (m s^{-1}), and (**c**) time series of sea surface temperature (°C) observed and derived during a selected 21-day TOGA COARE period, which are imposed in the model as the forcing in experiment COARE (After Gao et al. 2005)

Figure 4.2 shows time series of model domain-mean P_s, Q_{WVT}, Q_{WVF}, Q_{WVE}, and Q_{CM}. The variation of the surface rain rate basically follows the variation of vapor convergence associated with the imposed vertical velocity. Its high-frequency fluctuation is associated with the local vapor and hydrometeor changes. The calculations of variances that measure the fluctuations indicate that the variance of rain rate (0.18 mm^2 h^{-2}) is dominated by those of the local vapor change (0.11 mm^2 h^{-2}) and vapor convergence (0.12 mm^2 h^{-2}) as well as the local condensate change (0.03 mm^2 h^{-2}), whereas the magnitudes of other variances and covariances are relatively small (<0.01 mm^2 h^{-2}). The linear correlation coefficients between P_s and Q_{WVT}, P_s and Q_{WVF}, P_s and Q_{WVE}, and P_s and Q_{CM} are 0.54, 0.62, 0.01, and 0.16, respectively. There are 486 samples, and the critical correlation coefficient at the 1 % significance level in the student-t test is 0.12. Thus, the variation of the surface rain rate can be largely explained by the local vapor change and vapor convergence. The results suggest that Kuo's cumulus parameterization

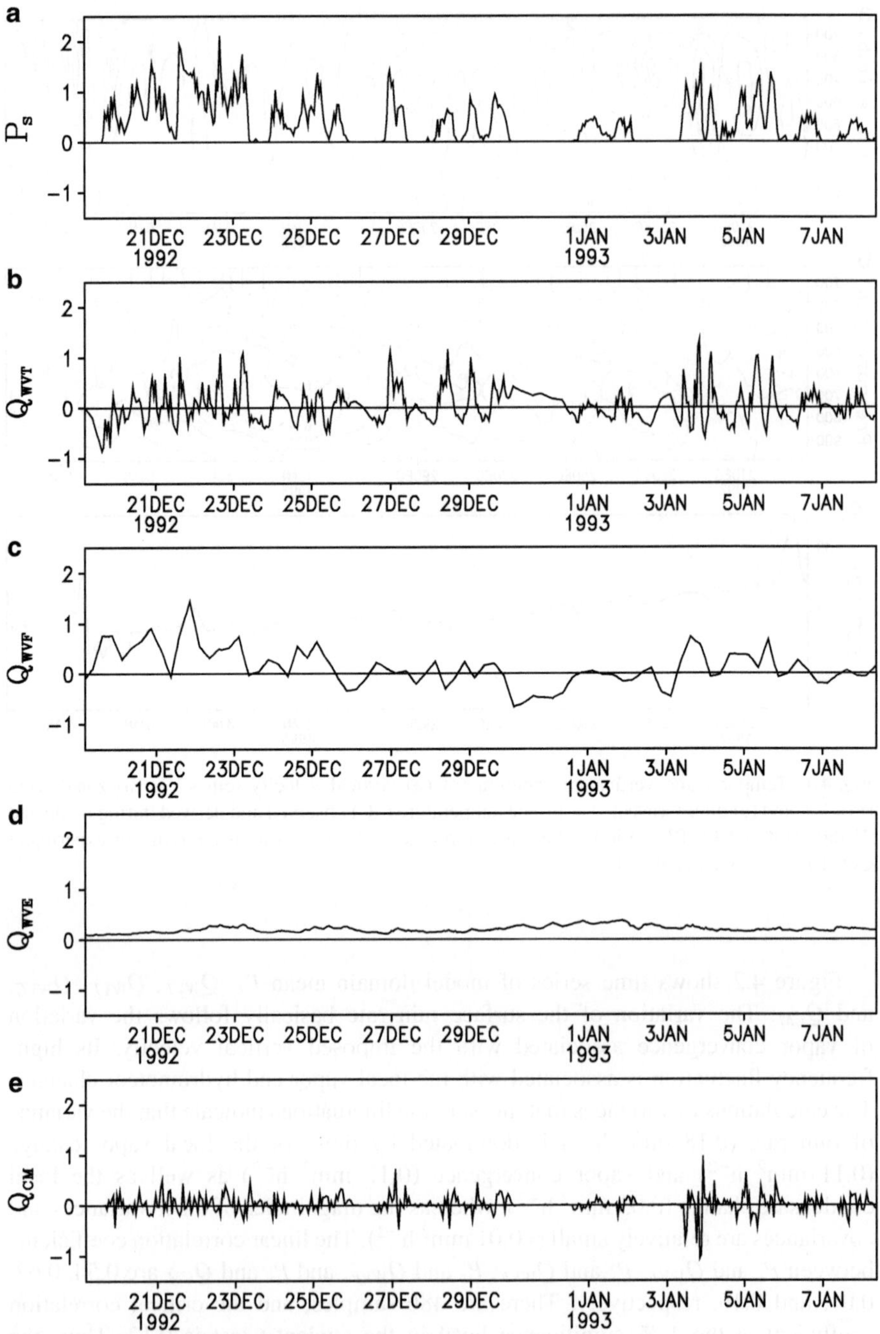

Fig. 4.2 Time series of domain-mean P_s in (**a**), Q_{WVT} in (**b**), Q_{WVF} in (**c**), Q_{WVE} in (**d**), and Q_{CM} in (**e**) during the 21-day integration in COARE (Unit is mm h^{-1})

scheme cannot predict the variation of the surface rain rate because Q_{WVT} and $Q_{WVF} + Q_{WVE}$ have similar variances, but their linear correlation coefficient is only −0.16, which is marginally statistically significant.

Q_{WV} and P_s are closely correlated with the coefficient of 0.92. Their RMS difference is 0.17 mm h^{-1}, which is much smaller than the standard deviation of P_s (0.43 mm h^{-1}). Thus, the variation of vapor sink (Q_{WV}) largely explains the variation of surface rain rate (P_s). The approximate balance between the surface rain rate and vapor sink guarantees reasonable simulations of the mean surface rain rate in the cloud-resolving model if the zonally uniform horizontal and vertical moisture advections imposed in the model are derived from moisture budgets.

The magnitudes of Q_{CM} are generally smaller than those of Q_{WV}, but they have similar magnitudes on 3 January 1993 when the negative Q_{CM} with the maximum negative value of up to −1 mm h^{-1} compensates for the overestimation of P_s with Q_{WV} significantly. Q_{WV} and Q_{CM} are negatively correlated, but their linear correlation coefficient is small (−0.33). The negative correlation in which a negative Q_{CM} associated with a positive Q_{WV} suggests that the increase of cloud hydrometeors is a result of vapor loss. In particular, Q_{WV} and Q_{CM} have the same magnitudes, but they have the opposite signs on 26 and 31 December 1992, indicating that the entire vapor sink supports the growth of cloud hydrometeors during the genesis of tropical convection.

Accuracy of the calculations of the surface rain rate with the vapor sink (in Kuo's scheme) can be evaluated with magnitudes of the cloud source/sink (Q_{CM}). Figure 4.3 shows the mean Q_{CM} (=P_s − Q_{WV}) versus the mean P_s. A positive Q_{CM} measures the underestimation of P_s with Q_{WV}, whereas a negative Q_{CM} denotes the overestimation of P_s. Maximum positive (1.2 mm h^{-1}) and negative (−0.8 mm h^{-1}) Q_{CM} occur around 1 mm h^{-1} of P_s. When P_s is small (0–0.2 mm h^{-1}), the amplitudes of Q_{CM} could be larger than P_s. The results show a significant contribution from the cloud source/sink in the calculations of surface rain rates and particularly in the estimation of small surface rain rates.

Since clouds consist of water and ice clouds, the cloud source/sink (Q_{CM}) can be further broken into water cloud ($Q_{CMW} = -\partial\,[q_2]\,/\partial t$, $q_2 = q_c + q_r$) and ice cloud ($Q_{CMI} = -\partial\,[q_3]\,/\partial t$, $q_3 = q_i + q_s + q_g$) sources/sinks. The linear correlation coefficient between Q_{CMI} and Q_{CM} (0.61) is larger than that between Q_{CMW} and Q_{CM} (0.4). The RMS difference between Q_{CMI} and Q_{CM} (0.15 mm h^{-1}) is smaller than that between Q_{CMW} and Q_{CM} (0.166 mm h^{-1}). This suggests that the variation of ice hydrometeors may contribute to the variation of cloud source/sink more than the variation of water hydrometeors does.

To detect precursor for development of surface rainfall, the phase differences between P_s and its contributors (Q_{WV} /Q_{CMW}/ Q_{CMI}) are analyzed by calculating lag correlation coefficients (Fig. 4.4). Lag correlation coefficients between P_s and Q_{WV} and between P_s and Q_{CMW} have maximum values of 0.92 and 0.21 at lag hour 0, respectively, indicating that the surface rain rate is in-phase with the vapor sink and water cloud sink in tropical deep convective regime. The lag correlation coefficient between P_s and Q_{CMI} has the maximum negative value (−0.23) at lag hour 1–2. The discussion of statistical significance for this phase lag can be found in Gao et al.

Fig. 4.3 Model domain-mean Q_{CM} ($P_s - Q_{WV}$) versus model domain-mean P_s in COARE. Unit is mm h^{-1} (After Gao et al. 2005)

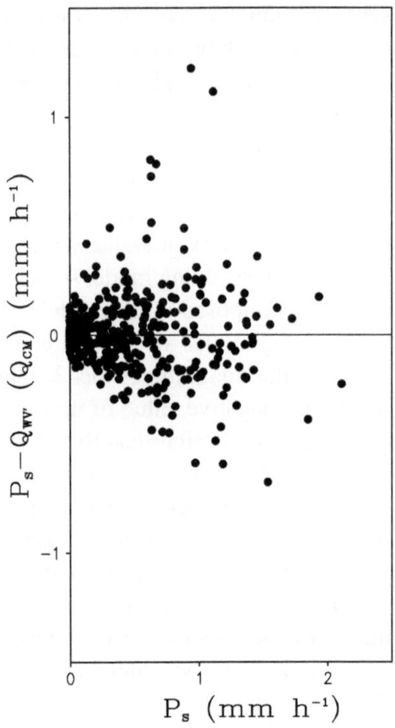

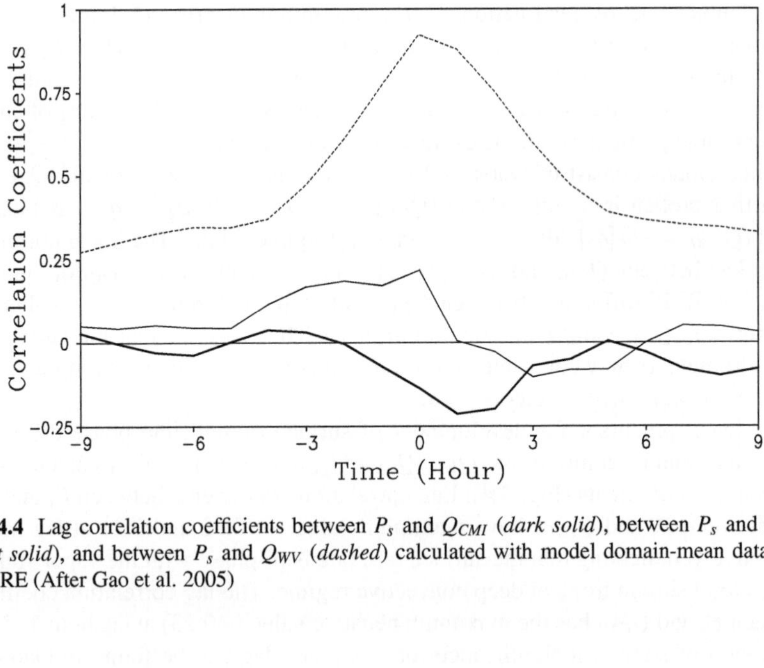

Fig. 4.4 Lag correlation coefficients between P_s and Q_{CMI} (*dark solid*), between P_s and Q_{CMW} (*light solid*), and between P_s and Q_{WV} (*dashed*) calculated with model domain-mean data from COARE (After Gao et al. 2005)

(2005). A positive lag time means that Q_{CMI} leads P_s by 1–2 h, while the negative correlation coefficient indicates that P_s and Q_{CMI} are out of phase. Since a negative Q_{CMI} denotes local ice hydrometeor gain, the ice clouds develop 1–2 h ahead of subsequent gain in the surface rainfall.

4.2 Time-Mean Surface Rainfall Processes

The time and model domain means of P_s, Q_{WVT}, Q_{WVF}, Q_{WVE}, and Q_{CM} are 0.37, 0.03, 0.12, 0.2, and 0.02 mm h^{-1}, respectively. This indicates that 8.1, 32.4, 54.1, and 5.4 % of P_s come from Q_{WVT}, Q_{WVF}, Q_{WVE}, and Q_{CM}, respectively. Thus, the surface evaporation rate and vapor convergence associated with the imposed vertical velocity mainly contribute to the time and model domain-mean surface rain rate. Since model domain-mean data include information from convective, raining stratiform, non-raining stratiform clouds, as well as clear-sky regions, Cui and Li (2006) applied the partitioning method to each grid point to determine the type (clear sky, raining stratiform, convective, or non-raining stratiform) and took the summations of grid points and other quantities (e.g., *IWP*, *LWP*, and the others) and divided them by the total zonal gird points (512) and the length of hourly data (486). They used the time-mean data to study the processes from different regions that contribute to the mean rainfall.

In COARE, convective clouds (4.3 %) occupy smaller areas than raining stratiform clouds (9.9 %) do (Table 4.1). In clear-sky regions, Q_{WVE} (0.090 mm h^{-1}) is balanced by Q_{WVT} (−0.064 mm h^{-1}) and Q_{WVF} (−0.029 mm h^{-1}), indicating that the surface evaporation is the only vapor source. One third of the mean evaporation rate is used to offset the vapor divergence associated with subsidence, whereas two thirds is used to moisten the atmosphere.

Table 4.1 Time means of fractional cloud coverage, *IWP*, *LWP*, P_s, Q_{WVT}, Q_{WVF}, Q_{WVE}, and Q_{CM} over clear-sky regions, raining stratiform regions, convective regions, and non-raining stratiform regions and their sums (model domain means) in COARE

	Clear-sky regions	Raining stratiform regions	Convective regions	Non-raining stratiform regions	Model domain mean
Fractional coverage (%)	47.1	9.9	4.3	38.7	100.0
IWP (mm)	0.000	0.080	0.027	0.038	0.145
LWP (mm)	0.000	0.083	0.077	0.011	0.171
P_s (mm h^{-1})	0.000	0.151	0.217	0.000	0.368
Q_{WVT} (mm h^{-1})	−0.064	0.197	−0.116	0.012	0.029
Q_{WVF} (mm h^{-1})	−0.029	−0.165	0.410	−0.097	0.120
Q_{WVE} (mm h^{-1})	0.090	0.024	0.014	0.075	0.203
Q_{CM} (mm h^{-1})	0.003	0.096	−0.091	0.009	0.017

After Cui and Li (2006)

Over raining stratiform regions, *IWP* (0.080 mm) and *LWP* (0.083 mm) are similar. P_s (0.151 mm h^{-1}), Q_{WVT} (0.197 mm h^{-1}), Q_{WVF} (−0.165 mm h^{-1}), and Q_{CM} (0.092 mm h^{-1}) have similar magnitudes, while Q_{WVE} (0.024 mm h^{-1}) is relatively small. Thus, about a half of magnitudes of the sum of local vapor loss and local hydrometeor loss/hydrometeor convergence loss is canceled out by the vapor divergence, whereas the other half feeds the stratiform rainfall.

Over convective regions, *IWP* (0.027 mm) is smaller than *LWP* (0.077 mm), suggesting that the water hydrometeors are more common than the ice hydrometeors. Q_{WVF} (0.410 mm h^{-1}) supports convective rainfall (P_s =0.217 mm h^{-1}), local atmospheric moistening ($Q_{WVT} = -0.116$ mm h^{-1}), and local cloud growth ($Q_{CM} = -0.091$ mm h^{-1}). Q_{WVE} (0.014 mm h^{-1}) is negligibly small. The small surface evaporation in convective and raining stratiform regions is also demonstrated by the analysis of moisture budget by Gao et al. (2006) in their 2D coupled ocean–cloud-resolving atmosphere model experiment.

Over non-raining stratiform regions, *IWP* (0.038 mm) is significantly larger than *LWP* (0.011 mm), indicating the dominance of ice hydrometeors. Q_{WVE} (0.075 mm h^{-1}) compensates vapor divergence ($Q_{WVF} = -0.097$ mm h^{-1}), local atmospheric drying (Q_{WVT} =0.012 mm h^{-1}), and local hydrometeor loss (Q_{CM} =0.009 mm h^{-1}). The vapor divergence associated with the subsidence has a larger magnitude than does the surface evaporation that leads to local atmospheric drying (Q_{WVT} =0.012 mm h^{-1}).

Thus, the mean surface evaporation flux comes mainly from the non-raining regions (clear-sky and non-raining stratiform regions), whereas the mean vapor convergence is mainly from the convective regions. Therefore, in additional to rainfall forced by the imposed upward motion, surface evaporation pumps water vapor into rainfall-free regions, and the water vapor is transported from rainfall-free regions into rainfall regions, which feeds rainfall.

4.3 Surface Rainfall Processes Associated with Individual Cloud

Since the surface rain rate at model grid is much larger than the mean value, the grid data (in the box in Fig. 4.5) are analyzed by Zhou et al. (2006). The box along the propagation of the major rainband in a life cycle of tropical convection has a zonal length of 18 km and a temporal length of 15 h. Hours 6–11, 12–18, and 19–20 represent the genesis, development, and decay of tropical convection, respectively, which are denoted by cases A, B, and C. The time and model domain-mean surface rain rates are 3.3 mm h^{-1} in case A, 15.9 mm h^{-1} in case B, and 1.9 mm h^{-1} in case C. Table 4.2 shows contributions of Q_{WV} and Q_{CM} to P_s in cases A, B, and C, respectively. In all three cases, Q_{WV} is positive, whereas Q_{CM} is negative. Calculations with the water vapor overestimate P_s by 38 % in case A, 23 % in case B, and 37 % in case C, respectively.

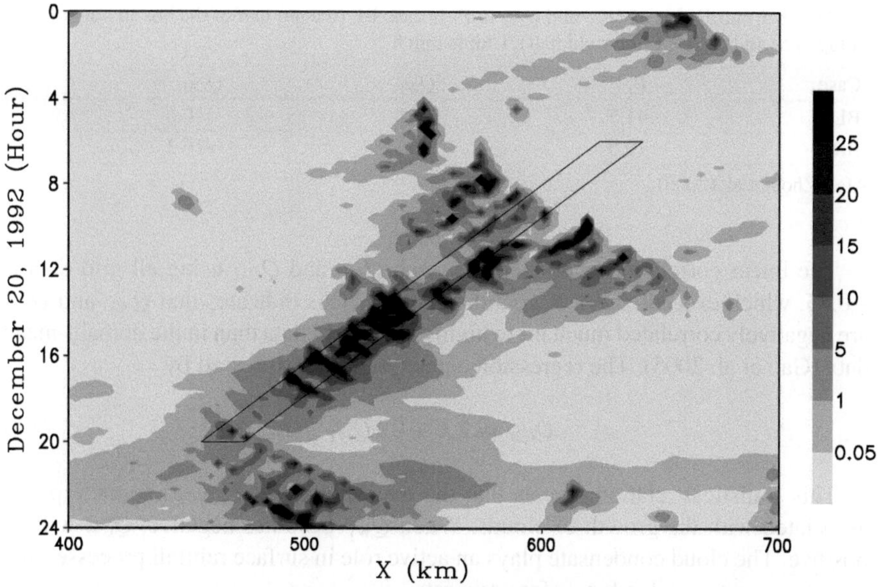

Fig. 4.5 Time evolution and horizontal distribution of surface rain rate (mm h^{-1}) within 400–700 km on 20 December 1992 simulated in COARE. The box between hours 6–20 (0600–2000 LST) is used to analyze the surface rainfall processes during the life cycle of convection (After Li et al. 2002)

Table 4.2 Contributions of Q_{WV} and Q_{CM} to P_s in case A (hours 6–11), case B (hours 12–18), and case C (hours 19–20) on 19–20 December 1992 in the box shown in Fig. 4.5

Case	P_s (mm h^{-1})	Q_{WV} (mm h^{-1})	Q_{CM} (mm h^{-1})
A	3.3	4.54	−1.24
B	15.9	19.6	−3.7
C	1.9	2.6	−0.7

After Zhou et al. (2006)

The analysis is further carried out in the left half (case BL) and the right half (case BR) of the box in hours 13–15 for mature convection. Maximum upward motion appears between 500 and 750 mb with the surface rain rate of 41.7 mm h^{-1} in case BL, whereas maximum upward motion occurs around 400 mb with the surface rain rate of 19.5 mm h^{-1} in case BR [see Fig. 6 in Li et al. (2002)]. Deep convective clouds are dominant in case BL, whereas anvil clouds are dominant in case BR. In case BL, P_s (41.7 mm h^{-1}) is contributed to by both Q_{WV} (31.1 mm h^{-1}) and Q_{CM} (10.6 mm h^{-1}) (Table 4.3). This suggests that the calculation with water vapor underestimates P_s by 25 % in deep convective clouds during the development of the convective system. It is interesting to note that in both BL and BR, calculations of surface rain rates with water vapor (Q_{WV}) are similar; the difference of surface rain rate (P_s) depends on the effects of cloud condensate (Q_{CM}).

Table 4.3 Contribution of Q_{WV} and Q_{CM} to P_s in case BL (the left half of the box in hours 13–15 in Fig. 4.5) and case BR (the right half). Unit is mm h^{-1}

Case	P_s	Q_{WV}	Q_{CM}
BL	41.7	31.1	10.6
BR	19.5	35.8	−16.3

After Zhou et al. (2006)

The linear correlation coefficient between Q_{WV} and Q_{CM} using all grid data is −0.85, which exceeds the 1 % confidence level. This indicates that Q_{WV} and Q_{CM} are negatively correlated much more highly in the grid data than in the domain-mean data (Gao et al. 2005). The regression equation can be expressed by

$$Q_{CM} = 2.6 - 0.7Q_{WV}. \tag{4.4}$$

This statistical relation reveals that the overestimate of P_s with water vapor is associated with the growth of clouds. When Q_{WV} becomes negative, Q_{CM} will be positive. The cloud condensate plays an active role in surface rainfall processes and assures a positive value for surface rain rate.

Q_{CM} is further broken into Q_{CMW} and Q_{CMI} for water and ice clouds, respectively. The variation of Q_{CM} is largely contributed by that of Q_{CMW}. Their regression equation can be expressed by

$$Q_{CM} = -0.17 + 1.02Q_{CMW} \cong Q_{CMW}. \tag{4.5}$$

The linear correlation coefficient is 0.91, which is much larger than the linear correlation coefficient calculated with the domain-mean data (0.4) (Gao et al. 2005). Gao et al. (2005) showed that the linear correlation coefficient between Q_{CM} and Q_{CMI} (0.61) is larger than that between Q_{CM} and Q_{CMW} (0.4) in the calculations with the domain-mean data, whereas Zhou et al. (2006) displayed that the results are opposite in the calculations with the grid data. This may be due to the fact that the vertical structures of vertical velocity for water and ice clouds can be separated in the grid data [e.g., Fig. 6 in Li et al. (2002)] and the large surface rain rate is associated with a large Q_{CM} as a result of a large Q_{CMW} (e.g., in case BL), whereas the domain-mean data shows that the maximum domain-mean upward motions appear in 350–450 mb and the surface rain rate and ice-cloud and water-cloud sources/sinks are positively correlated. Thus, contribution of cloud hydrometeors to surface rain rate comes mainly from the variation of water hydrometeors in tropical convection in the analysis of grid data.

References

Bannon PR (2002) Theoretical foundations for models of moist convection. J Atmos Sci 59: 1967–1982

Cui X, Li X (2006) Role of surface evaporation in surface rainfall processes. J Geophys Res 111. doi10.1029/2005JD006876, (c) American Geophysical Union. Reprinted with permission

Gao S, Cui X, Zhou Y, Li X (2005) Surface rainfall processes as simulated in a cloud resolving model. J Geophys Res 110. doi:10.1029/2004JD005467, (c) American Geophysical Union. Reprinted with permission

Gao S, Ping F, Cui X, Li X (2006) Short timescale air-sea coupling in the tropical deep convective regime. Meteorol Atmos Phys 93:37–44

Kuo HL (1965) On formation and intensification of tropical cyclones through latent heat release by cumulus convection. J Atmos Sci 22:40–63

Kuo HL (1974) Further studies of the parameterization of the influence of cumulus convection on large-scale flow. J Atmos Sci 31:1232–1240

Li X, Sui CH, Lau KM (2002) Dominant cloud microphysical processes in a tropical oceanic convective system: A 2-D cloud resolving modeling study. Mon Wea Rev 130:2481–2491, (c) American Meteorological Society. Reprinted with permission

Ooyama KV (1990) A thermodynamic foundation for modeling the moist atmosphere. J Atmos Sci 47:2580–2593

Ooyama KV (2001) A dynamic and thermodynamic foundation for modeling the moist atmosphere with parameterized microphysics. J Atmos Sci 58:2073–2102

Sui CH, Lau KM, Takayabu Y, Short D (1997a) Diurnal variations in tropical oceanic cumulus ensemble during TOGA COARE. J Atmos Sci 54:639–655

Sui CH, Li X, Lau KM, Adamec D (1997b) Multi-scale air-sea interactions during TOGA COARE. Mon Wea Rev 125:448–462

Takayabu YN, Lau KM, Sui CH (1996) Observation of a quasi-2-day wave during TOGA COARE. Mon Weather Rev 124:1892–1913

Webster PJ, Lukas R (1992) TOGA COARE: the Coupled Ocean–Atmosphere Response Experiment. Bull Am Meteorol Soc 73:1377–1416

Zhou Y, Cui X, Li X (2006) Contribution of cloud condensate to surface rain rate. Prog Nat Sci 16:967–973, (c) National Natural Science Foundation of China. Reprinted with permission

Chapter 5
Structures of Precipitation Systems I: Cloud-Content Analysis

Abstract In this chapter, interaction between water and ice clouds is examined through the analysis of tendency of a cloud ratio. The cloud ratio is defined as the ratio of ice water path to liquid water path. The tendency equation of the cloud ratio is derived from prognostic equations of mixing ratios of five cloud hydrometeors. The tendency of the cloud ratio is associated with conversion between the ice and liquid water paths, condensation and depositions, the evaporation of raindrops and rainfall. The cloud ratio is also used to develop a new rainfall separation method for studying structures of precipitation systems. The new rainfall partitioning method is compared to previous separation technique with the magnitude of rain rate through the analysis of cloud microphysical budgets associated with convective and stratiform rainfall.

Keywords Cloud ratio • Ice and liquid water paths • Rainfall partitioning method • Convective and stratiform rainfall • Cloud microphysical budget

In precipitation systems, the deep convection is usually associated with upward motions throughout the troposphere, producing liquid-phase clouds in the lower troposphere and ice-phase clouds in the upper troposphere, depending on air temperature. Due to mass continuity, the upward motions lead to strong divergence in the upper troposphere, which advects ice-phase hydrometeors out to form anvil clouds. The rainfalls from deep convection and surrounding anvil clouds are defined as convective and stratiform rainfall, respectively. To examine convective and stratiform rainfall, rainfall data are separated based on either the magnitude of radar signal in observational studies or rainfall intensity in numerical modeling studies after a pioneer research work done by Houze (1973). Since convective and stratiform rainfall clouds are dominated by water and ice hydrometeors, respectively, water and ice hydrometeors can be used to identify convective and stratiform rainfall. In this chapter, interaction between water and ice clouds in experiment COARE (Sui and Li 2005) will be discussed in Sect. 5.1. A convective-stratiform rainfall separation scheme developed by Sui et al. (2007) based on modeling information of water and ice clouds from COARE is discussed in Sect. 5.2.

© Springer International Publishing Switzerland 2016

X. Li, S. Gao, *Cloud-Resolving Modeling of Convective Processes*,
Springer Atmospheric Sciences, DOI 10.1007/978-3-319-26360-1_5

5.1 Interaction Between Water and Ice Clouds

To study the interaction between water and ice clouds, Sui and Li (2005) defined a cloud ratio (*CR*) as the ratio of the *IWP* (mass integration of sum of mixing ratios of cloud ice, snow, and graupel) to *LWP* (mass integration of sum of mixing ratios of cloud water and raindrops), which measures the importance of ice and water hydrometeors in precipitating clouds. Figure 5.1 displays x-z cross sections of the total hydrometeor mixing ratio and vertical profiles of zonal-mean total hydrometeor mixing ratio and vertical velocity at 0800 LST 20, 0000 LST 21, and 0600 LST 24 December 1992, respectively. The zonal-mean *CR* is 0.2 when several cloud clusters only extend to 400 mb, and the water clouds dominate at 0800 LST 20 December 1992, representing the development of convective clouds. The vertical velocities in the upper and lower troposphere have similar magnitudes. When clouds extend to 100 mb at 0000 21 December 1992 and the zonal-mean *CR* is 0.8, both ice and water clouds develop. There exists a strong upward motion above 600 mb which supports ice clouds. The zonal-mean *CR* is 2.8 at 0600 LST 24 December 1992 when anvil ice clouds exist and the ice hydrometeor mixing ratio is larger than water hydrometeor mixing ratio, denoting the development of stratiform clouds. A strong mean upward motion occurs in the upper troposphere, whereas a downward motion appears in the lower troposphere.

Cui et al. (2007) calculated the 21-day mean cloud hydrometeors and associated microphysical budgets in raining stratiform, convective, and non-raining stratiform regions in COARE (Fig. 5.2). In raining stratiform regions, the *IWP* (0.89 mm) is slightly smaller than the *LWP* (0.98 mm), and *CR* is 0.91. The magnitude of cloud ice (0.09 mm) only is about one third of the magnitude of cloud water (0.28 mm), whereas the magnitude of precipitation ice ($[q_s]+[q_g]=0.8$ mm) is slightly larger than that of precipitation water (0.7 mm). In convective regions, the *IWP* (0.6 mm) is much smaller than the *LWP* (1.72 mm), and *CR* is 0.35. The cloud ice (0.05 mm) is one order of magnitude smaller than the magnitude of cloud water (0.78 mm), and the magnitude of precipitation ice (0.55 mm) is smaller than that of precipitation water (0.94 mm). In non-raining stratiform regions, the *IWP* (0.12 mm) is much larger than the *LWP* (0.043 mm), and *CR* is 2.79. The cloud ice (0.018 mm) and cloud water (0.017 mm) have similar magnitudes, whereas the magnitude of precipitation ice (0.102 mm) is much larger than that of precipitation water (0.026 mm). Thus, both water and ice clouds are important in raining stratiform regions, whereas water and ice clouds are dominant in convective and non-raining stratiform regions, respectively.

In raining stratiform regions (Fig. 5.2a), the vapor deposition rate ($[P_{DEP}]+[P_{SDEP}]+[P_{GDEP}]=0.71$ mm h^{-1}) is less than half of the vapor condensation rate ($[P_{CND}]=1.64$ mm h^{-1}), and the ratio of the deposition rate to condensation rate is 0.43. In convective regions (Fig. 5.2b), the vapor deposition rate (0.62 mm h^{-1}) is one order of magnitude smaller than the vapor condensation rate (7.24 mm h^{-1}), and the ratio of the deposition rate to condensation rate is 0.09. In non-raining stratiform regions (Fig. 5.2c), the vapor deposition rate (0.078 mm h^{-1}) is much higher than

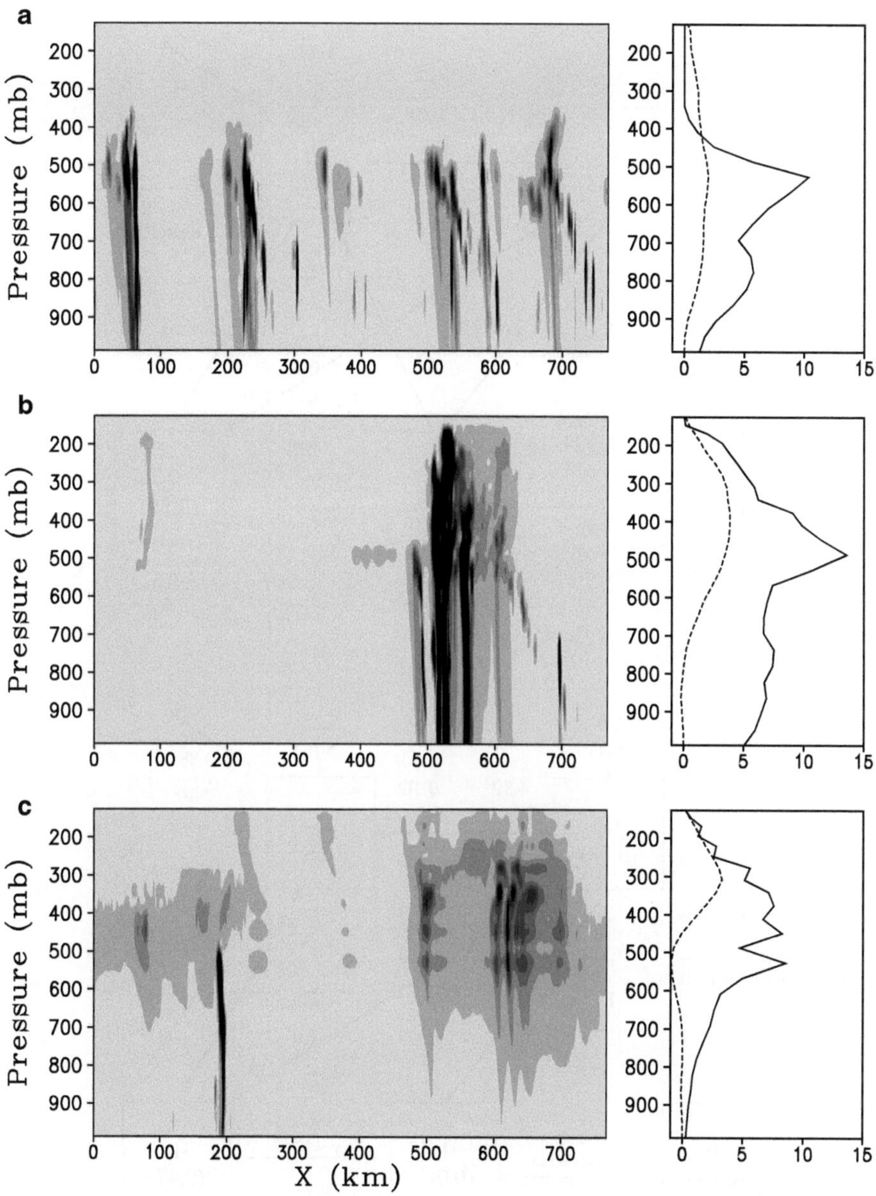

Fig. 5.1 Horizontal and vertical distributions of total hydrometeor mixing ratio (*left panels*) and vertical profiles of zonally averaged total hydrometeor mixing ratio (*solid*) and vertical velocity (*dashed*) (*right panels*) simulated in COARE at (**a**) 0800 LST 20, (**b**) 0000 LST 21, and (**c**) 0600 LST 24 December 1992. Units are 10^{-2} g kg^{-1} for total hydrometeor mixing ratio and cm s^{-1} for vertical velocity, respectively (After Sui and Li 2005)

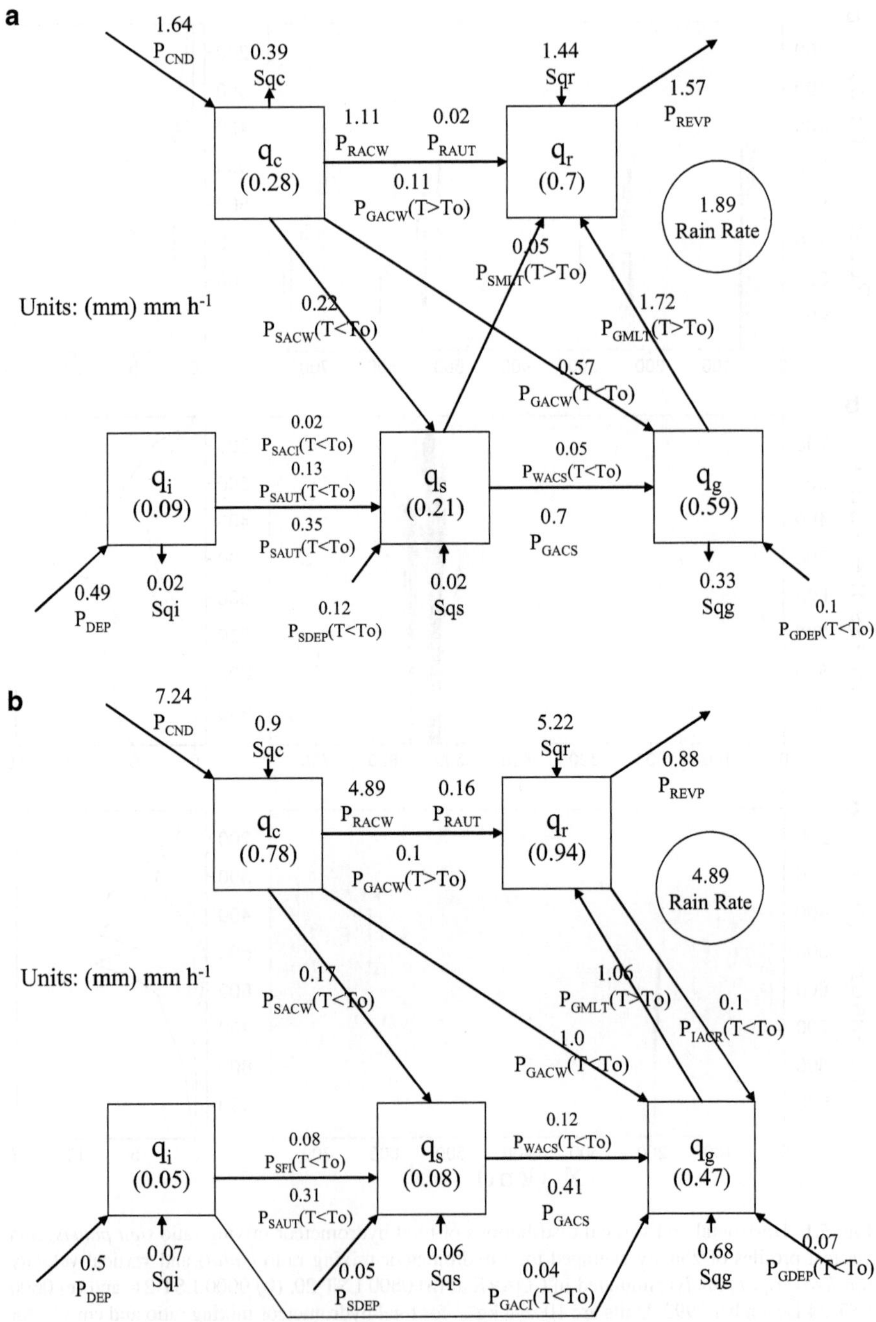

Fig. 5.2 21-day mean cloud microphysics budgets in (**a**) raining stratiform regions, (**b**) convective regions, and (**c**) non-raining stratiform regions simulated in COARE. Units for cloud hydrometeors and conversions are mm and mm h^{-1}, respectively (After Cui et al. 2007)

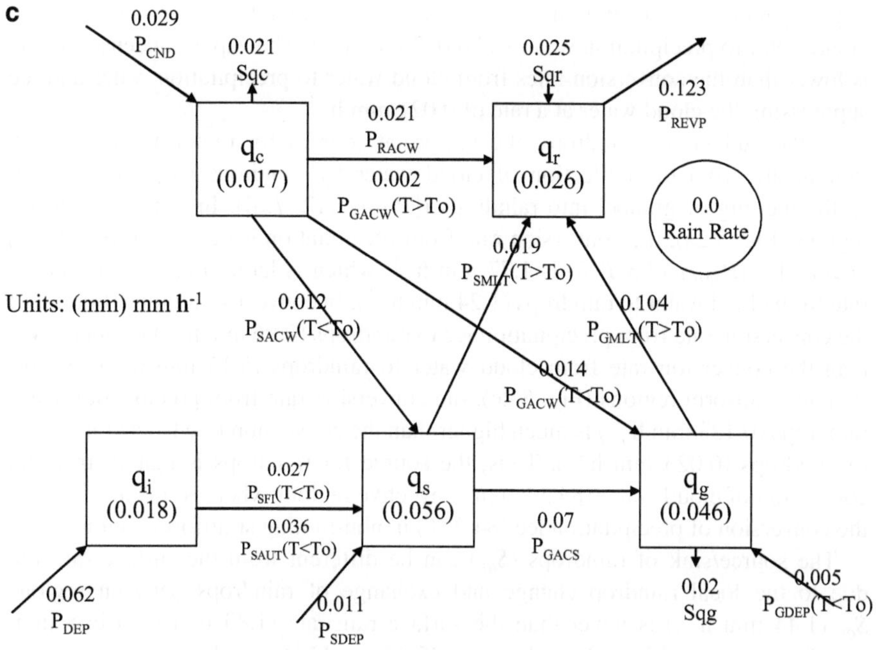

Fig. 5.2 (continued)

the vapor condensation rate (0.029 mm h^{-1}), and the ratio of the deposition rate to condensation rate is 2.7. These results indicate that a large *LWP* is associated with a large vapor condensation rate in convective regions, whereas a relatively large *IWP* is due to a large vapor deposition rate in non-raining stratiform regions.

The cloud water is converted into precipitation water mainly through the collection of cloud water by rain ($[P_{RACW}]$) and precipitation ice mainly through the accretion of cloud water by snow ($[P_{SACW}(T < T_o)]$) and graupel ($[P_{GACW}(T < T_o)]$). In raining stratiform regions (Fig. 5.2a), the conversion rate from cloud water to precipitation ice ($[P_{GACW}\ (T < T_o)] + [P_{SACW}\ (T < T_o)] = 0.79$ mm h^{-1}) is lower than the conversion rate from cloud water to precipitation water ($[P_{RACW}] + [P_{RAUT}] + [P_{GACW}\ (T > T_o)] = 1.24$ mm h^{-1}). About two thirds of cloud water becomes raindrops whereas about one third of cloud water is converted to precipitation ice. The vapor condensation rate is smaller than the conversion rates to precipitation water and ice, which leads to a reduction in cloud water at a rate of 0.39 mm h^{-1} (S_{qc}). In convective regions (Fig. 5.2b), the conversion rate from cloud water to precipitation ice (1.17 mm h^{-1}) is lower than the conversion rate from cloud water to precipitation water (5.15 mm h^{-1}). About four fifth of cloud water becomes raindrops whereas about one fifth of cloud water is converted to precipitation ice. The vapor condensation rate is larger than the conversion rates to precipitation water and ice, which enhances the cloud water at a rate of 0.9 mm h^{-1}. In non-raining stratiform regions (Fig. 5.2c), the conversion rate from cloud water to precipitation ice (0.026 mm h^{-1}) is slightly higher than the conversion rate from cloud water to precipitation water by (0.023 mm h^{-1}). The vapor condensation rate is lower than the conversion rates from cloud water to precipitation water and ice, suppressing the cloud water at a rate of 0.021 mm h^{-1}.

In the budgets of raindrops, the important sources for raindrops and surface rainfall are from the collection of cloud water by raindrops ($[P_{RACW}]$) as well as the melting of graupel into raindrops ($[P_{GMLT}\ (T > T_o)]$). In raining stratiform regions (Fig. 5.2a), the conversion rate from precipitation ice to raindrops ($[P_{GMLT}\ (T > T_o)] + [P_{SMLT}\ (T > T_o)]$) is 1.77 mm h^{-1}, which is larger than the conversion rate from cloud water to raindrops (1.24 mm h^{-1}). In convective regions (Fig. 5.2b), the conversion rate from precipitation ice to raindrops (0.96 mm h^{-1}) is much lower than the conversion rate from cloud water to raindrops (5.15 mm h^{-1}). In non-raining stratiform regions (Fig. 5.2c), the conversion rate from precipitation ice to raindrops (0.123 mm h^{-1}) is much higher than the conversion rate from cloud water to raindrops (0.023 mm h^{-1}). Thus, the source for raindrops is mainly from the conversion of cloud water (84.3 %) in convective regions whereas it is mainly from the conversion of precipitation ice (84.2 %) in non-raining stratiform regions.

The source/sink of raindrops (S_{qr}) can be different from the surface rain rate due to the local raindrop change and exchange of raindrops between regions. S_{qr} (1.44 mm h^{-1}) is lower than the surface rain rate (1.89 mm h^{-1}) in raining stratiform regions (Fig. 5.2a), whereas it (5.22 mm h^{-1}) is higher than the rain rate (4.89 mm h^{-1}) in convective regions (Fig. 5.2b). In non-raining stratiform regions (Fig. 5.2c), S_{qr} is 0.025 mm h^{-1} even though there is not any surface rainfall. Since the sum of S_{qr} should be the sum of the surface rain rates in three regions, the differences between S_{qr} and the surface rain rate in raining regions are mainly due to the exchange between raining stratiform regions and convective regions. In addition to the surface rainfall, the evaporation of raindrops ($[P_{REVP}]$) is another sink for the raindrops. The rain evaporation rate in raining stratiform regions (1.57 mm h^{-1}) is higher than in convective regions (0.88 mm h^{-1}) and in non-raining stratiform regions (0.123 mm h^{-1}).

The conversion between the *IWP* and *LWP* is mainly determined by the melting of precipitation ice ($[P_{GMLT}\ (T > T_o)] + [P_{SMLT}\ (T > T_o)]$) and the accretion of cloud water by precipitation ice ($[P_{GACW}\ (T < T_o)] + [P_{SACW}\ (T < T_o)]$) (Fig. 5.2). The conversion is from the *IWP* to *LWP* are 0.98 mm h^{-1} in raining stratiform regions, 0.097 mm h^{-1} in non-raining stratiform regions, and 0.21 mm h^{-1} in convective regions. This suggests that over the tropical deep convective regime, the exchange between the water and ice clouds enhances the *LWP*. Over raining stratiform regions, S_{qg} is –0.33 mm h^{-1}. This suggests that cloud microphysical processes suppress the graupel in raining stratiform regions. Over convective regions, S_{qg} is 0.68 mm h^{-1}. This indicates that cloud microphysical processes enhance the graupel in convective regions. In non-raining stratiform regions, S_{qg} (0.02 mm h^{-1}) is small.

CR is different over different regions as indicated in Fig. 5.3. Thus, the linear regression between *IWP* and *LWP* is calculated over three regions. The

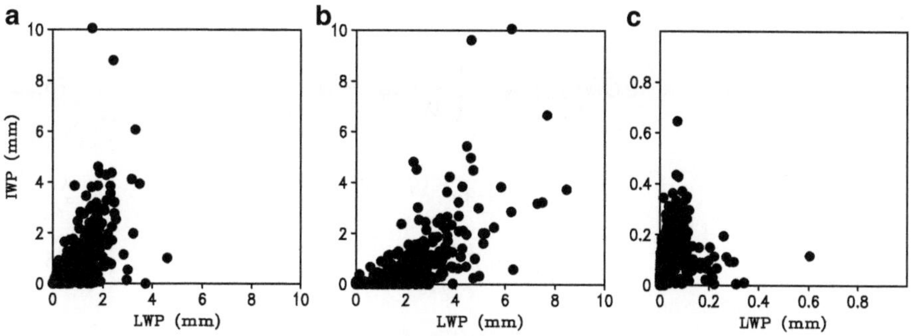

Fig. 5.3 *IWP* versus *LWP* in (**a**) raining stratiform regions, (**b**) convective regions, and (**c**) non-raining stratiform regions simulated in COARE (After Cui et al. 2007)

linear regression equations are $IWP = 0.88LWP$-0.01 in raining stratiform regions, $IWP = 0.49LWP$-0.21 in raining convective regions, and $IWP = 1.15LWP + 0.06$ in non-raining stratiform regions. Thus, the ratios of *IWP* to *LWP* (slopes of the regression equations) are 0.88 in raining stratiform regions, 0.49 in raining convective regions, and 1.15 in non-raining stratiform regions. The linear correlation coefficients are 0.61 in raining stratiform regions, 0.68 in raining convective regions, and 0.3 in non-raining stratiform regions. Thus, the linear relations over three regions are statistically significant. To statistically determine an upper limit in the ratio of the *IWP* to *LWP* over convective and raining stratiform clouds, the ratio of ($IWP_M + IWP_{SD}$) to ($LWP_M + LWP_{SD}$), where subscripts "M" and "SD" are time mean and standard deviation, respectively, is defined as the critical value for the upper limit, which is 0.54 for convective clouds, and 1.15 for raining stratiform clouds. Thus, the areas where $CR < 0.54$, $0.54 < CR < 1.15$, and $CR > 1.15$ are defined as the convective, raining stratiform, and non-raining stratiform regions, respectively.

The increase of *CR* shows development of ice clouds whereas the decrease of *CR* indicates the development of water clouds. Thus, Sui and Li (2005) further derived a tendency equation for *CR* by adding (1.7f)–(1.7g) and (1.7h)–(1.7j), respectively, and vertically integrate the resulting equations:

$$\frac{\partial LWP}{\partial t} = CONV_{LWP} - P_{qr} + [P_{CND}] + C(IWP, LWP) - [P_{REVP}], \tag{5.1}$$

$$\begin{aligned}\frac{\partial IWP}{\partial t} = {} & CONV_{IWP} + [P_{DEP}] + [P_{SDEP}] \\ & + [P_{GDEP}] - C(IWP, LWP) \\ & - P_{MLTS}(T > T_o) - P_{MLTG}(T > T_o) \\ & - P_{qs} - P_{qg},\end{aligned} \tag{5.2}$$

where

$$CONV_{LWP} = -\left[\frac{\partial}{\partial x} u\left(q_c + q_r\right)\right], \tag{5.3a}$$

$$CONV_{IWP} = -\left[\frac{\partial}{\partial x} u\left(q_i + q_s + q_g\right)\right], \tag{5.3b}$$

$$P_{qr} = \overline{\rho} w_{Tr} q_r \Big|_{z=0}, \tag{5.3c}$$

$$P_{qs} = \overline{\rho} w_{Ts} q_s \Big|_{z=0}, \tag{5.3d}$$

$$P_{qg} = \overline{\rho} w_{Tg} q_g \Big|_{z=0}, \tag{5.3e}$$

$$\begin{aligned} C\left(IWP, LWP\right) = & -\left[P_{SACW}\left(T < T_o\right)\right] - \left[P_{SFW}\left(T < T_o\right)\right] \\ & -\left[P_{GACW}\left(T < T_o\right)\right] - \left[P_{IHOM}\left(T < T_{oo}\right)\right] \\ & +\left[P_{IMLT}\left(T > T_o\right)\right] - \left[P_{IDW}\left(T_{oo} < T < T_o\right)\right] \\ & +\left[P_{RACS}\left(T < T_o\right)\right] - \left[P_{IACR}\left(T < T_o\right)\right] \\ & -\left[P_{GACR}\left(T < T_o\right)\right] - \left[P_{SACR}\left(T < T_o\right)\right] \\ & -\left[P_{GFR}\left(T < T_o\right)\right] + \left[P_{SMLT}\left(T > T_o\right)\right] \\ & +\left[P_{GMLT}\left(T > T_o\right)\right]. \end{aligned} \tag{5.3f}$$

C(IWP, LWP) is simplified as $[P_{GMLT}]-[P_{SACW}]-[P_{GACW}]$ for tropical convective systems (Li et al. 2002). Precipitation due to snow and graupel (P_{qs} and P_{qg}) are negligible, and P_{qr} accounts for surface rain rate (P_s) in tropical case here. $[P_{DEP}]+[P_{SDEP}]+[P_{GDEP}]=[\Sigma P_{DEP}]$. Therefore, (5.1, and 5.2) can be simplified as

$$\begin{aligned} \frac{\partial LWP}{\partial t} = & \; CONV_{LWP} - P_s + \left[P_{CND}\right] \\ & + C\left(IWP, LWP\right) - \left[P_{REVP}\right], \end{aligned} \tag{5.4a}$$

$$\frac{\partial IWP}{\partial t} = CONV_{IWP} + \left[\Sigma P_{DEP}\right] - C\left(IWP, LWP\right). \tag{5.4b}$$

Taking zonal mean on (5.4) leads to

$$\frac{\partial LWP}{\partial t} = -P_s + \left[P_{CND}\right] + C\left(IWP, LWP\right) - \left[P_{REVP}\right], \tag{5.5a}$$

$$\frac{\partial IWP}{\partial t} = \left[\Sigma P_{DEP}\right] - C\left(IWP, LWP\right). \tag{5.5b}$$

Take the time derivative of zonal-mean CR and use (5.5), the tendency equation of zonal-mean CR can be expressed by

$$\frac{\partial \ln CR}{\partial t} = CR_{CLI} + CR_{CNDDEP} + CR_{PSREVP} \tag{5.6}$$

$$CR_{CLI} = C\,(LWP, IWP)\left(\frac{LWP + IWP}{LWP \times IWP}\right), \tag{5.6a}$$

$$CR_{CNDDEP} = \frac{[\Sigma P_{DEP}]}{IWP} - \frac{[P_{CND}]}{LWP}, \tag{5.6b}$$

$$CR_{PSREVP} = \frac{P_{sfc}}{LWP} + \frac{[P_{REVP}]}{LWP}. \tag{5.6c}$$

The relation that $C(IWP, LWP) = -C(LWP, IWP)$ is used in the derivation of (5.6). Eq. (5.6) shows that the tendency of the zonal-mean CR is determined by the conversion between LWP and IWP through the melting of graupel and the accretion of cloud water by precipitation ice (CR_{CLI}), vapor condensation and deposition (CR_{CNDDEP}), rainfall, and evaporation of rain (CR_{PSREVP}). The accretion of cloud water by precipitation ice and vapor deposition enhances the development of ice clouds, and the rainfall and evaporation of rain suppress the development of water clouds increasing the cloud ratio whereas the melting of graupel and vapor condensation enhance the development of water clouds decreasing the cloud ratio.

Dominant responsible processes for the CR tendency may be different in different stages of cloud development, which can be identified by surface rain rates (Fig. 5.4). Two stages are categorized based on the hourly zonal-mean data. Since the time-mean surface rain rate during the integration is 0.37 mm h^{-1}, we choose $P_s < 0.3$ mm h^{-1} as the genesis/dissipating stages and $P_s > 0.3$ mm h^{-1} as the mature stage of tropical convection. In the genesis/dissipating stages of tropical convection, the linear correlation coefficients and RMS differences between lnCR tendency and its components are 0.13 and 2.31 h^{-1} for CR_{CLI}, 0.66 and 1.5 h^{-1} for CR_{CNDDEP}, and 0.14 and 3.33 h^{-1} for CR_{PSREVP}, respectively. Thus, the lnCR tendency is mainly determined by CR_{CNDDEP} (Fig. 5.4b in left panel). The lnCR tendency increases with increasing CR_{CNDDEP}, indicating that the vapor condensation and deposition rates associated with the upward motion determine the CR tendency and the variations of convective and stratiform clouds. The vapor deposition causes positive CR tendency and development of ice clouds, whereas the vapor condensation leads to negative CR tendency and growth of water clouds. In the mature stage of tropical convection, the linear correlation coefficients and RMS differences between the lnCR tendency and its components are 0.78 and 1.76 h^{-1} for CR_{CLI}, 0.01 and 2.81 h^{-1} for CR_{CNDDEP}, and 0.27 and 3.55 h^{-1} for CR_{PSREVP}, respectively. Thus, the lnCR tendency is mainly contributed to by CR_{CLI} (Fig. 5.4a in right panel). The lnCR tendency increases with increasing ($[P_{SACW}] + [P_{GACW}] - [P_{GMLT}]$). The CR decreases (increases) and clouds are more convective (stratiform) when the melting rate of graupel is larger (smaller) than the accretion rate of cloud water by precipitation ice in the mature stages of clouds.

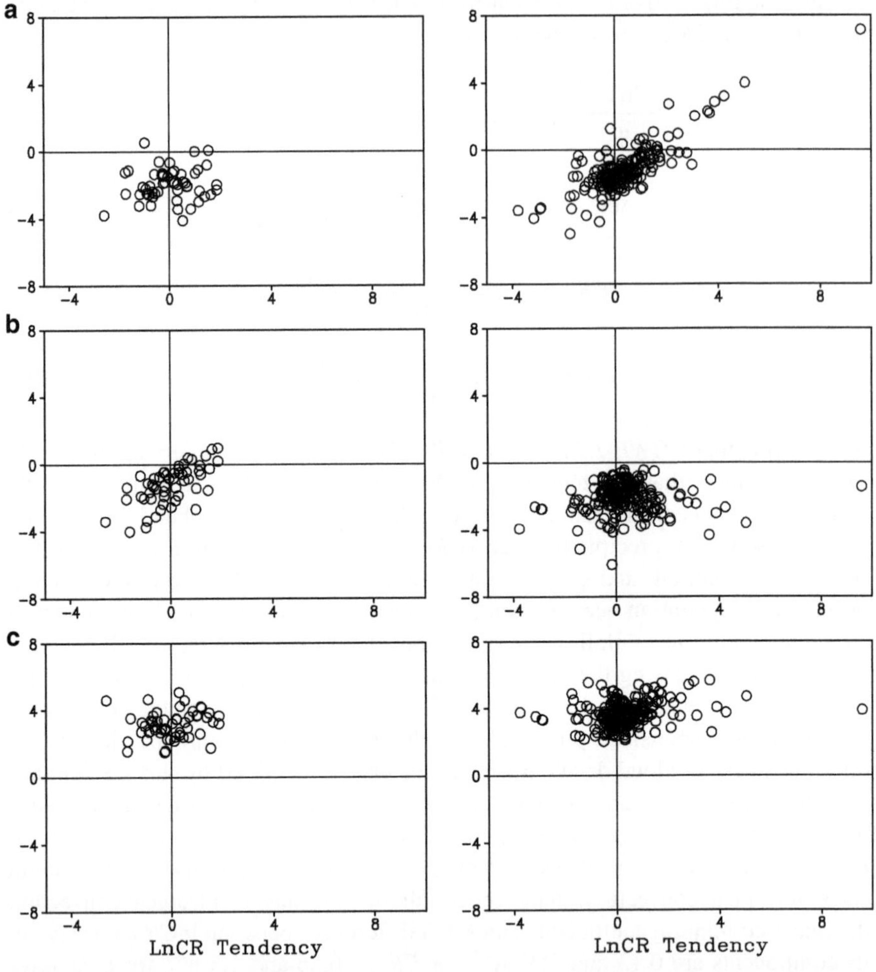

Fig. 5.4 ln*CR* tendency versus the contributions from (**a**) CR_{CLI}, (**b**) CR_{CNDDEP}, and (**c**) CR_{PSREVP} simulated in COARE when the zonal-mean surface rain rates are smaller than 0.3 mm h^{-1} in *left panel* and larger than 0.3 mm h^{-1} in *right panel*. Unit is h^{-1} (After Sui and Li 2005)

5.2 Convective-Stratiform Rainfall Separation Scheme by Cloud Content

Before a new scheme is developed using COARE data, we examine vertical structures of vertical velocity (Fig. 5.5), the contoured frequency by altitude diagrams (CFAD) of vertical velocity (Fig. 5.6), and cloud microphysical budgets (Fig. 5.2a, b) over convective and stratiform rainfall regions partitioned by the TS scheme [developed by Tao et al. (1993) and modified by Sui et al. (1994); Also see

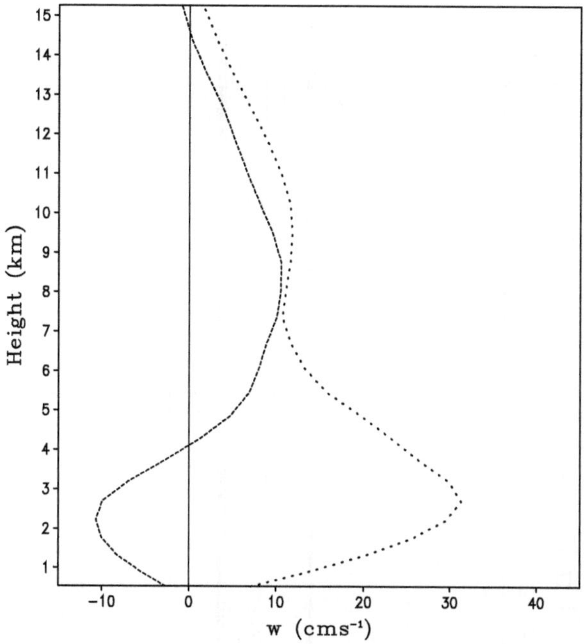

Fig. 5.5 Vertical profiles of vertical velocity (cm s-1) averaged in the stratiform (*dashed*) and convective rainfall regions partitioned by the TS scheme (After Sui et al. 2007)

Sect. 2.5], which serves as a reference for the development of the new separation scheme. In convective rainfall regions, mean upward motions occur in the entire troposphere with the maximum of 30 cm s^{-1} at 3 km (Fig. 5.5). In stratiform rainfall regions, mean downward motions appear below 4 km whereas mean upward motions occur above. The maximum upward and downward velocities are 10 cm s^{-1} at 8.5 km and −10 cm s^{-1} at 2.2 km, respectively. The CFADs of vertical velocity in the two regions reveal a broader distribution of vertical velocity in convective rainfall regions than that in stratiform rainfall regions (Fig. 5.6). The maximum vertical velocity is larger than 10 m s^{-1} in convective rainfall regions, and less than 5 m s^{-1} in stratiform rainfall regions.

Time-mean cloud microphysical budgets show that the melting rate of graupel to rain ($[P_{GMLT}]$) is more than twice larger than the collection rate of cloud water by rain ($[P_{RACW}]$) over stratiform rainfall regions (Fig. 5.2a). Thus, the melting of precipitation ice is a major source for production of stratiform rainfall. Over convective rainfall regions, $[P_{RACW}]$ is about five times larger than $[P_{GMLT}]$ (Fig. 5.2b) Thus, the collection of cloud water by rain is a major source for production of convective rainfall. The vapor condensation rate ($[P_{CND}]$) is much higher over convective rainfall regions than over stratiform rainfall regions, whereas deposition rate ($[P_{DEP} + P_{SDEP} + P_{GDEP}]$) are similar for both convective and stratiform rainfall regions, which is responsible for similar ice water paths (sum of mass integrated mixing ratios of cloud ice, snow, and graupel). The evaporation rate of raindrops ($[P_{REVP}]$) is higher over stratiform rainfall regions than over convective rainfall regions.

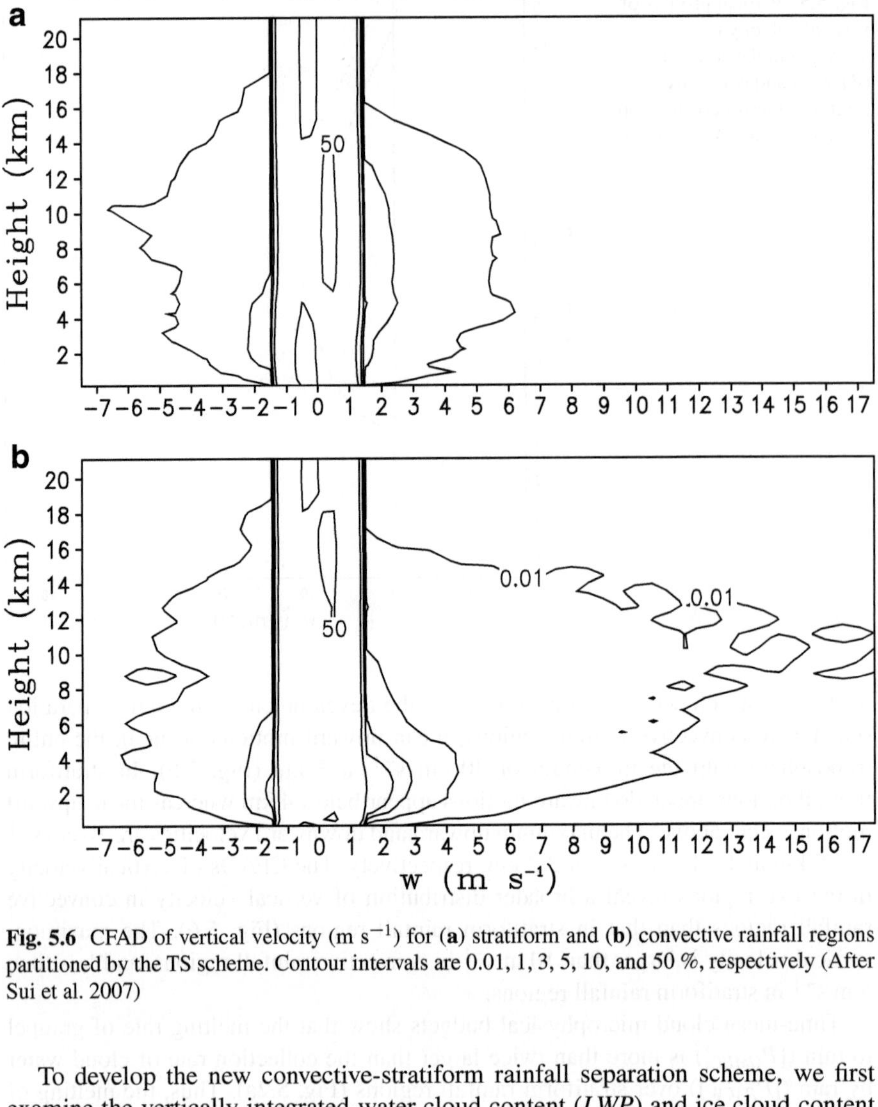

Fig. 5.6 CFAD of vertical velocity (m s^{-1}) for (**a**) stratiform and (**b**) convective rainfall regions partitioned by the TS scheme. Contour intervals are 0.01, 1, 3, 5, 10, and 50 %, respectively (After Sui et al. 2007)

To develop the new convective-stratiform rainfall separation scheme, we first examine the vertically integrated water cloud content (*LWP*) and ice cloud content (*IWP*) in the framework of a cloud model. *LWP* is the sum of mass-integrated mixing ratios of cloud water and raindrops, whereas *IWP* is the sum of mass-integrated mixing ratios of cloud ice, snow, and graupel. The *IWP* versus LWP are plotted in Fig. 5.7. It shows the different statistical distributions of *IWP* corresponding to small and large *LWP*, respectively. In case *LWP* is small (less than 5 mm), *IWP* could range from 0 to more than 30 mm where most cases of *IWP* greater than *LWP* occur. The features of such *IWP* distribution are indicative of strong ice cloud activities that can be regarded as the signals of the development of either graupels associated with convective updrafts or ice hydrometeors associated with stratiform ascents. One may

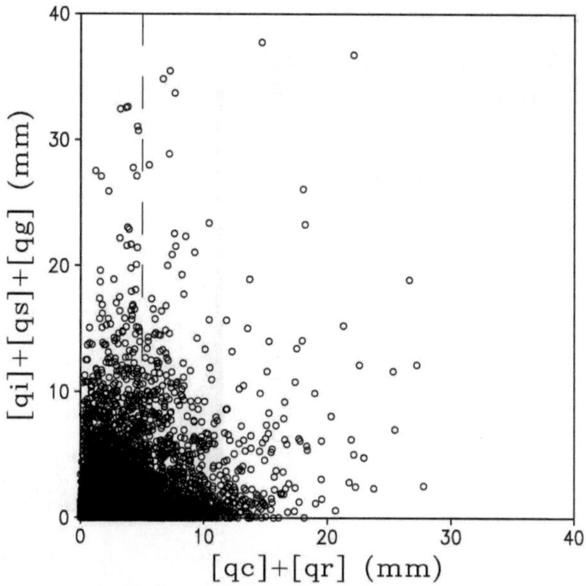

Fig. 5.7 Ice water path ($IWP = [q_i] + [q_s] + [q_g]$) versus liquid water path ($LWP = [q_c] + [q_r]$) (mm). The *dashed line* denotes $[q_c] + [q_r] = 5$ mm (After Sui et al. 2007)

distinguish the convective and stratiform ice clouds by the spatial variability of *IWP* (say *IWP* is one standard deviation above the average taken over an appropriate area), similar to the convective-stratiform rain separation methods developed for radar data analysis by previous studies (e.g, Churchill and Houze 1984; Steiner and Houze 1993; Steiner et al. 1995). For cases of large *LWP* (say, greater than 5 mm), *IWP* tends to decrease with increasing *LWP*, indicative of dominance of water hydrometeors in precipitating clouds. Therefore, the ratio of *IWP* to *LWP* (the cloud ratio) can be used to identify the convective and stratiform rainfall.

The cloud ratio as a function of surface rain rate is further examined in Fig. 5.8. For those grids of heavy surface rain rate (greater than 40 mm h^{-1}), signifying convective rainfall regions, the corresponding values of cloud ratio are near zero. On the other hand, for grids of light surface rain rate (smaller than 10 mm h^{-1}), indicating stratiform rainfall regions, values of cloud ratio are much greater than 1. For grids of surface rainfall rate of 10–40 mm h^{-1}, the corresponding cloud ratios vary within 0–1. The majority of these grids may represent a transition stage of convective development from vigorous convective elements to well-developed stratiform system.

To determine the criterion of cloud ratio for separation of convective rainfall from the others, we calculate linear correlation coefficients and RMS differences between fractional coverage of convective rainfall partitioned by the TS scheme and that by cloud ratio. The cloud ratio is 0.2 when the linear correlation coefficient and RMS difference reach a maximum (0.62) and a minimum (3.30 %), respectively. Therefore, rainfall is designated convective when the corresponding cloud ratio is smaller than 0.2.

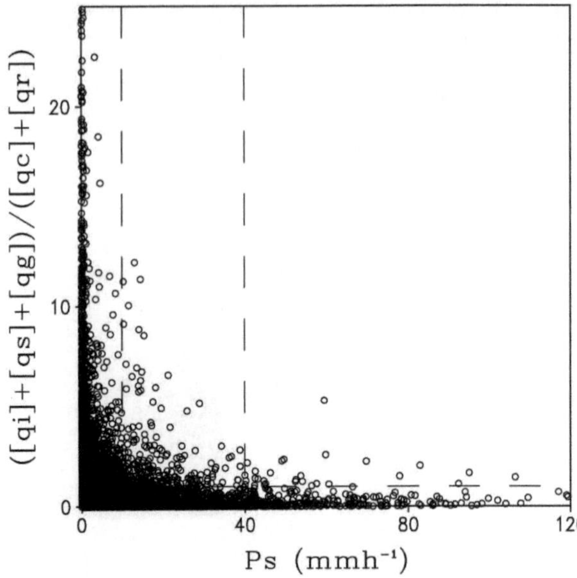

Fig. 5.8 The cloud ratio $(([q_i]+[q_s]+[q_g])/([q_c]+[q_r]))$ versus P_s. The *dashed lines* denote $([q_i]+[q_s]+[q_g])/([q_c]+[q_r])=1$, $P_s=10$, and $P_s=40$ (After Sui et al. 2007)

In addition, the criterion may still miss convective precipitation containing graupels within regions of strong updrafts. To detect convective portions of rainfall from grids of cloud ratio of larger than 0.2, we calculate mean and standard deviation of *IWP* over raining regions, which are 0.85 and 1.70 mm, respectively. We choose the mean plus one standard deviation of *IWP* (2.55 mm) as a threshold value. Those precipitating grids where *IWP* >2.55 mm are also classified as convective.

For remaining grids, we further separate them into stratiform rainfall regions and mixed rainfall regions by a threshold of the cloud ratio. This value may be determined based on the dominance of ice clouds in precipitation systems. Thus, we simply choose the cloud ratio >1 (*IWP* > *LWP*) as the criterion for classifying stratiform rainfall regions. The remaining precipitating grids are considered as mixed.

The time-mean fractional areas based on the new scheme are 3.8 % for convective rainfall regions, 3.2 % for mixed rainfall regions, and 4.2 % for stratiform rainfall regions. The corresponding profiles of time-mean vertical velocity averaged in the three regions are shown in Fig. 5.9. In convective rainfall regions, mean upward motions occur throughout the troposphere with peak values reaching 30 and 20 cm s^{-1} at 4 and 9 km, respectively. In stratiform rainfall regions, mean upward motions occur above 5.5 km while downward motions appear below. The maximum ascending motion reaches 10 cm s^{-1} at 9 km. In mixed rainfall regions, mean ascending motion appears above 3.5 km whereas mean descending motion occurs below. The maximum ascending motion is 10 cm s^{-1} at 5.5 km. The general features of Fig. 5.9 are similar to those of Fig. 5.5, if the profiles for stratiform and mixed rainfall regions in Fig. 5.9 are added for stratiform rainfall. The CFAD for

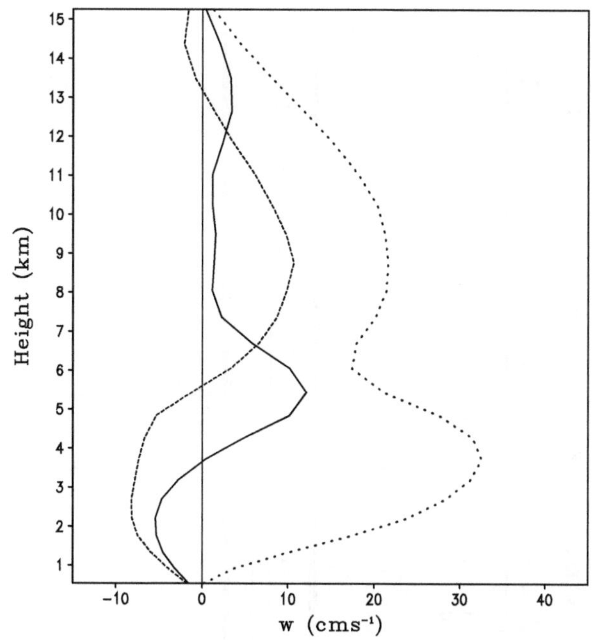

Fig. 5.9 Vertical profiles of vertical velocity (cm s^{-1}) averaged in the region of stratiform (*dashed*), mixed (*solid*), and convective (*dotted*) rainfall, partitioned by the new scheme (After Sui et al. 2007)

vertical velocity in the three regions shows a wide distribution of vertical velocity in convective rainfall regions, with maximum values exceeding 15 m s^{-1} (Fig. 5.10). In the designated stratiform rainfall regions, the distribution is narrow, with absolute values of vertical velocity confined within 5 m s^{-1}.

The mixed rainfall category represents a transition stage of convective development. It contains features more convective than stratiform like strong updraft and downdrafts below 6 km in the CFAD for vertical velocity (Fig. 5.10a, b). The cloud ratio of 0.46 also indicates that the mixed rainfall category is more convective than stratiform. These features are a consequence of the separation of clouds. But the usage of this proposed scheme in any applied or comparative sense may require one to put the mixed rainfall category into either the convective or stratiform categories. In such an occasion, the mixed rainfall category can be put into the stratiform rainfall category for the following reasons. First, from the roughly equal occurrence of stratiform rain (4.2 %) compared to convective rain (3.8 %), it would be more reasonable to add the mixed category (3.2 %) to the stratiform category to be consistent with the mesoscale structure in the model simulated convective systems. Second, the mean rain rate of the mixed category (2.59 mm h^{-1}) is much closer to the stratiform category (0.4 mm h^{-1}) than the convective category (6.19 mm h^{-1}). Finally, by comparing the vertical velocity profiles in Fig. 5.9, the vertical motions in mixed and stratiform rainfall regions are qualitatively similar, characterized by upward motions in the mid and upward troposphere and downward motions in the lower troposphere. This structure differs fundamentally from the structure seen in convective rainfall regions. From the above features and the theoretical point of

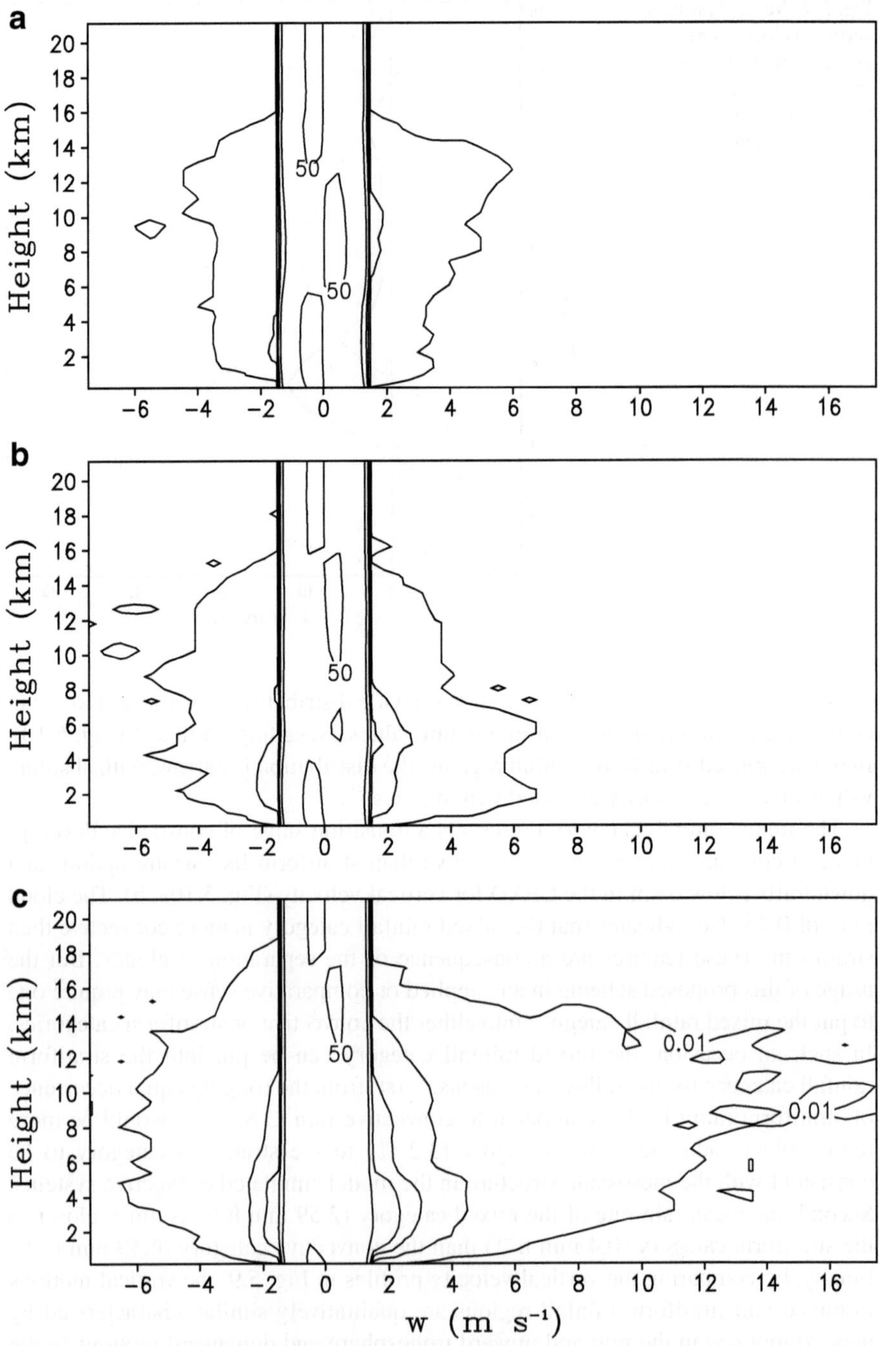

Fig. 5.10 CFAD of vertical velocity (m s^{-1}) for the region of (**a**) stratiform, (**b**) mixed, and (**c**) convective rain, partitioned by the new scheme. Contour intervals are 0.01, 1, 3, 5, 10, and 50 %, respectively (After Sui et al. 2007)

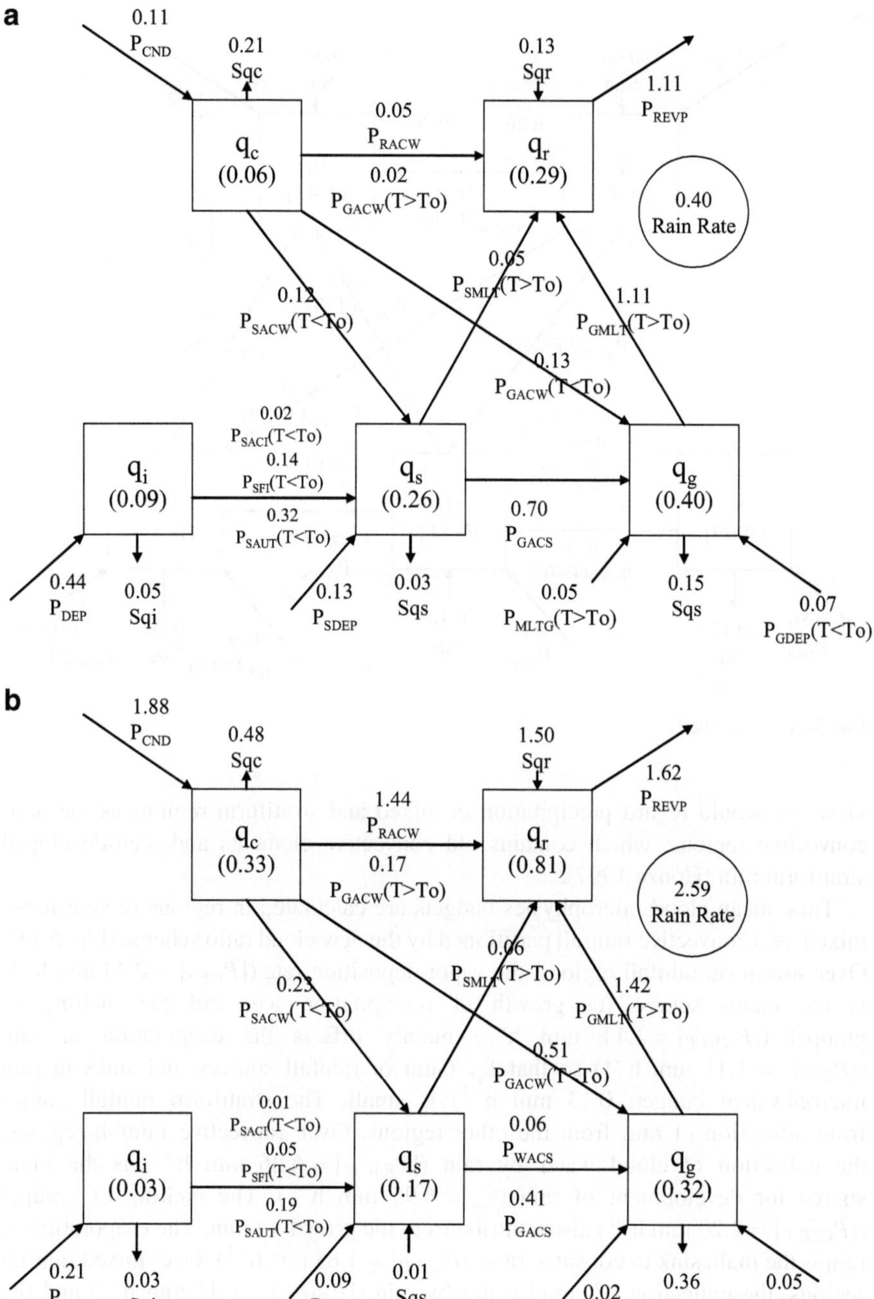

Fig. 5.11 Time mean cloud microphysics budgets averaged in (**a**) stratiform, (**b**) mixed, and (**c**) convective rainfall regions, partitioned by the new scheme. Unites are mm for cloud hydrometeors and mm h^{-1} for conversions (After Sui et al. 2007)

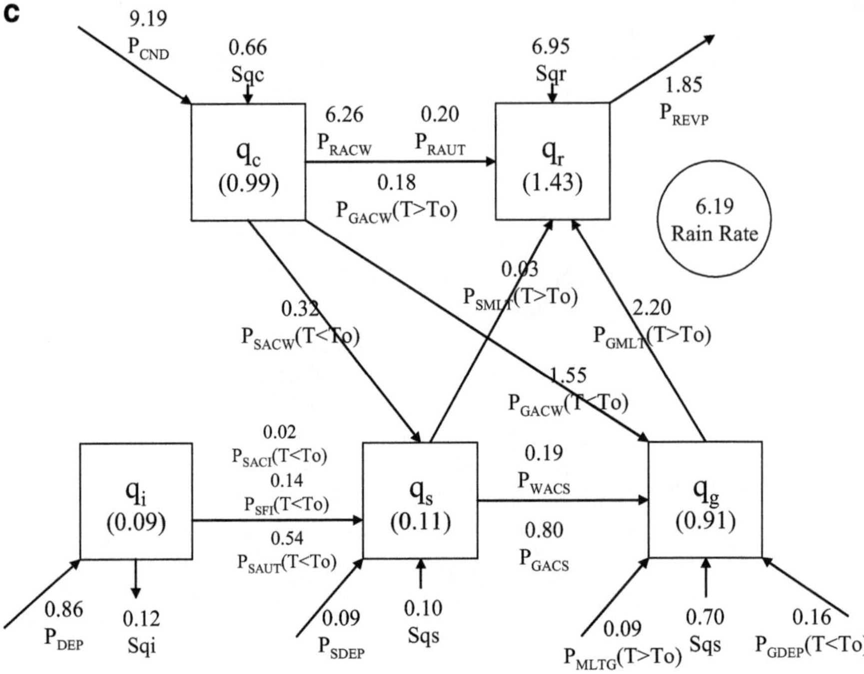

Fig. 5.11 (continued)

view, we would regard precipitation in mixed and stratiform regions as the non-convective regions, which contains old convective elements and well-developed stratiform rain (Houze 1997).

Time-mean cloud microphysics budgets are calculated in regions of stratiform, mixed, and convective rainfall partitioned by the new cloud ratio scheme (Fig. 5.11). Over stratiform rainfall regions, the vapor deposition rate ($[P_{DEP}] = 0.44$ mm h^{-1}) is the major source for growth of precipitation ice, and the melting of graupel ($[P_{GMLT}] = 1.11$ mm h^{-1}) mainly offsets the evaporation of rain ($[P_{REVP}] = 1.11$ mm h^{-1}) so that S_{qr} (sum of rainfall sources and sinks in rain microphysical budget; 0.13 mm h^{-1}) is small. The stratiform rainfall comes from advection of rain from the other regions. Over convective rainfall regions, the collection of cloud water by rain ($[P_{RACW}] = 6.26$ mm h^{-1}) is the main source for development of rain ($S_{qr} = 6.95$ mm h^{-1}). The melting of graupel ($[P_{GMLT}] = 2.20$ mm h^{-1}) also contributes to the growth of rain. The evaporation of rain is the main sink to consume rain ($[P_{REVP}] = 1.85$ mm h^{-1}). Over mixed rainfall regions, the collection of cloud water by rain ($[P_{RACW}] = 1.44$ mm h^{-1}) and the melting of graupel ($[P_{GMLT}] = 1.42$ mm h^{-1}) contribute equally to the development of rain ($S_{qr} = 1.50$ mm h^{-1}). The evaporation of rain is the main sink to consume rain ($[P_{REVP}] = 1.62$ mm h^{-1}).

The cloud microphysical budget over convective rainfall regions identified by the new scheme is similar to that by the TS scheme, and water cloud microphysical processes are dominant in the production of precipitation. Over stratiform rainfall regions, the budget based on the new scheme shows that ice microphysical processes dominate over water microphysical processes, whereas the budget based on the TS scheme shows that water and ice microphysical processes are equally important. Over mixed rainfall regions, the budget based on the new scheme indicates a comparable contribution from water and ice microphysical processes to rain production.

The existing partition methods for convective and stratiform clouds have some disadvantages. The separation between convective and stratiform clouds needs information from the neighboring grids (e.g., Churchill and Houze 1984) or some unconventional information such as vertical motion (e.g., Xu 1995), and fall spend of precipitation particles (e.g., Lang et al. 2003) (also see Sect. 2.5). In contrast, the new scheme does not require the information from neighboring grids. Cloud information is available from satellite retrievals. The disadvantage of the new scheme is that there is a singularity over the regions where water hydrometeor is completely absent.

References

Churchill DD, Houze RA Jr (1984) Development and structure of winter monsoon cloud clusters on 10 December 1978. J Atmos Sci 41:933–960

Cui X, Zhu Y, Li X (2007) Cloud microphysical properties in tropical convective and stratiform regions. Meteorol Atmos Phys 98:1–11

Houze RA Jr (1973) A climatological study of vertical transports by cumulus-scale convection. J Atmos Sci 30:1112–1123

Houze RA Jr (1997) Stratiform precipitation in region of convection: a meteorological paradox? Bull Am Meteorol Soc 78:2179–2196

Lang S, Tao WK, Simpson J, Ferrier B (2003) Modeling of convective-stratiform precipitation processes: sensitivity to partition methods. J Appl Meteorol 42:505–527

Li X, Sui CH, Lau KM (2002) Dominant cloud microphysical processes in a tropical oceanic convective system: a 2-D cloud resolving modeling study. Mon Weather Rev 130:2481–2491

Steiner M, Houze RA Jr (1993) Three-dimensional validation at TRMM ground truth sites: some early results from Darwin, Australia. In: 26th International conference on radar meteorology, Norman, OK, American Meteorological Society, 417–420

Steiner M, Houze RA Jr, Yuter SE (1995) Climatological characterization of three-dimensional storm structure from operational radar and rain gauge data. J Appl Meteorol 34:1978–2007

Sui CH, Li X (2005) A tendency of cloud ratio associated with the development of tropical water and ice clouds. Terr Atmos Ocean Sci 16:419–434, (c) Chinese Geoscience Union. Reprinted with permission

Sui CH, Lau KM, Tao WK, Simpson J (1994) The tropical water and energy cycles in a cumulus ensemble model. Part I: equilibrium climate. J Atmos Sci 51:711–728

Sui CH, Tsay CT, Li X (2007) Convective-stratiform rainfall separation by cloud content. J Geophys Res. doi:10.1029/2006JD008082, (c) American Geophysical Union. Reprinted with permission

Tao WK, Simpson J, Sui CH, Ferrier B, Lang S, Scala J, Chou MD, Pickering K (1993) Heating, moisture, and water budgets of tropical and midlatitude squall lines: comparisons and sensitivity to longwave radiation. J Atmos Sci 50:673–690
Xu KM (1995) Partitioning mass, heat, and moisture budgets of explicit simulated cumulus ensembles into convective and stratiform components. J Atmos Sci 52:1–23

Chapter 6
Structures of Precipitation Systems II: Budget Analysis

Abstract In this chapter, surface rainfall, cloud, and rain budgets are used to develop new rainfall partitioning methods for studying structures of precipitation systems. The rainfall is separated by atmospheric drying/moistening, water vapor, and hydrometeor convergence/divergence in surface rainfall budget. The rainfall types are combined based on vertical profiles of vertical velocity and major cloud microphysical processes to develop a new technique for separating convective and stratiform rainfall. The rainfall is partitioned by net condensation/evaporation and hydrometeor convergence/divergence in cloud budget and rain source/sink and rain hydrometeor convergence/divergence in rain budget. The relationship between rainfall separation methods is discussed. The analysis of mean data and grid-scale data is compared to highlight the scale dependence of rainfall separation.

Keywords Surface rainfall budget • Cloud budget • Rain budget • Rainfall separation method • Mean data • Grid-scale data

The rainfall results from the development of convective systems under large-scale dynamic, thermodynamic, and cloud microphysical conditions, which are governed by surface rainfall and cloud microphysical budgets. The rainfall can be partitioned into convective and stratiform components based on convective signals or rain intensity. The convective–stratiform separation rainfall based on information of liquid and ice water paths from experiment COARE is discussed in Chap. 5. The development of precipitation systems relies on water vapor processes, which is described by surface rainfall budget. The rainfall and water vapor are linked through cloud microphysical processes that are measured by cloud microphysical budget. Rain microphysical budget is directly responsible for the production of rainfall. The detailed structures of precipitation systems can be examined through budget analysis. Thus, the rainfall separation analyses by surface rainfall (Shen et al. 2010; Li et al. 2014), cloud (Li et al. 2011), and rain (Li and Shen 2013) budgets are, respectively, discussed using COARE data in Sects. 6.1, 6.2, and 6.3. The relationship between separation analyses is examined in Sect. 6.4. The scale dependence of rainfall separation analysis is discussed in Sect. 6.5.

© Springer International Publishing Switzerland 2016

X. Li, S. Gao, *Cloud-Resolving Modeling of Convective Processes*,
Springer Atmospheric Sciences, DOI 10.1007/978-3-319-26360-1_6

6.1 The Analysis of Surface Rainfall Budget

In surface rainfall budget (2.4), local change of water vapor (Q_{WVT}), water vapor convergence (Q_{WVF}), and hydrometeor change/convergence (Q_{CM}) can be positive or negative, whereas surface evaporation (Q_{WVE}) is positive. Following Shen et al. (2010), the grid-scale model simulation data are partitioned into eight rainfall types (TFM, TFm, tFM, tFm, TfM, Tfm, tfM, tfm) (see Table 6.1), where *T* and *t* represent local atmospheric drying ($Q_{WVT} > 0$) and moistening ($Q_{WVT} < 0$), respectively, F and f denote water vapor convergence ($Q_{WVF} > 0$) and divergence ($Q_{WVF} < 0$), respectively, and M and m represent hydrometeor loss/convergence ($Q_{CM} > 0$) and hydrometeor gain/divergence ($Q_{CM} < 0$), respectively. The rainfall type tfm is not analyzed because its rain rate is negligibly low compared to the seven other rainfall types.

The total rainfall is largely attributable to the rainfall associated with local atmospheric drying and hydrometeor loss and water vapor divergence in TfM (30.9 %) partly because this rainfall type covers large model domain (3.9 %) (Table 6.2). Model domain-mean surface rain rate of TfM is 0.036 mm h^{-1}, and associated rain rate averaged over TfM area (rain intensity; *RI*) is 2.833 mm h^{-1}. The rainfall is related to local atmospheric drying rate and hydrometeor loss/convergence, while water vapor divergence prevails. The water vapor divergence is associated with downward motions below 7 km that reaches its maximum around 2 km (Fig. 6.1); the downward motions produce downward water vapor mass flux with a maximum around 1.3 km (Fig. 6.2) and weak downward hydrometeor mass flux below 3 km (Fig. 6.3). The perturbation specific humidity of TfM is positive with a maximum above 1 km, whereas it is negative below 1 km (Fig. 6.4). The hydrometeor mixing

Table 6.1 Summary of rainfall types separated by surface rainfall budget

Type	Description
TFM	Local atmospheric drying, water vapor convergence, and hydrometeor loss/convergence
TFm	Local atmospheric drying, water vapor convergence, and hydrometeor gain/divergence
tFM	Local atmospheric moistening, water vapor convergence, and hydrometeor loss/convergence
tFm	Local atmospheric moistening, water vapor convergence, and hydrometeor gain/divergence
TfM	Local atmospheric drying, water vapor divergence, and hydrometeor loss/convergence
Tfm	Local atmospheric drying, water vapor divergence, and hydrometeor gain/divergence
tfM	Local atmospheric moistening, water vapor divergence, and hydrometeor loss/convergence
tfm	Local atmospheric moistening, water vapor divergence, and hydrometeor gain/divergence

T and t represent local atmospheric drying and moistening, respectively. F and f represent water vapor convergence and divergence, respectively. M and m represent hydrometeor loss/convergence and gain/divergence, respectively (After Shen et al. 2010)

Table 6.2 Fractional coverage (FC), percentage of rain amount over total rainfall amount (PRA), and model domain-mean surface rainfall budget (P_S, Q_{WVT}, Q_{WVF}, Q_{WVE}, and Q_{CM}) in TFM, TFm, tFM, tFm, TfM, Tfm, and tfM in COARE

	TFM	TFm	tFM	tFm	TfM	Tfm	tfM
FC	0.129	0.770	2.765	1.712	3.949	1.189	0.652
PRA	9.951	17.681	19.261	14.182	30.875	5.013	3.037
P_S	0.036	0.064	0.069	0.051	0.111	0.018	0.011
Q_{WVT}	0.010	0.063	−0.285	−0.208	0.511	0.152	−0.011
Q_{WVF}	0.012	0.118	0.258	0.409	−0.595	−0.099	−0.014
Q_{WVE}	0.000	0.003	0.007	0.005	0.011	0.003	0.002
Q_{CM}	0.014	−0.120	0.089	−0.156	0.184	−0.038	0.034
RI	27.814	8.273	2.511	2.986	2.822	1.523	1.683
PE	100	34.8	19.5	12.3	15.7	11.6	30.6

The rain intensity ($RI = P_S/FC \times 100$ %) and precipitation efficiency (*PE*) defined in surface rainfall budget are included in the table. Units are mm h^{-1} for surface rainfall budget and *RI* and % for *FC* and *PRA* and *PE* (After Li et al. 2014)

ratio of TfM has a maximum around 5 km (Fig. 6.5). The strong water divergence leads to the weak net evaporation. The evaporation of cloud water ($P_{CND} < 0$) is associated with the strong local atmospheric drying (Fig. 6.6e), which forms a large sink for cloud water along with the collection of cloud water by rain. The vapor depositions ($P_{DEP} + P_{SDEP} + P_{GDEP}$) cause the large melting rate of graupel (P_{GMLT}) as the major rainfall source (*Sqr*). The large melting rate of graupel also causes the sinks for ice hydrometeors.

The rainfalls associated with water vapor convergence make significant contributions to total rainfall. 10 %, 17.7 %, 19.3 %, and 14.2 % of the total rainfall come, respectively, from the rainfall associated with local atmospheric drying and hydrometeor loss/convergence (TFM), the local atmospheric drying and hydrometeor gain/divergence (TFm), the local atmospheric moistening and hydrometeor loss/convergence (tFM), and the local atmospheric moistening and hydrometeor gain/divergence (tFm) (Table 6.2). The large contributions of rainfall associated with the local atmospheric moistening may result from the larger fractional coverage (2.8 % in tFM and 1.7 % in tFm) compared to areas occupied by the rainfall associated with the local atmospheric drying in TFM (0.1 %) and TFm (0.8 %). Although they cover small areas, the rainfalls associated with local atmospheric drying have large averaged magnitudes (27.814 mm h^{-1} in TFM and 8.273 mm h^{-1} in TFm).

In TFM, local atmospheric drying, water vapor convergence, and hydrometeor loss/convergence have similar contributions (~30 %) to the rain rate. Although the weak downward motions occur near the surface, relatively strong upward motions with their maximum around 3.5 km produce the strong upward water vapor and hydrometeor mass fluxes (Figs. 6.1, 6.2, and 6.3). Maximum perturbation specific humidity appears at 2 km with the averaged value of 1.8 gkg^{-1} (Fig. 6.4), whereas maximum hydrometeor mixing ratio occurs below 5.5 km with the averaged value

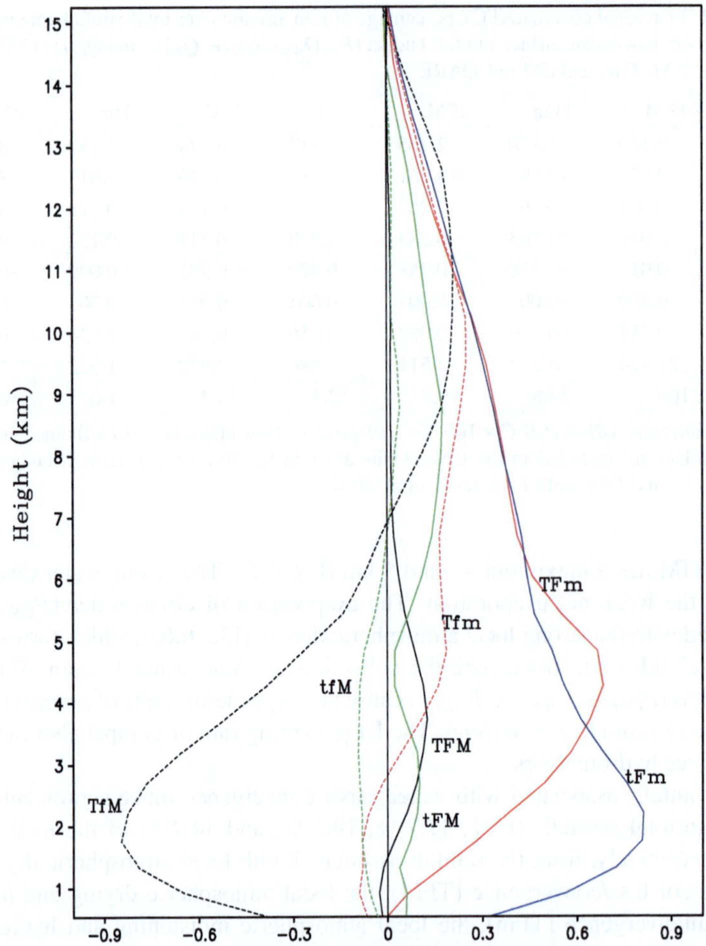

Fig. 6.1 Vertical profiles of mean vertical velocity (cm s^{-1}) in TFM (*black solid*), TFm (*red solid*), tFM (*green solid*), tFm (*blue solid*), TfM (*black dash*), Tfm (*red dash*), and tfM (*green dash*) in COARE. After Li et al. (2014)

of 1.2 gkg^{-1} (Fig. 6.5). The analysis of cloud microphysical budget in TFM shows that the vapor condensation is much larger than the vapor depositions ($P_{DEP} + P_{SDEP} + P_{GDEP}$) and the collection of water vapor by rain (P_{RACW}) is five times larger than the melting of graupel to rain (Fig. 6.6a). This suggests the dominance of liquid microphysical processes for the development of this rain type.

In TFm, the water vapor convergence is nearly offset by the hydrometeor gain/divergence, whereas rainfall is related to the local atmospheric drying (Table 6.2). The water vapor convergence rate is larger in TFm than in TFM because the upward motions in TFm occur throughout the troposphere and are

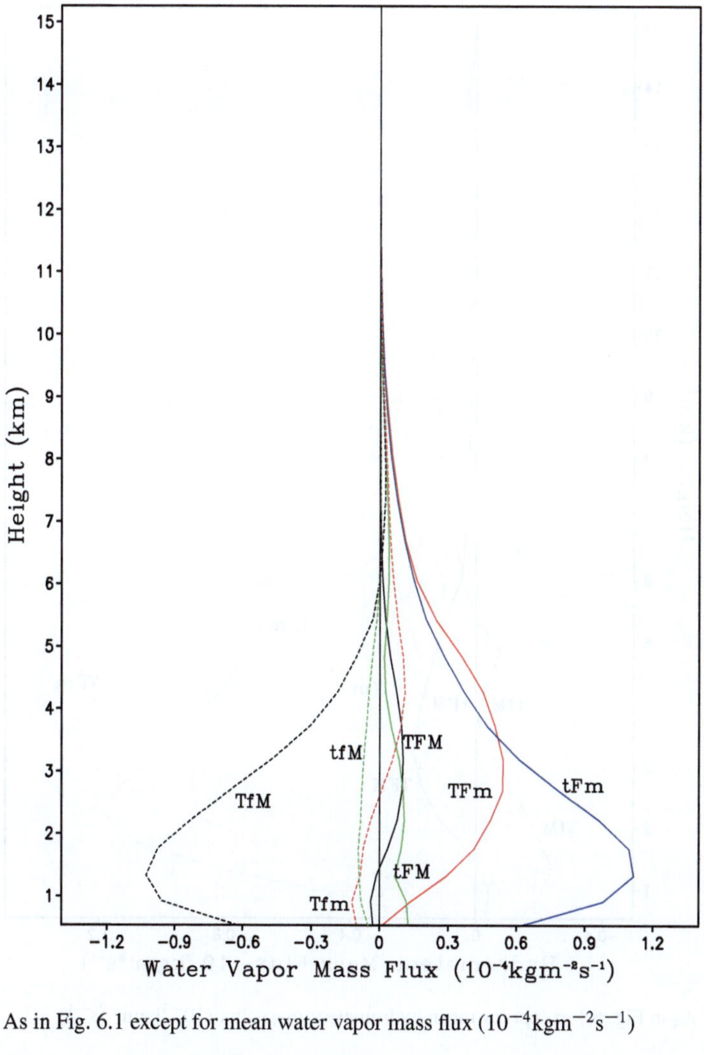

Fig. 6.2 As in Fig. 6.1 except for mean water vapor mass flux (10^{-4}kgm^{-2}s^{-1})

much stronger than those in TFM (Fig. 6.1). Upward motions in TFm generate the upward water vapor and hydrometeor mass fluxes with their peaks around 3 km (Figs. 6.2 and 6.3). Perturbation specific humidity and hydrometeor mixing ratio are much larger in TFm than in TFM (Figs. 6.4 and 6.5). The large net condensation is associated with the strong vapor condensation rate, which is almost one order of magnitude larger than the vapor deposition rates (Fig. 6.6b). As a result, the collection rate of cloud water by rain is about six times larger than the melting rate of graupel to rain and is a major rainfall source, which is larger than what is required by the surface rainfall. The large vapor condensation leads to the sources for all five cloud species ($Si > 0$, i = qc, qr, qi, qs, and qg). The comparison between

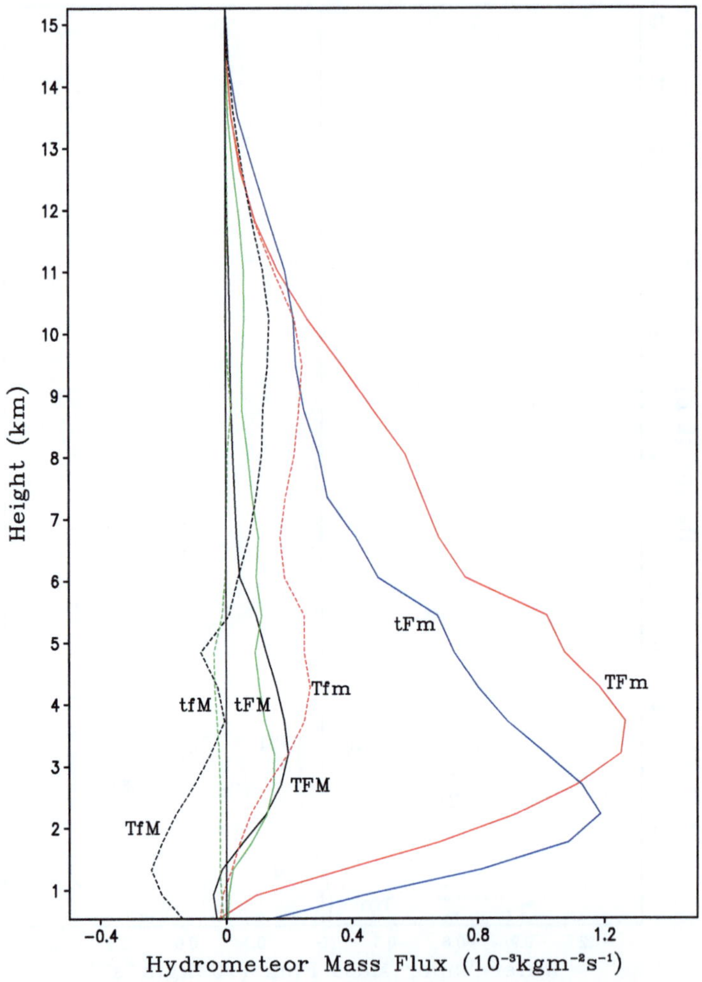

Fig. 6.3 As in Fig. 6.1 except for mean hydrometeor mass flux ($10^{-3}kgm^{-2}s^{-1}$)

TFM and TFm reveals that the rain intensity of TFM is lower than that of TFm if the cloud hydrometeor transport is excluded. This indicates the importance of the cloud hydrometeor transport in determining maximum rainfall.

In tFM, water vapor convergence largely moistens the local atmosphere (Table 6.2). As a result, the rainfall corresponds to the hydrometeor loss/convergence. The weak upward motions extend from the surface to the tropopause (Fig. 6.1) and yield weak upward water vapor and hydrometeor mass fluxes (Figs. 6.3 and 6.4). Perturbation specific humidity reaches the maximum of 0.015 gkg^{-1} between 2 km and 5 km (Fig. 6.4), and hydrometeor mixing ratio shows a maximum of 0.01 gkg^{-1} around 5 km (Fig. 6.5). The analysis of cloud microphysical budget shows that the vapor condensation rate is lower and is similar to the vapor deposition rates

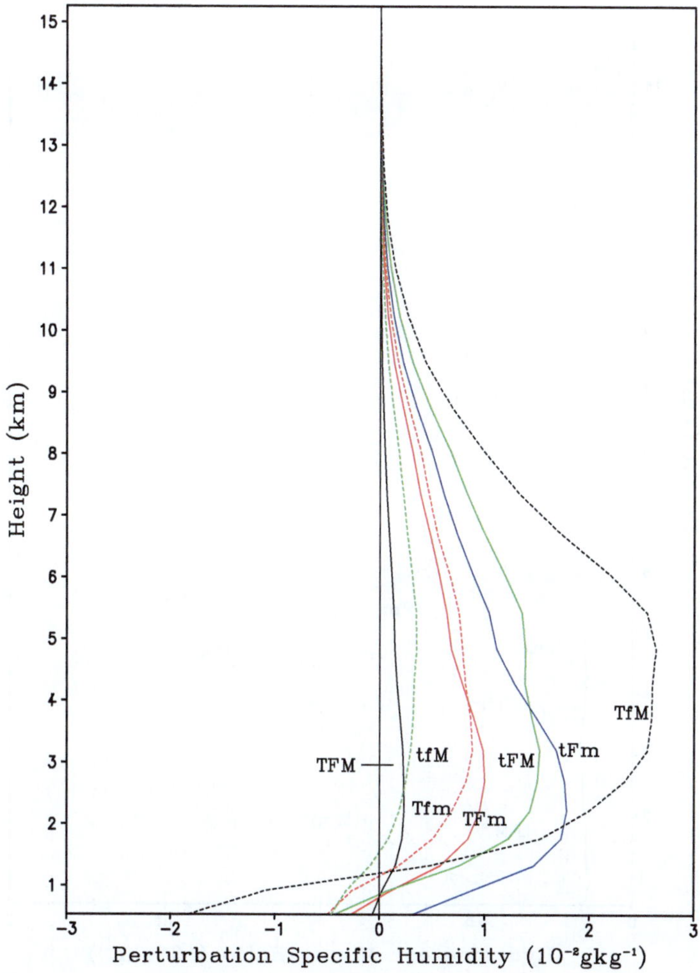

Fig. 6.4 As in Fig. 6.1 except for mean perturbation specific humidity (10^{-2} gkg^{-1})

(Fig. 6.6c). As a result, the melting rate of graupel to rain is slightly higher than the collection rate of cloud water by rain, indicating equal importance of liquid and ice microphysical processes for the development of this rain type. The weak vapor condensation and depositions cannot afford the collection and melting, forming large sinks for cloud water, cloud ice, and graupel.

In tFm, the water vapor convergence is the only source for the production of rainfall, local atmospheric moistening, and enhancement of hydrometeor concentration/hydrometeor divergence (Table 6.2). The upward motions occur through the troposphere while they reach their maximum around 2 km (Fig. 6.1). The upward motions are about 0.8 cm s^{-1}, which is the largest of upward motions among rainfall types. The strong lower-tropospheric upward motions lead to the strong upward

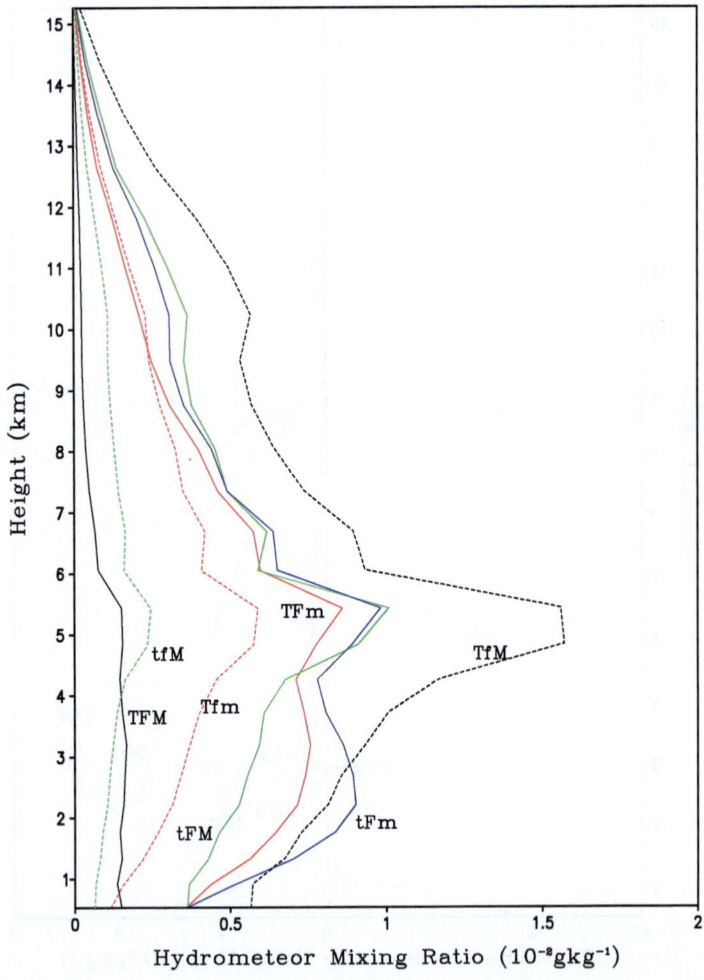

Fig. 6.5 As in Fig. 6.1 except for mean hydrometeor mixing ratio (10^{-2} gkg^{-1})

water vapor and hydrometeor mass fluxes around 1.5–2 km (Figs. 6.2 and 6.3). Maximum perturbation specific humidity is 0.018 gkg^{-1} at 2 km (Fig. 6.4), and maximum hydrometeor mixing ratio is about 0.008 gkg^{-1} around 2 km (Fig. 6.5). The vapor condensation is more than one order of magnitude larger than the vapor deposition rates, and the collection rate of cloud water by rain is about four times higher than the melting rate of graupel to rain in tFm (Fig. 6.6d). The comparison between TFm and tFm shows that the rain intensity of TFm is more than three times higher than that of tFm, although the water vapor convergence rate of TFm is more than 70 % less than that of tFm.

Although the rainfall of TfM is the largest contributor to the total rainfall, the other rainfall types associated with water vapor divergence are not important for the

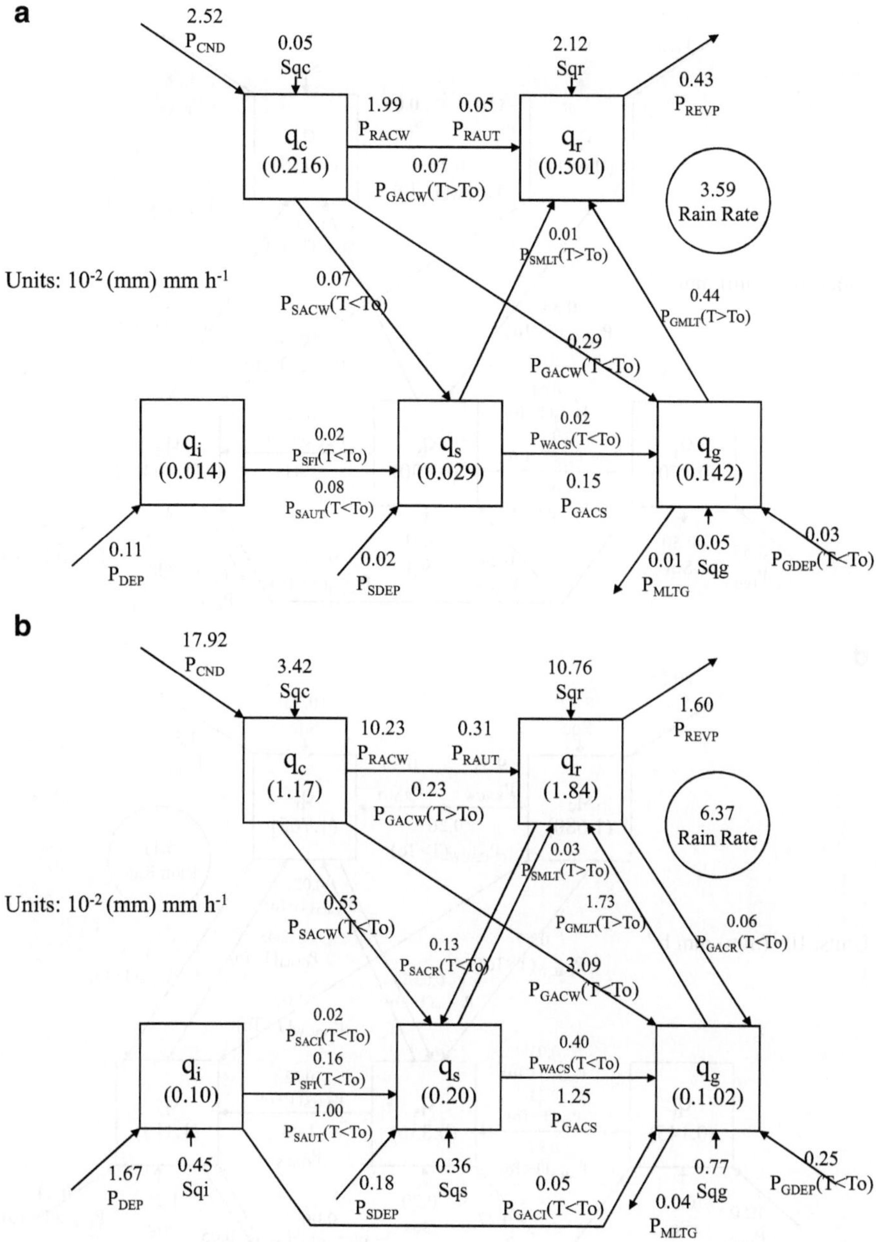

Fig. 6.6 Time-mean cloud microphysical budgets for (**a**) TFM, (**b**) TFm, (**c**) tFM, (**d**) tFm, (**e**) TfM, (**f**) Tfm, and (**g**) tfM. Units for cloud hydrometeors and conversions are 10^{-2} mm and 10^{-2} mm h^{-1}, respectively. Cloud microphysical conversion terms and their schemes can be found in Table 1.2

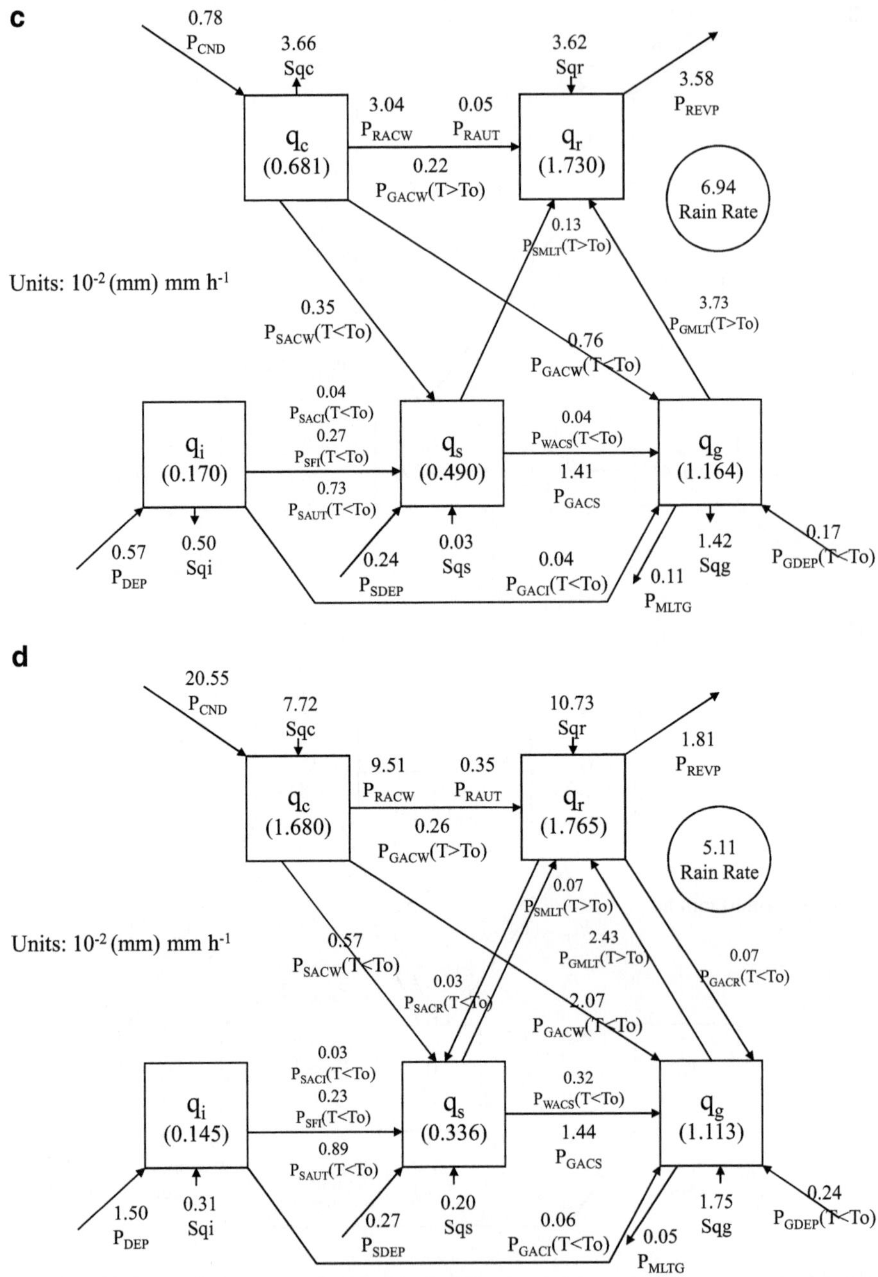

Fig. 6.6 (continued)

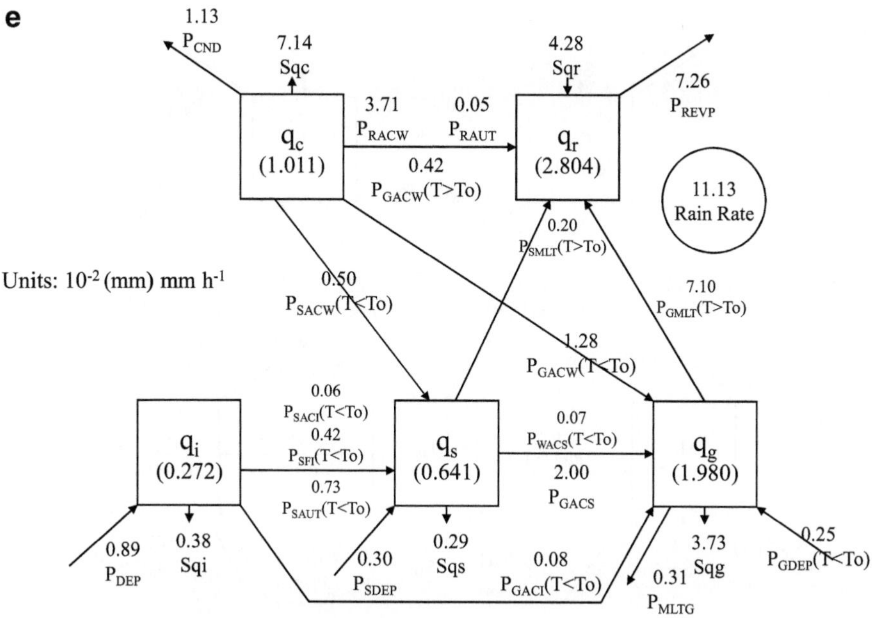

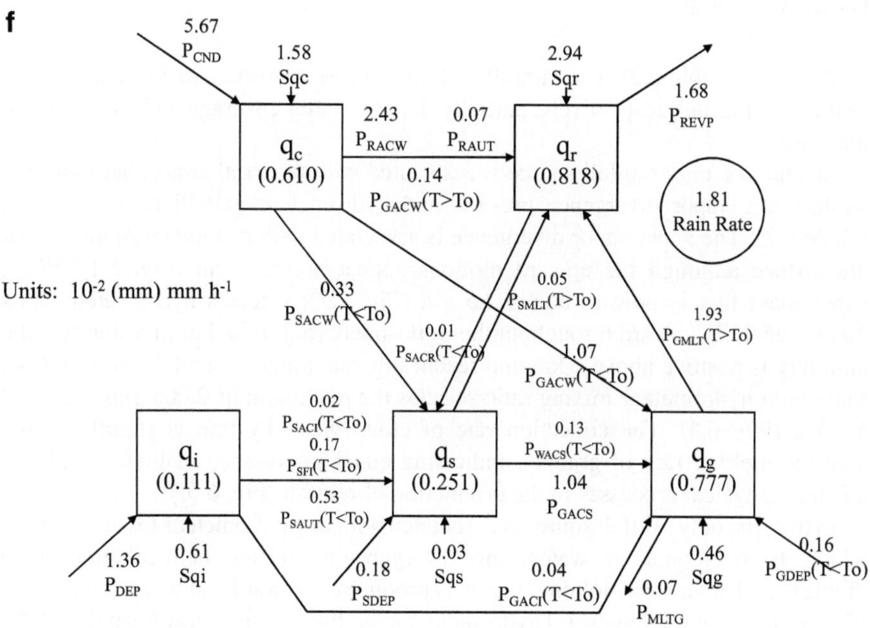

Fig. 6.6 (continued)

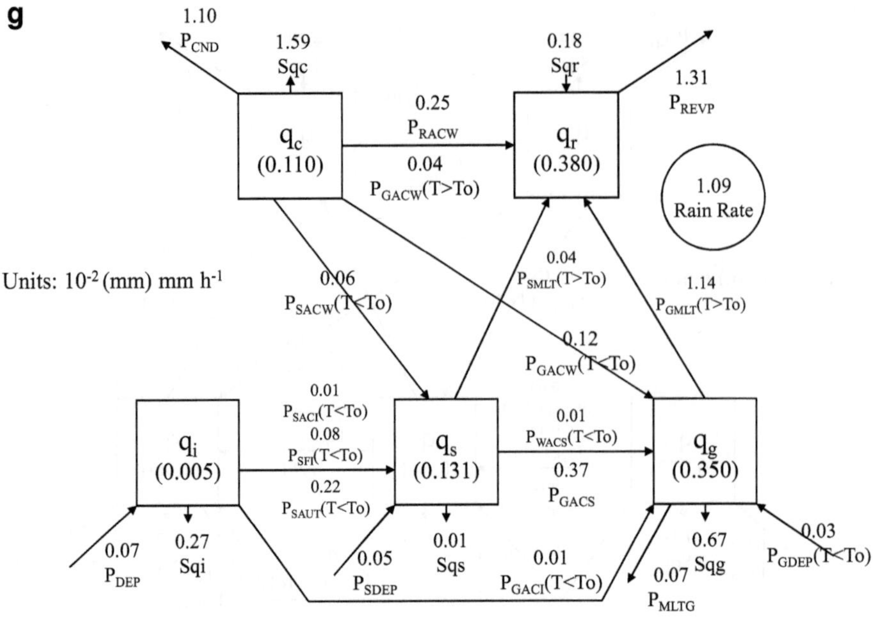

Fig. 6.6 (continued)

total rainfall (Table 6.2). The rainfall in Tfm and tfM contribute to 5.0 % and 3.0 % of the total rainfall, respectively, partially due to smaller coverage and weak rainfall intensity.

In Tfm, the only rainfall source is associated with the local atmospheric drying, while water vapor divergence prevails and hydrometeor gain/divergence occurs (Table 6.2). The water vapor divergence is associated with downward motions near the surface although the upward motions appear above 3 km (Fig. 6.1). Water vapor mass flux is upward above 2.5 km (Fig. 6.2), whereas hydrometeor mass flux is generally upward throughout the troposphere (Fig. 6.3). Perturbation specific humidity is positive above 1 km and reaches its maximum around 3 km (Fig. 6.4). Maximum hydrometeor mixing ratio reaches the maximum of 0.005 gkg^{-1} around 5.5 km (Fig. 6.5). The collection rate of cloud water by rain is slightly higher than the melting rate of graupel, indicating equal importance of both liquid and ice microphysical processes in the production of rainfall (Fig. 6.6f).

In tfM, the only rainfall source is associated with the hydrometeor loss/convergence, which also overcomes the water vapor divergence to moisten the local atmosphere (Table 6.2). Downward motions (Fig. 6.1) produce downward water vapor mass flux (Fig. 6.2). The weak downward hydrometeor mass flux occurs below 6 km (Fig. 6.3). Perturbation specific humidity is positive above 1.5 km (Fig. 6.4). The hydrometeor mixing ratio in tfM has its peak around 5 km (Fig. 6.5). The evaporation of cloud water occurs, and the melting rate of graupel is much higher than the collection rate of cloud water by rain, although the vapor deposition rates in tfM are much lower than those in TfM (Fig. 6.6g).

The analysis of vertical profile of vertical velocity and rain microphysical budget suggests that TFM, TFm, and tFm are rain types for convective rainfall, whereas TfM and tfM are rain types for stratiform rainfall. tFM and Tfm show similar rain microphysical budgets in which the collection rate of cloud water by rain and the melting rate of graupel to rain are nearly equal, although tFM and Tfm have the convergence and divergence of water vapor and cloud hydrometeor, respectively. These two rain types are similar to the mixed category discussed in Sect. 5.2 which is considered as a transition stage from the convective rainfall to the stratiform rainfall. The transition stage from the convective rainfall to the stratiform rainfall can be also seen from the vertical profiles in which tFM and Tfm show much weaker upward motions than TFm and tFm do and much weaker downward motions than TfM does (Fig. 6.1). Table 6.2 also shows that the fractional coverage of tFM is much larger than those of three other rain types associated with water vapor convergence. Thus, tFM + Tfm may be categorized into stratiform rainfall.

FTM + FTm + tFm and tFM + Tfm + TfM + tfM form convective and stratiform rainfall, respectively, in this study. TFM + TFm + tFm and tFM + Tfm + TfM + tfM are, respectively, compared to convective (CT93)–stratiform (ST93) rainfall separated by the TS scheme in Table 6.3. TFM + TFm + tFm and CT93 are associated with water vapor convergence and hydrometeor divergence, but they are related to the local atmospheric moistening and drying, respectively. Since the area of CT93 is larger than that of TFM + TFm + tFm, their rain intensities are

Table 6.3 Fractional coverage (FC) and contribution of surface rainfall budget (P_S, Q_{WVT}, Q_{WVF}, Q_{WVE}, and Q_{CM}) from (a) TFM + TFm + tFm and tFM + Tfm + TfM + tfM and (b) convective (CT93) and stratiform (ST93) rainfall (Tao et al. 1993) to model domain mean

(a)	TFM + TFm + tFm	tFM + Tfm + TfM + tfM
FC	2.611	8.555
P_S	0.151	0.209
Q_{WVT}	−0.135	0.367
Q_{WVF}	0.539	−0.450
Q_{WVE}	0.008	0.023
Q_{CM}	−0.262	0.269
RI	5.783	2.443

(b)	CT93	ST93
FC	4.391	8.890
P_S	0.234	0.127
Q_{WVT}	0.123	0.100
Q_{WVF}	0.223	−0.107
Q_{WVE}	0.015	0.021
Q_{CM}	−0.127	0.112
RI	5.329	1.429

The rain intensity (*RI*) is included in the table. Units are mm h^{-1} for surface rainfall budget and *RI* and % for FC (After Li et al. 2014)

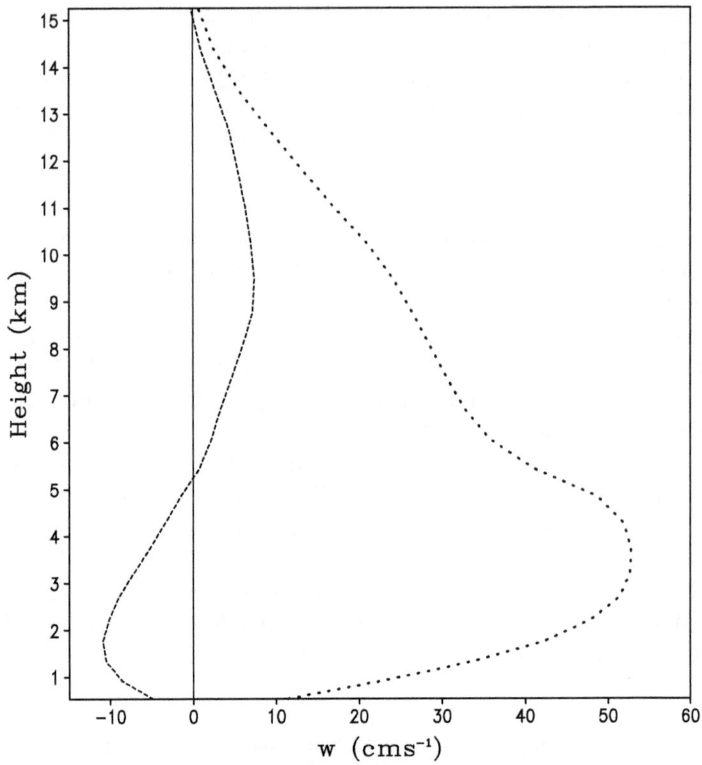

Fig. 6.7 Vertical profiles of vertical velocity (cm s^{-1}) averaged for tFM + Tfm + TfM + tfM (*dashed*) and TFM + TFm + tFm (*dotted*) (After Li et al. 2014)

similar. The maximum upward motions occur in the lower troposphere, but the magnitude of TFM + TFm + tFm (~50 cm s^{-1}) is much larger than that of CT93 (~30 cm s^{-1}) (Fig. 6.7; also see Fig. 5.5) because over 30 % of CT93 has downward motions in the lower troposphere. The collection rate of cloud water by rain is much higher than the melting rate of graupel to rain, and liquid microphysical processes are dominant in the precipitation of rainfall for TFM + TFm + tFm (Fig. 6.6) and CT93 (see Fig. 5.2b). tFM + Tfm + TfM + tfM and ST93 are mainly associated with hydrometeor convergence while the water vapor divergence largely causes the local atmospheric drying. The upward motions occur in the mid- and upper troposphere, while the downward motions appear in the lower troposphere for tFM + Tfm + TfM + tfM and ST93. The melting rate of graupel to rain is much higher than the collection of cloud water by rain and becomes the major rainfall source for tFM + Tfm + TfM + tfM (Fig. 6.6) and ST93 (see Fig. 5.2a). Although TFM + TFm + tFm and tFM + Tfm + TfM + tfM are qualitatively similar to CT93 and ST93, respectively, the current scheme provides a more in-depth analysis of the different rain types in terms of dynamics, thermodynamics, and cloud microphysics as to be presented in the following discussions.

Now we analyze convective–stratiform rainfall structures through the examination of surface rainfall budgets for rain types (Table 6.2). Over the area of TFM + TFm + tFm, Q_{WVT}, Q_{WVF}, and Q_{CM} are positive, and the rain rate is associated with local atmospheric drying, water vapor convergence, and hydrometeor convergence for TFM. The precipitation efficiency defined in surface rainfall budget is the ratio of rain rate to rain source including local atmospheric drying, water vapor convergence, surface evaporation, and hydrometeor convergence. Thus, the precipitation efficiency of TFM is 100 %. Since contribution from each rain type to total rainfall relies on rainfall area, the rain intensity is calculated by P_S over fractional coverage. The rain intensity of TFM is much higher than those of other rain types over convective rainfall regions. Thus, TFM is defined as maximum rainfall in the rainfall system. Both water vapor and hydrometeor convergences contribute about 40 % to the rain rate of TFM, respectively, indicating the crucially importance of cloud hydrometeor transport in the development of maximum rainfall. TFM is associated with hydrometeor convergence although hydrometeor divergence generally prevails over the area of TFM + TFm + tFm.

TFm and tFm show large cloud hydrometeor divergence rates while they cover moderate rainfall areas. In TFm, the cloud hydrometeor divergence is largely offset by the water vapor convergence, and rainfall is associated with the local atmospheric drying. Because water vapor convergence supports cloud hydrometeor divergence, the precipitation efficiency of TFm is 34.8 %. The rain intensity for TFm is much smaller than that of TFM. The comparison between TFM and TFm reveals that the rain intensity of TFM could be lower than that of TFm if the cloud hydrometeor transport is excluded. This again indicates the importance of the cloud hydrometeor transport in determining maximum rainfall. In tFm, the water vapor convergence rate is largest over convective rainfall regions (Table 6.2). The water vapor convergence to the area of this rain type is mainly used to moisten local atmosphere and support the cloud hydrometeor divergence to the other rain types. Thus, the precipitation efficiency of tFm is 12.3 %. The comparison between TFM and tFm reveals that maximum rain rate is not associated with maximum water vapor convergence since maximum water vapor convergence cannot fully used to produce rainfall in tFm.

Over the area of tFM + Tfm + TfM + tfM, surface rainfall budget in TfM shows that water vapor divergence is used to dry the local atmosphere, while the rainfall corresponds to the cloud hydrometeor convergence. The precipitation efficiency of TfM is about 16 %. The melting of graupel to rain is the major rain source of TfM (Fig. 6.6), and associated ice clouds occupy the largest area among the seven rain types. Due to the large rainfall area, the rain intensity only is less than 3 mm h^{-1}. The cloud hydrometeor convergence plays a crucial role in the production of rainfall in tfM. tfM covers much smaller rainfall area than TfM does. The precipitation efficiency of tfM is about 30 %. In Tfm, the local atmospheric drying is nearly offset by the divergence of water vapor and cloud hydrometeor, which produce precipitation less efficiently (precipitation efficiency of 11.6 %). tFM has cloud hydrometeor convergence while it occupies a large rainfall area. In tFM, water vapor convergence is almost used to moisten the local atmosphere. The rainfall corresponds to cloud hydrometeor convergence. Due to large rainfall area, the rain intensity for tFM is similar to that of TfM. The precipitation efficiency only is 19.5 %. The rainfall of tFM + Tfm + TfM + tfM mainly comes from TfM and tFM due to the fact that their fractional coverage is much larger than that of Tfm + tfM. The above analysis reveals that the rainfall separation based on surface rainfall budget is consistent with the rainfall separation based on the rain intensity, but it can provide in-depth dynamic, thermodynamic, and cloud microphysical structures of precipitation systems.

6.2 The Analysis of Cloud Microphysical Budget

The mass-integrated cloud microphysical budget (2.6) shows that surface rain rate is associated with the net condensation and hydrometeor change/convergence. The net condensation ($Q_{NC} > 0$) occurs when vapor condensation and depositions are larger than the evaporation of precipitation hydrometeors (rain, snow, and graupel), whereas the net evaporation ($Q_{NC} < 0$) appears when vapor condensation and depositions are smaller than the evaporation of precipitation hydrometeors. Thus, surface rainfall can be partitioned into three types based on cloud budget: CM, Cm, and cM, where C and c represent the net condensation ($Q_{NC} > 0$) and net evaporation ($Q_{NC} < 0$), respectively, and M and m represent hydrometeor loss/convergence ($Q_{CM} > 0$) and hydrometeor gain/divergence ($Q_{CM} < 0$), respectively (Table 6.4). A rainfall type is excluded in this study because net evaporation and hydrometeor gain/divergence do not produce rainfall.

To examine contribution of each rainfall type to total rainfall, percentage of rainfall amount of each rainfall type over total rainfall amount (PRA) is calculated and shown in Table 6.5a. The rainfall associated with net condensation and hydrometeor loss/convergence (CM) and the rainfall associated with net condensation and hydrometeor gain/divergence (Cm) have similar contributions to total rainfall (36–37 %), which are larger than the contribution (27.1 %) from the rainfall associated with net evaporation and hydrometeor loss/convergence (cM). Although CM (1.0 %) covers much smaller area over the entire model domain than Cm (3.7 %) does during the model integration, their time and model domain-mean rain rates are similar

Table 6.4 Summary of rainfall types partitioned by cloud budget

Type	Description
CM	Net condensation and hydrometeor loss/convergence
Cm	Net condensation and hydrometeor gain/divergence
cM	Net evaporation and hydrometeor loss/convergence
CM1	In CM, net condensation is larger than hydrometeor loss/convergence
CM2	In CM, net condensation is smaller than hydrometeor loss/convergence

C and c represent net condensation and net evaporation, respectively. M and m represent hydrometeor loss/convergence and gain/divergence, respectively (After Li et al. 2011)

Table 6.5 (a) Fractional coverage (FC), percentage of rain amount over total rainfall amount (PRA), and model domain-mean cloud budgets (P_S, Q_{NC}, Q_{CM}) for CM, Cm, and cM and (b) breakdown of CM into CM1 and CM2 in COARE

(a)	CM	Cm	cM
FC	1.032	3.674	6.463
PRA	36.003	36.876	27.122
P_S	0.130	0.133	0.098
Q_{NC}	0.065	0.447	−0.158
Q_{CM}	0.065	−0.314	0.256

(b)	CM1	CM2
FC	0.465	0.566
PRA	17.691	18.267
P_S	0.064	0.066
Q_{NC}	0.048	0.018
Q_{CM}	0.016	0.048

Units are % for FC and PRA and mm h^{-1} for cloud budget (After Li et al. 2011)

(0.13 mm h^{-1}). The mean net condensation and hydrometeor loss/convergence contribute equally to the mean surface rain rate in CM. CM is associated with the mean upward motions and upward water vapor and hydrometeor mass fluxes above 2 km and the mean downward motions and downward water vapor and hydrometeor mass fluxes near the surface (Figs. 6.8, 6.9, and 6.10). In CM, perturbation specific humidity is negative near the surface, whereas it is positive above 1 km (Fig. 6.11). Hydrometeor mixing ratio (sum of mixing ratio of five cloud species) slightly increases as height increases from the surface to 5.5 km, whereas it decreases as height increases above 5.5 km (Fig. 6.12). The analysis of cloud microphysical budget shows that the vapor condensation rate is much larger than the vapor depositions, and the collection rate of cloud water by rain is twice larger than the melting rate of graupel (Fig. 6.13a). This indicates the dominance of liquid microphysical processes in the production of rainfall for this rain type.

The mean net condensation is much larger in Cm (0.447 mm h^{-1}) than in CM (0.065 mm h^{-1}) and is largely offset by the mean hydrometeor gain/divergence (−0.314 mm h^{-1}), which causes similar mean rain rates in CM and Cm (Table 6.5). Cm is related to the mean upward motions and upward water vapor and hydrometeor mass fluxes throughout the troposphere (Figs. 6.8, 6.9, and 6.10), whose magnitudes are much larger than those associated with CM. The maxima for the mean upward motions and the mean upward water vapor and hydrometeor mass fluxes are 4–5 km, 2 km, and 3 km, respectively. Like CM, Cm has negative mean perturbation specific humidity near the surface. The mean positive perturbation specific humidity is much larger in Cm than in CM above the surface, and its maximum reaches around 2–3 km (Fig. 6.11). The mean hydrometeor mixing ratio is much larger in Cm than in CM, but the two rainfall types have similar vertical maxima (Fig. 6.12). The vapor condensation rate is slightly lower than the vapor deposition rates, and the melting

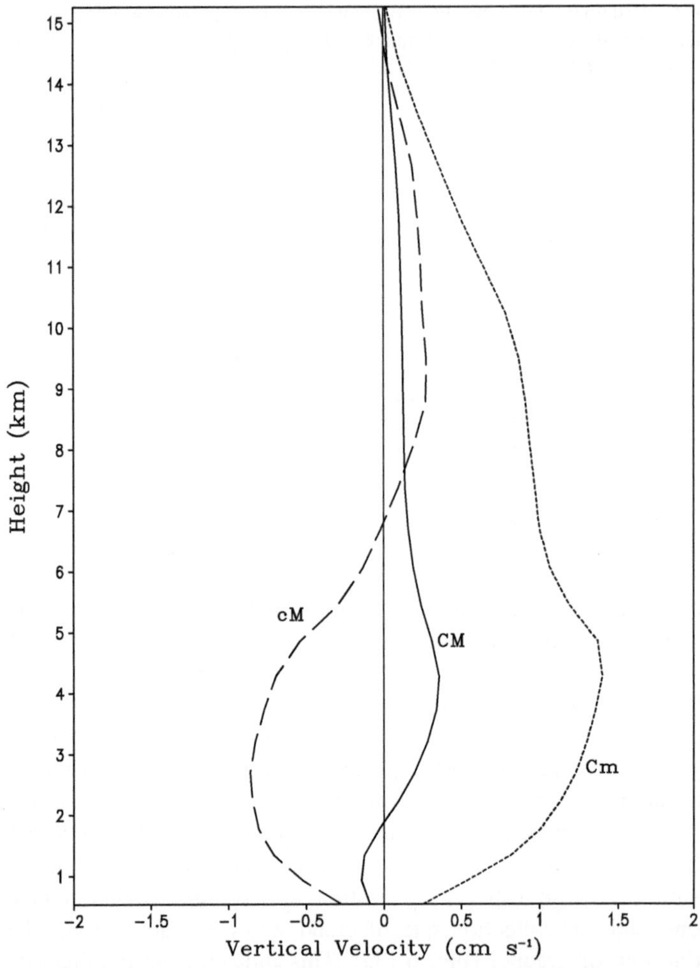

Fig. 6.8 Vertical profiles of time and model domain-mean vertical velocity (m s^{-1}) in CM (*solid*), Cm (*short dash*), and cM (*long dash*) in COARE (After Li et al. 2011)

rate of graupel is about three times larger than the collection rate of cloud water by rain, showing dominant ice microphysical processes (Fig. 6.13b). The large vapor condensation and depositions lead to large source for the growths of cloud water, cloud ice, and graupel.

The mean hydrometeor loss/convergence is much larger in cM (0.256 mm h^{-1}) than in CM (0.065 mm h^{-1}) and is largely offset by the mean net evaporation (−0.158 mm h^{-1}) (Table 6.5). Thus, the mean rain rate in cM is larger than those in the two other rainfall types. Significant contribution from cM indicates that the rainfall associated with hydrometeor loss/convergence plays an important role in regulating rainfall distribution. Fractional coverage of cM is nearly twice larger than

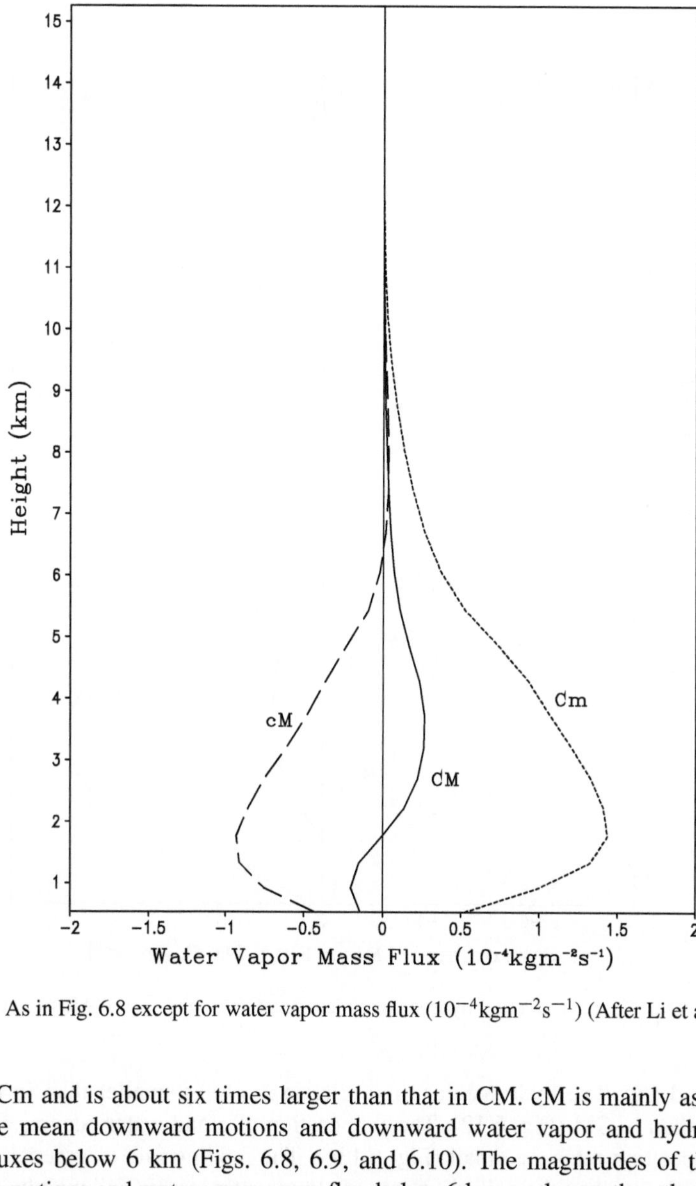

Fig. 6.9 As in Fig. 6.8 except for water vapor mass flux (10^{-4}kgm^{-2}s^{-1}) (After Li et al. 2011)

that in Cm and is about six times larger than that in CM. cM is mainly associated with the mean downward motions and downward water vapor and hydrometeor mass fluxes below 6 km (Figs. 6.8, 6.9, and 6.10). The magnitudes of the mean upward motions and water vapor mass flux below 6 km are larger than those above 6 km, whereas the magnitudes of the mean upward and downward hydrometeor mass fluxes are similar. The maxima for the mean downward motions and downward water vapor mass flux are 2.5 km and 1.5 km, respectively. The negative perturbation specific humidity in cM is larger than those in CM and Cm near the surface (Fig. 6.11). cM and Cm have similar magnitudes of positive perturbation specific humidity, but the maximum in cM (5 km) appears in higher altitude than that in Cm (3 km). The vertical structure of hydrometeor mixing ratio in cM is similar to those

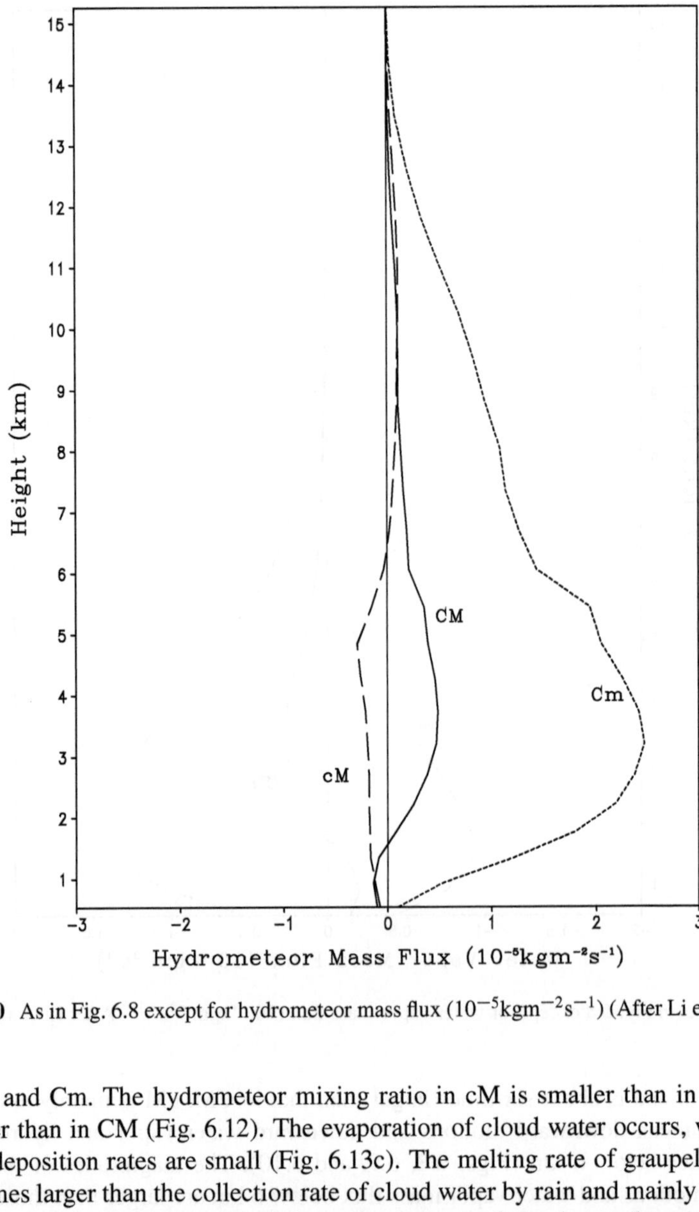

Fig. 6.10 As in Fig. 6.8 except for hydrometeor mass flux (10^{-5}kgm^{-2}s^{-1}) (After Li et al. 2011)

in CM and Cm. The hydrometeor mixing ratio in cM is smaller than in Cm, but is larger than in CM (Fig. 6.12). The evaporation of cloud water occurs, while the vapor deposition rates are small (Fig. 6.13c). The melting rate of graupel is about four times larger than the collection rate of cloud water by rain and mainly balances the evaporation of rain. As a result, the collection rate of cloud water by rain roughly accounts for the rainfall source. The analysis of cloud microphysical budgets reveals that the CM is more convective, while Cm and cM are more stratiform.

CM is associated with net condensation and hydrometeor loss/convergence. It can be broken down into two sub-rainfall types based on relative importance of net condensation and hydrometeor loss/convergence in CM. Net condensation is larger than hydrometeor loss/convergence in CM1, whereas it is smaller than hydrometeor

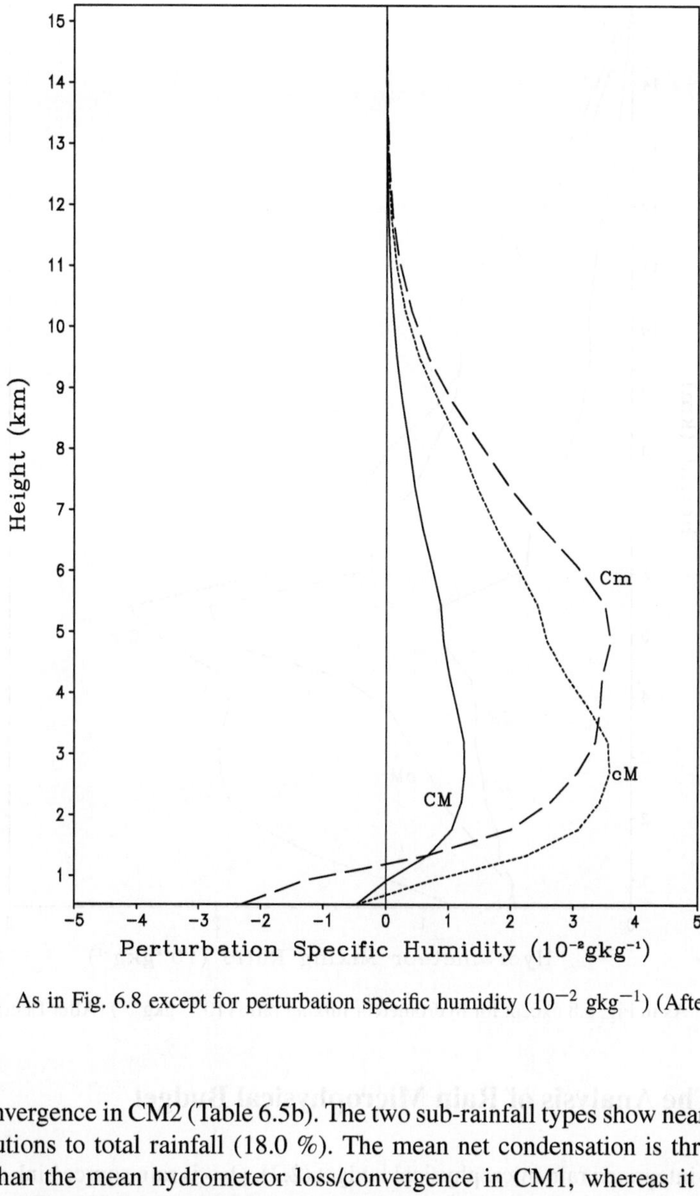

Fig. 6.11 As in Fig. 6.8 except for perturbation specific humidity (10^{-2} gkg^{-1}) (After Li et al. 2011)

loss/convergence in CM2 (Table 6.5b). The two sub-rainfall types show nearly equal contributions to total rainfall (18.0 %). The mean net condensation is three times larger than the mean hydrometeor loss/convergence in CM1, whereas it is about one thirds of the mean hydrometeor loss/convergence in CM2. CM2 covers slightly larger areas than CM1 does. The results suggest that hydrometeor loss/convergence and net condensation are equally important in CM.

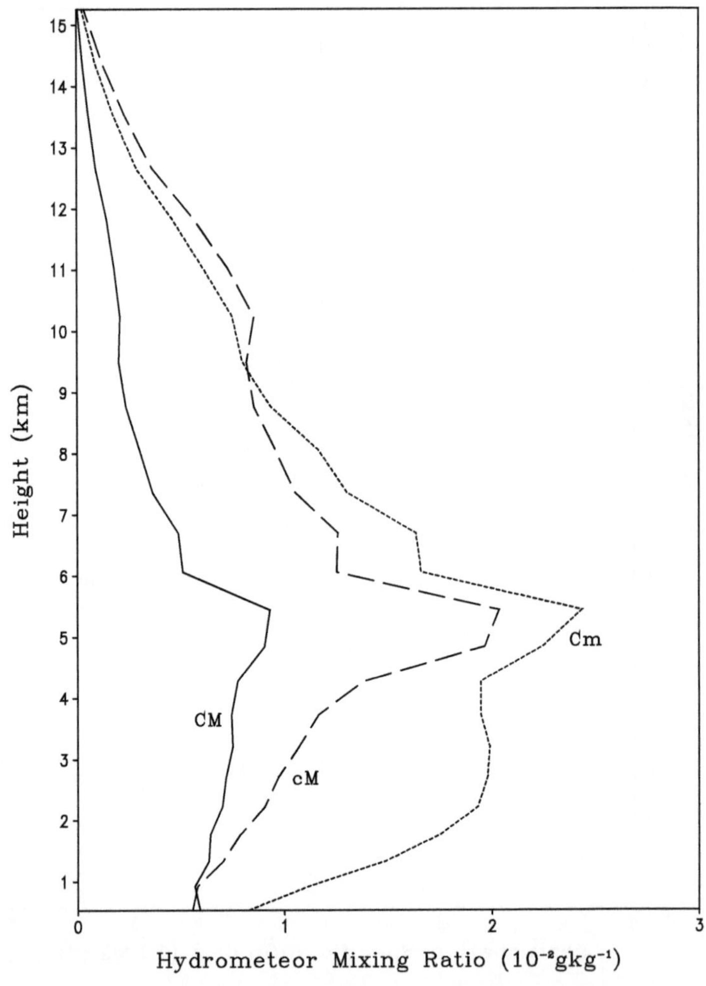

Fig. 6.12 As in Fig. 6.8 except for hydrometeor mixing ratio (10^{-2} gkg^{-1}) (After Li et al. 2011)

6.3 The Analysis of Rain Microphysical Budget

In mass-integrated rain microphysical budget (2.7), Q_{NC} is rain source/sink and Q_{RM} is local rain hydrometeor gain/loss and convergence/divergence. Here, R and r represent rain source ($Q_{NR} > 0$) and sink ($Q_{NR} < 0$), respectively. H and h represent rain hydrometeor loss/convergence ($Q_{RM} > 0$) and gain/divergence ($Q_{RM} < 0$), respectively. W and I denote water ($[P_{RAUT}] + [P_{RACW}] + [P_{GACW}] > [P_{SMLT}] + [P_{GMLT}]$) and ice ($[P_{RAUT}] + [P_{RACW}] + [P_{GACW}] < [P_{SMLT}] + [P_{GMLT}]$) microphysical processes dominated in production of rainfall, respectively.

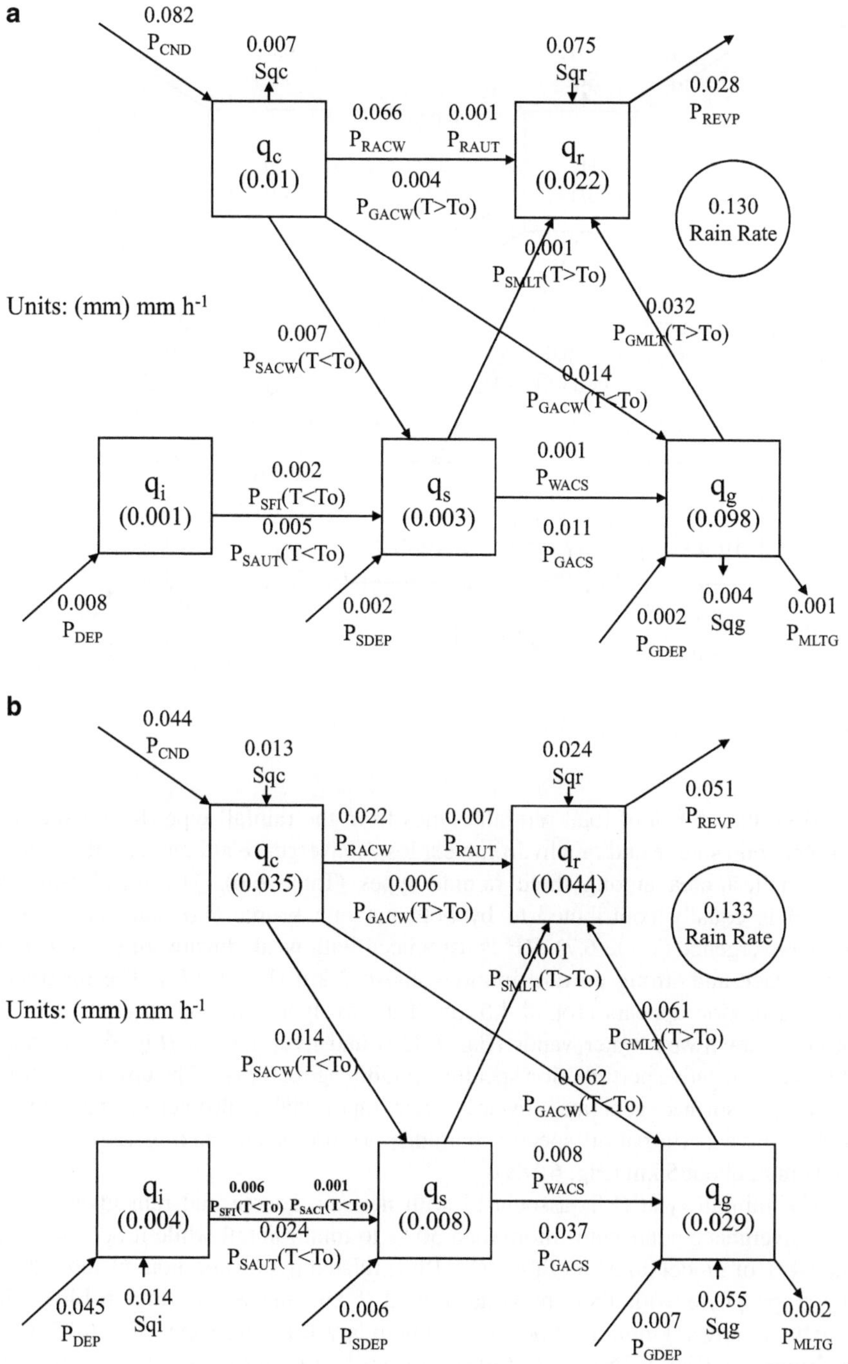

Fig. 6.13 Time-mean cloud microphysical budgets for (**a**) CM, (**b**) Cm, and (**c**) cM. Units for cloud hydrometeors and conversions are mm and mm h^{-1}, respectively. Cloud microphysical conversion terms and their schemes can be found in Table 1.2

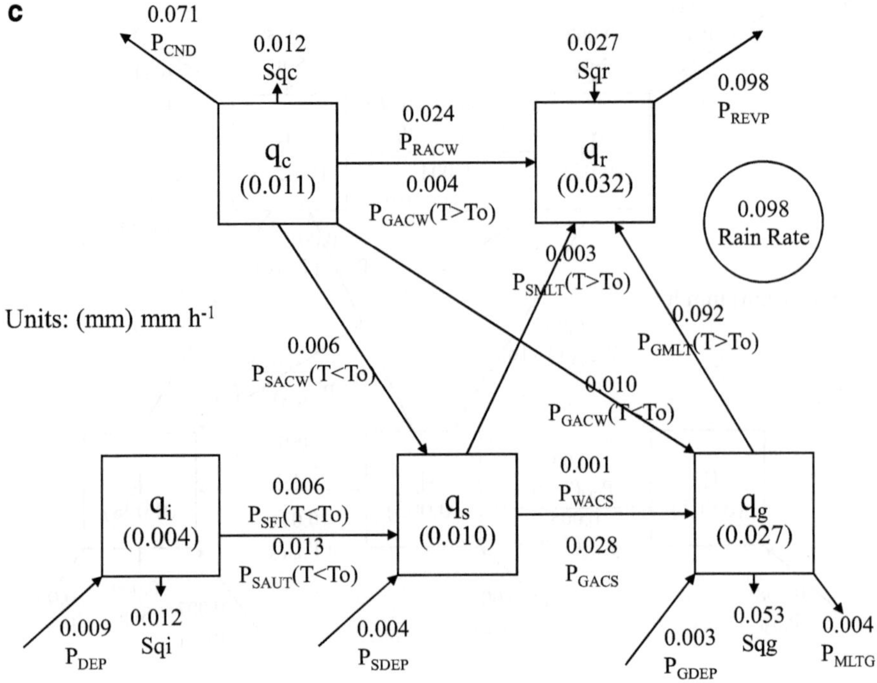

Fig. 6.13 (continued)

More than 50 % of total rainfall comes from the rainfall type (RH) associated with net rain source and rain hydrometeor loss/convergence although rainfall covers the smallest area among three rainfall types (Table 6.6). The mean rain rate of RH is equally contributed to by both net rain source and rain hydrometeor loss/convergence (Table 6.7). RH is associated with weak downward motions near the surface and strong upward motions above 2 km (Fig. 6.14a). The maximum upward motion appears around 4.5 km. The downward motions near the surface produce downward water vapor (Fig. 6.15a) and hydrometeor (Fig. 6.16a) mass fluxes and negative perturbation specific humidity (Fig. 6.17a). The upward motions above the surface generate upward water vapor and hydrometeor mass fluxes and positive perturbation specific humidity. Hydrometeor mixing ratio shows a maximum about 5 km (Fig. 6.18a).

The rainfall type (Rh) associated with net rain source and rain hydrometeor gain/divergence contributes more than 30 % to total rainfall while it occupies less than 4 % of model domain (Table 6.7). Rh is related to upward motions throughout the troposphere with their maxima around 2 km and 8 km (Fig. 6.14b). The maximum upward motions produce maximum upward water vapor (Fig. 6.15b) and hydrometeor (Fig. 6.16b) mass fluxes and positive perturbation specific humidity (Fig. 6.17b) around 2–3 km. The maximum hydrometeor mixing ratio occurs around 6 km (Fig. 6.18b).

Table 6.6 Summary of rainfall types separated by rain budget

Type	Description
RH	Net rain source and rain hydrometeor loss/convergence
RHW	Net rain source with dominant water microphysical processes and rain hydrometeor loss/convergence
RHI	Net rain source with dominant ice microphysical processes and rain hydrometeor loss/ convergence
Rh	Net rain source and rain hydrometeor gain/divergence
RhW	Net rain source with dominant water microphysical processes and rain hydrometeor gain/divergence
RhI	Net rain source with dominant ice microphysical processes and rain hydrometeor gain/divergence
rH	Net rain sink and rain hydrometeor loss/convergence
rHW	Net rain sink with dominant water microphysical processes and rain hydrometeor loss/convergence
rHI	Net rain sink with dominant ice microphysical processes and rain hydrometeor loss/convergence

R and r represent rain source ($Q_{NR}>0$) and sink ($Q_{NR}<0$), respectively. H and h represent rain hydrometeor loss/convergence ($Q_{RM}>0$) and gain/divergence ($Q_{RM}<0$), respectively. W and I denote water ($[P_{RAUT}]+[P_{RACW}]+[P_{GACW}]>[P_{SMLT}]+[P_{GMLT}]$) and ice ($[P_{RAUT}]+[P_{RACW}]+[P_{GACW}]<[P_{SMLT}]+[P_{GMLT}]$) microphysical processes dominated in production of rainfall, respectively (After Li and Shen 2013)

Table 6.7 Fractional coverage (FC), percentage of rain amount over total rainfall amount (PRA), and model domain-mean rain budget (P_S, Q_{NR}, Q_{RM}) for RH, Rh, and rH in COARE

	RH	Rh	rH
FC	2.120	3.625	5.425
PRA	54.079	30.769	15.152
P_S	0.195	0.111	0.055
Q_{NR}	0.095	0.284	−0.032
Q_{RM}	0.100	−0.173	0.087

Units are % for FC and PRA and mm h^{-1} for rain budget (After Li and Shen 2013)

About 15 % of total rainfall comes from the rainfall type (rH) associated with net rain sink and rain hydrometeor loss/convergence, which covers more than 5 % of model domain, the largest rainfall area among three rainfall types (Table 6.7). rH is associated with downward motions below 4.5 km and upward motions above 4.5 km, in which upward motions are weaker than downward motions (Fig. 6.14c). Strong downward motions generate strong downward water vapor mass flux below 4.5 km (Fig. 6.15c). Upward hydrometeor mass flux appears above 3 km, which are stronger than downward hydrometeor mass flux below 3 km (Fig. 6.16c). Surface perturbation specific humidity shows negative, while positive perturbation specific humidity reaches its maximum around 5 km (Fig. 6.17c). Maximum hydrometeor mixing ratio appears around 5.5 km (Fig. 6.18c).

In RH, the rainfall source comes from the collection of cloud water by rain and the melting of graupel, in which P_{RACW} is twice larger than P_{GMLT} (Fig. 6.19a).

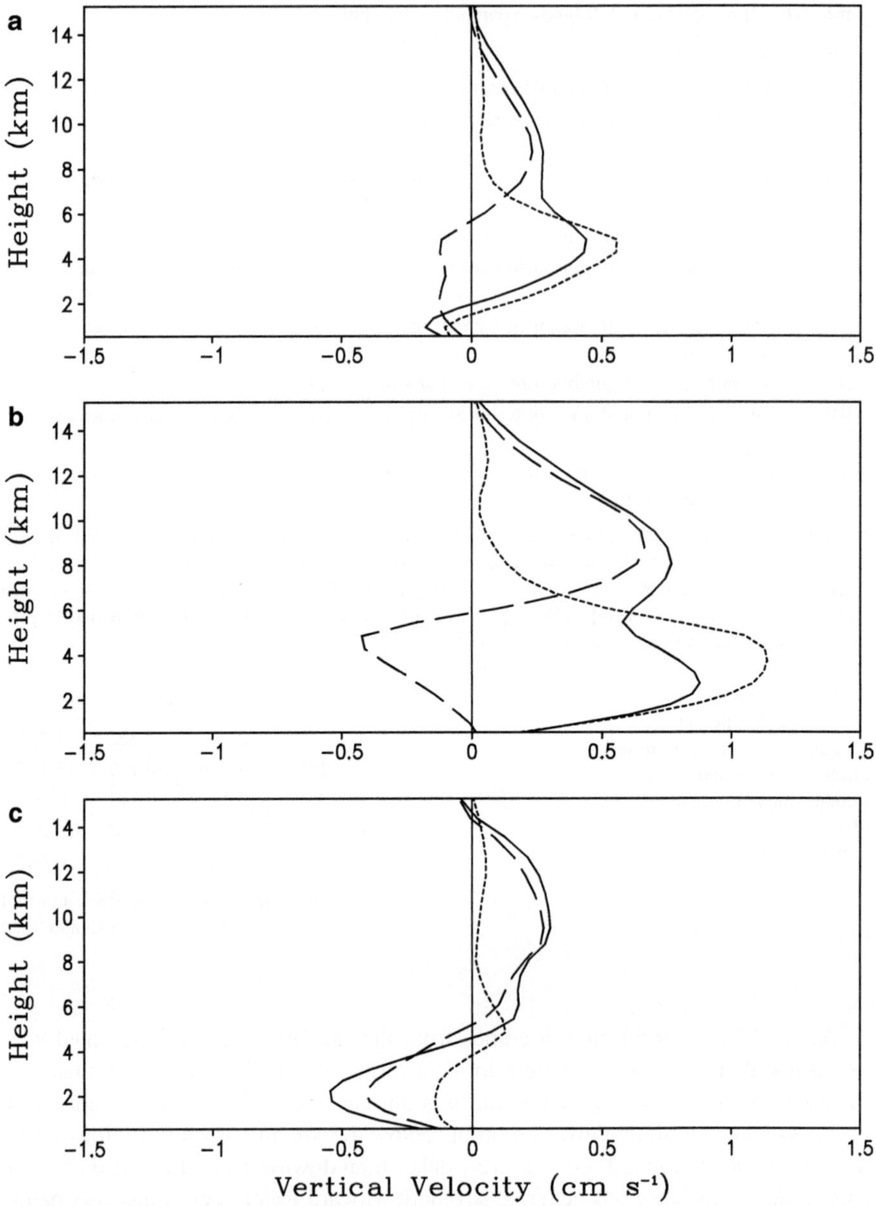

Fig. 6.14 Vertical profiles of time and model domain-mean vertical velocity (cm s^{-1}) in (**a**) RH (*solid*), RHW (*short dash*), and RHI (*long dash*); (**b**) Rh (*solid*), RhW (*short dash*), and RhI (*long dash*); and (**c**) rH (*solid*), rHW (*short dash*), and rHI (*long dash*) in COARE (After Li and Shen 2013)

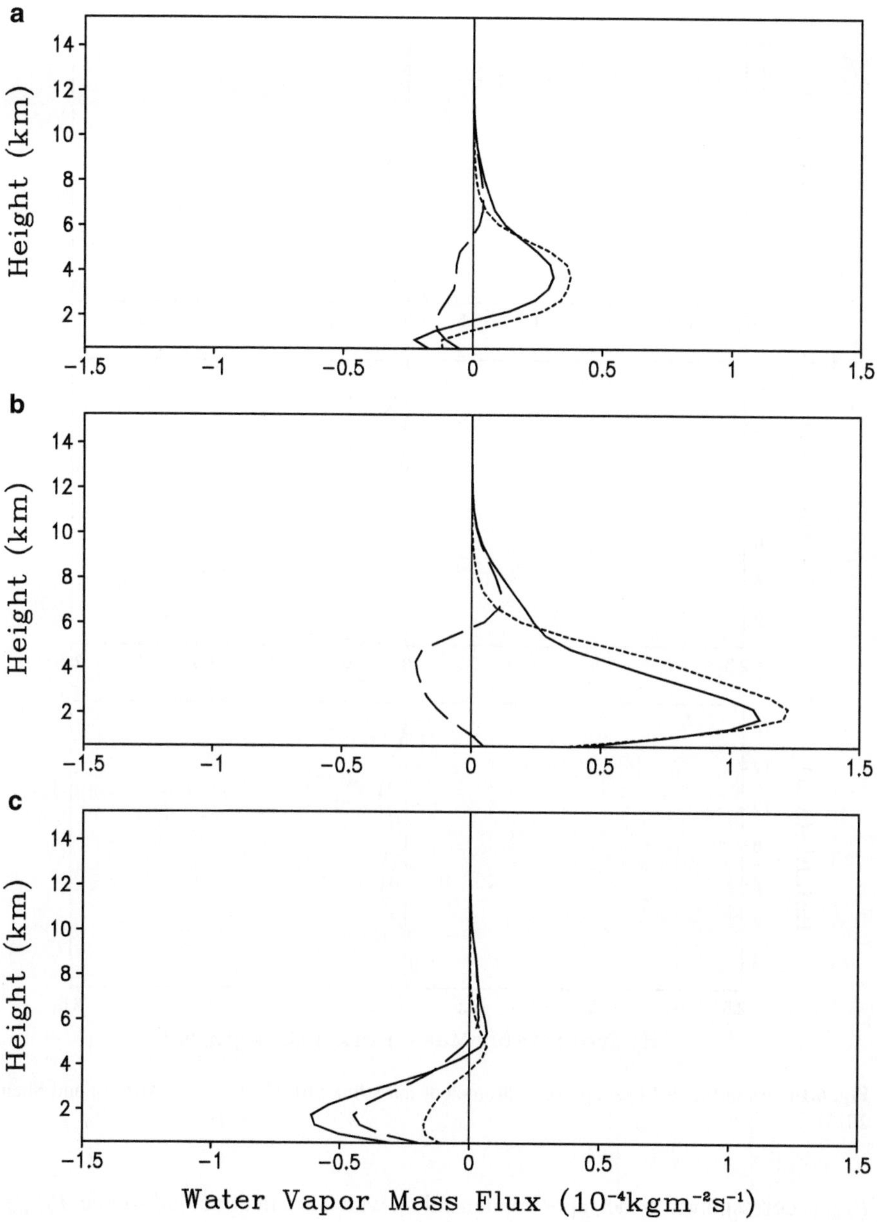

Fig. 6.15 As in Fig. 6.14 except for water vapor mass flux ($10^{-4}kgm^{-2}s^{-1}$) (After Li and Shen 2013)

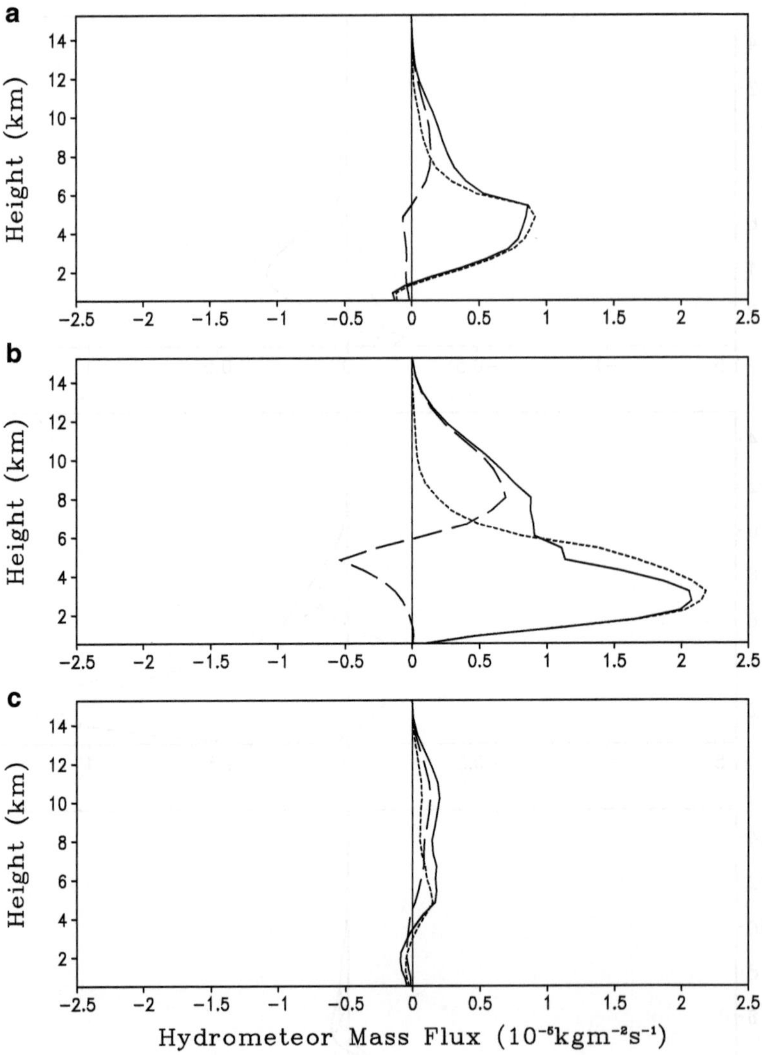

Fig. 6.16 As in Fig. 6.14 except for hydrometeor mass flux (10^{-5}kgm^{-2}s^{-1}) After Li and Shen 2013)

P_{RACW} corresponds to the vapor condensation, which is a major cloud source. P_{GMLT} is associated with the accretion of cloud water (P_{GACW}) and snow by graupel. The results reveal dominance of water microphysical processes in the production of RH. The cloud microphysical budget in Rh is similar to that in RH, but the budget magnitude of the former is about twice larger than that of the latter (Fig. 6.19b). The similarity of rainfall sources in RH and Rh suggests that both rain types are more convective. Although the rainfall source in Rh is much larger than that in RH, the rain rate of Rh is much smaller than that of RH due to the divergence of

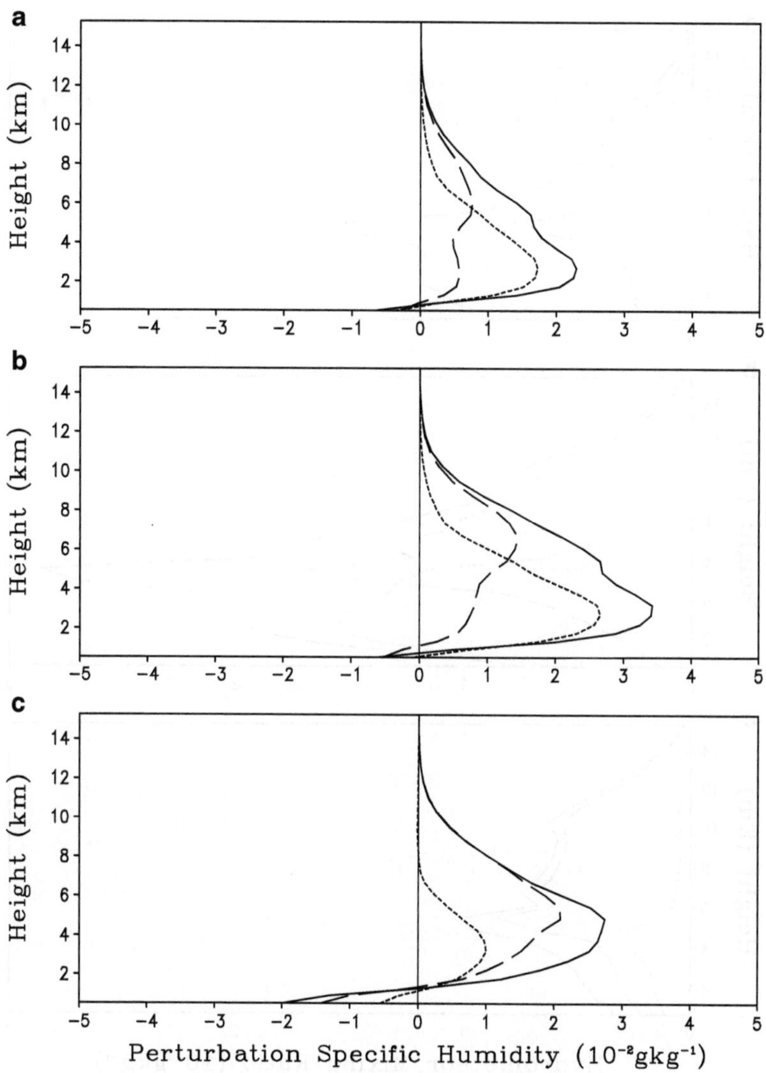

Fig. 6.17 As in Fig. 6.14 except for perturbation specific humidity (10^{-2} gkg^{-1}) (After Li and Shen 2013)

rain hydrometeor as indicated by Table 6.7. The calculation of rain intensity with the rain rate divided by fractional coverage reveals that the rain intensity of RH (9.2 mm h^{-1}) is about three times larger than that of Rh (3.1 mm h^{-1}) because Rh gets convergence of rain hydrometeor. Thus, high rainfall source is not necessary to generate high rain rate. In rH, the vapor condensation rate is much smaller than the vapor deposition rates (Fig. 6.19c). As a result, P_{GMLT} is four times larger than P_{RACW}. Ice microphysical processes are dominant, indicating more stratiform. The

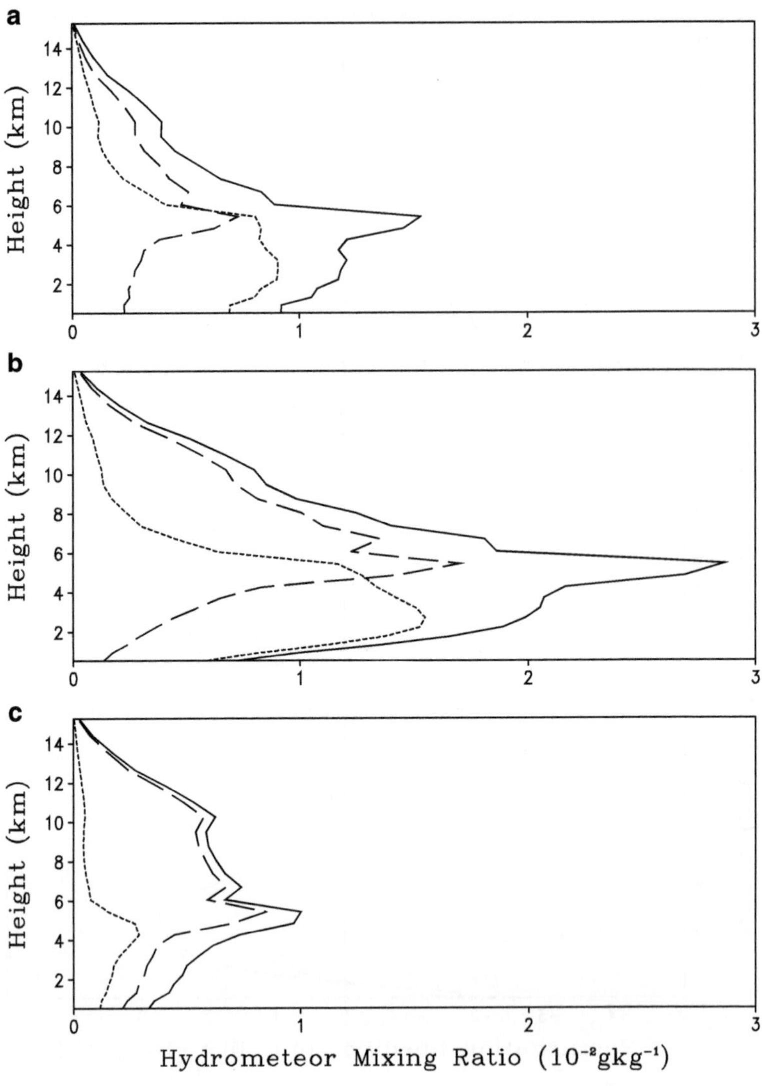

Fig. 6.18 As in Fig. 6.14 except for hydrometeor mixing ratio (10^{-2} gkg^{-1}) (After Li and Shen 2013)

magnitude of the evaporation of rain (P_{REVP}) is larger than those of P_{GMLT} and P_{RACW}, forming a net rainfall sink. In summary, RH and Rh are more convective when the collection rate of cloud water by rain is about twice larger than the melting of graupel, whereas rH is more stratiform as the melting of graupel is a dominant cloud microphysical process.

Among 54 % of rainfall contribution from RH, about 42 % of total rainfall comes from RH (RHW) associated with dominant water hydrometeor processes

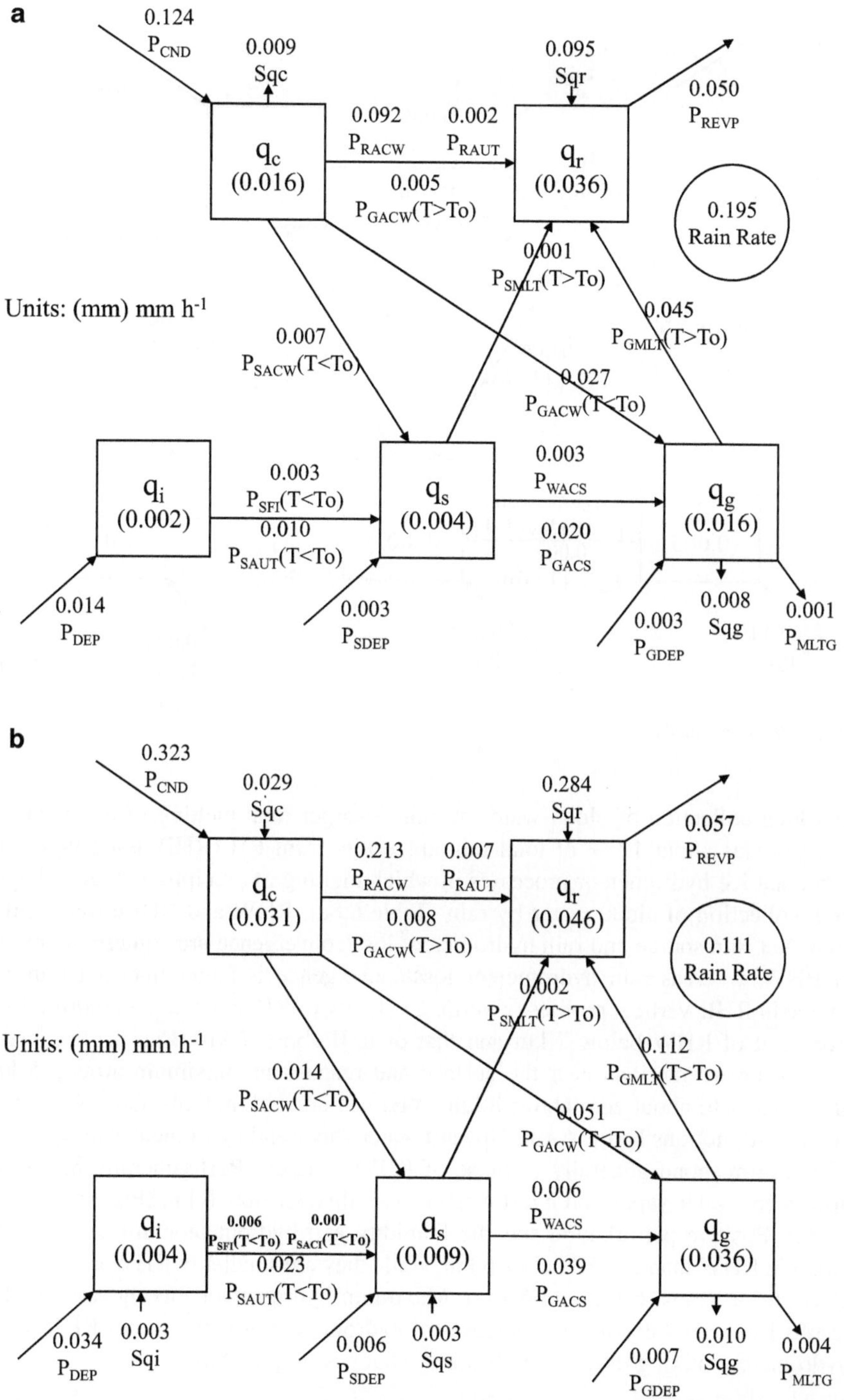

Fig. 6.19 Time-mean cloud microphysical budgets for (**a**) RH, (**b**) Rh, and (**c**) rH. Units for cloud hydrometeors and conversions are mm and mm h^{-1}, respectively. Cloud microphysical conversion terms and their schemes can be found in Table 1.2

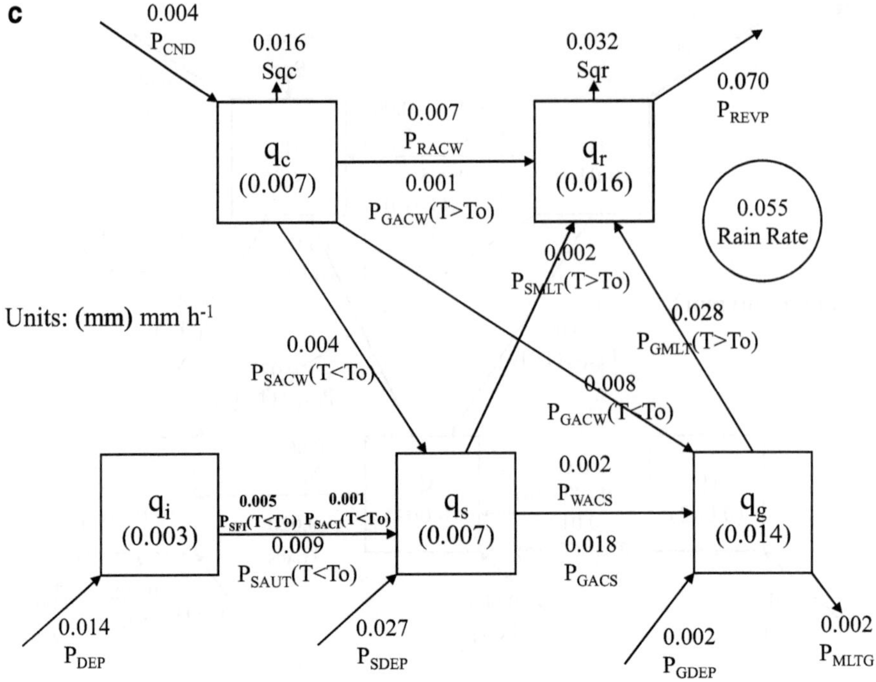

Fig. 6.19 (continued)

in which collection of cloud water by rain is larger than melting of precipitation ice, whereas about 12 % of total rainfall results from RH (RHI) associated with dominant ice hydrometeor processes in which melting of precipitation ice is larger than collection of cloud water by rain (Table 6.8a). RHW and RHI cover similar area. Net rain source and rain hydrometeor loss/convergence are equally important in RHW, whereas rain hydrometeor loss/convergence is larger than net rain net source in RHI. Vertical profiles of vertical velocity of RH show major contributions from that of RHW below 7 km and that of RHI above 7 km (Fig. 6.14a). RHW has downward motions near the surface and reaches its maximum around 5 km and reduces to about zero above 8 km. WHI has upward motions above 6 km and downward motions below 6 km. Upward water vapor and hydrometeor mass fluxes of RH correspond generally to those of RHW, whereas RHI generally has weak downward water vapor and hydrometeor mass fluxes below 6 km (Figs. 6.15a and 6.16a). Positive perturbation specific humidity and hydrometeor mixing ratio are larger in RHW than in RHI below 6 km, while they are smaller in RHW than in RHI above 6 km (Figs. 6.17a and 6.18a). Maximum hydrometeor mixing ratio of RH results from rapid downward decreases of hydrometeor mixing ratio of RHI, while hydrometeor mixing ratio of RHW barely changes (Fig. 6.18a).

Table 6.8 Breakdown of (a) RH into RHW and RHI, (b) RH into RhW and RhI, and (c) rH into rHW and rHI for fractional coverage (FC), percentage of rain amount over total rainfall amount (PRA), and model domain-mean rain budget (P_S, Q_{NR}, Q_{RM}) in COARE

(a)	RHW	RHI
FC	1.048	1.072
PRA	42.072	12.007
P_S	0.152	0.043
Q_{NR}	0.078	0.017
Q_{RM}	0.074	0.026

(b)	RhW	RhI
FC	1.511	2.114
PRA	25.672	5.096
P_S	0.093	0.018
Q_{NR}	0.212	0.071
Q_{RM}	−0.120	−0.053

(c)	rHW	rHI
FC	0.851	4.574
PRA	5.971	9.181
P_S	0.022	0.033
Q_{NR}	−0.007	−0.025
Q_{RM}	0.029	0.058

Units are % for FC and PRA and mm h^{-1} for rain budget (After Li and Shen 2013)

Among about 30 % contribution of Rh to total rainfall, about 26 % comes from RhW, whereas about 5 % comes from RhI (Table 6.8b), although RhI occupies more model domain than RhW does. Upward motions of Rh are associated with upward motions of RhW below 6.5 km and of RhI above 6.5 km (Fig. 6.14b). RhW has upward motions throughout the troposphere with maximum around 3.5 km, which generate upward water vapor and hydrometeor mass fluxes and positive perturbation specific humidity throughout the troposphere (Figs. 6.15b, 6.16, and 6.17b). RhI has downward motions below 6 km and upward motion above 6 km, which produce downward water vapor and hydrometeor mass fluxes below 6 km and upward water vapor and hydrometeor mass fluxes above 6 km (Figs. 6.15b and 6.16b). RhI generally has positive perturbation specific humidity with its maximum around 7 km, whereas RhW has maximum positive perturbation specific humidity around 3 km, although the former is weaker than the latter (Fig. 6.17b). Maximum hydrometeor mixing ratios of RhW and RhI occur, respectively, around 6 km and 3 km (Fig. 6.18b). Thus, maximum hydrometeor mixing ratio of Rh around 6 km results from that of RhI.

rHI contributes more to total rainfall than rHW because rHI occupies much more model domain than rHW does (Table 6.8c). Downward motions below 4.5 km and upward motions above 4.5 km in rH result primarily from those associated with rHI, while the vertical velocities of rHI are much stronger than those of rHW (Fig. 6.14c). Downward water vapor mass fluxes of rH are primarily contributed to by those of rHI (Fig. 6.15c). Both rHW and rHI have small magnitudes of hydrometeor mass fluxes compared to other rainfall types (Fig. 6.16c). Perturbation specific humidity of rHI mainly accounts for that of rH (Fig. 6.17c). Hydrometeor mixing ratio of rHI is much larger than that of rHW, in particular above 4 km (Fig. 6.18c).

The above analysis shows that vertical profiles of vertical velocity in RHI, RhI, and rH have downward motions in the lower troposphere (Fig. 6.14). The ice microphysical processes are dominant over water microphysical processes in RHI and RhI. The melting rate of graupel is much larger than the collection rate of cloud water by rain in rH (Fig. 6.19c). Thus, RHI, RhI, and rH are components of stratiform rainfall. RHW and RhW have upward motions throughout the troposphere with maxima in the lower troposphere (Fig. 6.14a, b) and dominant water microphysical processes, which are components of convective rainfall.

6.4 Relationship Between Separation Analyses

Table 6.9 shows rain rates separated by surface rainfall and cloud microphysical budgets. Because local atmospheric dry, water vapor convergence and surface evaporation contribute to net condensation in TFM and TFm, thus, the rain rates are 0.036 mm h^{-1} for TFM and CM and 0.064 mm h^{-1} for TFm and Cm. In tFM, net condensation occurs when the local atmospheric moistening rate is lower than water vapor convergence plus surface evaporation, whereas net evaporation appears when the local atmospheric moistening rate is higher than water vapor convergence plus surface evaporation. Thus, the rain rate of tFM is the sum of rain rates of CM and cM. Since local hydrometeor gain/divergence (tFm and Tfm) occurs, water vapor convergence plus surface evaporation are higher than local atmospheric moistening, forming net condensation and producing rainfall in tFm, and local atmospheric drying plus surface evaporation are higher than water vapor divergence in Tfm, developing net condensation and rainfall in Tfm. Thus, the rain rates are 0.051 mm h^{-1} for tFm and Cm and 0.018 mm h^{-1} for Tfm and Cm. When local hydrometeor loss/convergence appears in TfM, local atmospheric drying plus surface evaporation overcome water vapor divergence to develop net condensation in CM, while they cannot overcome water vapor divergence to generate net evaporation in cM. As a result, the rain rate of TfM is the sum of the

Table 6.9 Rain rate partitioned by surface rainfall budget versus rain rate separated by cloud budget

	CM	Cm	cM
TFM	0.036	0.0	0.0
TFm	0.0	0.064	0.0
tFM	0.046	0.0	0.023
tFm	0.0	0.051	0.0
TfM	0.048	0.0	0.063
Tfm	0.0	0.018	0.0
tfM	0.0	0.0	0.011

Unit is mm h^{-1}

Table 6.10 Rain rate partitioned by surface rainfall budget versus rain rate separated by rain budget

	RH	Rh	rH
TFM	0.032	0.003	0.001
TFm	0.021	0.041	0.002
tFM	0.047	0.010	0.013
tFm	0.014	0.035	0.002
TfM	0.068	0.012	0.031
Tfm	0.007	0.009	0.001
tfM	0.005	0.001	0.005

Unit is mm h^{-1}

rain rates of CM and cM. In tfM, magnitudes of local atmospheric moistening and water vapor divergence are largely higher than that of surface evaporation. Thus, tfM and cM have similar rain rates.

Table 6.10 shows the rain rate partitioned by surface rainfall and rain microphysical budgets. The rain rate of TFM is largely associated with the rain rate of RH. This implies that local atmospheric drying, water vapor convergence, and surface evaporation account for the net rain source and local hydrometeor loss/convergence occurs mainly over rain area. The rain rates of TFm and tFm are primarily related to the rain rate of Rh partly because local hydrometeor gain/divergence occurs over rain area and local atmospheric drying, water vapor convergence, and surface evaporation contribute to the net rain source in TFm and water vapor convergence plus surface evaporation overcome local atmospheric moistening to form the net rain source in tFm. The rain rates of TFm and tFm are also associated with the rain rate of RH. This suggests that rain hydrometeor loss/convergence occurs while ice hydrometer moistening/divergence appears. The rain rates of tFM and TfM are primarily associated with the rain rate of RH because local hydrometeor has loss/convergence over rain area and water vapor convergence the net rain sources are associated with water vapor convergence in tFM and local atmospheric drying in TfM. The rain rate of TfM is also related to that of rH when the net rain sink responds to water vapor divergence. The rain rate of Tfm is similar to those of RH and Rh, indicating that local atmospheric drying accounts for the net rain source. The rain rate of tfM is similar to those of RH and rH, suggesting that water hydrometeor loss/convergence occurs.

Table 6.11 reveals the rain rates separated by cloud and rain microphysical budgets. The rain rate of CM is mainly associated with that of RH. The net condensation is the net rain source and hydrometeor loss/convergence appears over rain area. The rain rate of Cm is the sum of the rain rates of RH and Rh. The net condensation is the net rain source. Rh has more contribution to the rain rate of Cm than RH does. The hydrometeor gain/divergence occurs over rain area. The rain rate of cM is the sum of rain rates of RH and rH. The rain hydrometeor loss/convergence leads to the production of rainfall.

The partitioning analysis of convective–stratiform rainfall using surface rainfall-derived scheme shows that convective rainfall includes 31.7 % of rainfall associated with water vapor divergence (Table 6.12a). The convective rainfall is usually defined as the rainfall associated with water vapor convergence caused by strong

Table 6.11 Rain rate partitioned by cloud microphysical budget versus rain rate separated by rain budget

	RH	Rh	rH
CM	0.104	0.017	0.009
Cm	0.043	0.085	0.005
cM	0.048	0.009	0.041

Unit is mm h^{-1}

Table 6.12 Percentage (%) of rainfall types partitioned by (a) surface rainfall, (b) cloud, and (c) rain budgets over rainfall amount over convective and raining stratiform regions

(a)	TFM	TFm	tFM	tFm	TfM	Tfm	tfM
Convective	13.82	22.71	15.11	16.65	25.28	5.01	1.41
Stratiform	2.80	8.38	26.93	9.62	41.22	5.01	6.04

(b)	CM	Cm	cM
Convective	40.9	44.4	14.7
Stratiform	26.9	22.9	50.1

(c)	RH	Rh	rH
Convective	58.7	43.9	6.3
Stratiform	45.5	22.9	31.5

After Shen et al. (2010) and Li and Shen (2013)

updrafts. Thus, the convective–stratiform rainfall separation scheme may not have capability to exclude the rainfall associated with water vapor divergence when the only magnitude of convective signal is used in the scheme. Only 52.27 % of stratiform rainfall is associated with water vapor divergence, but 77 % of stratiform rainfall corresponds to dynamic hydrometeor advection. Thus, dynamic hydrometeor advection associated with updrafts originated from convective core may be the major characteristics of stratiform rainfall.

When convective–stratiform rainfall is evaluated with microphysics-derived schemes, more than 85 % of convective rainfall is associated with net condensation in cloud microphysical budget and net rain source in rain microphysical budget. In contrast, only 50 % and 68.6 % of stratiform rainfall correspond to net condensation in cloud microphysical budget and net rain source in rain microphysical budget, respectively. 77.1 % of stratiform rainfall is related to dynamic hydrometeor advection for both microphysics-derived schemes. Thus, stratiform rainfall may be defined as the rainfall associated with dynamic hydrometeor advection.

6.5 Scale-Dependent Analysis: Time-Mean Data Versus Grid-Scale Data

The budget analysis shows that rainfall source could be underestimated due to the offset between different processes when the averages are taken temporally and spatially. For example, temporal evolution and spatial distribution of rainfall processes may lead to an offset between net condensation and net evaporation and between hydrometeor advections in averaged calculation of cloud budget, implying temporal and spatial scale dependence of rainfall partitioning analysis. Large temporal and spatial fluctuations of precipitation may cause significant differences between temporally and spatially averaged calculations in cloud microphysical budgets associated with production of precipitation. In this section, time-mean and grid-scale data are, respectively, used to evaluate sensitivity of rainfall analysis to data average.

Table 6.13a shows the surface rainfall-derived partitioning analysis using time-mean and grid-scale rainfall data. The four rainfall types with water vapor convergence account for about 61 % of the total rainfall in the calculation of grid-scale data, which is significantly smaller than that (86.6 %) calculated from the model domain-mean simulation data. For each mean rainfall type, the largest rainfall contributions come from TfM with a range of 25.4–44.1 % as revealed by the calculations of grid-scale rainfall data. Thus, for mean rainfall types associated

Table 6.13 Percentage of rain amount over total rainfall amount (PRA) calculated using model domain-mean simulation data for mean rainfall types separated based on (a) surface rainfall and (b) cloud and (c) rain budgets and PRA of rainfall types calculated using grid-scale simulation data for each mean rainfall type

		Rainfall type separated by time-mean data						
(a)		TFM	TFm	tFM	tFm	TfM	Tfm	tfM
Mean data		26.572	34.789	16.627	8.631	6.207	4.301	2.387
Rainfall type separated by grid-scale data	TFM	12.264	9.567	11.117	7.767	7.282	6.721	4.376
	TFm	15.675	21.716	14.375	17.990	15.895	19.338	7.330
	tFM	19.016	19.390	22.063	16.600	19.181	15.170	18.153
	tFm	12.174	16.168	11.294	17.166	12.667	20.108	10.091
	TfM	33.983	25.403	33.589	31.165	37.194	27.628	44.101
	Tfm	3.321	5.292	4.551	6.159	4.642	8.164	10.502
	tfM	3.567	2.465	3.012	3.153	3.138	2.871	5.446

		Rainfall type separated by time-mean data		
(b)		CM	Cm	cM
Mean data		50.553	48.207	1.240
Rainfall type separated by grid-scale data	CM	38.906	33.243	24.943
	Cm	31.117	43.233	24.513
	cM	29.978	23.524	50.544

		Rainfall type separated by time-mean data		
(c)		RH	Rh	rH
Mean data		53.736	46.087	0.178
Rainfall type separated by grid-scale data	RH	59.748	47.537	36.419
	Rh	25.134	37.423	9.056
	rH	15.118	15.040	54.526

Unit is %. After Shen et al. (2010)

with water vapor convergence, a large amount of rainfall is related to the rainfall from the regions with water vapor divergence. The cloud microphysics-derived separation analysis reveals that although mean rainfall types come primarily from same grid-scale rainfall types, other grid-scale rainfall types contribute significantly to mean rainfalls (Table 6.13b). For example, for mean rainfall type CM, 38.9 % of mean rainfall results from same grid-scale rainfall type. The other grid-scale rainfall types Cm and cM contribute 31.1 % and 30.0 % to the mean rainfall of CM, respectively. For mean rainfall type Rh, separated by rain microphysical budget, the largest contribution to the mean rainfall comes from the grid-scale rainfall type RH (47.5 %), while the grid-scale rainfall type of Rh only contributes 37.4 % to the mean rainfall (Table 6.13c). The analyses indicate that the contribution of each rainfall type to the total rainfall in the analysis of grid-scale data could be significantly different from that determined by the analysis of model domain-mean data. Therefore, the analysis of precipitation statistics is temporal and spatial scale dependent.

References

Li X, Shen X (2013) Rain microphysical budget over tropical deep convective regime. J Meteorol Soc Jpn 91:801–815, (c) Meteorological Society of Japan. Reprinted with permission

Li X, Shen X, Liu J (2011) A partitioning analysis of tropical rainfall based on cloud budget. Atmos Res 102:444–451. (c) Elsevier. Reprinted with permission

Li X, Zhai G, Gao S, Shen X (2014) A new convective-stratiform rainfall separation scheme. Atmos Sci Lett 15:245–251, (c) the Royal Meteorological Society. Reprinted with permission

Shen X, Wang Y, Zhang N, Li X (2010) Precipitation and cloud statistics in the deep tropical convective regime. J Geophys Res doi: 10.1029/2010JD014481, (c) American Geophysical Union. Reprinted with permission

Tao WK, Simpson J, Sui CH, Ferrier B, Lang S, Scala J, Chou MD, Pickering K (1993) Heating, moisture, and water budgets of tropical and midlatitude squall lines: comparisons and sensitivity to longwave radiation. J Atmos Sci 50:673–690

Chapter 7
Tropical Cloud Clusters

Abstract In this chapter, kinetics and spatial structure of cloud clusters during TOGA COARE are discussed based on the analysis of cloud-resolving model simulation data. The merging processes of cloud clusters are examined in terms of circulations and the composite analysis of total hydrometeor mixing ratio, zonal wind and vertical velocity, temperature and specific humidity perturbations, and surface rainfall budget. A rainfall event during the South China Sea Monsoon Experiment is also analyzed to study the evolution of cloud clusters associated with the development of precipitation system.

Keywords Cloud cluster • Cloud merger • Surface rainfall budget • Composite analysis

Observational studies using satellite measurements have revealed that cloud cluster groups (super cloud clusters) and individual cloud clusters embedded within them propagate in different directions and have different evolution and spatial distributions (e.g., Nakazawa 1988; Lau et al. 1991; Sui and Lau 1992). This hierarchical cloud structures and behaviors can be simulated with 2D cloud-resolving model (e.g., Peng et al. 2001; Ping et al. 2008). The kinetics and spatial structures of cloud clusters are discussed in this chapter based on Ping et al. (2008). The cloud clusters with different propagations also can merge, and merging processes can affect intensity and spatial structures of merged cloud clusters. The composite analysis of vertical structure of temperature and water vapor anomalies, cloud hydrometeors, and surface rainfall processes before and after merging is compared in this chapter. The surface rainfall processes associated with the development of cloud clusters by Wang et al. (2007) are also discussed in this chapter.

7.1 Introduction

Lau et al. (1991) analyzed IR radiance measurements at cloud top from the Japanese geostationary meteorological satellite (GMS) and showed the observed cloud clusters with a hierarchy of collective motions at time scales of 1 day,

© Springer International Publishing Switzerland 2016

X. Li, S. Gao, *Cloud-Resolving Modeling of Convective Processes*,
Springer Atmospheric Sciences, DOI 10.1007/978-3-319-26360-1_7

2–3 days, and 10–15 days. The 1–15-day time scale is closely related to the intraseasonal oscillation, and their super cloud clusters propagate eastward along the equator from the Indian Ocean to the western Pacific all around the global tropics. The cloud clusters embedded in the super cloud clusters have the 2–3-day time scale and propagate in the opposite direction of the super cloud clusters. The diurnal time scale is significant in the cloud clusters, in which the signals are more pronounced over the continent than over the open ocean. Sui and Lau (1992) analyzed the First GARP Global Experiment IIIb circulation data along with the Japanese GMS-1 IR data to study the atmospheric multiscale variabilities over the tropical western Pacific during the 1979 Northern Hemisphere. Two intraseasonal oscillations propagate eastward from the Indian Ocean to the western Pacific. Over the western Pacific warm pool, the intraseasonal oscillations develop and move barely with the enhanced rotational circulations. The intraseasonal oscillations interact with regional and synoptic-scale systems such as monsoon circulations. The intraseasonal oscillations also excite 2–4-day disturbances. Sui and Lau also found that the diurnal signal becomes strong when the intraseasonal oscillation loses the intensity, whereas the opposite is true.

The hierarchical cloud structures and behaviors have been given much attention in the meteorological research community for recent decades. Various numerical models have been employed to search the physical processes controlling their formation, development, and propagation (e.g., Lau et al. 1989; Numaguti and Hayashi 1991; Chao and Lin 1994; Yano et al. 1995). Chao and Lin (1994) simulated hierarchical cloud patterns with a 2D hydrostatic model and found that the simulations of hierarchical cloud clusters are sensitive to spatial-mean flows and cumulus parameterization schemes. The spatial-mean flows are critical in organizing patterns for cloud clusters, and the cumulus parameterization schemes are crucial to their formation. Peng et al. (2001) conducted a 2D cloud-resolving experiment over a large domain, in which there is a warm pool surrounded by cold pools that mimics the equatorial western Pacific, to examine the genesis and evolution of hierarchical cloud clusters. The model used an open lateral boundary. The simulated cloud clusters have the typical horizontal scale of a few hundred kilometers. The new cloud clusters are generated at the leading edge of a propagating cold-air pool. The largest cloud cluster has a horizontal scale of 3000 km that consists of four cloud clusters that have a life cycle of 12–36 h. The condensational heating associated with the constituent cloud clusters initiates an overall tropospheric-deep gravity wave. The cumulative cluster-induced wave effects lead to the development of new cloud clusters.

7.2 Kinetics and Spatial Structures of Cloud Clusters

Ping et al. (2008) analyzed kinematics and spatial structures of tropical cloud clusters using the 2D simulation data in COARE. There are four eastward-propagating rainbands as shown in Fig. 7.1 (cloud clusters A–D; CCA, CCB, CCC,

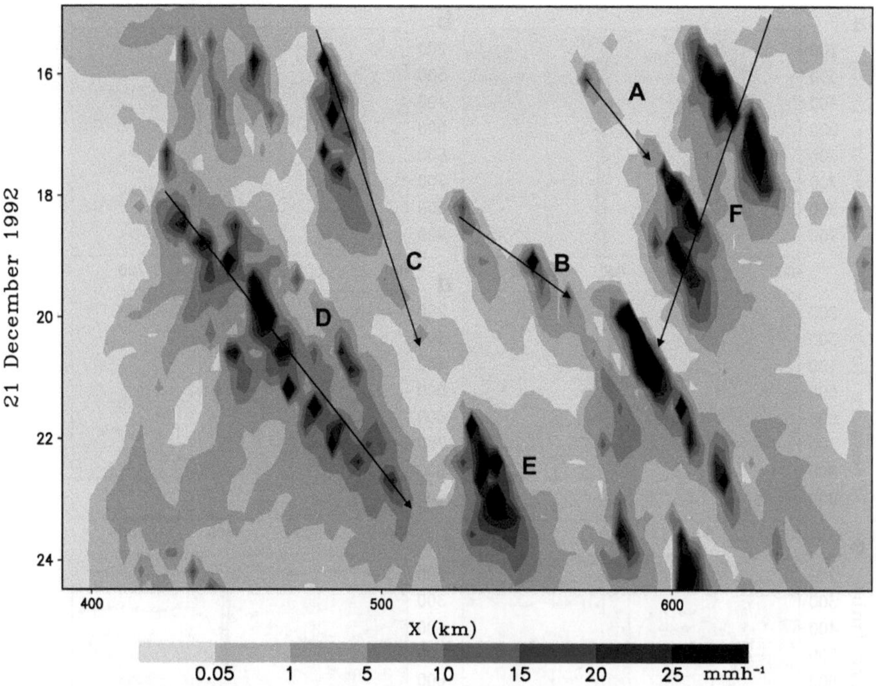

Fig. 7.1 Temporal evolution and zonal distribution of surface rain rate (mm h^{-1}) simulated in COARE on 21 December 1992. Arrows indicate propagation directions of cloud clusters A–F (After Ping et al. 2008)

CCD hereafter), a westward-propagating rainband (cloud cluster F; CCF hereafter), and a newly formed rainband (cloud cluster E; CCE hereafter). CCD shows that the individual cloud propagates westward while the cloud clusters move eastward. CCF exhibits that the individual cloud moves eastward while the cloud clusters propagate westward. This is consistent with the observations reported by Nakazawa (1988) and model simulations reported by Peng et al. (2001). CCA and CCB merge into CCF. CCC weakens and dissipates while it moves eastward, whereas CCD undergoes genesis, development, and weakening during its eastward-propagating course. CCE forms between the weakening CCD and CCF.

At 1600 LST, CCF and CCA extend to upper and mid-troposphere, respectively (Fig. 7.2a). Both share a negative vorticity circulation centered at 550 mb and 590 km. CCF is supported by an upward motion in the eastern part of the circulation, whereas CCA is suppressed by a weak downward motion in the western part. The two clusters move closer as the circulation associated with the clusters weakens and propagates upward at 1700 LST (Fig. 7.2b). At 1800 LST, CCA merges with CCF around 600 km, west of the previous CCF and east of the previous CCA (Fig. 7.2c). The merging CCF extends to 100 mb. Meanwhile, CCB forms and is 70–80 km away from CCF. A circulation linking CCF and CCB is developing above 500 mb. At 1900 LST, with the weakening of circulation, the clouds extend upward to form

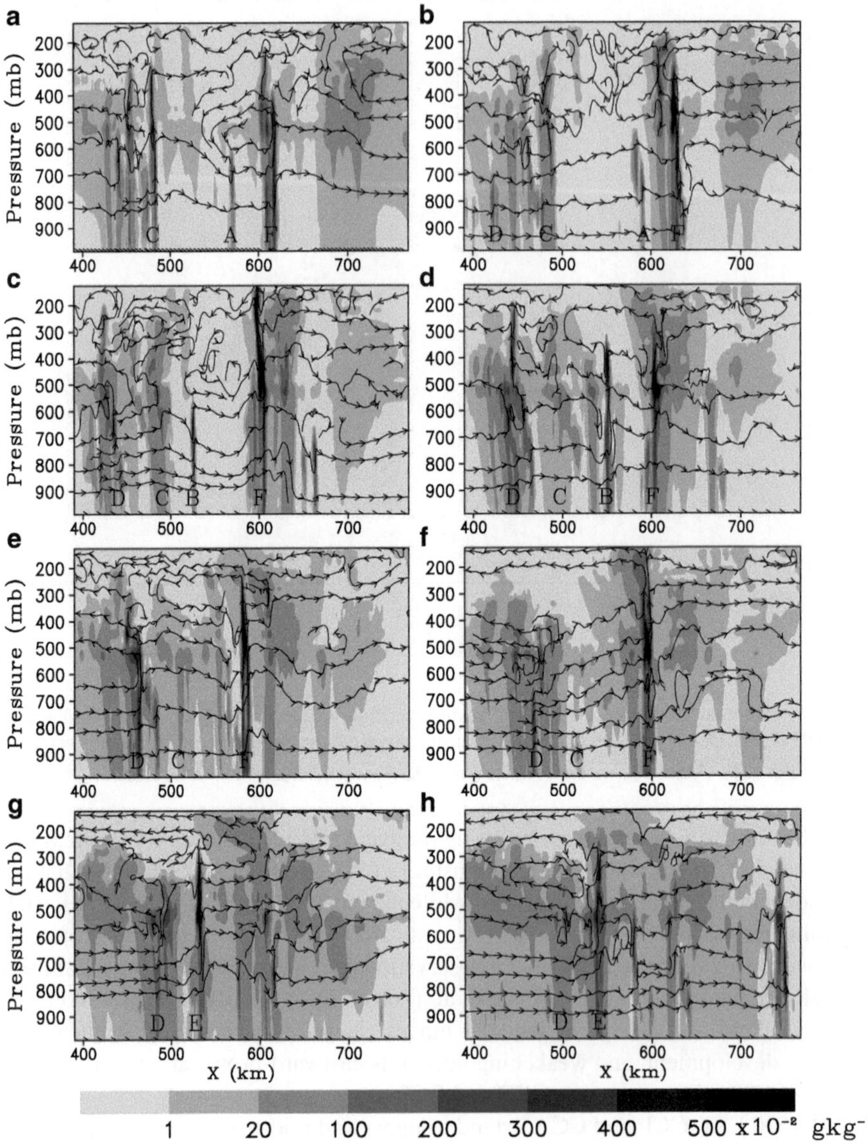

Fig. 7.2 Zonal–vertical cross sections of streamlines and total hydrometeor mixing ratio (*background shading*) for cloud clusters A–F simulated in COARE in 1-h interval from 1600 to 2300 LST 21 December 1992 (After Ping et al. 2008)

ice clouds while moving eastward (Fig. 7.2d). At 2000 LST, CCB merges with CCF around 580 km while the circulation associated with the cells disappears (Fig. 7.2e). The downward circulation develops around CCF at 2100 LST (Fig. 7.2f), which leads to its dissipation at 2200 LST (Fig. 7.2g).

7.3 Cloud Merger

The cloud merger has been observed and studied for decades (e.g., Malkus 1954; Simpson and Woodley 1971; Leavy and Houze 1979). Simpson (1980) showed the important roles of cumulus downdrafts and associated cold outflows in the cloud merging process. Tao and Simpson (1984) conducted 48 experiments with a 2D multicell model and found a total of 14 cloud merging cases. One of the 14 cloud merging cases (the case with a merger of two cells in different cycle stages) is very similar to cloud clusters A and B here except that the weak cell has a significant eastward propagation in this study, whereas it barely moves in their study. Tao and Simpson showed that the downdraft and cold outflow associated with the strong cell induce the formation of a new cell between the two old cells. The weak old cell eventually merges with the new cell while the strong old cell dissipates into an anvil cloud. In contrast, CCF and CCA/CCB move together, causing the merger.

To examine the impacts of merging processes in CCF, composites of CCF before merging (CCF1) and after merging (CCF2) are calculated using the data of 1700 and 1900 LST and of 1800 and 2000 LST, respectively (Figs. 7.3 and 7.4). For the composite calculations, the maximum surface rain rate is defined as the center point ($x = 0$), and the area of interest extends 16.5 km to the west and east of this point for each hour of the cluster's existence. The composite is temporally averaged over this period. Both clusters are limited to the zonal size of 15 km. CCF2 (4 g kg^{-1}) has a larger total hydrometeor mixing ratio (q_l) than CCF1 (2.5 g kg^{-1}) does. CCF2 displays smaller zonal scales of perturbations than does CCF1 in the total hydrometeor mixing ratio, updrafts and downdrafts, and temperature and specific humidity perturbations. The weak westerly winds advect the maximum total hydrometeor mixing ratio to the east while CCF moves westward, suggesting that westerly winds may be accounted for eastward propagation but may not be responsible for the propagation of cloud clusters. While maximum negative Q_{CMI} occurs at -3 km in CCF1 and -1.5 km in CCF2 (Figs. 7.3f and 7.4f), maximum negative Q_{CMW} appears at -6 km in CCF1 and 4.5 km in CCF2. The positive Q_{WV} mainly balances the maximum negative Q_{CMW} while the surface rain rate is small in CCF1 (Fig. 7.3f). This suggests that the maximum negative Q_{CMW} at -6 km in CCF1 is the growth of a new water cloud. The positive Q_{WV} mainly balances P_s and maximum negative Q_{CMW} in CCF2 (Fig. 7.4f). This indicates that the maximum negative Q_{CMW} at 4.5 km in CCF2 is the growth of the existing cloud after merging. Thus, the development of the new cloud at the west of the cloud cluster in CCF1 may explain the westward movement of cloud cluster group, whereas the advection of the maximum total hydrometeor mixing ratio by the westerly winds may account for the eastward propagation of individual cloud clusters inside CCF2. An organized downdraft associated with CCF1 in the lower troposphere causes negative specific humidity perturbations near surface. The positive specific humidity perturbations associated with CCF2 near the surface indicate that the environment has a more favorable water vapor condition for the development of clouds after merging, which leads to larger Q_{WV} in CCF2 than in CCF1 (see Figs. 7.3f and 7.4f).

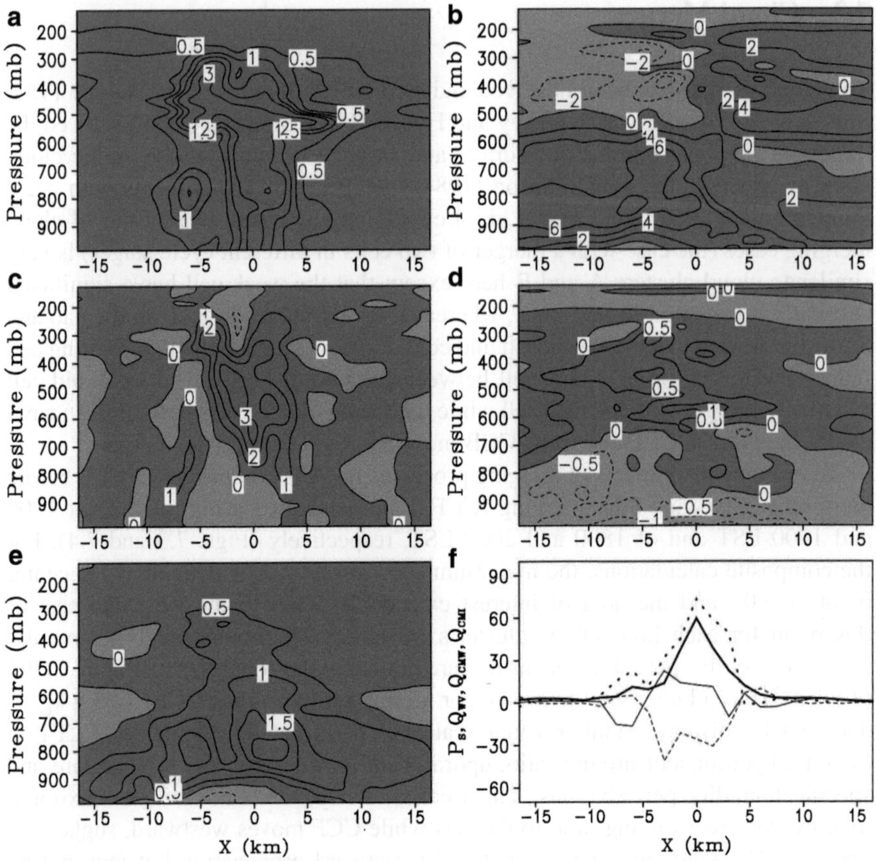

Fig. 7.3 Composites of zonal and vertical distributions of (**a**) total hydrometeor mixing ratio ($g\ kg^{-1}$), (**b**) zonal wind ($m\ s^{-1}$), (**c**) vertical velocity ($m\ s^{-1}$), (**d**) temperature perturbation (°C), (**e**) specific humidity perturbation ($g\ kg^{-1}$), and (**f**) P_s (*dark solid*), Q_{WV} (*dot*), Q_{CMW} (*light solid*), and Q_{CMI} (*dashed*) ($mm\ h^{-1}$) for cloud cluster F1 simulated in COARE (After Ping et al. 2008)

7.4 Surface Rainfall Processes Associated with Cloud Clusters

To examine surface rainfall processes associated with the development of cloud clusters, the results by Wang et al. (2007) are discussed here. Wang et al. (2007) conducted a 2D cloud-resolving model simulation with the forcing of meridionally uniform vertical velocity, meridional wind, along with thermal and moisture advection calculated over the area of 116–117°E, 16–23°N, during the South China Sea Monsoon Experiment (SCSMEX). The model simulations show many similarities to the radar observations. The rainband associated with the convection remains at a very stable position throughout its life cycle in the northern SCS region. The

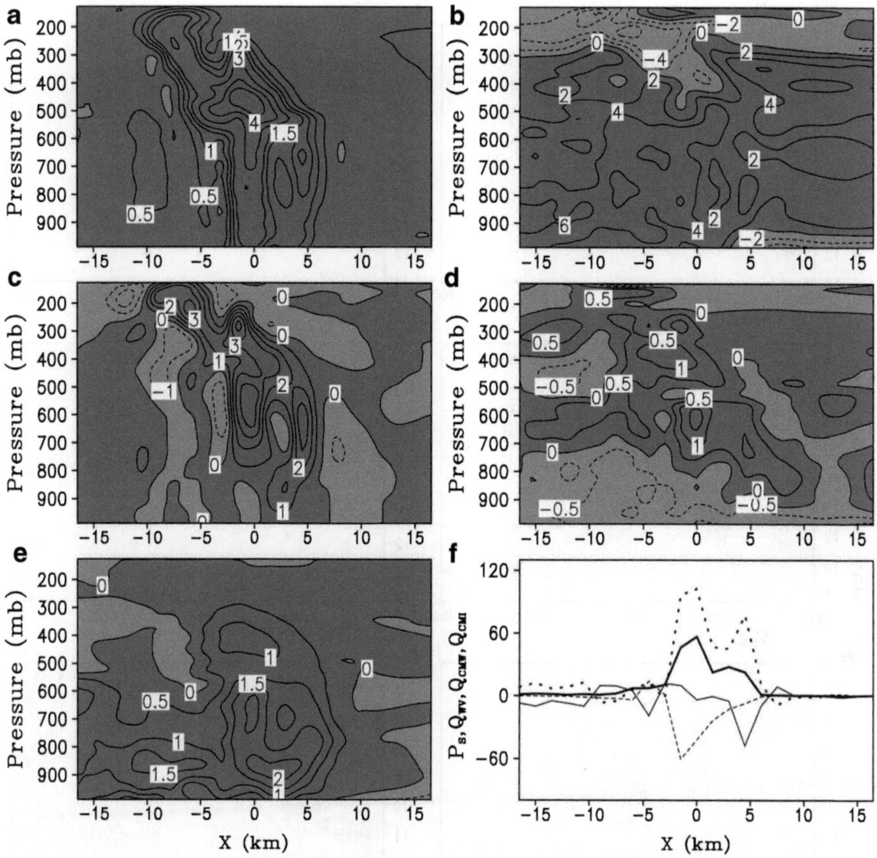

Fig. 7.4 As in Fig. 7.3 except for cloud cluster F2 (After Ping et al. 2008)

reflectivity pattern exhibits a straight upward structure with little tilt. The positions of the convective, transition, and stratiform regions produced by the model are consistent with observations. The major difference from the observations is that the model tends to overestimate the magnitude of updraft. As a result, the maximum reflectivity generated by the model appears at an elevated altitude.

Figure 7.5a shows that a major rainband initiates around 670 km after hour 5, barely moves, and intensifies quickly, reaching up to 21.9 mm h^{-1} at hour 7 (also see Table 7.1). The rainband weakens quickly after hour 8 and it maintains light rain until hour 13. Meanwhile, new rainbands form around 660, 653, 647, and 640 km at hour 7, 8, 9, and 10, respectively. Thus, rainbands propagate southward while the individual rainband barely moves. The surface rain rate is contributed by the local vapor change, vapor convergence, and the local cloud change/hydrometeor convergence, whereas the surface evaporation flux is much smaller than the other rainfall processes and negligible (Fig. 7.5b–e). The surface rainfall is always located

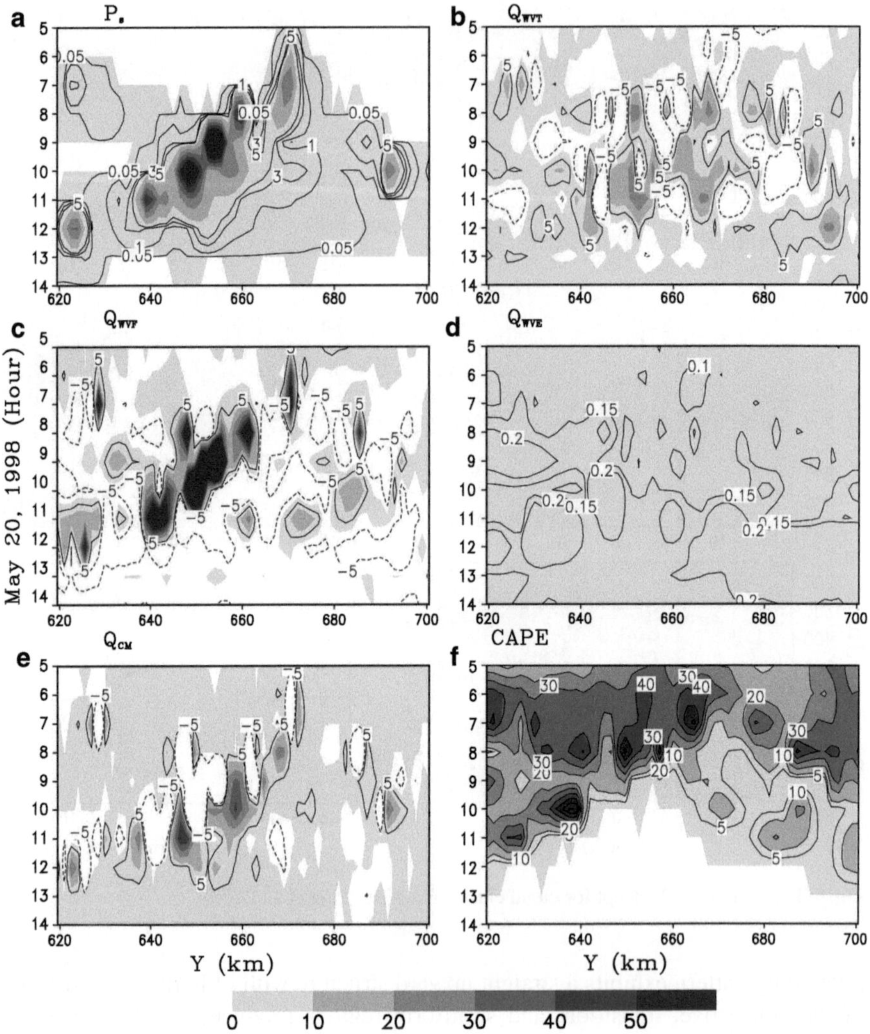

Fig. 7.5 Temporal and horizontal distribution of (**a**) P_s, (**b**) Q_{WVT}, (**c**) Q_{WVF}, (**d**) Q_{WVE}, (**e**) Q_{CM}, and (**f**) *CAPE* on 20 May 1998. Contour intervals are 0.05, 1, 3, and 5 mm h^{-1} for P_s; -5, 5 mm h^{-1} for Q_{WVT}, Q_{WVF}, and Q_{CM}; 0.05, 0.1, 0.15, and 0.2 mm h^{-1} for Q_{WVE}; and 5, 10, 20, 30, 40, 50, and 60×10 J kg^{-1} for *CAPE*. The units are mm h^{-1} for P_s, Q_{WVT}, Q_{WVF}, Q_{WVE}, and Q_{CM} and 10 J kg^{-1} for *CAPE* (After Wang et al. 2007)

to the north of large amounts of the *CAPE*, and the *CAPE* becomes small after the passage of rainbands (Fig. 7.5f), indicating the release of unstable energy for the development of convection. As an example, the rainband around 670 km from hour 5 to hour 9 will be analyzed.

The averaged surface rainfall budget during its preformation stage at 0500 LST (Table 7.1) shows that local atmospheric moistening ($Q_{WVT} = -2.6$ mm h^{-1})

Table 7.1 P_s, Q_{WVT}, Q_{WVF}, Q_{WVE}, and Q_{CM} (mm h^{-1}) along a life span of convection averaged in 669–672 km from 0500 LST to 0900 LST 19 May 1998

Stage	LST	P_s	Q_{WVT}	Q_{WVF}	Q_{WVE}	Q_{CM}
Preformation	0500	0	−2.6	2.4	0.2	0
Formation	0600	6.7	−2.3	16.0	0.2	−7.2
Mature	0700	21.9	−7.2	30.8	0.2	−1.9
Weakening	0800	12.1	11.7	−11.8	0.2	12.0
Dissipating	0900	0.2	2.8	−4.5	0.2	1.7

After Wang et al. (2007)

and vapor convergence ($Q_{WVF} = 2.4$ mm h^{-1}) nearly cancel each other out. This indicates that the vapor convergence moistens the atmosphere so that surface rainfall as well as clouds does not occur. During its formation stage at 0600 LST, the rain rate ($P_s = 6.7$ mm h^{-1}) is mainly determined by vapor convergence ($Q_{WVF} = 16.0$ mm h^{-1}) and hydrometeor convergence ($Q_{CM} = -7.2$ mm h^{-1}). Thus, the vapor convergence enhances both surface rainfall and cloud hydrometeors. During its mature phase at 0700 LST, the rain rate ($P_s = 21.9$ mm h^{-1}) is mainly determined by vapor convergence ($Q_{WVF} = 30.8$ mm h^{-1}) and local atmospheric moistening ($Q_{WVT} = -7.2$ mm h^{-1}). Thus, the vapor convergence is the only source that is responsible for the surface rainfall. During its weakening stage at 0800 LST, the rain rate ($P_s = 12.1$ mm h^{-1}) is determined by local vapor change ($Q_{WVT} = 11.7$ mm h^{-1}), vapor convergence ($Q_{WVF} = -11.8$ mm h^{-1}), and hydrometeor convergence ($Q_{CM} = 12.0$ mm h^{-1}). Thus, the local vapor and hydrometeor loss overcome the vapor divergence to support the surface rainfall. During its dissipating stage at 0900 LST, the rain rate ($P_s = 0.2$ mm h^{-1}) is small because of a large cancellation among local vapor change ($Q_{WVT} = 2.8$ mm h^{-1}), vapor convergence ($Q_{WVF} = -4.5$ mm h^{-1}), and hydrometeor convergence ($Q_{CM} = 1.7$ mm h^{-1}).

References

Chao WC, Lin SJ (1994) Tropical intraseasonal oscillation, super cloud clusters, and cumulus convection schemes. J Atmos Sci 51:1282–1297

Lau KM, Peng L, Sui CH, Nakazawa T (1989) Super cloud clusters, westerly wind bursts, 30–60 day oscillations, and ENSO: a unified view. J Meteorol Soc Jpn 67:205–219

Lau KM, Nakazawa T, Sui CH (1991) Observations of cloud cluster hierarchy over the tropical western Pacific. J Geophys Res 96:3197–3208

Leavy CA, Houze RA Jr (1979) The structure and evolution of convection in a tropical cloud cluster. J Atmos Sci 36:437–457

Malkus JS (1954) Some results of a trade cumulus cloud investigation. J Meteor 11:220–237

Nakazawa T (1988) Tropical super clusters within intraseasonal variations over the western Pacific. J Meteorol Soc Jpn 66:823–839

Numaguti A, Hayashi YY (1991) Behavior of cumulus activity and the structures of circulations in an "aqua planet" model. Part I. The structure of the super cloud clusters. J Meteorol Soc Jpn 69:541–561

Peng L, Sui CH, Lau KM, Tao WK (2001) Genesis and evolution of hierarchical cloud clusters in a two-dimensional cumulus-resolving model. J Atmos Sci 58:877–895
Ping F, Luo Z, Li X (2008) Kinematics, cloud microphysics, and spatial structures of tropical cloud clusters: A two-dimensional cloud-resolving modeling study. Atmos Res 88:323–336, (c) Elsevier. Reprinted with permission
Simpson J (1980) Downdrafts as linkages in dynamic cumulus seeding effects. J Appl Meteorol 19:477–487
Simpson J, Woodley WL (1971) Seeding cumulus in Florida: new 1970 results. Science 172:117–126
Sui CH, Lau KM (1992) Multi-scale phenomena in the tropical atmosphere over the western Pacific. Mon Weather Rev 120:407–430
Tao WK, Simpson J (1984) Cloud interactions and merging: numerical simulations. J Atmos Sci 41:2901–2917
Wang JJ, Li X, Carey L (2007) Evolution, structure, cloud microphysical and surface rainfall processes of a monsoon convection during the South China Sea monsoon experiment. J. Atmos. Sci 64:360–379. (c) American Meteorological Society. Reprinted with permission
Yano JI, McWilliams JC, Moncrieff MW, Emanuel RA (1995) Hierarchical tropical cloud systems in an analog shallow-water model. J Atmos Sci 52:1723–1742

Chapter 8
Cloud-Radiative and Microphysical Processes

Abstract In this chapter, cloud-radiative processes are examined through the comparison between the sensitivity experiments with varied and fixed cloud single-scattering albedo and asymmetric factor. Cloud microphysical processes are investigated through the comparison between the sensitivity experiments with and without schemes of depositional growth of snow from cloud ice. The effects of ice clouds on rainfall are also analyzed based on the sensitivity experiment study. Heat and water vapor budgets are analyzed to study condensation, associated release of latent heat and large-scale forcing, and their relationship. The lag correlation analysis is applied to the budgets of available potential and kinetic energy to identify dominant physical processes that are responsible for the phase difference between unstable energy and rainfall.

Keywords Cloud-radiative and microphysical processes • Depositional growth of snow from cloud ice • Condensation • Large-scale forcing • Phase difference • Convective available potential energy • Perturbation kinetic energy

Convective systems affect vertical heat and water vapor distributions through radiative and cloud microphysical processes. Due to fine spatial resolution, cloud-resolving models resolve individual cloud and simulate cloud-radiative and microphysical processes through solar and IR radiative parameterization and prognostic cloud microphysics parameterization. Cloud microphysical processes include vapor condensation and deposition, autoconversion, evaporation, collection, accretion, riming, and melting. Cloud–radiation interaction processes can be simulated by solar and IR radiative parameterization coupling with cloud single-scattering albedo and asymmetry factor associated with cloud hydrometeor mixing ratios. In this chapter, radiative and cloud microphysical processes associated with the development of tropical convection are intensively discussed based on Li et al. (1999, 2002a, b, 2005), Li (2006), and Gao et al. (2006a, b).

© Springer International Publishing Switzerland 2016

X. Li, S. Gao, *Cloud-Resolving Modeling of Convective Processes*,
Springer Atmospheric Sciences, DOI 10.1007/978-3-319-26360-1_8

8.1 Radiative Processes

Cloud–radiation interaction can be examined by a comparison between two experiments during selective 7-day period of TOGA COARE. In one, cloud single-scattering albedo and asymmetry factor varied with clouds and environmental thermodynamic conditions; in the other, they were fixed at 0.99 and 0.843, respectively (Li et al. 1999). A comparison of solar radiation calculations between the two experiments showed that the experiment with the varying single-scattering albedo and asymmetry factor had stronger solar radiation absorption by ice clouds in the upper troposphere than did the experiment with the constant single-scattering albedo and asymmetry factor. The difference in temperatures between the two experiments further showed that the temperature was 2 °C warmer around 200 mb in the experiment with variable single-scattering albedo and asymmetry factor than in the experiment with values that were kept constant.

A statistical analysis of the clouds and surface rain rates revealed that stratiform (convective) clouds contributed to 33 (67) % of the total rain in the experiment with the variable cloud optical properties and 40 (60) % in the experiment with the constant cloud optical properties. The fractional cover by stratiform clouds increased from 64 % in the experiment with the variations to 70 % in the experiment with the constants. These sensitivity tests showed the cloud–radiation interaction process for stabilizing the atmosphere in which the change in the vertical heating gradient by solar radiation due to variations of cloud optical properties stabilizes the middle and upper troposphere and contributes to the reduction of stratiform clouds that further stabilizes the cloud system by reducing infrared cloud-top cooling and cloud-base warming.

Li et al. (2005) examined the effects of precipitating cloud–radiation interactions on thermodynamic states by conducting the two experiments (COAREN and COARE). Both experiments are forced by the same TOGA COARE forcing and integrated for 10 days. The difference between COAREN and COARE is that COAREN excludes the precipitation–radiation interaction by simply setting mixing ratios of precipitation water (raindrops) and ice (snow and graupel) to zero in the calculation of radiation.

Within the first three days of integration, the temperature differences for COAREN–COARE are mostly positive except in late 21 December 1992 when negative temperature differences occur (Fig. 8.1a in the left panel). From late 22 December 1992 on, the temperature differences below 200 mb level become persistently negative with minimum values ranging from −1.5 to −2 °C. This indicates that the exclusion of precipitation–radiation interaction causes a cold bias in the troposphere. The differences in specific humidity for COAREN–COARE are persistently negative below 500 mb during the integrations, with minimum around −1 to −1.5 g kg^{-1} (Fig. 8.1a in the left panel). The significant atmospheric drying begins on 23 December 1992.

To examine the physical processes responsible for the temperature and water vapor differences, each term of (2.1) is integrated with time for COAREN and

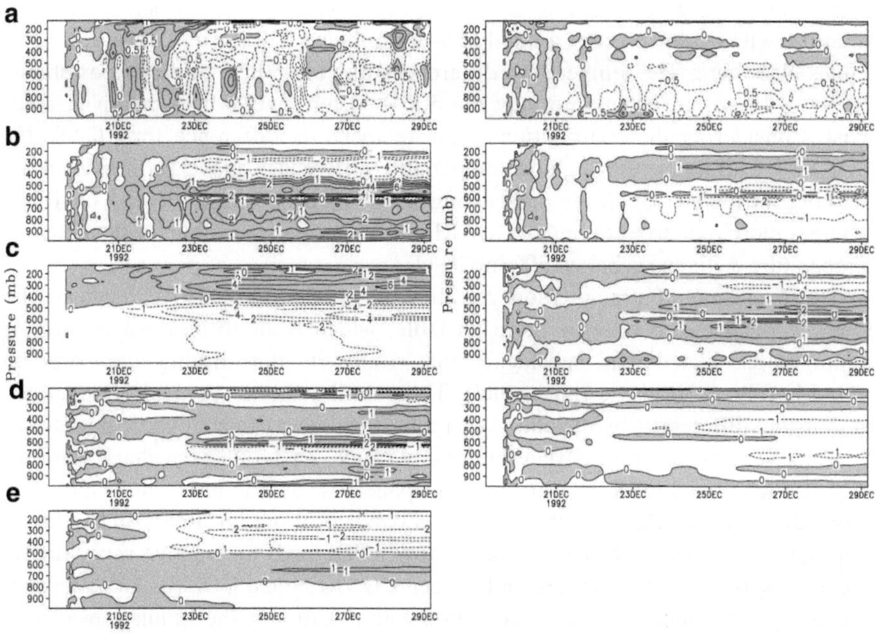

Fig. 8.1 Temporal and vertical distribution of (**a**) model domain-mean temperature differences (°C) between COAREN and COARE (COAREN–COARE) and differences due to (**b**) condensational heating, (**c**) radiative heating, (**d**) convergence of vertical heat flux, and (**e**) vertical temperature advection in the left panel and (*a*) differences in the mean specific humidity (g kg^{-1}) for COAREN–COARE and the differences due to (*b*) condensation, (*c*) convergence of vertical moisture flux, and (*d*) vertical moisture advection in the right panel. Positive differences are shaded (After Li et al. 2005)

COARE separately and then took the difference for COAREN–COARE to obtain the corresponding temperature differences due to condensational heating, radiative heating, convergence of vertical heat flux, vertical temperature advection (Figs. 8.1b–e in the left panel), and water vapor differences due to condensation, convergence of vertical moisture flux, and vertical moisture advection (Figs. 8.1b–d in the right panel), respectively. The same horizontal temperature and moisture advections are imposed in both experiments that do not contribute to the temperature and moisture differences.

The temperature differences due to radiation are negative below 500 mb, and positive above, with the maximum and minimum values reaching 6 °C around 325 mb and −6 °C around 575 mb, respectively, at the end of the integrations (Fig. 8.1c in the left panel). The radiation-induced differences in thermal stratification cause a more stable layer above 500 mb in COAREN than in COARE. This corresponds to less ice hydrometeors in COAREN than in COARE (also see Fig. 8.1b in the right panel), which leads to less latent heating above 500 mb in COAREN from 23 December 1992 on (Fig. 8.1b in the left panel). The temperature differences due to the convergence of vertical heat flux (Fig. 8.1d in the left panel) show a banded structure with negative zones around 650–800 mb and 550 mb and mostly positive values elsewhere. The temperature differences due to vertical advection as shown in Fig. 8.1e are all negative except the 500–800 mb layer where positive values exist. Considering the contribution to the temperature difference for COAREN–COARE by all terms, the negative temperature differences above 500 mb level are mainly contributed to by less cloud heating as a result of more stability in COAREN, whereas more radiative cooling in COAREN is directly responsible for the negative temperature differences below 500 mb level.

The exclusion of precipitation–radiation interaction in COAREN causes the lower troposphere to become more unstable, which leads to less evaporation of raindrops associated with subsidence that results in more drying, compared to COARE (Fig. 8.1b in the right panel). The contributions to vertical distributions of moisture differences by the other two terms, the convergence of vertical moisture flux and vertical moisture advection, largely cancel each other out due to an out-of-phase relation (Fig. 8.1c, d in the right panel). Thus, the less rain evaporation associated with large-scale subsidence as a result of more instability in the lower troposphere increases water vapor more slowly that leads to negative vapor differences. In summary, the comparison between COAREN and COARE indicates that exclusion of the interaction of precipitation with radiation in the simulations causes a significant cooling and drying bias in the troposphere.

8.2 Cloud Microphysical Processes

Cloud microphysical properties in deep convective regions significantly differ from those in stratiform regions. Along with the westward propagation and during the development of tropical convection, the area-mean vertical velocity profiles exhibit the major ascending motion below 500 mb in the left half of the cloud and the maximum ascending motion between 300 and 500 mb in the right half [Li et al. (2002b)], indicating that the left half of the cloud undergoes the deep convective development whereas the anvil cloud grows in the right half (see the box in Fig. 4.5). The surface rainfall is much larger in the left half than in the right half (Fig. 8.2). The *LWP* (9.0 mm) is much larger than *IWP* (4.8 mm) in the left half whereas *LWP* (8.1 mm) and *IWP* (8.4 mm) have similar amounts in the right half. The analysis of raindrop budget reveals the collection of cloud water by raindrops. (P_{RACW}) is a major process that is responsible for the surface rainfall, and thus, the water hydrometeor processes are dominant in the deep convective clouds in the left half. Both the collection of cloud water by raindrops and the melting of precipitation ice into raindrops (P_{GMLT}) are responsible for the surface rainfall in the anvil clouds in the right half.

The P_{SFI} is one of the main processes to consume cloud ice in the growth of snow, in particular, in the deep convective clouds (e.g., Fig. 8.2). Hsie et al. (1980) modified the work of Orville and Kopp (1977) that was based on the equation of

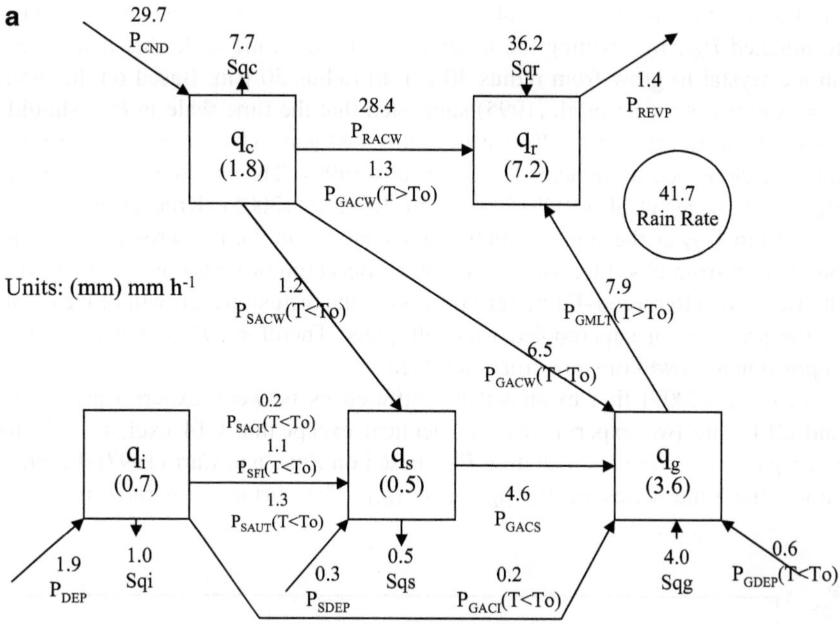

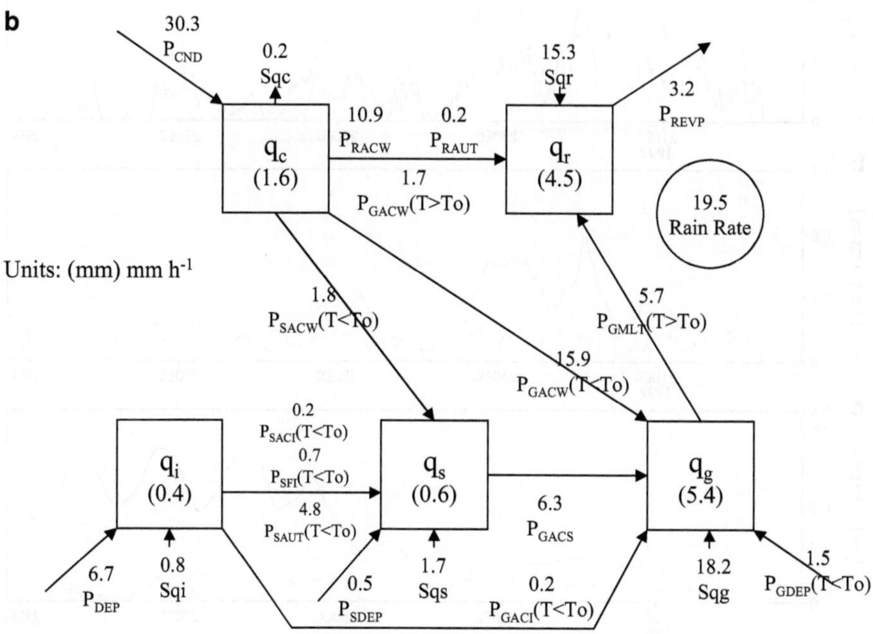

Fig. 8.2 Cloud microphysical budgets averaged within hours 13–15 in (**a**) the *left half* and (**b**) the *right half* of the box in Fig. 4.5. Units for cloud hydrometeor and conversions are mm and mm h^{-1}, respectively (After Li et al. 2002b)

the rate of growth of ice crystals by deposition proposed by Koenig (1971) and formulated P_{SFI} by dividing the mixing ratio by the time scale that is needed for an ice crystal to grow from radius 40 μm to radius 50 μm. Based on the aircraft observations, Krueger et al. (1995) suggested that the time scale in P_{SFI} should be for a crystal to grow from 40 μm to radius 100 μm, which increases the mixing ratio of cloud ice as indicated in Li et al. (1999). The modified formulation of P_{SFI} by Krueger et al. (1995) is used by Li et al. (2005). Krueger et al. (1995) referred to P_{SFI} as the snow formation associated with the Bergeron process (snow production from cloud ice via the growth of Bergeron process embryos). However, the Bergeron (Bergeron–Findeisen) process is the diffusional growth of ice crystals in the presence of supercooled water droplets. Therefore, P_{SFI} is defined as the depositional growth of snow from cloud ice.

Li et al. (2005) first examined the differences between experiments COARE and C14. The two experiments are identical except that C14 excludes 14 cloud microphysical processes including P_{SFI} based on Zhao and Carr (1997). Figure 8.3 shows the time series of P_s, $[q_c]$, and $[q_i]$. The surface rain rates in the two

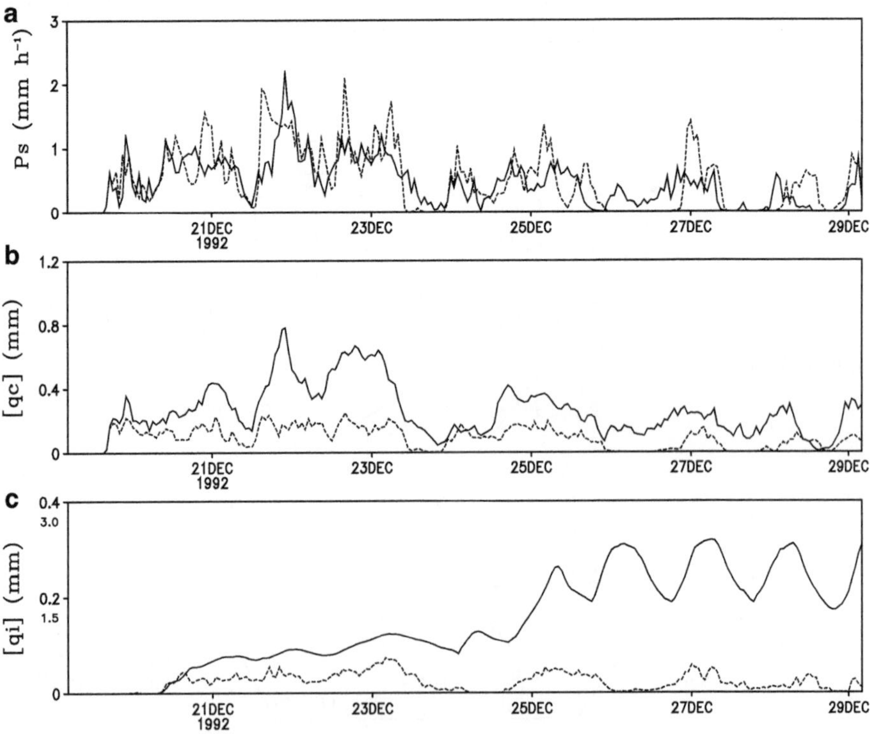

Fig. 8.3 Time series of (**a**) model domain-mean surface rain rate (P_s) and mass-integrated mixing ratios of (**b**) cloud water ($[q_c]$) and (**c**) cloud ice ($[q_i]$) simulated in COARE (*dashed*) and C14 (*solid*). The plotting scales for $[q_i]$ are 0–0.4 mm for COARE and 0–3 mm for C14, respectively. Units are mm h^{-1} for P_s and mm for $[q_c]$ and $[q_i]$ (After Li et al. 2005)

experiments have similar temporal evolution because of the same imposed vertical velocity. However, $[q_c]$ and $[q_i]$ are much larger in C14 than in COARE, in particular, $[q_i]$ in C14 is more than one order of magnitude larger than in COARE.

The anomalous $[q_i]$ causes a large cooling above the ice clouds due to the reflection of the solar radiation and a large warming below due to the energy trapping of the longwave radiation and positive difference of water vapor fields between C14 and COARE. The 10-day mean fractional covers of convective, raining stratiform, and non-raining stratiform clouds in C14 are 7.1 %, 15.6 %, and 65.6 %, respectively, whereas those in COARE are 4.8 %, 12.8 %, and 52.4 %, respectively. The result indicates a respective increase of 48 %, 22 %, and 25 % in the fractional coverage of convective, raining stratiform, and non-raining stratiform clouds from COARE to C14. C14 confirms unrealistic simulation of cloud ice. The budgets of $[q_i]$ in COARE and C14 are then analyzed to identify the cause of the unrealistic $[q_i]$ simulation in C14.

The time series of vertically integrated budgets of $[q_i]$ (1.8e) in COARE and C14 are shown, respectively, in Fig. 8.4a, b. In COARE, $[P_{DEP}]$ is nearly balanced by $[P_{SAUT}]$ and $[P_{SFI}]$ (Fig. 8.4a). Thus, $[S_{qi}]$ is very small. $[P_{SFI}]$ is smaller than $[P_{SAUT}]$, but it is an important sink of cloud ice. In the first five days of the integration

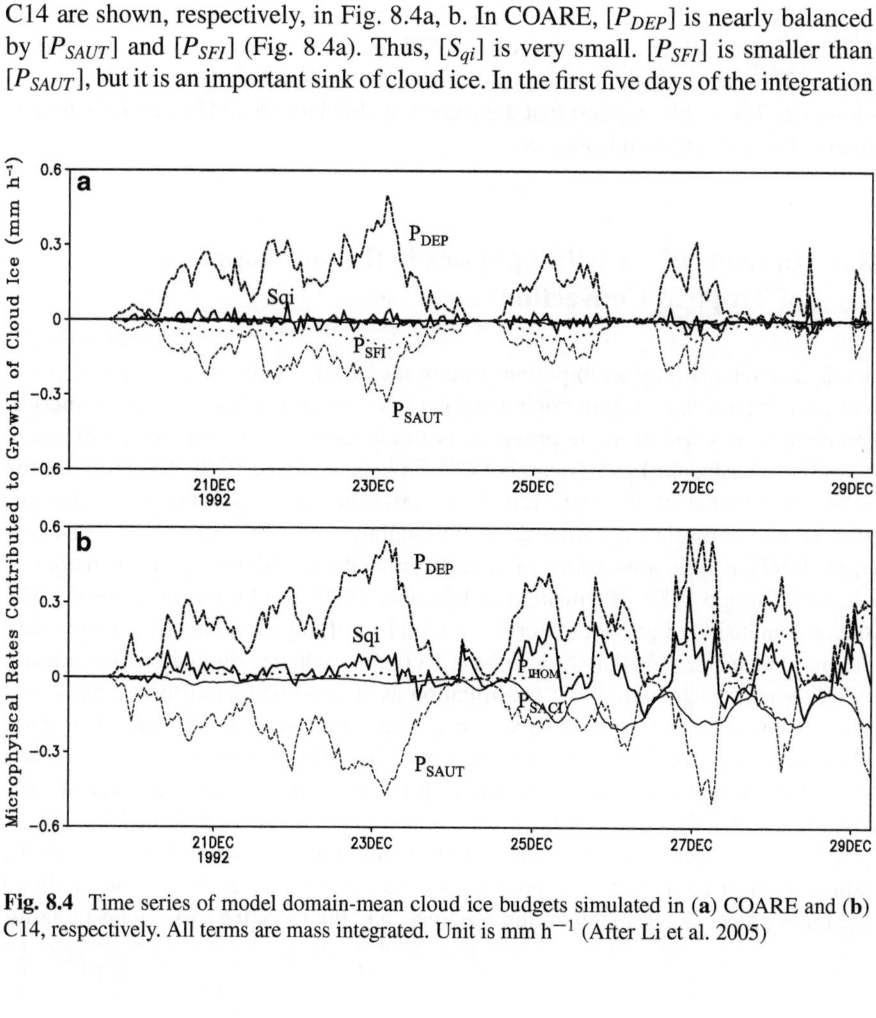

Fig. 8.4 Time series of model domain-mean cloud ice budgets simulated in (**a**) COARE and (**b**) C14, respectively. All terms are mass integrated. Unit is mm h^{-1} (After Li et al. 2005)

of C14, $[P_{DEP}]$ and $[P_{SAUT}]$ cancel each other out in large part, but there is a considerable magnitude of $[S_{qi}]$ contributed by $[P_{IHOM}]$, which leads to the initial anomalous growth of cloud ice (Fig. 8.4b). The rapid decrease of the magnitude of $[P_{SAUT}]$ cannot offset $[P_{DEP}]$ so that a large $[S_{qi}]$ is responsible for the rapid increase of $[q_i]$ in the late evening of 24 December and early morning of 25 December. In the second five days of the integration, $[P_{IHOM}]$ and $[P_{SACI}]$ become as important as $[P_{DEP}]$ and $[P_{SAUT}]$ in the budget of $[q_i]$. A comparison of budgets of $[q_i]$ between COARE and C14 indicates that $[P_{SFI}]$ is the important sink of cloud ice in COARE, but it is excluded in C14, which may cause anomalous growth of cloud ice in C14. Thus, P_{SFI} is included in the additional experiment CSFI that is conducted to further examine the cause of the difference between C14 and COARE.

The temperature differences in the upper troposphere are significantly reduced in CSFI–COARE compared to those in C14–COARE. Also notice that the positive and negative temperature differences appear alternately. The differences of water vapor exhibit a slightly positive trend with most of the positive differences at less than 1 g kg^{-1}. The fractional cloud covers in CSFI also display significant reduction from C14, and their 10-day mean values in CSFI become similar to those in COARE. Thus, CSFI confirms the crucial role of P_{SFI} as the sink of $[q_i]$ in the balance of cloud ice. The results suggest that the budget of cloud ice should be carefully treated to avoid unrealistic cloud calculations.

8.3 Impacts of Ice Microphysics in the Development of Tropical Convection

Ice hydrometeors play an important role in the development of tropical convective systems. Inclusion of ice microphysical parameterization schemes in the numerical simulations has led to the improvements in cloud-resolving simulations of squall lines (e.g., Yoshizaki 1986; Nicholls 1987; Fovell and Ogura 1988; Tao and Simpson 1989; McCumber et al. 1991; Tao et al. 1991), in cloud-resolving and mesoscale simulations of tropical cyclones (e.g., Willoughby et al. 1984; Lord et al. 1984; Liu et al. 1997), and in simulations of general circulation models (e.g., Ramanathan et al. 1983; Slingo 1987; Heymsfield and Donner 1990; and Fowler et al. 1996). The major conclusion of previous short-term cloud-resolving modeling (less than 1 day) studies on ice effects is that the ice phase is crucial for better simulation of stratiform development, such as the light precipitation associated with stratiform clouds over the trailing region of convective systems. The main results from recent long-term cloud-resolving modeling (more than 1 week) studies reveal that (1) the effects of cloud microphysics on temperature and water vapor profiles are statistically significant and the temperature in the upper troposphere is modified by microphysical processes associated with anvil clouds (Grabowski et al. 1999); (2) the simulations of cloud-radiative properties have been improved due to the modified ice microphysical parameterization schemes, and the radiative flux, cloud-radiative forcing, and albedo are sensitive to the effective radius of ice particles (Wu et al. 1999); (3) the convective systems simulated without ice microphysics have a reduced stratiform component, a larger propagation speed, and a shorter life cycle than do those simulated with ice microphysics (Grabowski and Moncrieff 2001; Grabowski 2003); (4) the models both with and without ice microphysics simulated similar propagation speeds and life cycles of convective systems (Tao and Simpson 1989); and (5) the model without ice microphysics simulated a much weaker and less organized convective system than does the model with ice microphysics in a weak unstable environment, whereas ice microphysics has important effects on the system-scale structures in a strong unstable environment, indicating that the environmental thermodynamics is a major factor in regulating ice effects on tropical convection (Liu and Moncrieff 1997).

Gao et al. (2006b) conducted an experiment (COAREW) that excludes ice microphysics and compared COAREW with COARE that includes ice microphysics. The *LWP* is larger in COAREW than in COARE. This indicates that exclusion of ice microphysics in the simulation enhances water hydrometeors. To explain the difference in *LWP* between the two experiments, 21-day mean cloud microphysics budgets in both experiments are calculated and shown in Fig. 8.5. $[q_c]$ is much larger in COAREW (0.173 mm) than in COARE (0.066 mm) whereas $[q_r]$ is slightly smaller in COAREW (0.091 mm) than in COARE (0.107 mm). Thus, a larger *LWP* in COAREW is mainly caused by a larger amount of cloud water. In COARE, $[P_{CND}]$ (0.48 mm h^{-1}) is nearly balanced by $[P_{RACW}]$ (0.33 mm h^{-1}) and $[P_{GACW}(T<T_O)]$ (0.09 mm h^{-1}). In COAREW, the accretion rate of cloud water by precipitation ice is excluded. Only $[P_{RACW}]$ (0.39 mm h^{-1}) and $[P_{RAUT}]$ (0.02 mm h^{-1}) consume cloud water, which leads to the growth of cloud water ($[S_{qc}] = 0.03$ mm h^{-1}).

Surface rain rate is smaller in COAREW (0.29 mm h^{-1}) than in COARE (0.36 mm h^{-1}). In COAREW, the vapor condensation rate ($[P_{CND}] = 0.44$ mm h^{-1}) supports the surface rain rate and the evaporation rate of rain ($[P_{REVP}] = 0.12$ mm h^{-1}) through the collection rate of cloud water by raindrops ($[P_{RACW}] = 0.39$ mm h^{-1}) and the autoconversion rate from cloud water to raindrops ($[P_{RAUT}] = 0.02$ mm h^{-1}), while it enhances cloud water persistently with a rate of 0.03 mm h^{-1} ($[S_{qc}]$). In COARE, the vapor condensation rate ($[P_{CND}] = 0.48$ mm h^{-1}) supports the surface rain rate and the evaporation rate of rain ($[P_{REVP}] = 0.22$ mm h^{-1}) directly through the collection rate of cloud water by raindrops ($[P_{RACW}] = 0.33$ mm h^{-1}) and the autoconversion rate from cloud water to raindrops ($[P_{RAUT}] = 0.01$ mm h^{-1}). Meanwhile, vapor deposition rates ($[P_{DEP}] + [P_{SDEP}] + [P_{GDEP}] = 0.13$ mm h^{-1}) and accretion rates of cloud water by precipitation ice ($[P_{SACW}] + [P_{GACW}(T<T_O)]$ $= 0.03 + 0.09$ mm h^{-1}) support the surface rain rate and the evaporation rate of rain through melting rates of precipitation ice to rain ($[P_{SMLT}] + [P_{GMLT}] = 0.01 +$ 0.22 mm h^{-1}). Thus, the larger surface rain rate and larger evaporation rate of rain in COARE stem from vapor deposition processes and conversion from precipitation ice to rain.

Model domain-mean surface rain rates in both experiments show similar evolution (Fig. 8.6). The vapor convergence (Q_{WVF}) in both experiments exhibits the same

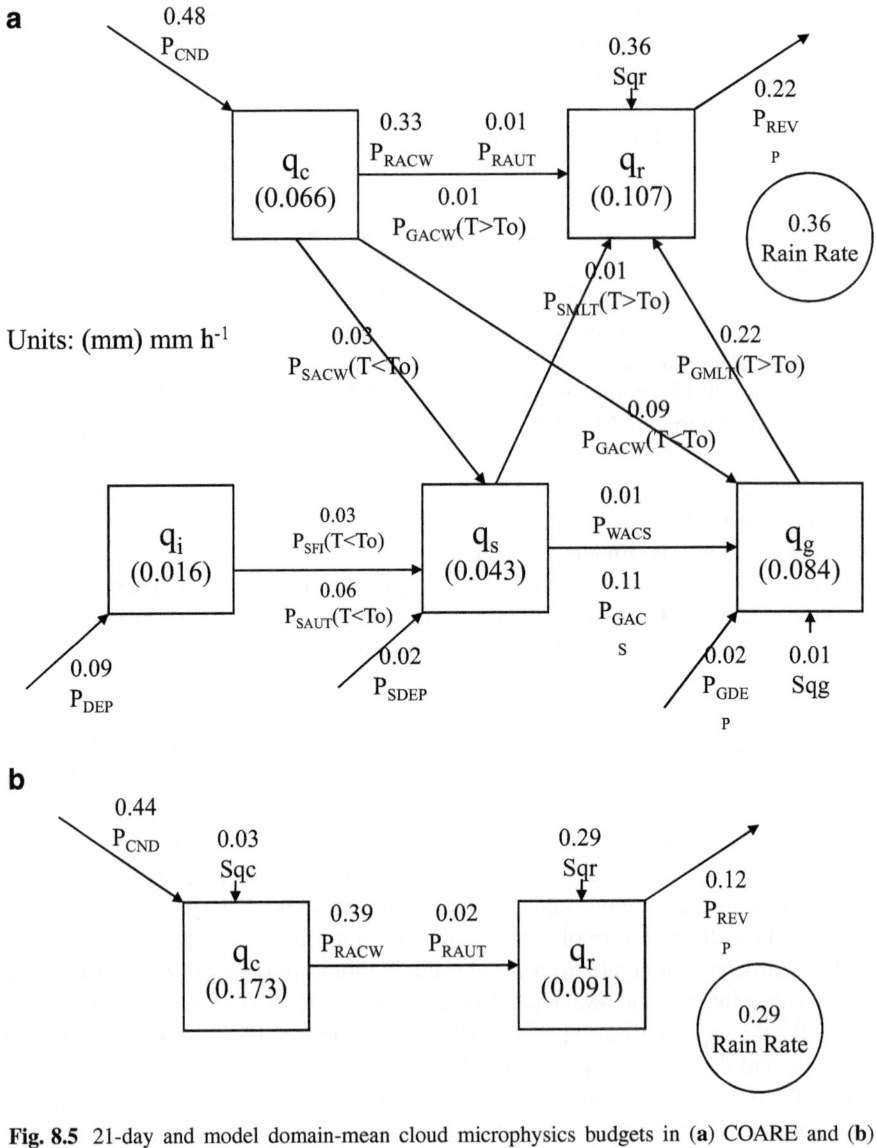

Fig. 8.5 21-day and model domain-mean cloud microphysics budgets in (**a**) COARE and (**b**) COAREW. Units for cloud hydrometeors and conversions are mm and mm h^{-1}, respectively (After Gao et al. 2006b)

magnitudes and evolution due to similar effects of the mean ascent on the moisture profiles (Fig. 8.6c). The surface evaporation fluxes (Q_{WVE}) in both experiments are small (Fig. 8.6d). Thus, the differences in local vapor (Q_{WVT}) and hydrometeor (Q_{CM}) changes (Fig. 8.6b, e) are responsible for the differences in magnitudes. In COARE, the 21-day means of P_s, Q_{WVT}, Q_{WVF}, Q_{WVE}, and Q_{CM} are 0.36, 0.04,

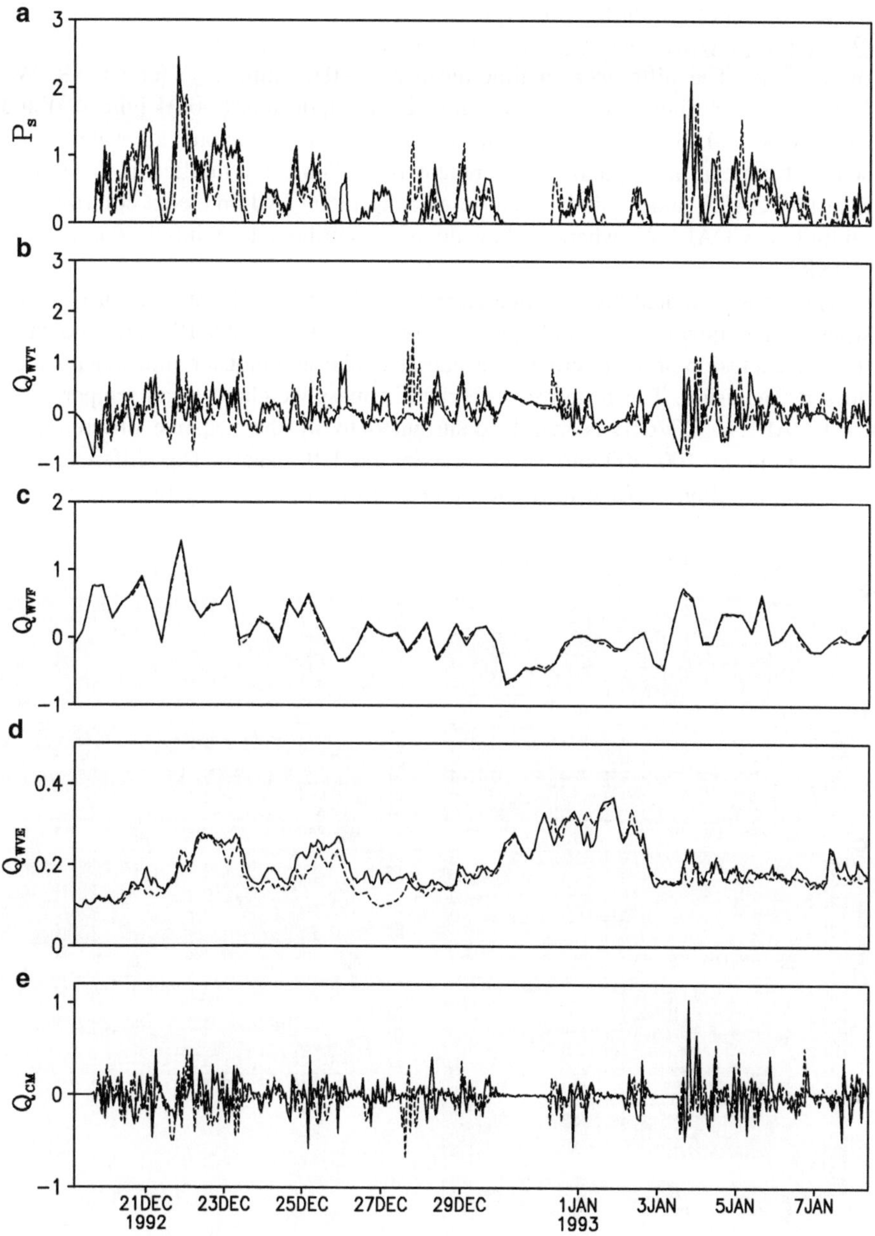

Fig. 8.6 Time series of model domain-mean (**a**) P_s, (**b**) Q_{WVT}, (**c**) Q_{WVF} , (**d**) Q_{WVE}, and (**e**) Q_{CM} during the 21-day integrations. Solid and dashed lines denote COARE and COAREW, respectively. Unit is mm h^{-1} (After Gao et al. 2006b)

0.12, 0.20, and 0.0 mm h^{-1}, respectively. In COAREW, the 21-day means of P_s, Q_{WVT}, Q_{WVF}, Q_{WVE}, and Q_{CM} are 0.29, 0.02, 0.11, 0.19, and −0.03 mm h^{-1}, respectively. The difference in time-mean P_s (−0.07 mm h^{-1}) for COAREW–COARE is contributed to by the differences in vapor sink (−0.04 mm h^{-1}) and cloud sink (−0.03 mm h^{-1}) partially because a smaller amount of water vapor source from the atmospheric environment is available for rainfall in COAREW than in COARE and partially because existing cloud hydrometeors contribute to rainfall in COAREW, whereas they do not contribute to rainfall processes in COARE.

The mass-weighted mean temperature is colder in COAREW than in COARE whereas the amount of PW is larger in COAREW than in COARE. This indicates that the exclusion of ice microphysics causes colder and moister atmospheres than does its inclusion. The negative difference in mass-weighted mean temperatures for COAREW–COARE is mainly contributed to by the negative difference of temperature in 100–400 mb (Fig. 8.7a in the left panel). The differences in heat budgets (Figs. 8.7b–e in the left panel) reveal that the negative temperature

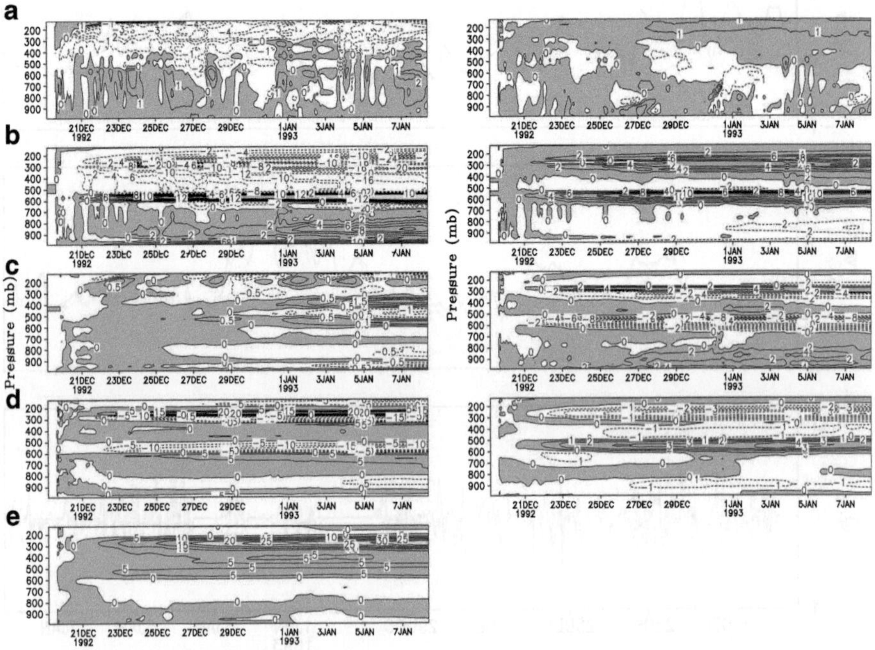

Fig. 8.7 Temporal and vertical distribution of (**a**) model domain-mean temperature differences (°C) for COAREW–COARE and differences due to (**b**) condensational heating, (**c**) radiative heating, (**d**) convergence of vertical heat flux, and (**e**) vertical temperature advection in the left panel and (*a*) differences of the mean specific humidity (g kg^{-1}) for COAREW–COARE and the differences due to (*b*) condensation, (*c*) convergence of vertical vapor flux, and (*d*) vertical vapor advection in the right panel. Positive differences are shaded (After Gao et al. 2006b)

difference in the upper troposphere is mainly due to the negative difference in condensational heating since ice-cloud heating is excluded in COAREW. The difference in convergence of vertical heat flux is largely cancelled out by the difference in vertical temperature advection in the upper troposphere whereas the difference in radiative heating is relatively small compared to the differences in the other thermal processes.

The positive difference in *PW* for COAREW–COARE is mainly contributed to by the positive difference in specific humidity in 100–300 mb (Fig. 8.7a in the right panel). The differences in water vapor budgets (Figs. 8.7b–d in the right panel) show that the positive difference in specific humidity in the upper troposphere is mainly due to the positive difference in condensation in the upper troposphere, which has a larger magnitude than the negative differences in vertical vapor advection, as well as in convergence of vertical vapor flux do.

Note that the imposed vertical velocity that is derived from observational data during TOGA COARE includes effects of ice clouds. To examine the sensitivity of effects of ice clouds to imposed vertical velocity, the zero vertical velocity is imposed in the experiments with (SST29) and without (SST29NIM) ice clouds. The two experiments are integrated to quasi-equilibrium states by Ping et al. (2007) (see Sect. 14.4 for detailed discussions of experiments). The similarities include the fact that the exclusion of ice microphysics yields a cold temperature bias and enhances cloud water amount. The difference is that the exclusion of ice microphysics in the simulation with the zero vertical velocity produces a dry bias whereas the exclusion of ice microphysics in the simulation with the nonzero vertical velocity generates a weak moist bias.

8.4 Condensation, Associated Heating, and Large-Scale Forcing

Gao et al. (2006a) analyzed the mass-weighted mean heat budget and *PW* budget using hourly model domain-mean data simulated in COARE. The linear regression equations between $\partial < \overline{T} > /\partial t$ and $< \overline{Q}_{cn} > /c_p - < \pi \overline{w}^o \partial \overline{\theta}/\partial z >$ in the heat budget (Fig. 8.8a in the left panel) and between $\partial \left[\overline{q}_v\right] /\partial t$ and $-\overline{P}_s - \left[\overline{w}^o \partial \overline{q}_v/\partial z\right]$ (here $\left[\overline{S}_{qv}\right] \cong \overline{P}_s$) in the PW budget (Fig. 8.8b in the left panel) can be, respectively, expressed by

$$\frac{\partial < \overline{T} >}{\partial t} = -0.02 + 0.9\left(\frac{< \overline{Q}_{cn} >}{c_p} - < \pi \overline{w}^o \frac{\partial \overline{\theta}}{\partial z} >\right), \tag{8.1a}$$

$$\frac{\partial \left[\overline{q}_v\right]}{\partial t} = 0.27 + \left(-\overline{P} - \left[\overline{w}^o \frac{\partial \overline{q}_v}{\partial z}\right]\right). \tag{8.1b}$$

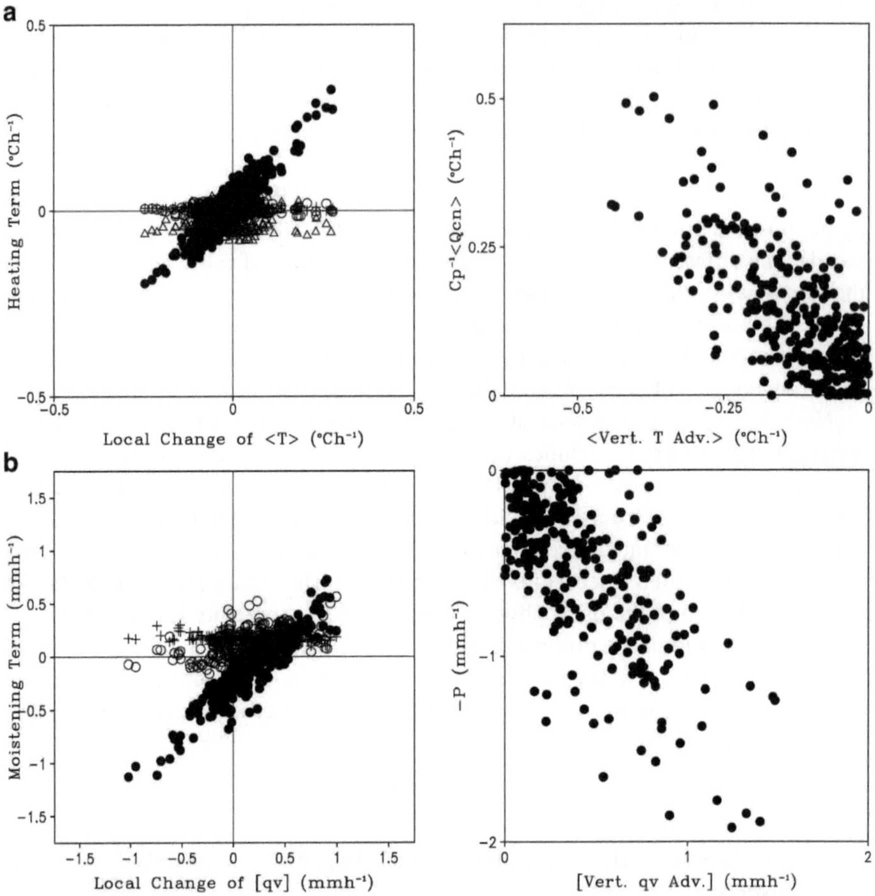

Fig. 8.8 Left panel: (**a**) components of model domain and mass-weighted mean heat budget versus local change of mass-weighted mean temperature ($°C\ h^{-1}$) and (**b**) components of model domain-mean *PW* budget versus local *PW* change ($mm\ h^{-1}$) in rainy conditions simulated in COARE. Closed dots denote $<\overline{Q}_{cn}>/c_p - <\pi\overline{w}^o\partial\overline{\theta}/\partial z>$ in (*a*) and $-\overline{P}_s - \left[\overline{w}^o\partial\overline{q}_v/\partial z\right]$ in (*b*). Open dots represent $-<\overline{u}^o\partial\overline{T}^o/\partial x>$ in (*a*) and $-\left[\overline{u}^o\partial\overline{q}_v{}^o/\partial x\right]$ in (*b*). Symbols "x" are $\overline{H}_s$ in (*a*) and $\overline{E}$ in (*b*). Symbols delta denote $<\overline{Q}_R>/c_p$ in (*a*). Right panel: (*a*) $<\overline{Q}_{cn}>/c_p$ versus $-<\pi\overline{w}^o\partial\overline{\theta}/\partial z>$ ($°C\ h^{-1}$) and (*b*) $-\overline{P}_s$ versus $-\left[\overline{w}^o\partial\overline{q}_v/\partial z\right]$ ($mm\ h^{-1}$) simulated in COARE (After Gao et al. 2006a)

The linear correlation coefficient for both equations is 0.9, which is statistically significant. Thus, the local heat change is largely determined by the sum of condensational heating and vertical thermal advection, whereas the local moisture change is mainly controlled by the sum of precipitation and vertical moisture advection.

Further analysis of the relations between $<\overline{Q}_{cn}>/c_p$ and $-<\pi\overline{w}^o\partial\overline{\theta}/\partial z>$ and between $-\overline{P}_s$ and $-[\overline{w}^o\partial\overline{q}_v/\partial z]$, respectively, shows that the latent heat of condensation and vertical potential temperature advection in the mass-weighted mean heat budget (Fig. 8.8a in the right panel) and the precipitation and vertical moisture advection in the *PW* budget (Fig. 8.8b in the right panel) have the same orders of magnitudes, although they have opposite signs. This indicates that in the first-order approximation, $< \overline{Q}_{cn} > /c_p - < \pi \overline{w}^o \partial \overline{\theta} / \partial z > \approx 0$ and $-\overline{P}_s - [\overline{w}^o \partial \overline{q}_v / \partial z] \approx 0$.

To explain the balances between the latent heat of condensation and vertical potential temperature advection in the heat budget and between the precipitation and vertical moisture advection in the *PW* budget in the lowest-order approximation and smaller variations of local temperature and *PW* in raining conditions, $< \overline{Q}_{cn} > /c_p - < \pi \overline{w}^o \partial \overline{\theta} / \partial z >$ versus $< \overline{Q}_{cn} > /c_p$ and $-\overline{P}_s - [\overline{w}^o \partial \overline{q}_v / \partial z]$ versus $\overline{P}_s$ are plotted in Fig. 8.9. $< \overline{Q}_{cn} > /c_p - < \pi \overline{w}^o \partial \overline{\theta} / \partial z >$ is negative (positive) whereas $-\overline{P}_s - [\overline{w}^o \partial \overline{q}_v / \partial z]$ is positive (negative) for the weak (strong) convection with small (large) values of $< \overline{Q}_{cn} > /c_p$ and $\overline{P}_s$. Thus, vertically advective moistening and cooling have larger (smaller) magnitudes than the condensation and associated latent heat so that the local vapor gains (losses) and local heat losses (gains) for the weak (strong) convection.

Li (2006) studied the cloud microphysical responses to the COARE-derived large-scale forcing by analyzing the 50-day coupled ocean–cloud-resolving atmosphere model simulation. Figure 8.10 displays upward and downward motions with similar amplitudes imposed in the model with large amplitudes in the upper troposphere and strong easterly winds with strong vertical shears in the upper troposphere and with 4-day oscillations in the lower troposphere during the selected 10-day weak-forcing period. From 15 to 17 November 1992, the sum of the mean *IWP* and *LWP* is smaller than 0.2 mm (Fig. 8.11). $[q_c]$ and $[q_r]$ have similar magnitudes that are smaller than 0.1 mm. $[q_g]$ is occasionally large around late 16 and 17 November 1992 whereas $[q_i]$ and $[q_s]$ are negligibly small. Thus, water clouds are dominant during this period. From 18 November 1992 on, $IWP + LWP$ increases its magnitudes to 0.4 mm. $[q_r]$ has significant enhancements and becomes twice as large as $[q_c]$ is. $[q_g]$ has large increases with the amplitudes of larger than 0.15 mm. A large increase in $[q_g]$ is associated with a large upward motion with the maximum magnitudes around 200 mb. $[q_i]$ and $[q_s]$ also show increases and they are as large as $[q_g]$ is around 19 November 1992.

To examine the enhancement of raindrops around 18 November 1992, the budgets of raindrops are analyzed based on the daily-mean data on 15 and 18 November, respectively. On 15 November, $[P_{GMLT}(T > T_o)]$ (0.0053 mm h^{-1}) is more than one order of magnitude smaller than P_s (0.101 mm h^{-1}) and $[P_{RACW}]$ (0.1081 mm h^{-1}) because $[q_g]$ (0.0016 mm) is more than one order of magnitude smaller than $[q_r]$ (0.0204 mm) and $[q_c]$ (0.0266 mm). $[P_{RACW}]$ is nearly balanced by P_s, and -$[P_{REVP}]$ (−0.0273 mm h^{-1}) contributes to $\partial [q_r] / \partial t$ (−0.0084 mm h^{-1}), which suppresses raindrops. On 18 November, $[P_{GMLT}(T > T_o)]$ (0.0936 mm h^{-1}) becomes as large as P_s (0.156 mm h^{-1}) and $[P_{RACW}]$ (0.1554 mm h^{-1}) are because $[q_g]$ (0.0406 mm) grows to the magnitude of $[q_r]$ (0.0403 mm). $[P_{RACW}]$ is nearly balanced by P_s, and $[P_{GMLT}(T > T_o)]$ overcomes -$[P_{REVP}]$ (−0.0763 mm h^{-1}), producing a

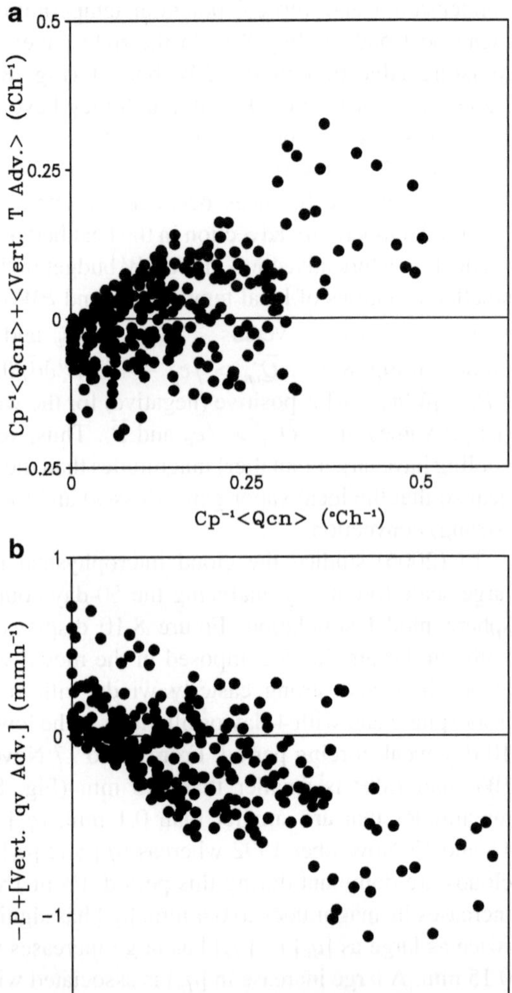

Fig. 8.9 (**a**) $<\overline{Q}_{cn}>/c_p - <\pi\overline{w}^o\partial\overline{\theta}/\partial z>$ versus $<\overline{Q}_{cn}>/c_p$ (°C h^{-1}) and (**b**) $-\overline{P}_s - \left[\overline{w}^o\partial\overline{q}_v/\partial z\right]$ versus $\overline{P}_s$ (mm h^{-1}) simulated in COARE (After Gao et al. 2006a)

positive $\partial\,[q_r]/\partial t$ (0.025 mm h^{-1}). Thus, the graupel enhances raindrops through its melting in the tropical convection. The enhancement of raindrops results from the growth of graupel on 18 November. The analysis of the budget of $[q_g]$ (1.7j and 1.8g) shows that the local growth of $[q_g]$ is mainly supported by $[P_{GACW}(T<T_o)]$ (0.0688 mm h^{-1}). $[P_{GACS}]$ (0.0244 mm h^{-1}) plays a second role.

Convective responses to the strong forcing are studied by analyzing the simulation data from 17 to 27 December 1992 (see vertical velocity and zonal wind in Fig. 4.1). The maxima of $IWP+LWP$ correspond to the maxima of upward motions (Fig. 8.12a). Although the upward motion (14 cm s^{-1}) around the early

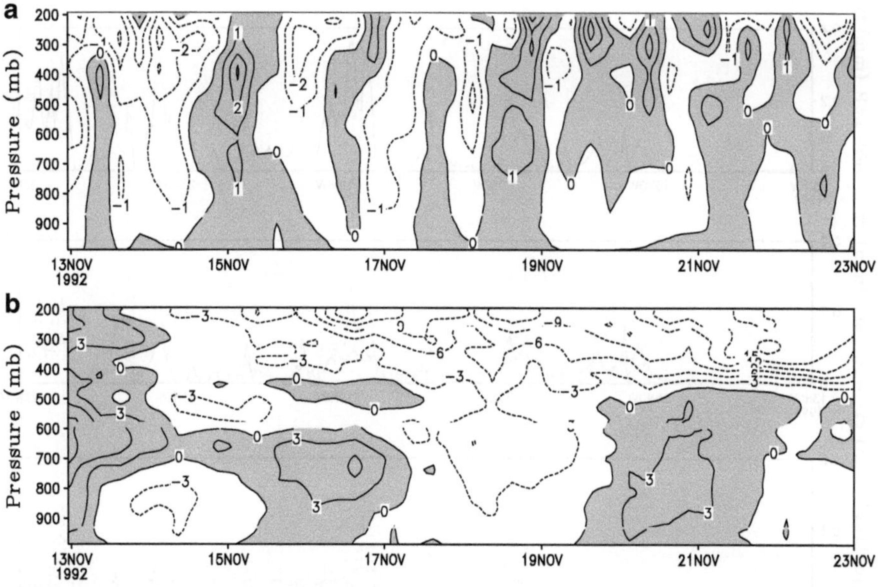

Fig. 8.10 Temporal and vertical distribution of (**a**) vertical velocity (cm s^{-1}) and (**b**) zonal wind (m s^{-1}) from 13 to 22 November 1992. Upward motion in (*a*) and westerly winds in (*b*) are shaded (After Li 2006)

morning of 23 December is stronger than that (10 cm s^{-1}) around the early morning of 25 December, the vertical extent of the former (e.g., contour of 2 cm s^{-1}) is smaller than that of the latter. Thus, their magnitudes of $IWP + LWP$ are about the same (~2 mm). $IWP + LWP$ is largely contributed to by the precipitation water and ice. The amplitudes of $[q_r]$ are larger than those of $[q_c]$ (Fig. 8.12b) whereas the magnitudes of $[q_g]$ are larger than those of $[q_i]$ and $[q_s]$ (Fig. 8.12c). It is interesting to notice that $[q_s]$ increases significantly around the late night of 22 December and early morning of 23 December 1992.

To examine the enhancement of snow around the late night of 22 December 1992, the daily-mean budget of snow on 22 December is compared with the budget on 21 December. $\partial [q_s] / \partial t$ switches to a positive value on 22 December (0.0058 mm h^{-1}) from a negative value on 21 December (−0.0275 mm h^{-1}). The positive $\partial [q_s] / \partial t$ is mainly contributed to by a significant increase of $[P_{SAUT}]$ from 0.1428 mm h^{-1} on 21 December to 0.2273 mm h^{-1} on 22 December and an increase of $[P_{SFW}]$ from 0.0808 mm h^{-1} on 21 December to 0.1241 mm h^{-1} on 22 December, which are in turn due to the increase of $[P_{DEP}]$ from 0.2546 mm h^{-1} on 21 December to 0.3621 mm h^{-1} on 22 December. The increased deposition rate of vapor corresponds to the intensified vertical velocity center in the upper troposphere. The local growth of $[q_s]$ also causes the local enhancement of $[q_g]$ on 22 December, in which $[P_{GACS}]$ (0.4412 mm h^{-1}) plays a primary role and $[P_{GACW}(T < T_o)]$ (0.2377 mm h^{-1}) is a second factor.

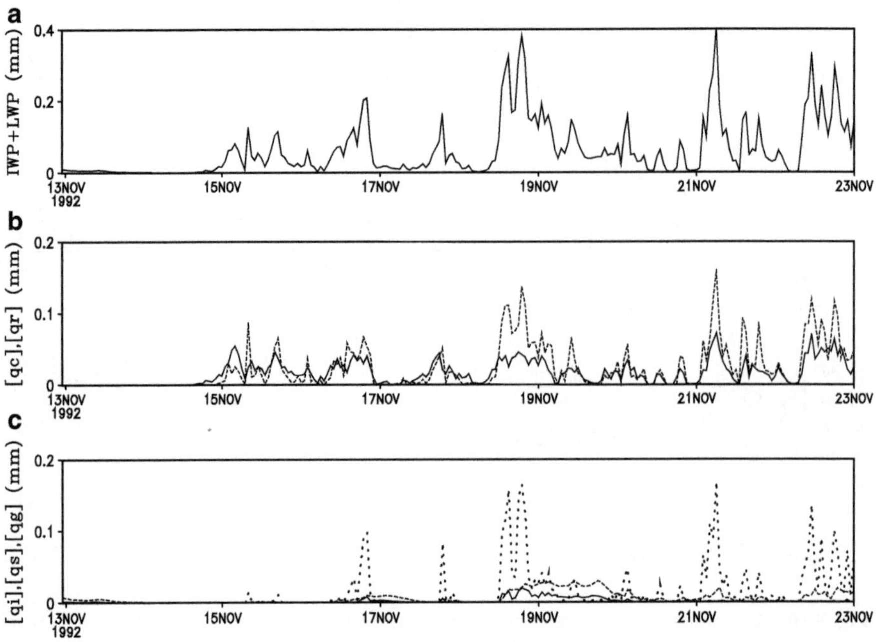

Fig. 8.11 Time series of $IWP + LWP$ in (**a**) $[q_c]$ (*solid*) and $[q_r]$ (*dashed*) in (**b**) and $[q_i]$ (*solid*), $[q_s]$ (*dashed*), and $[q_g]$ (*dot*) in (**c**) simulated in coupled experiment from 13 to 22 November 1992. Unit is in mm (After Li 2006)

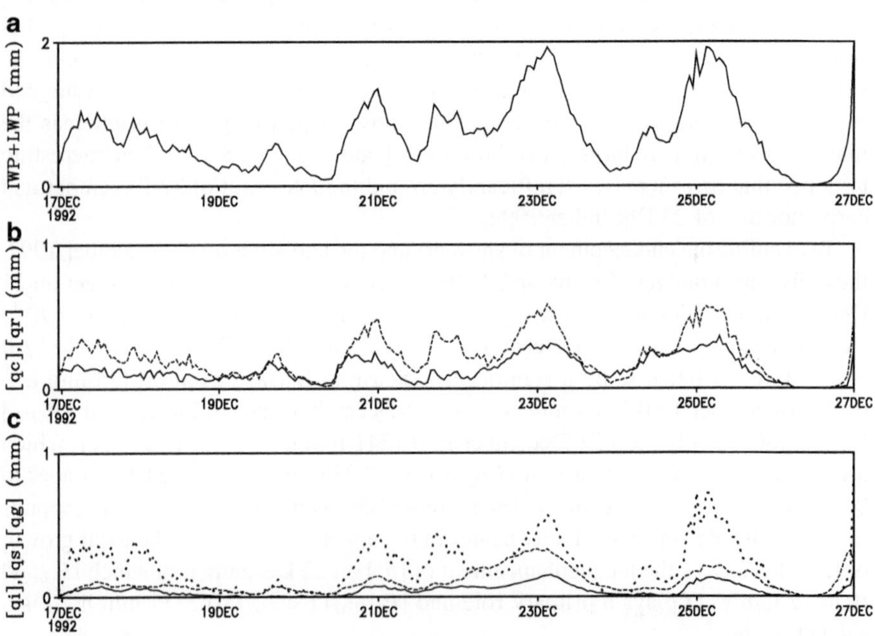

Fig. 8.12 Same as in Fig. 8.11 except from 17 to 26 December 1992 (After Li 2006)

8.5 Phase Relation Between Unstable Energy and Surface Rainfall

Tropical convection occurs as a result of instability in the environment. The large-scale environment provides favorable thermal and moisture conditions for the occurrence and development of convection. In return, it is affected by the vertical redistribution of temperature, moisture, and momentum as a result of the convection. Such an interaction allows us to use environmental conditions to estimate properties of the convection such as the precipitation. Since the environmental time scales (a few days and longer) are much longer than the convective time scales (a few hours or less), the rate of production of available potential energy by the large-scale processes is nearly balanced by the rate of consumption of the available potential energy by the convection (Manabe and Strickler 1964). This quasi-equilibrium concept is the basic premise of the cumulus parameterization scheme proposed by Arakawa and Schubert (1974). A decrease in the *CAPE*, that measures the thermal and moisture conditions of the environment, often coincides with the development of convection so that the *CAPE* and rain rate are negatively correlated (e.g., Thompson et al. 1979; Cheng and Yanai 1989; Wang and Randall 1994; Xu and Randall 1998). The phase relation between the *CAPE* and rainfall is due to the coupling between the environmental dynamic and thermodynamic fields (Cheng and Yanai 1989).

The phases of *CAPE* and rainfall could be different because it takes time for clouds to develop. This phase difference can be included by relaxing the quasi-equilibrium assumption in cumulus parameterization (e.g., Betts and Miller 1986; Randall and Pan 1993). The minimum *CAPE* typically occurs a few hours after the maximum rainfall. Such a phase lag was also demonstrated by Xu and Randall (1998) in their 2D cloud-resolving model simulations. Xu and Randall (1998) interpreted the maximum phase lag as the adjustment time scale from disequilibrium to equilibrium states in the presence of time-varying large-scale forcing. Since the *CAPE* is calculated in a Lagrangian framework and the relevant equations cannot be derived in that framework, the physical processes responsible for the phase difference between the *CAPE* and the surface rain rate cannot be examined. Potential and kinetic energy in the Eulerian framework represent the *CAPE* and surface rain rate, respectively, in the Lagrangian framework. Li et al. (2002a) derived a set of equations for conversions between the moist available potential energy and kinetic energy in the Eulerian framework (see Sect. 2.3). Their equations were demonstrated to be the same as those derived by Lorenz (1955) in the absence of moisture.

The analysis of lag correlation between the *CAPE* and surface rain rate shows that maximum *CAPE* leads maximum surface rain rate by 3 h whereas minimum *CAPE* lags maximum surface rain rate by 2 h. The calculations of lag correlation coefficients between the mean moist available potential energy ($\overline{P}$) and layer-mean vertical velocity as well as between perturbation kinetic energy (K') and layer-mean vertical velocity display that minimum $\overline{P}$ lags maximum upward motion by 6 h whereas maximum K' lags maximum upward motion by 1–2 h. Thus, the phase relation between $\overline{P}$ and K' is similar to that between the *CAPE* and surface rain rate, which lays down the foundation to use the moist energetics framework to explain the physical processes associated with the phase difference between the *CAPE* and surface rain rate.

The further calculations between $\partial\overline{P}/\partial t$ and layer-mean vertical velocity and between $\partial K'/\partial t$ and layer-mean vertical velocity reveal that minimum $\partial\overline{P}/\partial t$ lags maximum ascending motion by 3 h whereas maximum $\partial K'/\partial t$ leads maximum ascending motion by 1–2 h. To elucidate dominant physical processes that are responsible for the phase differences, lag correlation coefficients between each term of $\partial\overline{P}/\partial t$ in (2.10) and $\partial P'/\partial t$, $\partial\overline{P}/\partial t$ and each term of $\partial P'/\partial t$ in (2.11), each term of $\partial P'/\partial t$ and $\partial K'/\partial t$ in (2.12), and $\partial P'/\partial t$ and each term of $\partial K'/\partial t$ are calculated. The term $C_v\left(\overline{K},\overline{P}\right)$ of $\partial\overline{P}/\partial t$ associated with the vertical thermal and vapor advections and the term $G_R(P')$ of $\partial P'/\partial t$ associated with covariance between the radiation and h' are major players in the determination of phase relation between $\partial\overline{P}/\partial t$ and $\partial P'/\partial t$. The term $C_{cn}(P')$ of $\partial P'/\partial t$ associated with the covariance between the depositional heating and h' and the term $C(P',K')$ of $\partial K'/\partial t$ associated with the covariance between perturbation vertical velocity and temperature are major contributors to the phase relation between $\partial P'/\partial t$ and $\partial K'/\partial t$.

The phase relations between the convection and its environment are summarized in the following way. The imposed large-scale downward motion yields a growth of $\overline{P}$ by the associated vertical advective warming $[C_v\left(\overline{K},\overline{P}\right) > 0]$, building the favorable environment for occurrence of convection. The near-simultaneous occurrence of maximum $\partial K'/\partial t$, K', and imposed large-scale upward motion implies that convection is phase locked with the large-scale forcing. The life cycle of the simulated convective events (about 9 h) is much shorter than the time scales of imposed large-scale forcing (longer than the diurnal cycle). In the convective events, maximum $\partial K'/\partial t$ leads maximum $\partial P'/\partial t$ by about 3 h through perturbation cloud heating $[C_{cn}(P')]$ and the vertical heat transport by perturbation circulations $[C(P',K')]$. Maximum $\partial K'/\partial t$ also leads maximum K' by about 3 h, indicating that 3 h is the time required by convection to reach the maximum strength. Minimum $\partial P'/\partial t$ leads minimum $\partial\overline{P}/\partial t$ by about 1 h through perturbation radiative processes $[G_R(P')]$ and the large-scale vertical advective cooling $[C_v\left(\overline{K},\overline{P}\right)]$. Consequently, maximum $\partial K'/\partial t$ leads minimum $\partial\overline{P}/\partial t$ by 4–5 h, which is about half of the convective life span.

References

Arakawa A, Schubert WH (1974) Interaction of a cumulus cloud ensemble with the large-scale environment. Part I. J Atmos Sci 31:674–701

Betts AK, Miller MJ (1986) A new convective adjustment scheme. Part II: single column tests using GATE wave, BOMEX, ATEX and arctic air-mass data sets. Q J Roy Meteorol Soc 112:692–709

Cheng MD, Yanai M (1989) Effects of downdrafts and mesoscale convective organization on the heat and moisture budgets of tropical cloud cluster. Part III: effects of mesoscale convective organization. J Atmos Sci 56:3028–3042

Fovell RG, Ogura Y (1988) Numerical simulation of a midlatitude squall line in two dimensions. J Atmos Sci 45:3846–3879

Fowler LD, Randall DA, Rutledge DA (1996) Liquid and ice cloud microphysics in the CSU general circulation model. Part I: model description and simulated microphysical processes. J Climate 9:489–529

Gao S, Ping F, Li X (2006a) Tropical heat/water vapor quasi-equilibrium and cycle as simulated in a 2D cloud resolving model. Atmos Res 79:15–29, (c) Elsevier. Reprinted with permission

Gao S, Ran L, Li X (2006b) Impacts of ice microphysics on rainfall and thermodynamic processes in the tropical deep convective regime: a 2D cloud-resolving modeling study. Mon Wea Rev 134:3015–3024, (c) American Meteorological Society. Reprinted with permission

Grabowski WW (2003) Impact of ice microphysics on multiscale organization of tropical convection in two-dimensional cloud-resolving simulations. Q J Roy Meteorol Soc 129:67–81

Grabowski WW, Moncrieff MW (2001) Large-scale organization of tropical convection in two-dimensional explicit numerical simulations. Q J Roy Meteorol Soc 127:445–468

Grabowski WW, Wu X, Moncrieff MW (1999) Cloud-resolving model of tropical cloud systems during phase III of GATE. Part III: effects of cloud microphysics. J Atmos Sci 56:2384–2402

Heymsfield AJ, Donner LJ (1990) A scheme for parameterization ice-cloud water content in general circulation models. J Atmos Sci 47:1865–1877

Hsie EY, Farley RD, Orville HD (1980) Numerical simulation of ice phase convective cloud seeding. J Appl Meteorol 19:950–977

Koenig LR (1971) Numerical modeling of ice deposition. J Atmos Sci 28:226–237

Krueger SK, Fu Q, Liou KN, Chin HNS (1995) Improvement of an ice-phase microphysics parameterization for use in numerical simulations of tropical convection. J Appl Meteorol 34:281–287

Li X (2006) Cloud microphysical and precipitation responses to a large-scale forcing in the tropical deep convective regime. Meteorol Atmos Phys 94:87–102

Li X, Sui CH, Lau KM, Chou MD (1999) Large-scale forcing and cloud-radiation interaction in the tropical deep convective regime. J Atmos Sci 56:3028–3042

Li X, Sui CH, Lau KM (2002a) Interactions between tropical convection and its environment: an energetics analysis of a 2-D cloud resolving simulation. J Atmos Sci 59:1712–1722

Li X, Sui CH, Lau KM (2002b) Dominant cloud microphysical processes in a tropical oceanic convective system: a 2-D cloud resolving modeling study. Mon Wea Rev 130:2481–2491, (c) American Meteorological Society. Reprinted with permission

Li X, Sui CH, Lau KM, Tao WK (2005) Tropical convective responses to microphysical and radiative processes: a 2D cloud-resolving modeling study. Meteorol Atmos Phys 90:245–259

Liu C, Moncrieff MW (1997) Dynamic influence of microphysics in tropical squall lines: a numerical study. Mon Weather Rev 125:2193–2210

Liu Y, Zhang DL, Yau MK (1997) A multiscale numerical study of hurricane Andrew (1992). Part I: explicit simulation and verification. Mon Wea Rev 125:3073–3093

Lord SJ, Willoughby HE, Piotrowicz JM (1984) Role of a parameterized ice phase microphysics in an axisymmetric, nonhydrostatic tropical cyclone model. J Atmos Sci 41:2836–2848

Lorenz EN (1955) Available potential energy and the maintenance of the general circulation. Tellus 7:157–167

Manabe S, Strickler RF (1964) Thermal equilibrium of the atmosphere with a convective adjustment. J Atmos Sci 21:361–385

McCumber M, Tao WK, Simpson Penc JR, Soong ST (1991) Comparison of ice –phase microphysical parameterization schemes using numerical simulations of tropical convection. J Appl Meteorol 30:985–1004

Nicholls ME (1987) A comparison of the results of a two-dimensional numerical simulation of a tropical squall line with observations. Mon Weather Rev 115:3055–3077

Orville HD, Kopp FJ (1977) Numerical simulation of the life history of a hailstorm. J Atmos Sci 34:1596–1618

Ping F, Luo Z, Li X (2007) Microphysical and radiative effects of ice clouds on tropical equilibrium states: a two-dimensional cloud-resolving modeling study. Mon Weather Rev 135:2794–2802

Ramanathan V, Pitcher EJ, Malone RC, Blackmon ML (1983) The response of a spectral general circulation model to refinements in radiative processes. J Atmos Sci 40:605–630
Randall DA, Pan DM (1993) Implementation of the Arakawa-Schubert cumulus parameterization with a prognostic closure. In: Emanuel KA, Raymond DJ (ed) The representation of cumulus convection in numerical models of the atmosphere, Meteor Monogr 46 American Meteorological Society
Slingo JM (1987) The development and verification of a cloud prediction scheme for the ECMWF model. Q J Roy Meteorol Soc 113:899–927
Tao WK, Simpson J (1989) Modeling study of a tropical squall-type convective line. J Atmos Sci 46:177–202
Tao WK, Simpson J, Soong ST (1991) Numerical simulation of a subtropical squall line over the Taiwan Strait. Mon Weather Rev 119:2699–2723
Thompson RM Jr, Payne SW, Recker EE, Reed RJ (1979) Structure and properties of synoptic-scale wave disturbances in the intertropical convergence zone of the eastern Atlantic. J Atmos Sci 36:53–72
Wang J, Randall DA (1994) The moist available energy of a conditionally unstable atmosphere. Part II: further analysis of GATE data. J Atmos Sci 51:703–710
Willoughby HE, Jin HL, Lord SJ, Piotrowicz JM (1984) Hurricane structure and evolution as simulated by an axisymmetric, non-hydrostatic numerical model. J Atmos Sci 41:1169–1186
Wu X, Hall WD, Grabowski WW, Moncrieff MW, Collins WD, Kiehl JT (1999) Long-term evolution of cloud systems in TOGA COARE and their interactions with radiative and surface processes. Part II: effects of ice microphysics on cloud-radiation interaction. J Atmos Sci 56:3177–3195
Xu KM, Randall DA (1998) Influence of large-scale advective cooling and moistening effects on the quasi-equilibrium behavior of explicitly simulated cumulus ensembles. J Atmos Sci 55:896–909
Yoshizaki M (1986) Numerical simulations of tropical squall-line clusters: two-dimensional model. J Meteorol Soc Jpn 64:469–491
Zhao Q, Carr FH (1997) A prognostic cloud scheme for operational NWP models. Mon Weather Rev 125:1931–1953

Chapter 9
Thermal Effects of Doubled Carbon Dioxide on Rainfall

Abstract The increase in carbon dioxide has thermal effects on rainfall through the change in vertical stratification and moist effects on rainfall via the increase in ocean surface evaporation. In this chapter, the thermal aspect of rainfall responses to doubled carbon dioxide is examined by conducting cloud-resolving model experiments. The sensitivity experiments are conducted in pre-summer rainfall event around summer solstices and in tropical rainfall event around winter solstices. The radiative effects of water and ice clouds and effects of ice clouds on rainfall responses to doubled carbon dioxide are investigated. The cloud microphysical budget and vertical profiles of heat budget are analyzed to study dominant physical processes that link cloud-radiative and microphysical effects, doubled carbon dioxide, and rainfall.

Keywords Doubled carbon dioxide • Rainfall • Cloud-radiative and microphysical processes • Cloud budget • Heat budget

The Fourth Assessment Report from the Intergovernmental Panel on Climate Change (IPCC 2007) shows an increase in precipitation over land as a result of an increase in atmospheric water vapor flux caused by atmospheric warming. The increase in precipitation (Fields et al. 1993; Li et al. 2011) may be associated with the increase in tropospheric temperature through the enhancement of greenhouse gases (IPCC 2001). The increase in carbon dioxide has thermal effects on precipitation through the change in thermal stratification associated with the change in radiation and moist effects on precipitation through the enhanced water vapor associated with the accelerated surface evaporation. The increase in water vapor eventually causes the increase in global precipitation (Allen and Ingram 2002). The cloud-resolving model alone can be only used to study the thermal effects of increased carbon dioxide on precipitation. The change in carbon dioxide may lead to the change in solar heating and the solar heating rates around summer and winter solstice could have the largest difference; the pre-summer heavy rainfall event that occurred around the summer solstice (0200 LST 3 June – 0200 LST 8 June 2008) and the tropical rainfall event during TOGA COARE around the winter solstice (1000 LST 19 December – 1000 LST 24 December 1992) are chosen, and their corresponding large-scale forcing data are shown in Figs. 9.1 and 4.1,

© Springer International Publishing Switzerland 2016

X. Li, S. Gao, *Cloud-Resolving Modeling of Convective Processes*,
Springer Atmospheric Sciences, DOI 10.1007/978-3-319-26360-1_9

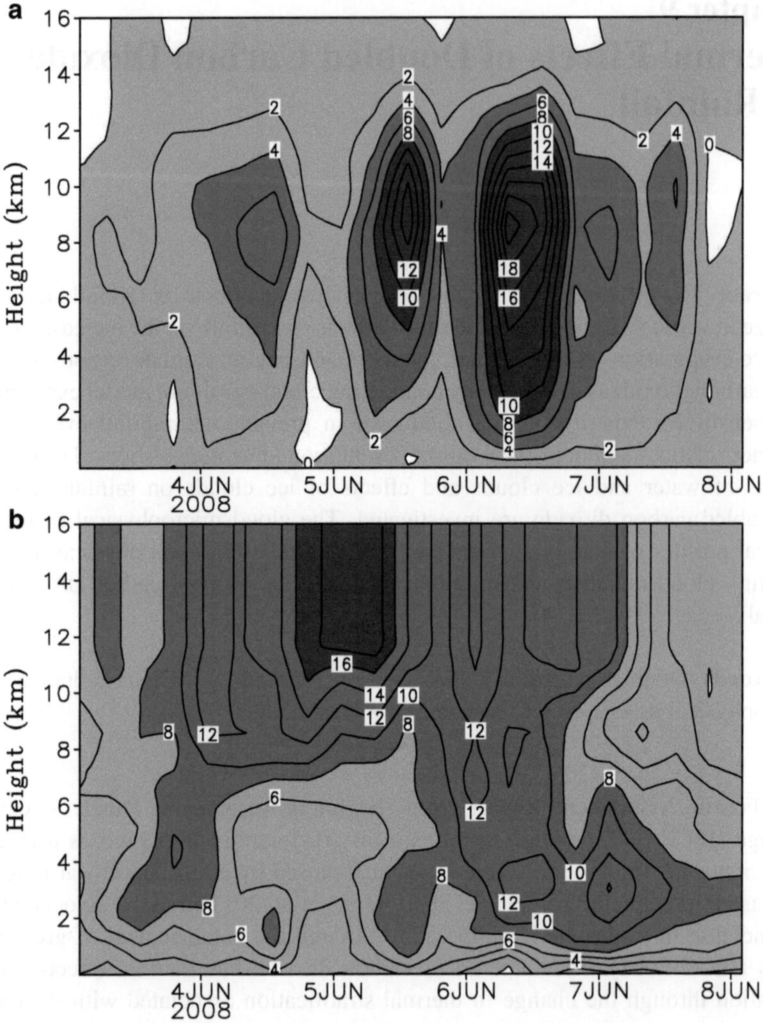

Fig. 9.1 Temporal and vertical distribution of (**a**) vertical velocity (cm s^{-1}) and (**b**) zonal wind (m s^{-1}) imposed in the P experiments. Ascending motion in (**a**) and westerly wind in (**b**) are *shaded*

respectively. The thermal effects of doubled carbon dioxide on rainfall are discussed in Sect. 9.1 through the comparison between experiments P and P2 for pre-summer rainfall event and C and C2 for tropical rainfall event (Li et al. 2014a). Because of the thermal effects of doubled carbon dioxide on precipitation through clouds, a series of sensitivity experiments are carried out and compared to the control experiments to study the effects of doubled carbon dioxide on rainfall responses to radiative processes of water and ice clouds and microphysical processes of ice clouds, respectively, in Sect. 9.2 (Li et al. 2014b), 9.3, and 9.4. The control and sensitivity experiments are summarized in Table 9.1. The data are averaged in 5 days over model domain in this chapter.

9.1 Rainfall Responses to Doubled Carbon Dioxide

The doubled carbon dioxide reduces the mean surface rain rate from P to P2 (Table 9.2). The comparison in the mean cloud budgets between P and P2 shows that the reduced mean rain rate is associated with the weakened mean net condensation and decreased mean hydrometeor loss. The decrease in mean hydrometeor loss corresponds to the reduction in mean cloud ice loss through the weakened melting of cloud ice to cloud water (P_{IMLT}). To explain the effects of doubled carbon dioxide on cloud microphysics, the mean heat budgets (2.1a) are compared between the two

Table 9.1 Experiment designs. X = (P, P2, C, C2), where P and P2 are pre-summer rainfall event, C and C2 are tropical rainfall event, C and C have current carbon dioxide concentration, and P2 and C2 have doubled carbon dioxide concentration

	Radiative effects of water clouds	Radiative effects of ice clouds	Ice clouds
X	Included	Included	Included
XNWR	Excluded	Included	Included
XNIR	Included	Excluded	Included
XNCR	Excluded	Excluded	Included
XNIM	Included	Excluded	Excluded

Table 9.2 Cloud microphysical budgets (P_S, Q_{NC}, and Q_{CM}) averaged for 5 days over model domain in (a) P, PNWR, PNIR, PNCR, PNIM; (b) PNWR-P, PNCR–PNIR, PNIR-P, PNCR–PNWR, PNIM–PNIR; (c) P2, PNWR2, PNIR2, PNCR2, PNIM2; and (d) PNWR2-P2, PNCR2–PNIR2, PNIR2-P2, PNCR2–PNWR2, PNIM2–PNIR2

(a)	P	PNWR	PNIR	PNCR	PNIM
P_S	1.364	1.322	1.326	1.358	1.168
Q_{NC}	1.333	1.347	1.326	1.344	1.280
Q_{CM}	0.031	−0.025	0.0	0.014	−0.113

(b)	PNWR-P	PNCR–PNIR	PNIR-P	PNCR–PNWR	PNIM–PNIR
P_S	−0.042	0.032	−0.038	0.036	−0.158
Q_{NC}	0.014	0.018	−0.007	−0.003	−0.046
Q_{CM}	−0.056	0.014	−0.031	0.039	−0.113

(c)	P2	PNWR2	PNIR2	PNCR2	PNIM2
P_S	1.331	1.357	1.380	1.362	1.190
Q_{NC}	1.323	1.313	1.317	1.312	1.291
Q_{CM}	0.008	0.045	0.063	0.050	−0.101

(d)	PNWR2-P2	PNCR2–PNIR2	PNIR2-P2	PNCR2-PNWR2	PNIM2–PNIR2
P_S	0.026	−0.018	0.049	0.005	−0.190
Q_{NC}	−0.010	−0.005	−0.006	−0.001	−0.026
Q_{CM}	0.037	−0.013	0.055	0.005	−0.164

Unit is mm h^{-1} (After Liu et al. (2014) and Li et al. 2014a, b)

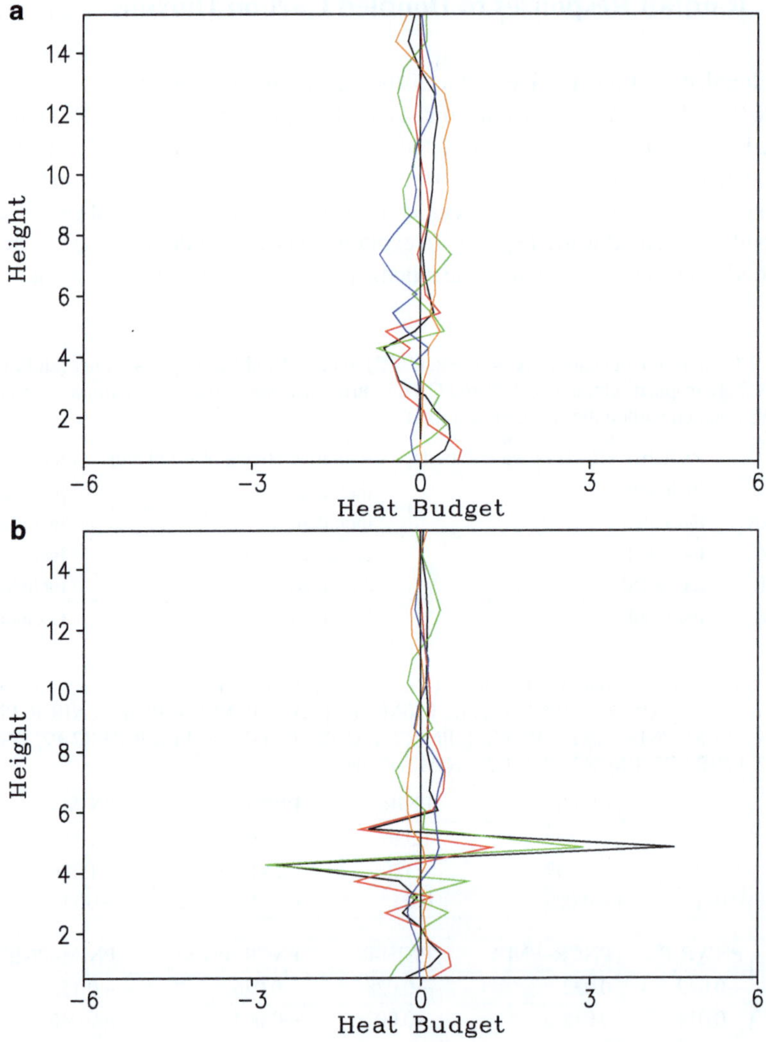

Fig. 9.2 Vertical profiles of differences between (**a**) P and P (P2-P) and (**b**) C2 and C (C2-C) for local temperature change (*black*), condensational heating (*red*), convergence of vertical heat flux (*green*), vertical temperature advection (*blue*), and radiation (*orange*) averaged for 5 days and model domain. Unit is °C d^{-1} (After Li et al. 2014a)

experiments. The difference in the mean radiative cooling results mainly from that in the mean infrared radiative cooling because the difference in the mean solar radiative heating is much smaller than that in the mean infrared radiative cooling (Fig. 9.3). The mean infrared radiative cooling decreases from P to P2 below 13 km, whereas it strengthens from P to P2 above 13 km (Fig. 9.3a). A reduced mean latent heat from 2 to 5.5 km is associated with the weakened mean net condensation (Fig. 9.2a).

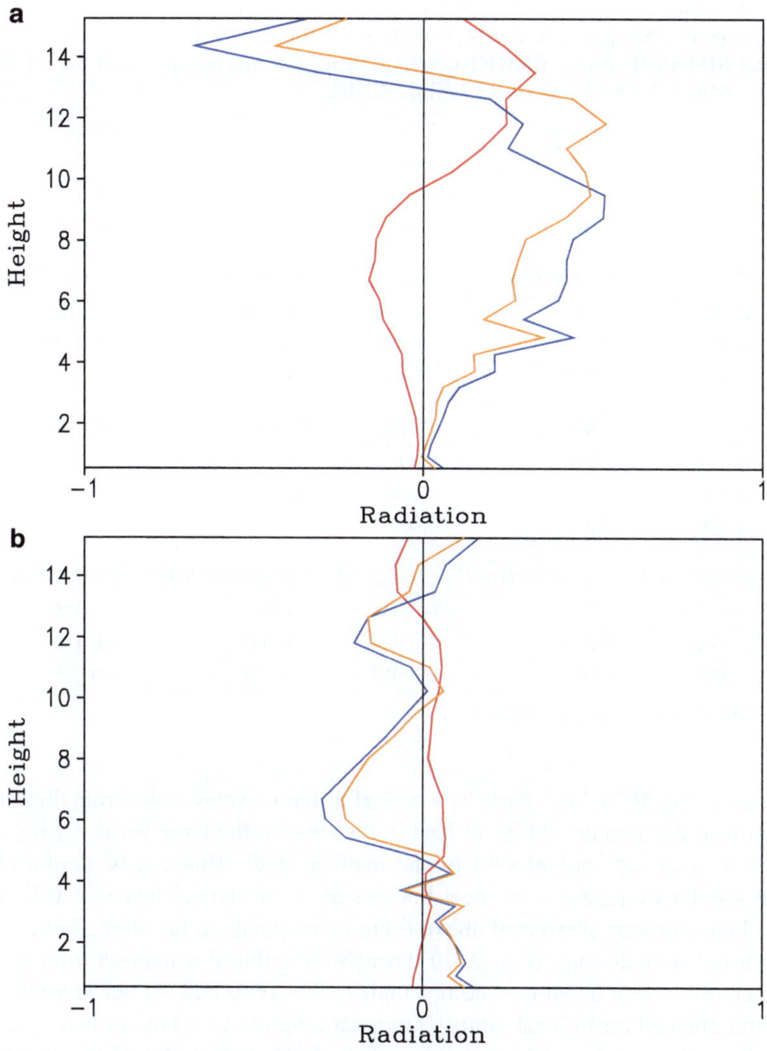

Fig. 9.3 As in Fig. 9.2 except for radiation (*orange*) and its components: solar radiative heating (*red*) and infrared radiative cooling (*blue*) (After Li et al. 2014a)

The reduced mean latent heat corresponds to the decreased mean infrared radiative cooling through the increase in saturation mixing ratio. The weakened melting of cloud ice is related to the enhanced mean infrared radiative cooling above 13 km.

The doubled carbon dioxide increases the mean rain rate from C to C2 (Table 9.3). The enhanced mean rainfall is associated with the strengthened mean net condensation. The analysis of the mean heat budgets (Fig. 9.2b) reveals the enhanced mean latent heat above 6 km and near surface associated with the intensified mean net condensation. Note a large difference in heat budget between 3

Table 9.3 Cloud microphysical budgets (P_S, Q_{NC}, and Q_{CM}) averaged for 5 days over model domain in (a) C, CNWR, CNIR, CNCR, CNIM; (b) CNWR-C, CNCR–CNIR, CNIR-C, CNCR–CNWR, CNIM–CNIR; (c) C2, CNWR2, CNIR2, CNCR2, CNIM2; and (d) CNWR2-C2, CNCR2–CNIR2, CNIR2-C2, CNCR2–CNWR2, CNIM2–CNIR2

(a)	C	CNWR	CNIR	CNCR	CNIM
P_S	0.918	0.919	0.909	0.927	0.648
Q_{NC}	0.877	0.877	0.907	0.910	0.753
Q_{CM}	0.041	0.042	0.002	0.017	−0.105

(b)	CNWR-C	CNCR–CNIR	CNIR-C	PNCR–PNWR	CNIM–CNIR
P_S	0.001	0.018	−0.009	0.008	−0.261
Q_{NC}	0.000	0.003	0.030	0.033	−0.154
Q_{CM}	0.001	0.015	−0.039	−0.025	−0.107

(c)	C2	CNWR2	CNIR2	CNCR2	CNIM2
P_S	0.940	0.916	0.902	0.923	0.667
Q_{NC}	0.915	0.892	0.894	0.900	0.738
Q_{CM}	0.025	0.024	0.008	0.023	−0.071

(d)	CNWR2-C2	CNCR2–CNIR2	CNIM2-C2	CNCR2–CNWR2	CNIM2–CNIR2
P_S	−0.024	0.021	−0.038	0.007	−0.235
Q_{NC}	−0.023	0.006	−0.021	0.008	−0.156
Q_{CM}	−0.001	0.015	−0.017	−0.001	−0.079

Unit is mm h^{-1} (After Li et al. 2014a, b)

and 6 km in Fig. 9.2b. Such large heat budget difference may result from the vertical shift of heat divergence and latent heat in response to the large vertical wind shear (Fig. 4.1b) and the vertical shift of the melting level. This can be demonstrated by the similar averages of the heat budgets at these vertical levels in C2 and C. The enhanced mean latent heat above 6 km corresponds to the strengthened mean infrared radiative cooling (Fig. 9.3b) through the reduced saturation mixing ratio. The increased mean latent heat corresponds to the weakened surface sensible heat. The strengthened mean local atmospheric warming above 6 km is associated with the enhanced mean latent heat that corresponds to the intensified mean infrared radiative cooling.

To determine the greenhouse effect on radiative heating in the mean heat budget, we calculated the radiation flux at the top of the model atmosphere (42 hPa) based on (2) in Lau et al. (1994). The radiation flux at the top of the model atmosphere can be written as

$$N_{toa} = SW_{CR} - \sigma {T_S}^4 + G_a + C_l + C_S, \tag{9.1}$$

where subscript "toa" denotes top of the model atmosphere, "CR" for clear-sky conditions, SW_{CR} is the net absorbed solar radiation averaged over clear-sky regions, $\sigma {T_S}^4$ is the longwave radiation emitted by the surface with $\sigma = 5.67 \times 10^{-8}$

Table 9.4 The difference in the radiation flux at the top of the model atmosphere between P2 and P and C2 and C averaged in 5 days over model domain

	SW_{CR}	$-\sigma T_S{}^4$	G_a	C_l	C_S
P2-P	0.577	0.0	2.374	2.916	10.621
C2-C	3.145	0.0	7.782	−7.414	−13.364

Unit is W m^{-2}

Wm^{-2}K^{-4}, G_a is the atmospheric greenhouse effect, and C_l and C_S are the longwave and shortwave cloud forcing, respectively. Table 9.4 shows that G_a has the same positive values for P2-P and C2-C, whereas C_l and C_S have the opposite signs for P2-P and C2-C. This indicates that the cloud-radiative forcing determines the signs of radiative heating in the mean heat budget.

9.2 Effects of Doubled Carbon Dioxide on Rainfall Responses to Radiative Processes of Water Clouds

In the presence of radiative effects of ice clouds, the exclusion of radiative effects of water clouds reduces the mean pre-summer rain rate from P to PNWR during summer solstice (Table 9.2), whereas it barely changes the mean tropical rain rate from C to CNWR during winter solstice (Table 9.3) in the current carbon dioxide concentration. When carbon dioxide is doubled, the removal of radiative effects of water clouds increases the mean pre-summer rain rate from P2 to PNWR2, whereas it decreases the mean tropical rain rate from C2 to CNWR2. Doubled carbon dioxide changes the mean pre-summer rainfall from the decrease for PNWR-P to the increase for PNWR2-P2, whereas it switches the mean tropical rainfall from an unchanged rainfall for CNWR-C to the decrease for CNWR2-C2.

In the current carbon dioxide concentration, the decrease in the mean pre-summer rain rate caused by the exclusion of radiative effects of water clouds is associated with the mean hydrometeor change from loss in P to gain in PNWR (Table 9.2), but the mean tropical rain rate is not sensitive to radiative effects of water clouds because the mean net condensation and hydrometeor barely change in C and CNWR (Table 9.3). In the doubled carbon dioxide concentration, the increase in the mean pre-summer rain rate resulting from the removal of radiative effects of water clouds corresponds to the strengthened mean hydrometeor loss, but the reduction in the mean tropical rain rate resulting from the elimination of radiative effects of water clouds corresponds to the decrease in the mean net condensation.

Liu et al. (2014) analyzed the difference in Q_{CM} between P and PNWR through the calculations of its components of five cloud species associated with pre-summer torrential rainfall [also see (2.6b, 2.6c, 2.6d, 2.6e, 2.6f, and 2.6g)]. While both experiments have similar Q_{CMR} and Q_{CMS}, the difference in Q_{CM} results from those in Q_{CMC}, Q_{CMI}, and Q_{CMG} (Table 9.5). The vapor condensation rate (P_{CND}) is larger

Table 9.5 Breakdown of Q_{CM} to Q_{CMC}, Q_{CMR}, Q_{CMI}, Q_{CMS}, and Q_{CMG} averaged for 5 days over model domain in (a) P, PNWR, PNIR, PNCR, PNIM; (b) PNWR-P, PNCR–PNIR, PNIR-P, PNCR–PNWR, PNIM–PNIR; (c) P2, PNWR2, PNIR2, PNCR2, PNIM2; and (d) PNWR2-P2, PNCR2–PNIR2, PNIR2-P2, PNCR2–PNWR2, PNIM2–PNIR2

(a)	P	PNWR	PNIR	PNCR	PNIM
Q_{CM}	0.031	−0.025	0.0	0.014	−0.113
Q_{CMC}	−0.001	−0.018	−0.013	−0.010	−0.120
Q_{CMR}	−0.005	−0.005	0.010	0.024	0.011
Q_{CMI}	0.021	−0.003	−0.003	−0.006	0.000
Q_{CMS}	−0.004	−0.003	−0.006	0.000	0.000
Q_{CMG}	0.021	0.004	0.011	0.006	0.000

(b)	PNWR-P	PNCR–PNIR	PNIR-P	PNCR–PNWR	PNIM–PNIR
Q_{CM}	−0.056	0.014	−0.031	0.039	−0.113
Q_{CMC}	−0.017	0.003	−0.012	0.008	−0.107
Q_{CMR}	0.000	0.014	0.015	0.029	0.001
Q_{CMI}	−0.024	−0.003	−0.024	−0.003	0.003
Q_{CMS}	0.001	0.006	−0.002	0.003	0.006
Q_{CMG}	−0.017	−0.005	−0.010	0.002	−0.011

(c)	P2	PNWR2	PNIR2	PNCR2	PNIM2
Q_{CM}	0.008	0.045	0.063	0.050	−0.101
Q_{CMC}	−0.002	0.006	0.013	−0.011	−0.108
Q_{CMR}	0.023	0.041	0.051	0.057	0.008
Q_{CMI}	−0.002	−0.005	−0.003	−0.006	0.000
Q_{CMS}	−0.009	−0.002	0.001	0.007	0.000
Q_{CMG}	−0.002	0.004	0.001	0.003	0.000

(d)	PNWR2-P2	PNCR2–PNIR2	PNIR2-P2	PNCR2–PNWR2	PNIM2–PNIR2
Q_{CM}	0.037	−0.013	0.055	0.005	−0.164
Q_{CMC}	0.008	−0.024	0.015	−0.017	−0.121
Q_{CMR}	0.018	0.006	0.028	0.016	−0.043
Q_{CMI}	−0.003	−0.003	−0.001	−0.001	0.003
Q_{CMS}	0.007	0.006	0.010	0.009	−0.001
Q_{CMG}	0.006	0.002	0.003	−0.001	−0.001

Unit is mm h^{-1} (After Li et al. 2014b)

in PNWR than in P forming a source for cloud water in PNWR (Fig. 9.4). The collection rate of cloud water by rain (P_{RACW}) and the accretion rate of cloud water by graupel [$P_{GACW}(T < T_O)$] are larger in P than in PNWR as a result of the melting of cloud ice to cloud water (P_{IMLT}) in P. Compared to those in P, more vapor condensation in PNWR may be related to more water vapor due to less consumption of water vapor as indicated by less rainfall. Compared to those in PNWR, the melting of cloud ice to cloud water is the microphysical process that is responsible

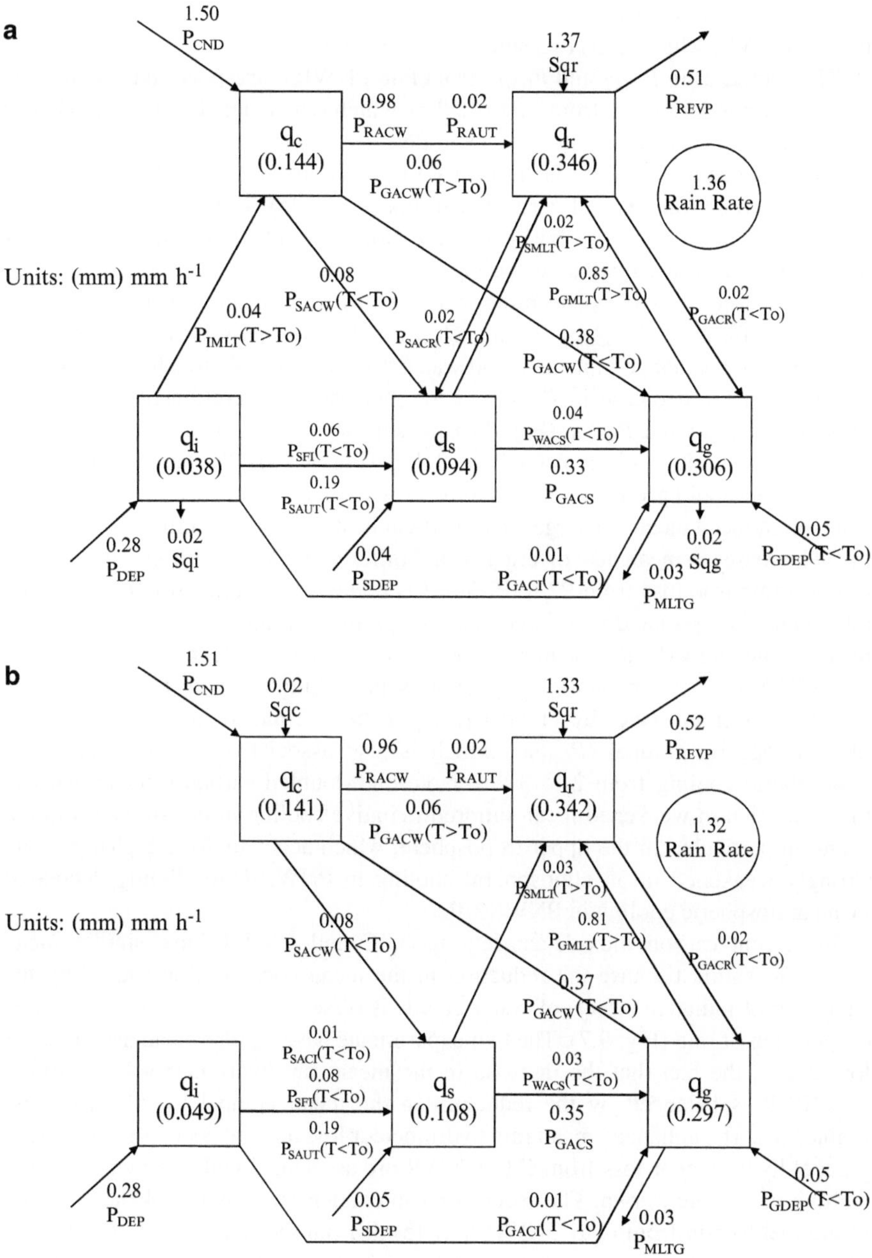

Fig. 9.4 Time-mean cloud microphysics budgets in (**a**) P and (**b**) PNWR. Units for cloud hydrometeors and conversions are mm and mm h^{-1}, respectively. Cloud microphysical conversion terms and their schemes can be found in Table 1.2. $T_0 = 0\ °C$ (After Liu et al. 2014)

for the sink for cloud ice in P. The melting of graupel to rain (P_{GMLT}) is larger in P than in PNWR, which leads to a sink for graupel in P.

The decreases in P_{IMLT} and P_{GMLT} from P to PNWR correspond to the enhanced local atmospheric cooling from 2.5 to 4.5 km, which is associated with the weakened convergence of vertical heat flux from P to PNWR around 4 km and the strengthened heat divergence and reduced latent heat P to PNWR from 2.5 to 4 km in response to intensified convergence of vertical heat flux (Fig. 9.4a). The decrease in infrared radiative cooling from P to PNWR is small. Note that radiation tendency is always determined by infrared radiative cooling as shown in Fig. 9.3.

When carbon dioxide concentration is doubled, the difference in Q_{CM} results from the differences in Q_{CMR}, Q_{CMC}, Q_{CMS}, and Q_{CMG} between P2 and PNWR2 (Table 9.5). The change in mean cloud water from a gain in P2 to a loss in PNWR2 results from the decrease in P_{CND} and the increases in accretion of cloud water by snow (P_{SACW}) and P_{RACW} (Fig. 9.6), which may be associated with weakened mean atmospheric cooling from 2 to 3 km (Fig. 9.5b). The suppressed mean atmospheric cooling corresponds mainly to the weakened release of mean latent heat and reduced mean convergence of vertical heat flux, which further corresponds to the increases in mean heat divergence and infrared radiative cooling. The increase in mean raindrop loss from P2 to PNWR2 is related to the reduction in mean rain rate, which is associated with large mass integration mixing ratio of raindrops as a result of the strengthened mean rain source. The enhanced mean rain source from P2 to PNWR2 corresponds to the increases in melting of snow to rain (P_{SMLT}), accretion of cloud water by graupel [$P_{GACW}(T > T_0)$], and reduction in accretion of raindrops by graupel (P_{GACR}), which is also associated with reduced mean atmospheric cooling from 2 to 3 km. Note that doubled carbon dioxide leads to the slowdown in weakened mean infrared radiative cooling in pre-summer rainfall event, in particular, in the upper troposphere, which accounts for the change from strongly weakened mean atmospheric cooling in PNWR-P to slightly decreased mean atmospheric cooling in PNWR2-P2.

In current carbon dioxide concentration, C and CNWR have similar mean net condensation because the reduction in the mean condensation caused by the exclusion of radiative effects of water clouds is offset by the decrease in the mean evaporation of rain (Fig. 9.7). The two experiments have similar mean hydrometeor loss due to the fact that the increase in the mean rain hydrometeor loss from C to CNWR is balanced by the reduction in the mean graupel hydrometeor loss (Table 9.6). The enhanced mean rain hydrometeor loss and the decrease in the mean graupel hydrometeor loss from C to CNWR are associated with the reduced mean melting of graupel to rain. The reduced mean condensation and weakened melting of graupel to rain from C to CNWR may result from the strengthened mean local atmospheric cooling between 3 and 4.5 km, which is associated with the reduced convergence of vertical heat flux, weakened release of the mean latent heat, and enhanced mean infrared radiative cooling (Fig. 9.8a).

When carbon dioxide concentration is doubled, the weakened mean net condensation from C2 to CNWR2 results from the decreases in the mean vapor condensation and deposition to cloud ice (Fig. 9.9), which is associated with the

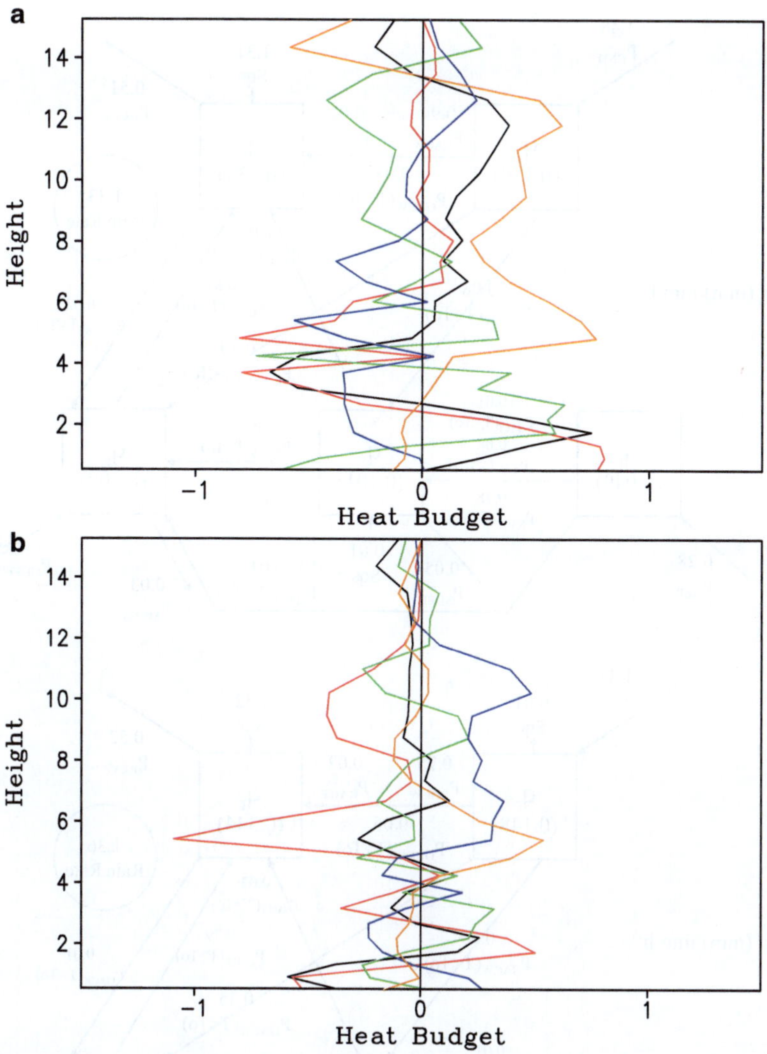

Fig. 9.5 Vertical profiles of differences between (**a**) PNWR and P (PNWR-P) and (**b**) PNWR2 and P2 (PNWR2-P2) for local temperature change (*black*), condensational heating (*red*), convergence of vertical heat flux (*green*), vertical temperature advection (*blue*), and radiation (*orange*) averaged for 5 days and model domain. Unit is °C d^{-1} (After Liu et al. 2014)

suppressed release of the mean latent heat between 2 and 4.5 km and above 5.5 km (Fig. 9.8b). The weakened release of the mean latent heat corresponds to the enhanced mean convergence of vertical heat flux between 2 and 4.5 km and the reduced mean infrared radiative cooling above 5.5 km. The decrease in the mean raindrop hydrometeor loss from C2 to CNWR2 and the change in the mean cloud water and graupel from a gain in C2 to a loss in CNWR2 lead to the similar mean

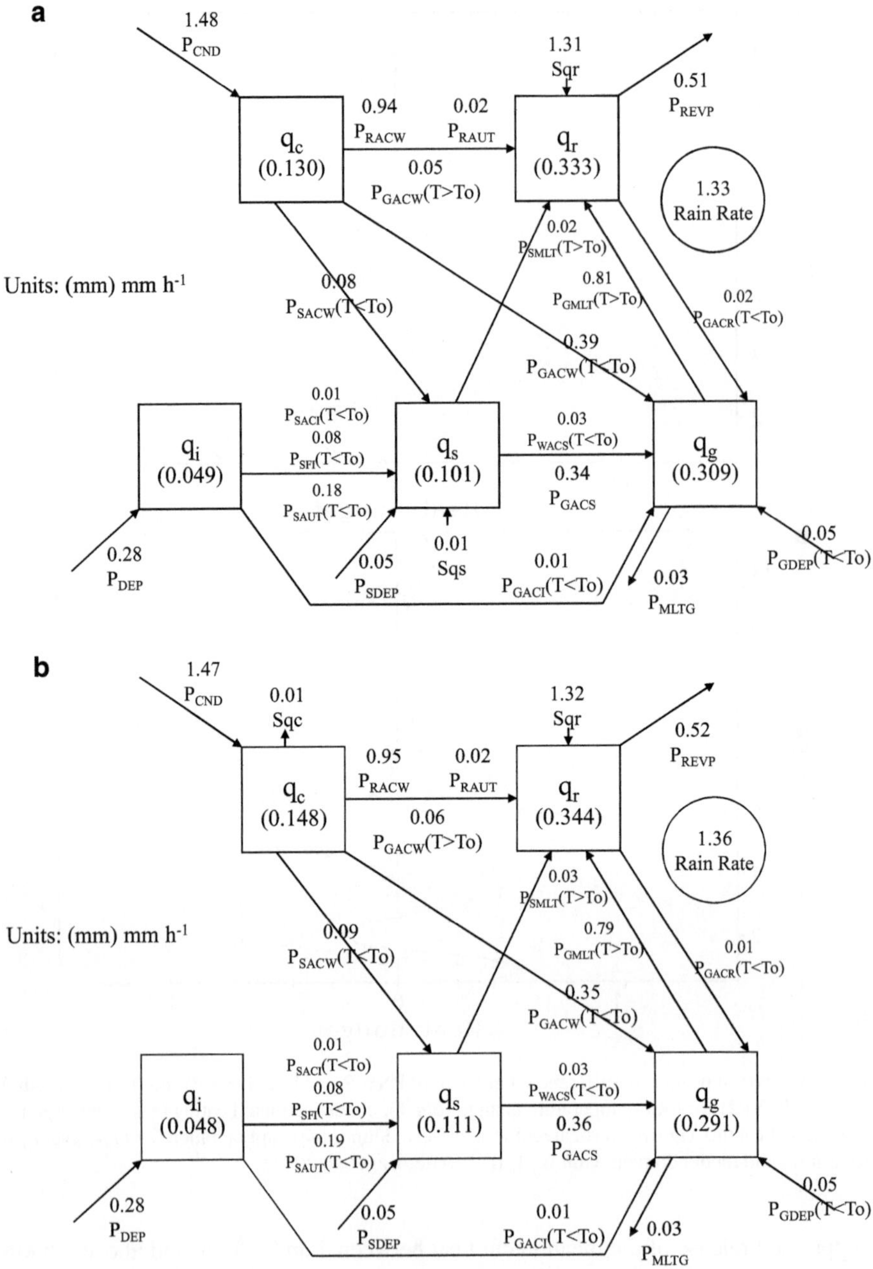

Fig. 9.6 As in Fig. 9.4 except for those in (**a**) P2 and (**b**) PNWR2 (After Li et al. 2014b)

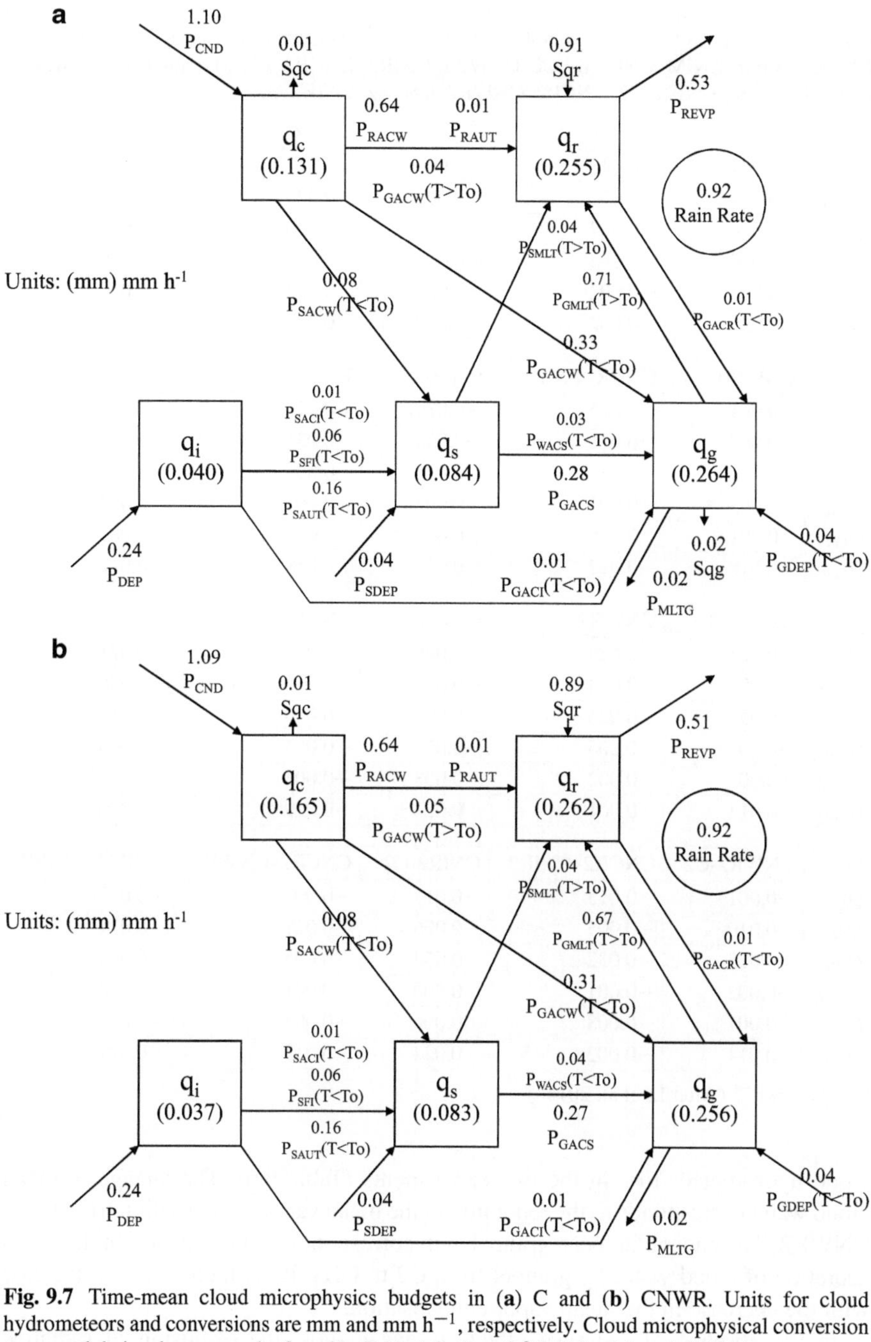

Fig. 9.7 Time-mean cloud microphysics budgets in (**a**) C and (**b**) CNWR. Units for cloud hydrometeors and conversions are mm and mm h^{-1}, respectively. Cloud microphysical conversion terms and their schemes can be found in Table 1.2. $T_0 = 0$ °C (After Li et al. 2014b)

Table 9.6 Breakdown of Q_{CM} to Q_{CMC}, Q_{CMR}, Q_{CMI}, Q_{CMS}, and Q_{CMG} averaged for 5 days over model domain in (a) C, CNWR, CNIR, CNCR, CNIM; (b) CNWR-C, CNCR–CNIR, CNIR-C, CNCR–CNWR, CNIM–CNIR; (c) C2, CNWR2, CNIR2, CNCR2, CNIM2; and (d) CNWR2-C2, CNCR2–CNIR2, CNIR2-C2, CNCR2–CNWR2, CNIM2–CNIR2

(a)	C	CNWR	CNIR	CNCR	CNIM
Q_{CM}	0.041	0.042	0.002	0.017	−0.105
Q_{CMC}	0.014	0.011	−0.009	−0.020	−0.129
Q_{CMR}	0.010	0.029	−0.016	0.020	0.025
Q_{CMI}	0.000	0.003	−0.003	−0.004	0.000
Q_{CMS}	−0.003	−0.002	−0.002	0.003	0.000
Q_{CMG}	0.020	0.002	0.032	0.020	0.000

(b)	CNWR-C	CNCR–CNIR	CNIR-C	PNCR–PNWR	CNIM–CNIR
Q_{CM}	0.001	0.015	−0.039	−0.025	−0.107
Q_{CMC}	−0.003	−0.011	−0.023	−0.031	−0.120
Q_{CMR}	0.019	0.036	−0.026	−0.009	0.041
Q_{CMI}	0.003	−0.001	−0.003	−0.007	0.003
Q_{CMS}	0.001	0.005	0.001	0.005	0.002
Q_{CMG}	−0.018	−0.012	0.012	0.018	−0.032

(c)	C2	CNWR2	CNIR2	CNCR2	CNIM2
Q_{CM}	0.025	0.024	0.008	0.023	−0.071
Q_{CMC}	−0.006	0.004	−0.012	−0.017	−0.086
Q_{CMR}	0.050	0.025	−0.002	0.030	0.016
Q_{CMI}	−0.003	−0.005	−0.003	−0.004	0.000
Q_{CMS}	−0.003	−0.002	0.003	−0.005	0.000
Q_{CMG}	−0.013	0.001	0.021	0.019	0.000

(d)	CNWR2-C2	CNCR2–CNIR2	CNIR2-C2	CNCR2–CNWR2	CNIM2–CNIR2
Q_{CM}	−0.001	0.015	−0.017	−0.001	−0.079
Q_{CMC}	0.010	−0.005	−0.006	−0.021	−0.074
Q_{CMR}	−0.025	0.032	−0.052	0.005	0.018
Q_{CMI}	−0.002	−0.001	0.000	0.001	0.003
Q_{CMS}	0.001	−0.008	0.006	−0.003	−0.003
Q_{CMG}	0.014	−0.002	0.034	0.018	−0.021

Unit is mm h^{-1} (After Li et al. 2014b)

cloud hydrometeor loss in the two experiments (Table 9.6). The change in mean cloud water corresponds to the reduction in the mean vapor condensation from C2 to CNWR2. The change in mean graupel gain corresponds to the decrease in the mean accretion of cloud water by graupel from C2 to CNWR2, which is also associated with the decrease in the mean vapor condensation.

In the absence of radiative effects of ice clouds, the exclusion of radiative effects of water clouds increases the mean rainfall through the enhancement in

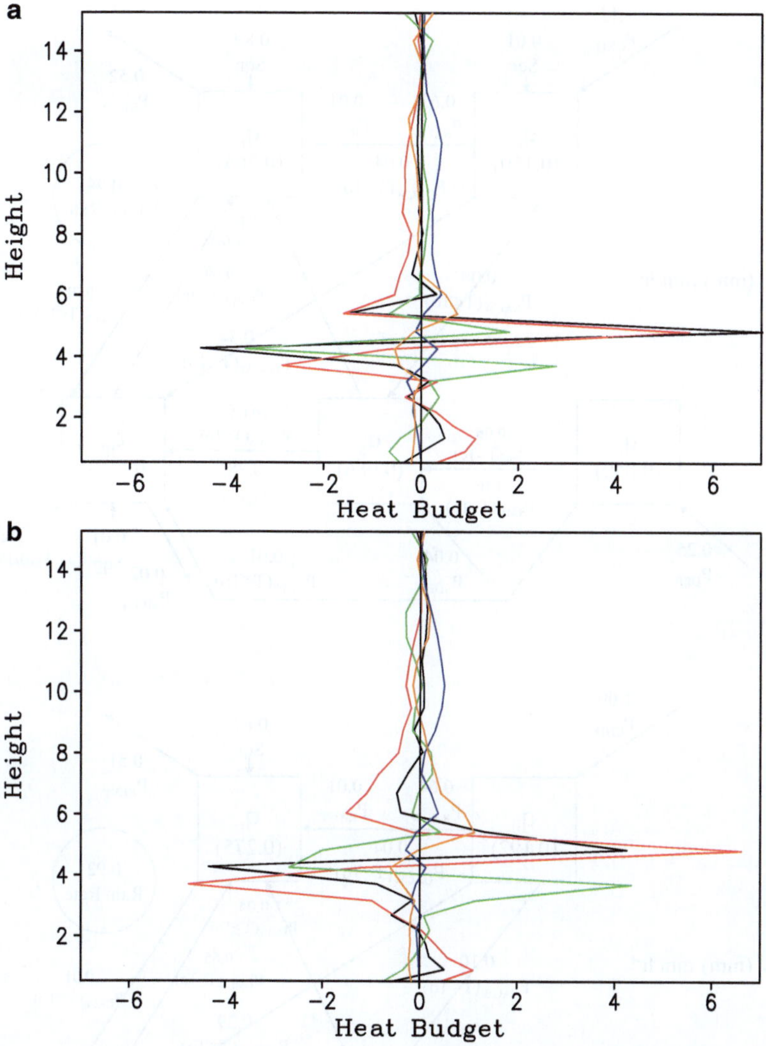

Fig. 9.8 Vertical profiles of differences between (**a**) CNWR and C (CNWR-C) and (**b**) CNWR2 and C2 (CNWR2-C2) for local temperature change (*black*), condensational heating (*red*), convergence of vertical heat flux (*green*), vertical temperature advection (*blue*), and radiation (*orange*) averaged for 5 days and model domain. Unit is °C d^{-1} (After Li et al. 2014b)

the mean net condensation and the mean hydrometeor loss from PNIR to PNCR in pre-summer rainfall event and the intensification in the mean hydrometeor loss from CNIR to CNCR in tropical rainfall event under the current carbon dioxide concentration (Tables 9.2 and 9.3). When carbon dioxide concentration is doubled, the removal of radiative effects of water clouds decreases the mean

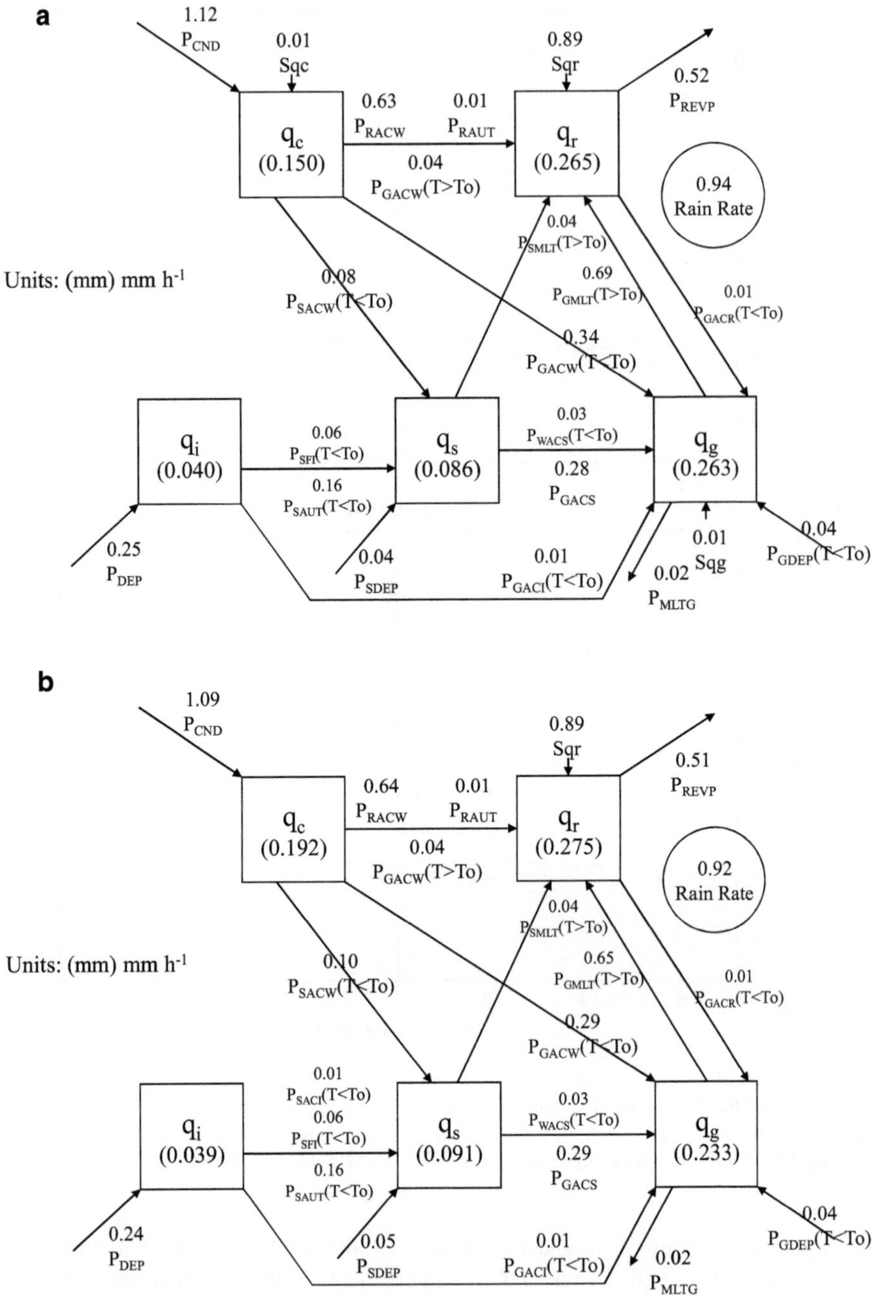

Fig. 9.9 As in Fig. 9.7 except for those in (**a**) C2 and (**b**) CNWR2 (After Li et al. 2014b)

pre-summer rainfall through the suppressed mean hydrometeor loss from PNIR2 to PNCR2, but it enhances the mean tropical rainfall through the strengthened mean hydrometeor loss from CNIR2 to CNCR2. Doubled carbon dioxide changes the mean pre-summer rainfall from the increase for PNCR–PNIR to the decrease for PNCR2–PNIR2, whereas it barely changes the increases in mean tropical rainfall for CNWR-C and CNWR2-C2.

The increase in the mean net condensation from PNIR to PNCR results from the weakened mean evaporation of rain (Fig. 9.10). The strengthened mean hydrometeor loss is caused by the enhanced mean rain hydrometeor loss (Table 9.5), which is due to the increase in the mean rainfall (Fig. 9.10). The strengthened mean rainfall is associated with the increase in mass-integrated rain hydrometeor mixing ratio from PNIR to PNCR, which corresponds to the increase in *Sqr*. The increased *Sqr* is also related to the reduced mean evaporation of rain. Since the evaporation of rain is determined by the difference between specific humidity and saturation specific humidity (Rutledge and Hobbs 1983), the reduced mean evaporation of rain may be related to the increased saturation specific humidity. The increased saturation specific humidity can be demonstrated by the strengthened mean local atmospheric warming from 1.5 to 4 km, which is associated with the enhanced release of the mean latent heat, corresponding to the strengthened mean infrared radiative cooling (Fig. 9.11a).

The reduced mean hydrometeor loss from PNIR2 to PNCR2 results from the mean cloud water loss in PNIR2 to the mean cloud water gain in PNCR2 (Table 9.5), which is associated with the reduction in P_{RACW} (Fig. 9.12). The weakened P_{RACW} corresponds to the reduced P_{CND} and reduced release of the mean latent heat below 3 km (Fig. 9.11b). The suppressed release of the mean latent heat and the enhanced mean infrared radiative cooling are partly offset by enhanced mean convergence of vertical heat flux and reduced mean heat divergence.

The strengthened mean hydrometeor loss from CNIR to CNCR is related to the mean rain hydrometeor change from a gain in CNIR to a loss in CNCR (Table 9.6), which corresponds to the reduced mean *Sqr* (Fig. 9.13). The weakened mean *Sqr* is mainly associated with the suppressed P_{GMLT} caused by the enhanced mean local atmospheric warming around 5 km (Fig. 9.14a). The strengthened mean local atmospheric warming is determined by the increased release of the mean latent heat.

The strengthened mean hydrometeor loss from CNIR2 to CNCR2 results from the change in mean rain hydrometeor from a gain in CNIR2 to a loss in CNCR2 (Table 9.6). The change in mean rain hydrometeor corresponds to the enhanced mean rainfall caused by the increased mass-integrated rain mixing ratio, which is associated with the enhanced mean collection of cloud water by rain caused by the enhanced mean vapor condensation (Fig. 9.15). Like CNIR and CNCR, the enhanced mean vapor condensation is caused by the enhanced mean infrared radiative cooling (Fig. 9.14b).

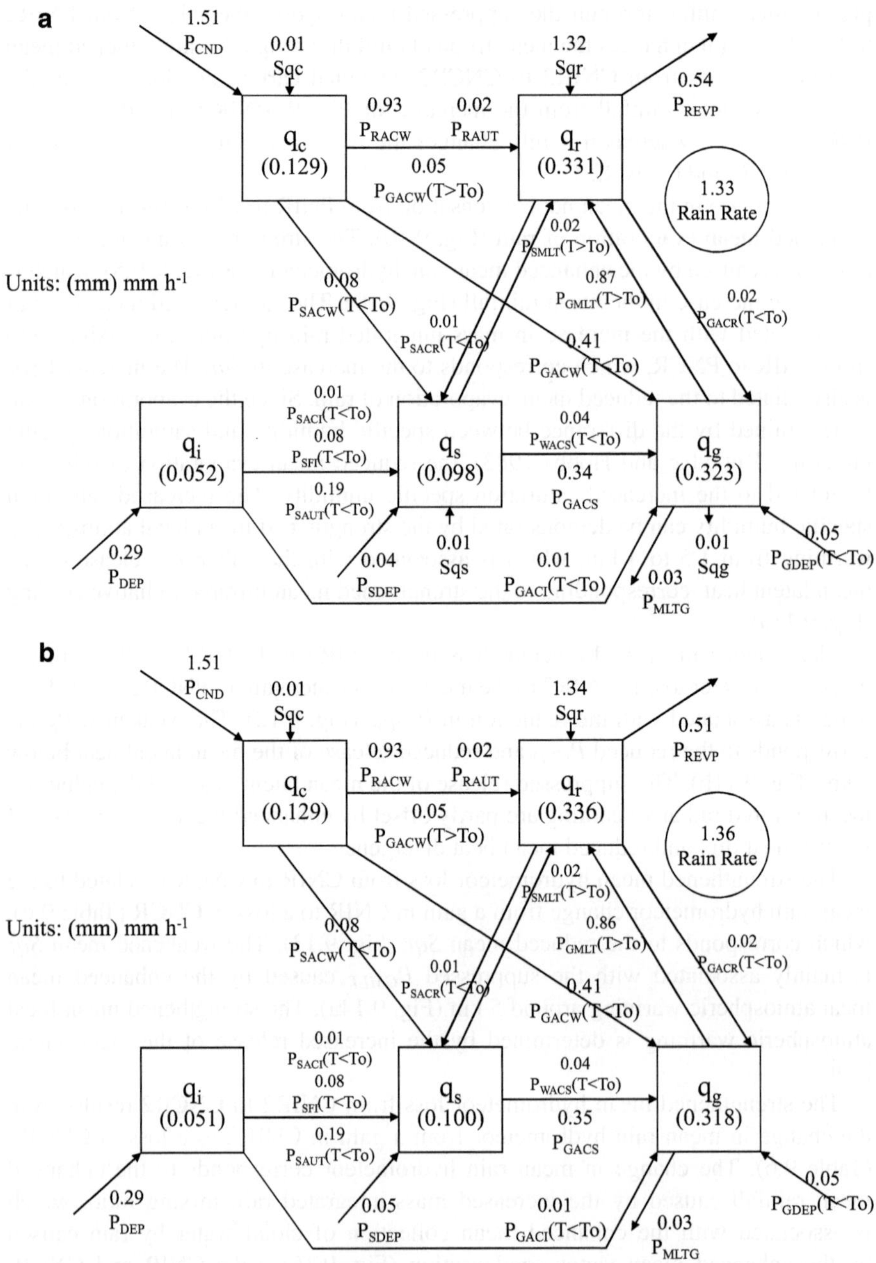

Fig. 9.10 Time-mean cloud microphysics budgets in (**a**) PNIR and (**b**) PNCR. Units for cloud hydrometeors and conversions are mm and mm h^{-1}, respectively. Cloud microphysical conversion terms and their schemes can be found in Table 1.2. $T_0 = 0\ °C$ (After Li et al. 2014b)

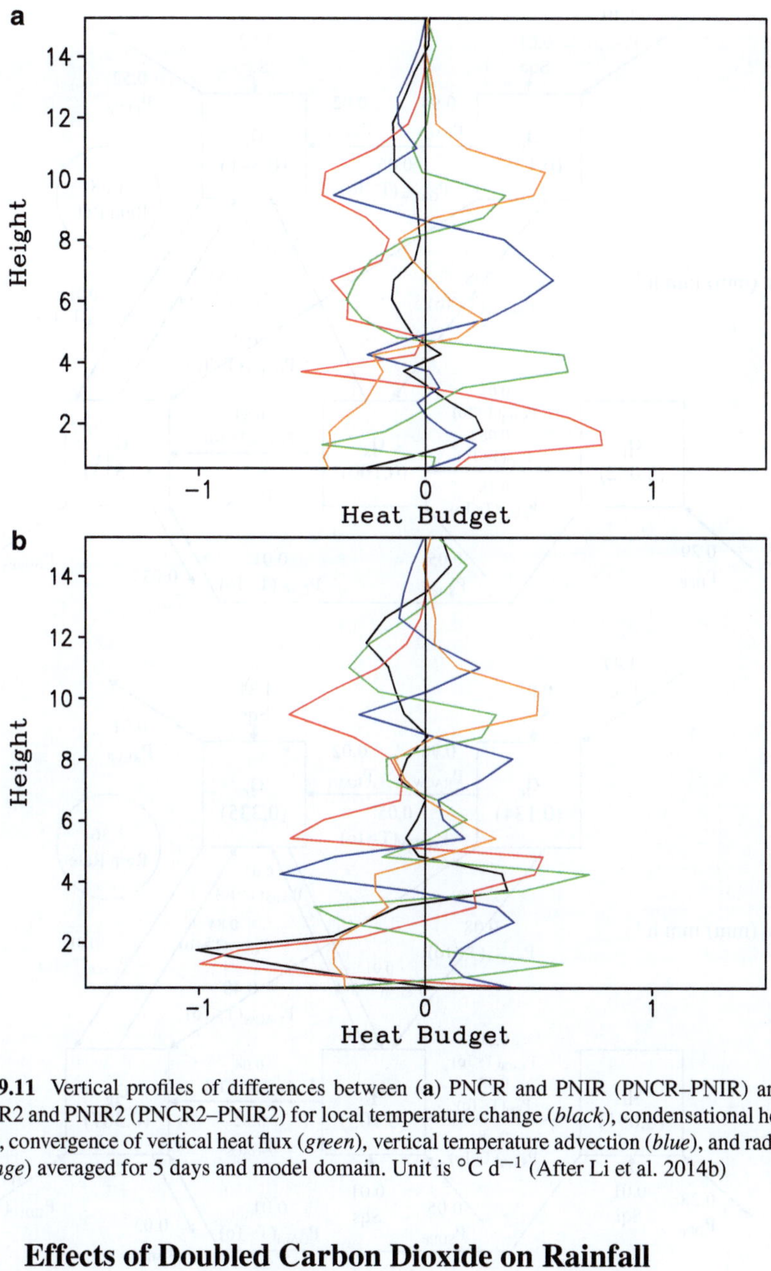

Fig. 9.11 Vertical profiles of differences between (**a**) PNCR and PNIR (PNCR–PNIR) and (**b**) PNCR2 and PNIR2 (PNCR2–PNIR2) for local temperature change (*black*), condensational heating (*red*), convergence of vertical heat flux (*green*), vertical temperature advection (*blue*), and radiation (*orange*) averaged for 5 days and model domain. Unit is °C d^{-1} (After Li et al. 2014b)

9.3 Effects of Doubled Carbon Dioxide on Rainfall Responses to Radiative Processes of Ice Clouds

In the presence of radiative effects of water clouds, the exclusion of radiative effects of ice clouds decreases the mean pre-summer rain rate from P to PNIR through the suppressed hydrometeor loss (Table 9.2), whereas it barely changes the

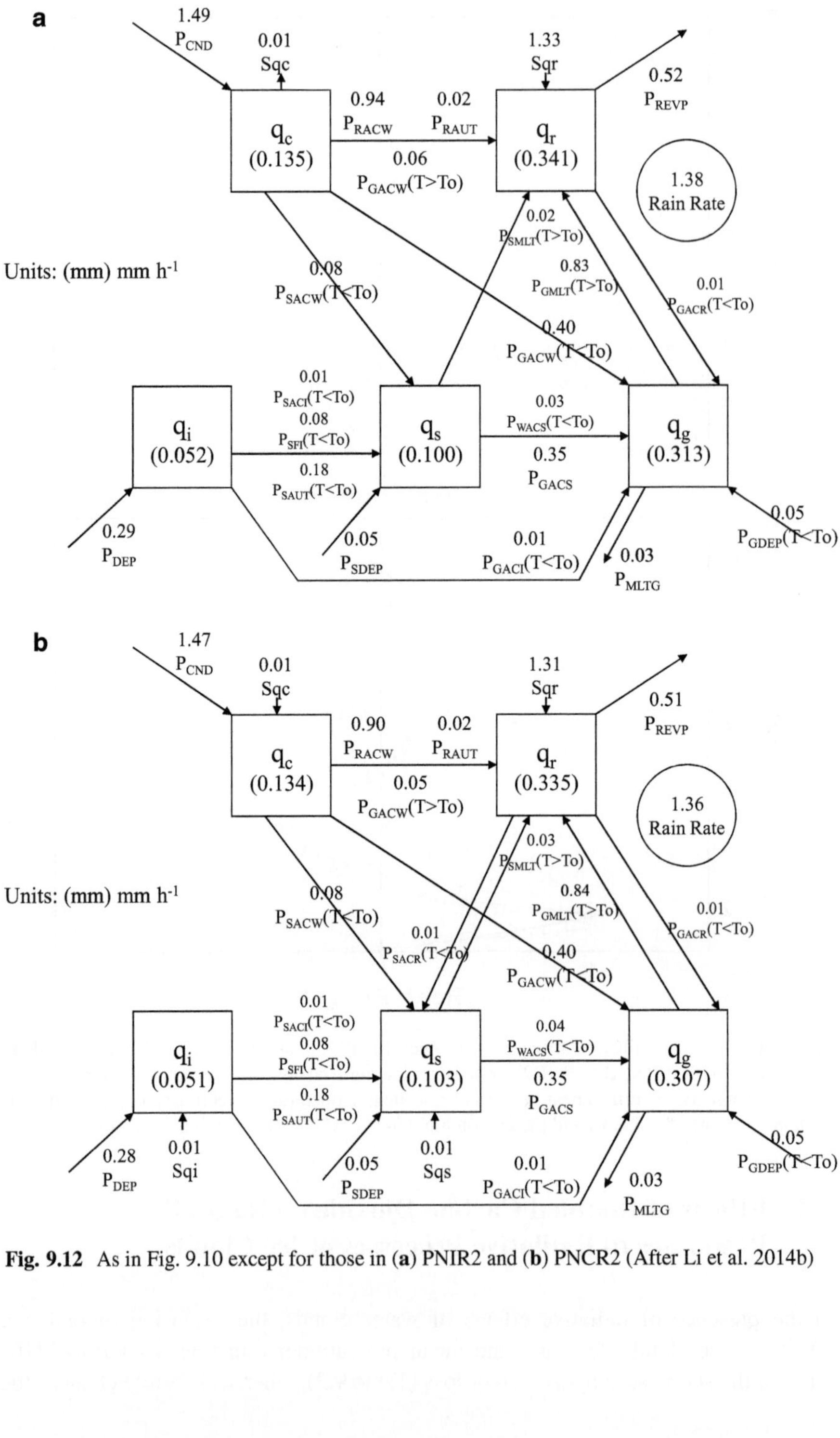

Fig. 9.12 As in Fig. 9.10 except for those in (**a**) PNIR2 and (**b**) PNCR2 (After Li et al. 2014b)

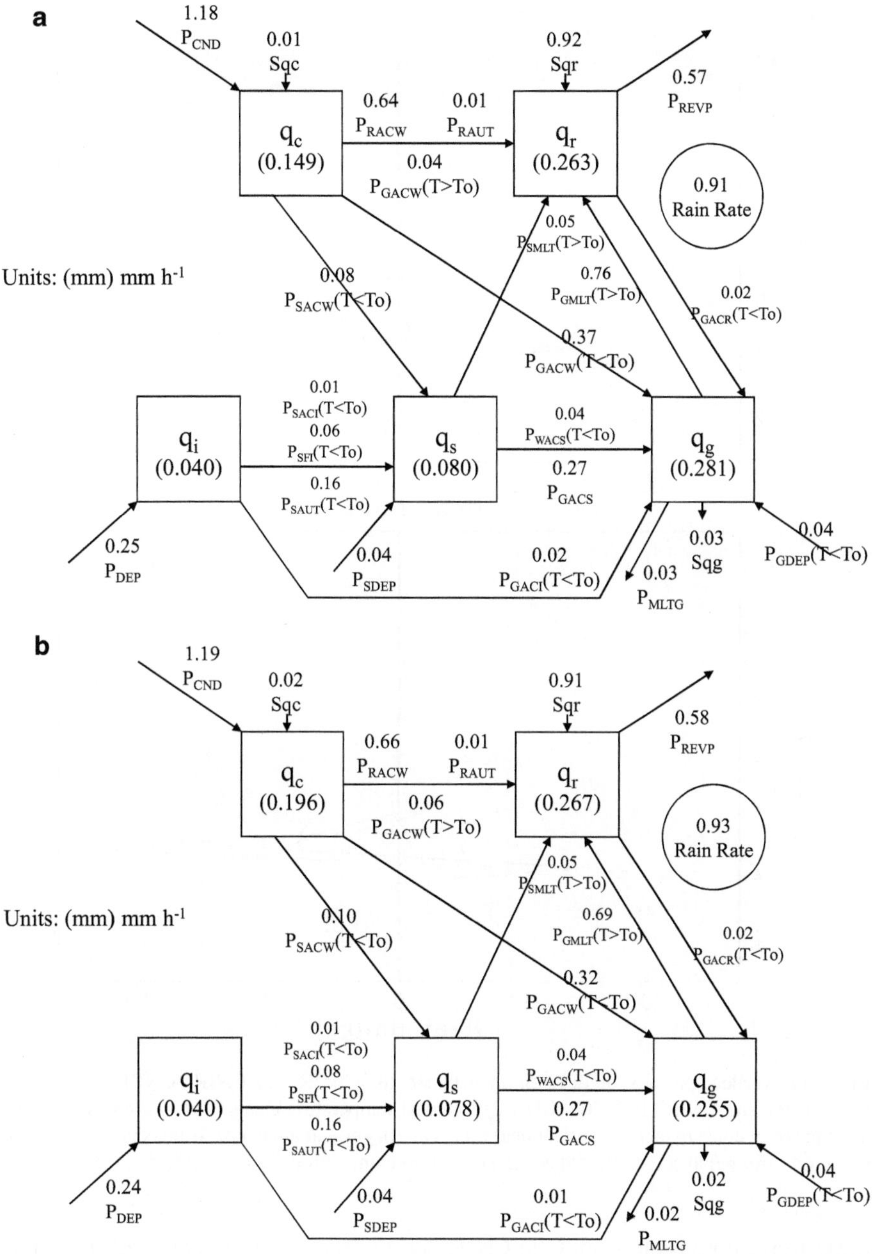

Fig. 9.13 Time-mean cloud microphysics budgets in (**a**) CNIR and (**b**) CNCR. Units for cloud hydrometeors and conversions are mm and mm h^{-1}, respectively. Cloud microphysical conversion terms and their schemes can be found in Table 1.2. $T_0 = 0$ °C (After Li et al. 2014b)

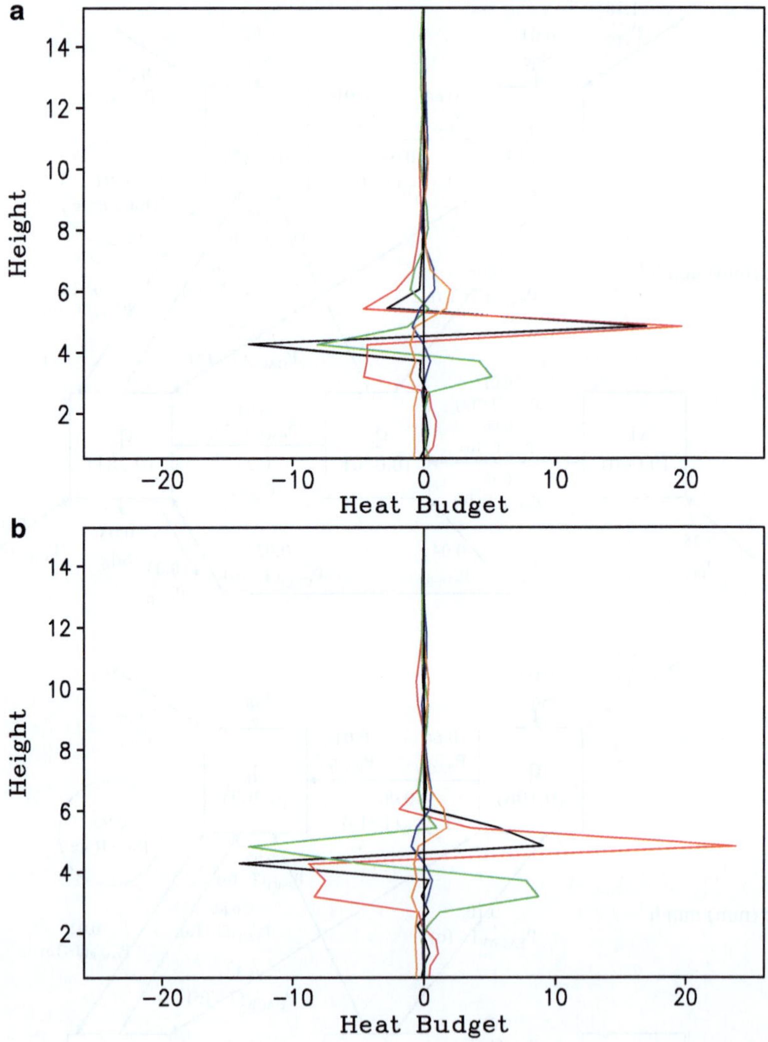

Fig. 9.14 Vertical profiles of differences between (**a**) CNCR and CNIR (CNCR–CNIR) and (**b**) CNCR2 and CNIR2 (CNCR2–CNIR2) for local temperature change (*black*), condensational heating (*red*), convergence of vertical heat flux (*green*), vertical temperature advection (*blue*), and radiation (*orange*) averaged for 5 days and model domain. Unit is °C d^{-1}. (After Li et al. 2014b)

mean tropical rain rate from C to CNIR because the enhanced net condensation is offset by the suppressed hydrometeor loss (Table 9.3) in the current carbon dioxide concentration. When carbon dioxide is doubled, the removal of radiative effects of ice clouds increases the mean pre-summer rain rate from P2 to PNIR2 via the enhanced hydrometeor loss, whereas it decreases the mean tropical rain rate from C2 to CNIR2 through the reductions in net condensation and hydrometeor loss.

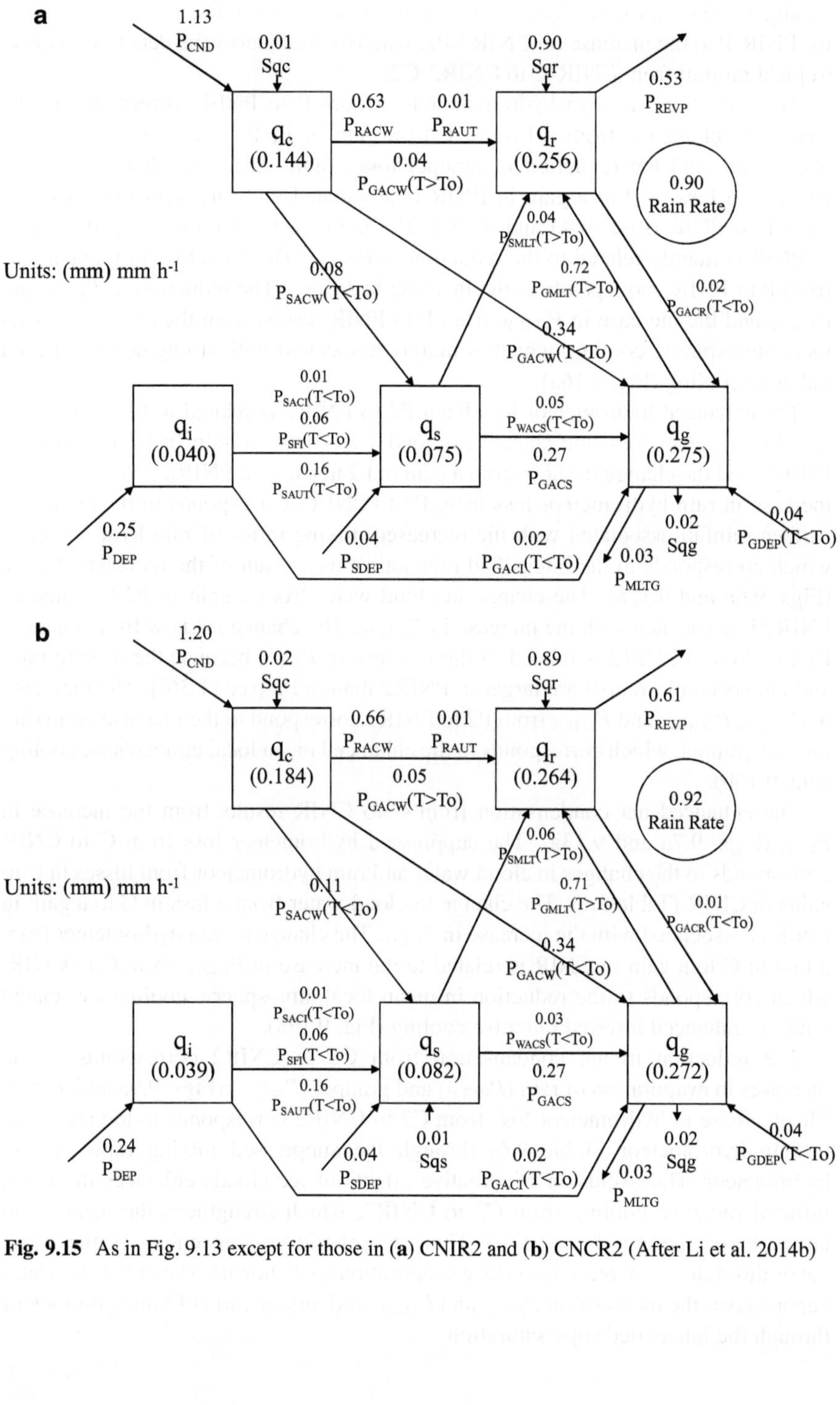

Fig. 9.15 As in Fig. 9.13 except for those in (**a**) CNIR2 and (**b**) CNCR2 (After Li et al. 2014b)

Doubled carbon dioxide changes the mean pre-summer rainfall from the decrease for PNIR-P to the increase for PNIR2-P2, whereas it enhances the decrease in mean tropical rainfall from CNIR-C to CNIR2-C2.

The reduction in mean hydrometeor loss from P to PNIR corresponds to the change in cloud ice from a loss in P to a gain in PNIR, the increase in cloud water gain, and the reduction in graupel loss (Table 9.5). The change in cloud ice from a loss in P to a gain in PNIR is associated with the reduction in P_{IMLT} from P to PNIR (Figs. 9.4a and 9.10a). The increase in cloud water gain from P to PNIR is mainly related to the reduction in P_{RACW}. The reduction in graupel loss from P to PNIR corresponds to the increase in P_{GACW}. The reduction in P_{IMLT} and P_{RACW} and the increase in P_{GACW} from P to PNIR results from the enhanced mean local atmospheric cooling, which is mainly associated with strengthened infrared radiative cooling (Fig. 9.16a).

The enhanced hydrometeor loss from P2 to PNIR2 is related to the increase in rain hydrometeor loss, the change in cloud water from a gain in P2 to a loss in PNIR2, and the change in snow from a gain in P2 to a loss in PNIR2 (Table 9.5). The increase in rain hydrometeor loss from P2 to PNIR2 corresponds to the enhanced surface rainfall associated with the increased mixing ratios of rain hydrometeors, which corresponds to the intensified rain source as a result of the increased P_{GMLT} (Figs. 9.6a and 9.12a). The change in cloud water from a gain in P2 to a loss in PNIR2 is associated with the increase in P_{GACW}. The change in snow from a gain in P2 to a loss in PNIR2 is related to the increase in P_{GACS} because the mixing ratio and fall speed of graupel are larger in PNIR2 than in P2 [see (1.36)]. The increases in P_{GMLT}, P_{GACW}, and P_{GACS} from P2 to PNIR2 correspond to the increase in mixing ratio of graupel, which corresponds to the enhanced mean local atmospheric cooling (Fig. 9.16b).

The enhanced net condensation from C to CNIR results from the increase in P_{CND} (Figs. 9.7a and 9.13a). The suppressed hydrometeor loss from C to CNIR corresponds to the changes in cloud water and rain hydrometeor from losses in C to gains in CNIR (Table 9.6). The change in cloud water from a loss in C to a gain in CNIR is associated with the increase in P_{CND}. The change in rain hydrometeor from a loss in C to a gain in CNIR is related to the increase in P_{GMLT} from C to CNIR, which corresponds to the reduction in mean local atmospheric cooling associated with the enhanced infrared radiative cooling (Fig. 9.17a).

The reduction in net condensation from C2 to CNIR2 corresponds to the increases in evaporation of rain (P_{REVP}) and graupel (P_{MLTG}) (Figs. 9.9a and 9.15a). The decrease in hydrometeor loss from C2 to CNIR2 corresponds to the reduction in rain hydrometeor (Table 9.6) through the suppressed mixing ratio of rain hydrometeor. The exclusion of radiative effects of ice clouds enhances the mean infrared radiative cooling from C2 to CNIR2, which strengthens the mean local atmospheric cooling (Fig. 9.17b). The atmospheric cooling enhances the water vapor through the increase in surface evaporation (not shown). The enhanced water vapor causes the increases in P_{REVP} and P_{MLTG} and mixing ratio of rain hydrometeor through the intensified supersaturation.

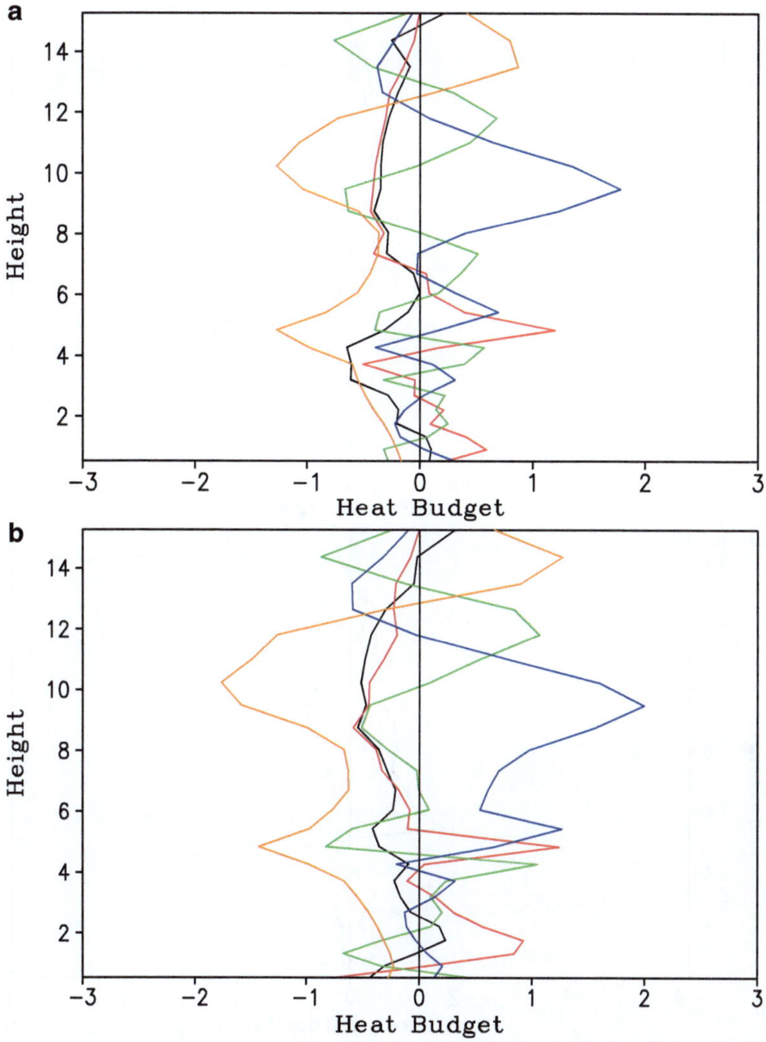

Fig. 9.16 Vertical profiles of differences between (**a**) PNIR and P (PNIR-P) and (**b**) PNIR2 and P2 (PNIR2-P2) for local temperature change (*black*), condensational heating (*red*), convergence of vertical heat flux (*green*), vertical temperature advection (*blue*), and radiation (*orange*) averaged for 5 days and model domain. Unit is °C d^{-1}

In the absence of radiative effects of water clouds, the exclusion of radiative effects of ice clouds increases the mean pre-summer rainfall through the change in mean hydrometeor from a gain in PNWR to a loss in PNCR (Table 9.2), whereas it barely changes the mean tropical rainfall from CNWR to CNCR because the enhanced net condensation is offset by the suppressed hydrometeor loss and the

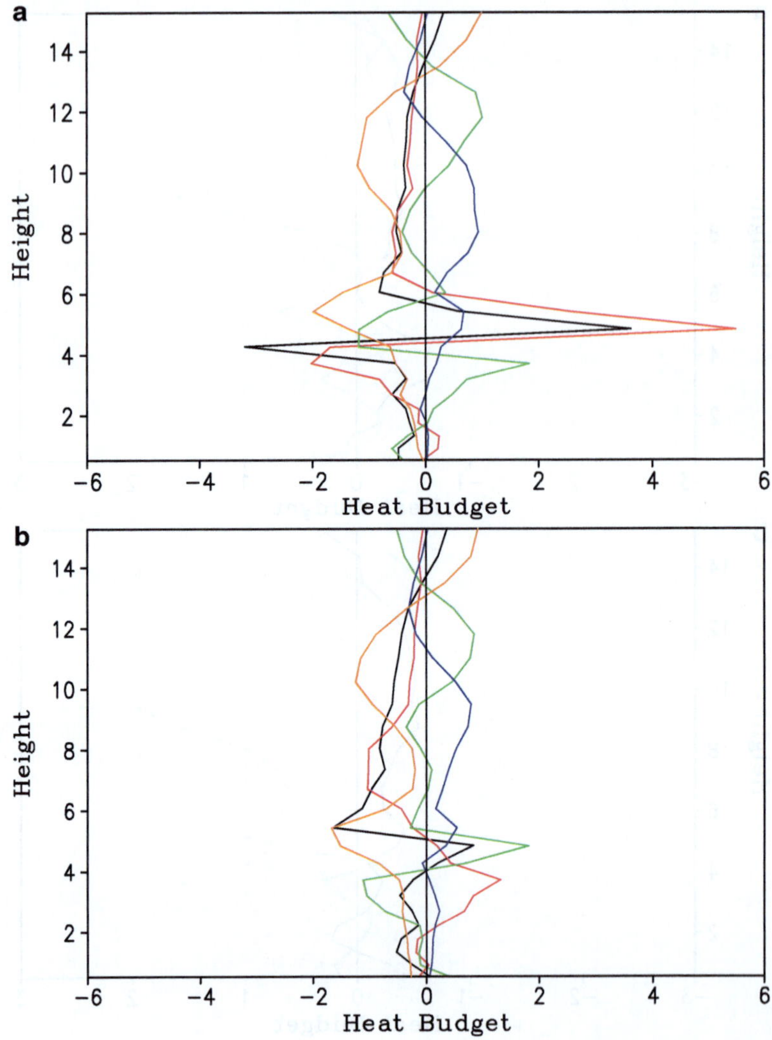

Fig. 9.17 Vertical profiles of differences between (**a**) CNIR and C (CNIR-C) and (**b**) CNIR2 and C2 (CNIR2-C2) for local temperature change (*black*), condensational heating (*red*), convergence of vertical heat flux (*green*), vertical temperature advection (*blue*), and radiation (*orange*) averaged for 5 days and model domain. Unit is °C d^{-1}

intensification in the mean hydrometeor loss from CNIR to CNCR (Table 9.3) under the current carbon dioxide concentration. When carbon dioxide concentration is doubled, the removal of radiative effects of ice clouds barely changes the mean pre-summer and tropical rainfall due to similar net condensation and hydrometeor loss in PNWR2 to PNCR2 and in CNWR2 to CNCR2. Doubled carbon dioxide weakens the increase in mean pre-summer rainfall from PNCR–PNWR to PNCR2–PNWR2, whereas the mean tropical rainfall is insensitive to radiative processes of ice clouds and carbon dioxide concentration in CNCR, CNWR, CNCR2, and CNWR2.

The change in mean hydrometeor from a gain in PNWR to a loss in PNCR corresponds to the change in rain hydrometeor from a gain in PNWR to a loss in PNCR (Table 9.5) through the enhanced surface rainfall associated with the increased rain source (Figs. 9.4b and 9.10b). The increased rain source from PNWR to PNCR corresponds to the strengthened P_{GMLT} through the enhanced mixing ratio of graupel, which is associated with the strengthened mean local atmospheric cooling as a result of the enhanced infrared radiative cooling (Fig. 9.18a).

The similar mean net condensation rates in PNWR2 and PNCR2 correspond to similar vapor condensation and deposition and evaporation of rain (Figs. 9.6b and 9.12b). The hydrometeor loss in PNWR2 is similar to that in PNCR2 because the change in cloud water from a loss in PNWR2 to a gain in PNCR2 is offset by the enhanced rain hydrometeor loss from PNWR2 toPNCR2 (Table 9.5). The change in cloud water from a loss in PNWR2 to a gain in PNCR2 corresponds mainly to the weakened P_{RACW}. The enhanced rain hydrometeor loss from PNWR2 toPNCR2 is related to the reduced rain source associated with the weakened P_{RACW}. The reduced P_{RACW} from PNWR2 toPNCR2 corresponds to the decreases in mixing ratios of cloud water and rain hydrometeor, which are associated with the enhanced mean local atmospheric cooling as a result of the strengthened mean infrared radiative cooling (Fig. 9.18b).

The enhanced net condensation from CNWR to CNCR corresponds to the increased P_{CND} (Figs. 9.7b and 9.13b). The suppressed hydrometeor loss from CNWR to CNCR is associated with the change in cloud water from a loss in CNWR to a gain in CNCR (Table 9.6) through the increased P_{CND}. The increased P_{CND} from CNWR to CNCR is related to the enhanced release of mean latent heat in response to the strengthened mean infrared radiative cooling (Fig. 9.19a).

The mean net condensation rates are similar in CNWR2 and CNCR2 because the enhanced P_{CND} from CNWR2 to CNCR2 is offset by the increased P_{REVP} (Figs. 9.9b and 9.15b). The rates of hydrometeor loss are similar in CNWR2 and CNCR2 because the change in cloud water from a loss in CNWR2 to a gain in CNCR2 is canceled by the increase in graupel loss from CNWR2 to CNCR2 (Table 9.6). The change in cloud water from a loss in CNWR2 to a gain in CNCR2 corresponds to the increased P_{CND} from CNWR2 to CNCR2. The increase in graupel loss from CNWR2 to CNCR2 corresponds to the strengthened P_{GMLT}. The enhanced P_{CND} is associated with the increased release of mean latent heat around 5 km, which corresponds to the intensified mean infrared radiative cooling (Fig. 9.19b). The increase in P_{REVP} results from the weakened saturation specific humidity associated with the strengthened mean infrared radiative cooling. The strengthened P_{GMLT} results from the increase in mixing ratio of graupel from CNWR2 to CNCR2 associated with the enhanced mean infrared radiative cooling.

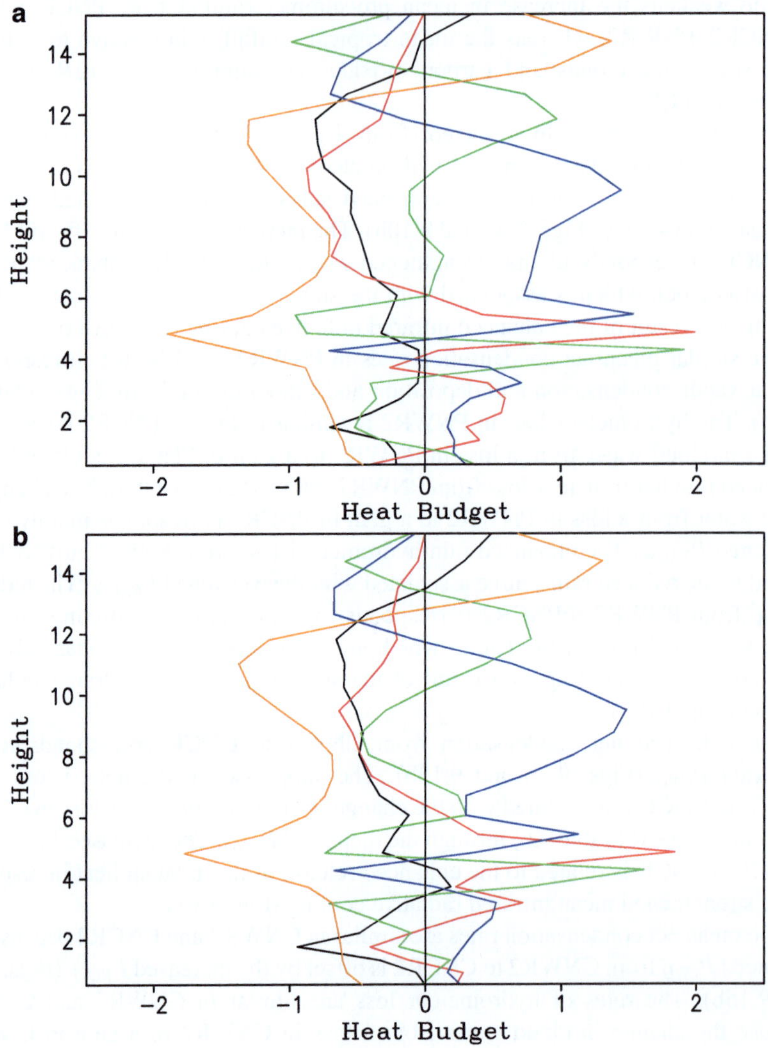

Fig. 9.18 Vertical profiles of differences between (**a**) PNCR and PNWR (PNCR–PNWR) and (**b**) PNCR2 and PNWR2 (PNCR2–PNWR2) for local temperature change (*black*), condensational heating (*red*), convergence of vertical heat flux (*green*), vertical temperature advection (*blue*), and radiation (*orange*) averaged for 5 days and model domain. Unit is °C d^{-1}

9.4 Effects of Doubled Carbon Dioxide on Rainfall Responses to Microphysical Processes of Ice Clouds

Doubled carbon dioxide enhances the reduction in mean pre-summer rainfall from PNIM–PNIR to PNIM2–PNIR2 because it increases the mean hydrometeor loss from PNIR to PNIR2 but decreases mean hydrometeor gain (Table 9.2). The

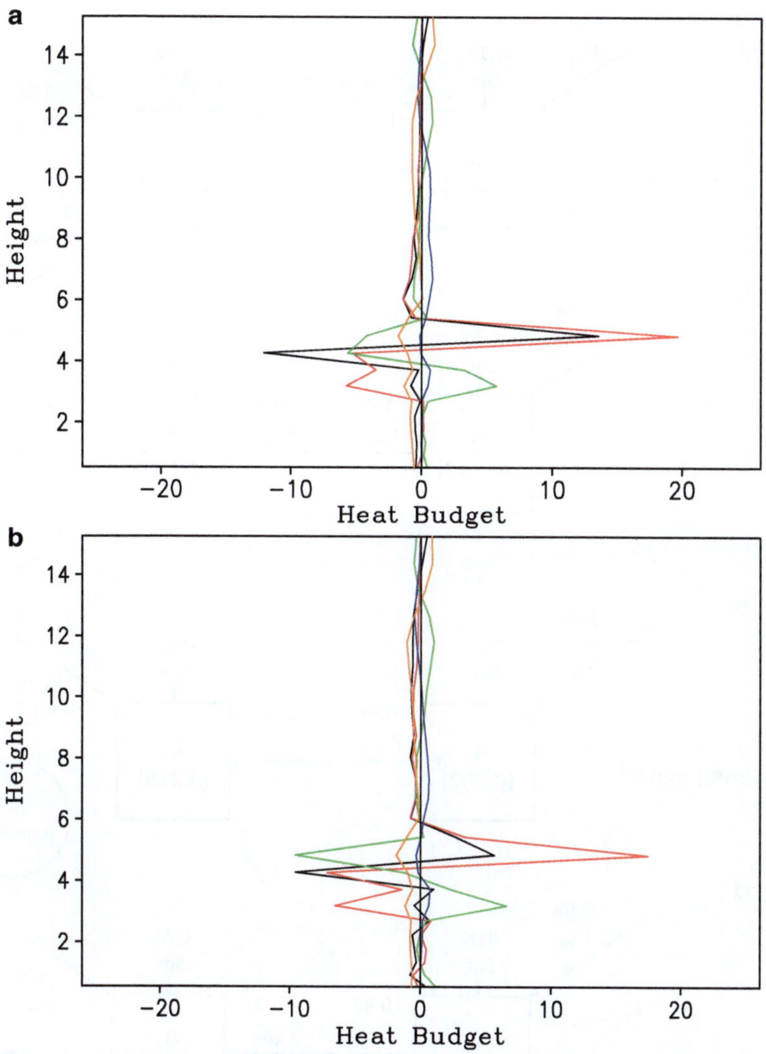

Fig. 9.19 Vertical profiles of differences between (**a**) CNCR and CNWR (CNCR–CNWR) and (**b**) CNCR2 and CNWR2 (CNCR2–CNWR2) for local temperature change (*black*), condensational heating (*red*), convergence of vertical heat flux (*green*), vertical temperature advection (*blue*), and radiation (*orange*) averaged for 5 days and model domain. Unit is °C d^{-1}

increase in mean hydrometeor loss from PNIR to PNIR2 corresponds to the change in mean cloud water from a gain in PNIR to a loss in PNIM and the increase in mean raindrop loss from PNIR to PNIR2 (Table 9.5). The change in mean cloud water from a gain in PNIR to a loss in PNIM is associated with the weakened P_{CND} from PNIR to PNIR2 (Figs. 9.10a and 9.12a). The increase in mean raindrop loss

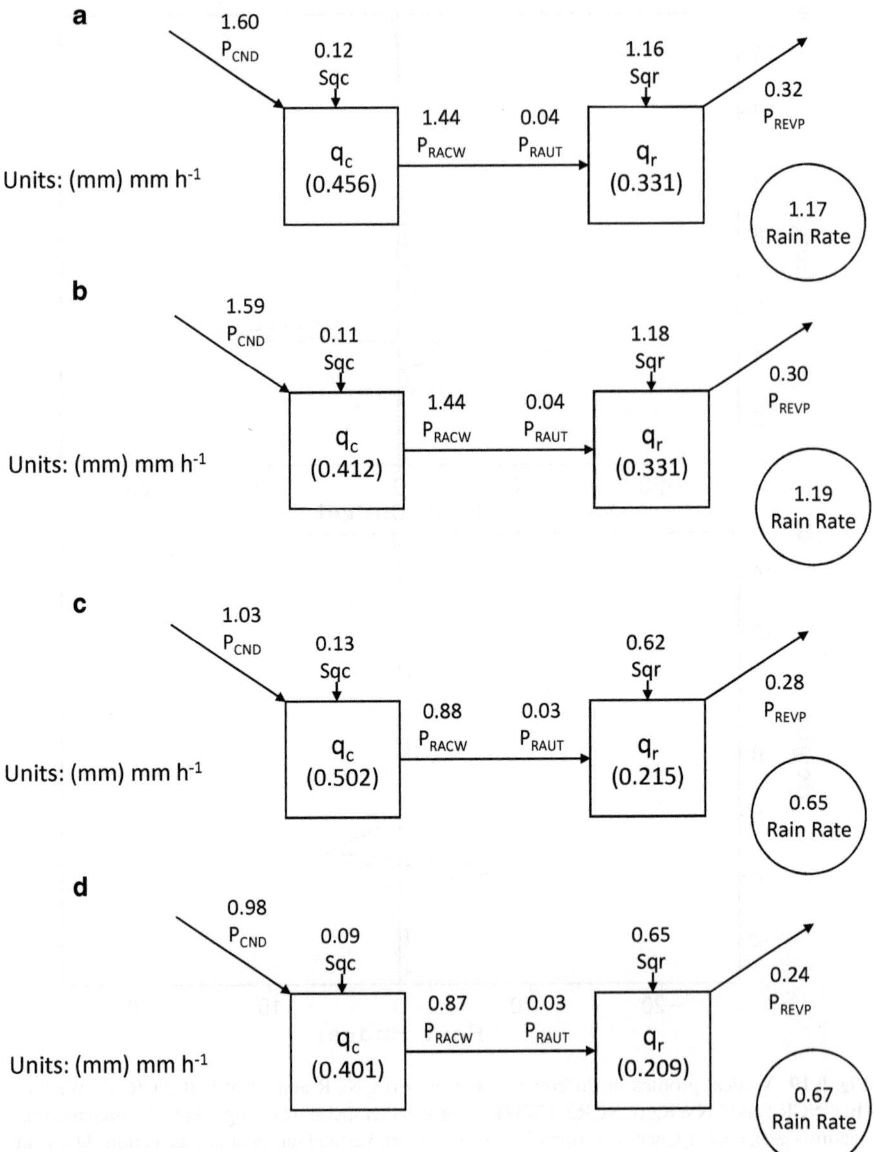

Fig. 9.20 Time-mean cloud microphysics budgets in (**a**) PNIM, (**b**) PNIM2, (**c**) CNIM, and (**d**) CNIM2. Units for cloud hydrometeors and conversions are mm and mm h^{-1}, respectively. Cloud microphysical conversion terms and their schemes can be found in Table 1.2. $T_0 = 0\ °C$

from PNIR to PNIR2 is related to the enhanced P_{RACW} due to the strengthened P_{CND}. The decrease in mean hydrometeor gain corresponds to the reduction in mean cloud water gain (Table 9.5) through the weakened P_{CND} (Figs. 9.20a, b).

Doubled carbon dioxide suppresses the decrease in mean tropical rainfall from CNIM–CNIR to CNIM2–CNIR2 through the weakened mean hydrometeor gain from CNIM to CNIM2 (Table 9.3). The weakened mean hydrometeor gain from CNIM to CNIM2 corresponds to the increased Sqr via the reduced P_{REVP} (Figs. 9.20c, d).

References

Allen MR, Ingram WJ (2002) Constraints on future changes in climate and the hydrological cycle. Nature 419:224–232

Fields PA, Graham JB, Rosenblatt RH, Somero GN (1993) Effects of expected global change on marine faunas. Trends Ecol Evol 8:361–367

Intergovernmental Panel for Climate Change (IPCC) (2001) Climate change 2001: the scientific basis. Cambridge University Press, New York

Intergovernmental Panel for Climate Change (IPCC) (2007) Climate change 2007. Switzerland, Geneva

Lau KM, Sui CH, Chou MD, Tao WK (1994) An inquiry into the cirrus cloud thermostat effect for tropical sea surface temperature. Geophys Res Lett 21:1157–1160

Li HM, Feng L, Zhou TJ (2011) Multi-model projection of July–August climate extreme changes over China under CO_2 doubling. Part I: precipitation. Adv Atmos Sci 28:433–447

Li X, Shen X, Liu J (2014a) Effects of doubled carbon dioxide on rainfall responses to the large-scale forcing: a two-dimensional cloud resolving modeling study. Adv Atmos Sci 31:525–531

Li X, Li T, Lou L (2014b) Effects of doubled carbon dioxide on rainfall responses to radiative processes of water clouds. J Meteor Res 28:1114–1126

Liu J, Shen X, Li X (2014) Radiative effects of water clouds on heat, cloud microphysical and surface rainfall budgets associated with pre-summer torrential rainfall. Terr Atmos Ocean Sci 25:41–50, (c) Chinese Geoscience Union. Reprinted with permission

Rutledge SA, Hobbs RV (1983) The mesoscale and microscale structure and organization of clouds and precipitation in midlatitude cyclones. Part VIII: a model for the "seeder-feeder" process in warm-frontal rainbands. J Atmos Sci 40:1185–1206

Chapter 10
Convective, Moist, and Dynamic Vorticity Vectors

Abstract In this chapter, convective, moist, and dynamic vorticity vectors are defined as the cross products of relative vorticity vector and gradients of equivalent potential temperature and specific humidity and wind vector, respectively. These vorticity vectors and cloud hydrometeor are highly correlated. The tendency equations of these vorticity vectors are derived to show the linkage between dynamic, thermodynamic, and cloud microphysical processes. These vorticity vectors are compared in the two- and three-dimensional frameworks.

Keywords Convective vorticity vector • Moist vorticity vector • Dynamic vorticity vector

As one of the most important dynamic/thermodynamic parameters, potential vorticity has been studied to enhance the understanding of the genesis and development of weather systems for more than six decades since it was first introduced by Ertel (1942). Potential vorticity is conserved in a frictionless, adiabatic flow in a dry atmosphere. Later, moist potential vorticity was introduced by replacing potential temperature with the equivalent potential temperature. Moist potential vorticity is conserved in frictionless moist adiabatic processes. Many studies have contributed to understanding the dry and moist potential vorticity associated with dynamic and thermodynamic processes in the genesis and development of weather systems (e.g., Bennetts and Hoskins 1979; Emanuel 1979; Danielsen and Hipskind 1980; Thorpe 1985; Hoskins and Berrisford 1988; Xu 1992; Montgomery and Farrell 1993; Cao and Cho 1995; Cho and Cao 1998; Gao et al. 2002). Helicity, as an important dynamic concept, has been applied to the study of convective storms in recent decades (e.g., Lilly 1986; Droegemeier et al. 1993; Tan and Wu 1994) since it was introduced by Betchov (1961). However, potential vorticity and helicity cannot be applied to the analysis in the 2D framework.

Convective (***CVV***), moist (***MVV***), and dynamic (***DVV***) vorticity vectors are introduced in the 2D framework by Gao et al. (2004, 2005) and are further studied in the 3D framework by Gao et al. (2007) and Gao (2007). How well do these vorticity vectors represent convective signals? What dominant physical processes determining the variations of these vorticity vectors? What are the differences between the 2D and 3D vorticity vectors? What are the differences between ***CVV***

© Springer International Publishing Switzerland 2016

X. Li, S. Gao, *Cloud-Resolving Modeling of Convective Processes*,
Springer Atmospheric Sciences, DOI 10.1007/978-3-319-26360-1_10

and moist potential vorticity and between ***DVV*** and helicity in the study of 3D convection? These questions will be discussed based on Gao et al. (2004, 2005, 2007) and Gao (2007) in this chapter.

10.1 Convective Vorticity Vector

The 3D ***CVV*** can be expressed by

$$\mathbf{CVV} = \frac{\boldsymbol{\xi} \times \nabla\theta_e}{\overline{\rho}} = C_x\mathbf{i} + C_y\mathbf{j} + C_z\mathbf{k}, \tag{10.1a}$$

where $C_x = \zeta_y \partial\theta_e/\overline{\rho}\partial z - \zeta_z \partial\theta_e/\overline{\rho}\partial y$, $C_y = \zeta_z \partial\theta_e/\overline{\rho}\partial x - \zeta_x \partial\theta_e/\overline{\rho}\partial z$, $C_z = \zeta_x \partial\theta_e/\overline{\rho}\partial y - \zeta_y \partial\theta_e/\overline{\rho}\partial x$; $\zeta_x = \partial w/\partial y - \partial v/\partial z$, $\zeta_y = \partial u/\partial z - \partial w/\partial x$, $\zeta_z = \partial v/\partial x - \partial u/\partial y$; θ_e is the equivalent potential temperature; u, v, w are zonal, meridional, and vertical components of wind; planetary vorticity is excluded since it is much smaller than vertical component of relative vorticity.

The ***CVV*** in the 2D *x*–*z* frame can be written by

$$\mathbf{CVV} = P_x\mathbf{i} + P_z\mathbf{k}, \tag{10.1b}$$

where $P_x = \zeta_y \partial\theta_e/\overline{\rho}\partial z$, $P_z = -\zeta_y \partial\theta_e/\overline{\rho}\partial x$. The ***CVV*** has zonal and vertical components in the 2D *x*–*z* frame.

Zonally averaged and mass-integrated P_x, P_z, and the total hydrometeor mixing ratio were calculated to examine their relationship. They are denoted by $[P_x]$, $[P_z]$, and *IWP*+*LWP*, respectively, and are shown in Fig. 10.1. The time evolution of $[P_z]$ is in phase with that of *IWP*+*LWP* as indicated by their correlation coefficient of 0.81, though more fluctuations occur in $[P_z]$ than in *IWP*+*LWP*. $[P_x]$, however, does not follow the time evolution of *IWP*+*LWP*, and their correlation coefficient is only 0.18. Thus, the tendency of the vertical component of ***CVV*** is closely associated with the variation of tropical convection and its tendency equation will be derived next.

The zonal and vertical momentum equations can be expressed by

$$\frac{\partial u}{\partial t} = -u\frac{\partial u}{\partial x} - w\frac{\partial u}{\partial z} - c_p\frac{\partial\left(\overline{\theta}\pi\right)}{\partial x}, \tag{10.2a}$$

$$\frac{\partial w}{\partial t} = -u\frac{\partial w}{\partial x} - w\frac{\partial w}{\partial z} - c_p\frac{\partial\left(\overline{\theta}\pi\right)}{\partial z} + B, \tag{10.2b}$$

where $B = g\left(\theta'/\theta_b + 0.61q_v - q_l\right)$ is the buoyancy force.

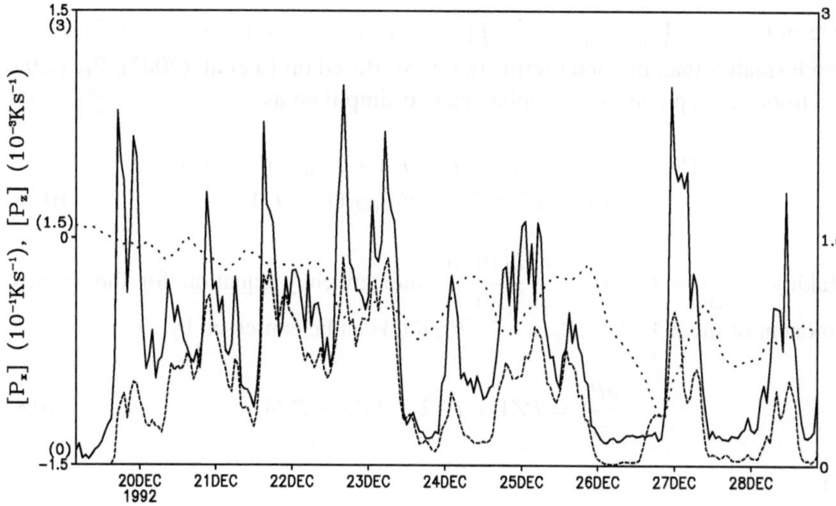

Fig. 10.1 Time series of domain averaged, mass-integrated zonal ($[P_x]$, 10^{-1}Ks^{-1}, *dot line*), and vertical ($[P_z]$, 10^{-3}Ks^{-1}, *solid line*) components of the 2D *CVV*, and *IWP*+*LWP* (mm, *dashed line*) in the 10-day 2D cloud-resolving model simulation during TOGA COARE. The plotting scales of $[P_x]$ and $[P_z]$ and *IWP* + *LWP* are −0.15 to 0.15 Ks^{-1}, 0–3 × 10^{-3} Ks^{-1}, and 0–3 mm, respectively (Gao et al. 2004)

Taking $\frac{\partial}{\partial z}(10.2a) - \frac{\partial}{\partial x}(10.2b)$ and applying mass continuity equation, the tendency equation for the vorticity in an elastic approximation is expressed by

$$\frac{\partial}{\partial t}\left(\frac{\zeta_y}{\overline{\rho}}\right) = -u\frac{\partial}{\partial x}\left(\frac{\zeta_y}{\overline{\rho}}\right) - w\frac{\partial}{\partial z}\left(\frac{\zeta_y}{\overline{\rho}}\right) - \frac{1}{\overline{\rho}}\frac{\partial B}{\partial x}. \tag{10.3}$$

The tendency equations for potential temperature and specific humidity are

$$\frac{\partial \theta}{\partial t} = -u\frac{\partial \theta}{\partial x} - w\frac{\partial \theta}{\partial z} + \frac{Q_{cn}}{\pi c_p} + \frac{Q_R}{\pi c_p}, \tag{10.4a}$$

$$\frac{\partial q_v}{\partial t} = -u\frac{\partial q_v}{\partial x} - w\frac{\partial q_v}{\partial z} - S_{qv}, \tag{10.4b}$$

Taking $\frac{1}{\theta} \times$ (10.4a) + $\frac{L_v}{c_p T} \times$ (10.4b) and defining the equivalent potential temperature θ_e as $\theta \exp(L_v q_v / c_p T)$ yields a tendency equation for equivalent potential temperature,

$$\frac{\partial \theta_e}{\partial t} = -u\frac{\partial \theta_e}{\partial x} - w\frac{\partial \theta_e}{\partial z} + \frac{L_f}{c_p}\frac{P_{18}\theta_e}{T} + \frac{1}{c_p}\frac{Q_R \theta_e}{T}. \tag{10.5}$$

The term $\left(-\frac{L_v}{c_p}\frac{\theta_e q_v}{T^2}\left(\frac{\partial T}{\partial t}+\frac{\partial T}{\partial x}+w\frac{\partial T}{\partial z}\right)\right)$ is omitted in the derivation of (10.5) since it is much smaller than the other terms in (10.5). Based on Li et al. (2002), P_{18} (2.9a) in the tropical deep convective regime can be simplified as

$$P_{18} = P_{DEP} - P_{MLTG}\,(T > T_o) + P_{SACW}\,(T < T_o) \\ -P_{SMLT}\,(T > T_o) - P_{GMLT}\,(T > T_o)\,. \tag{10.5a}$$

Taking $-\dfrac{\partial\theta_e}{\partial x}\times(10.3) - \dfrac{\zeta_y}{\overline{\rho}}\dfrac{\partial(10.5)}{\partial x}$, the tendency equation for the vertical component of the 2D ***CVV*** ($P_z = -\zeta_y\partial\theta_e/\overline{\rho}\partial x$) can be expressed by

$$\frac{\partial P_z}{\partial t} = PZ1 + PZ2 + PZ3 + PZ4, \tag{10.6}$$

where

$$PZ1 = \left(-u\frac{\partial P_z}{\partial x} - w\frac{\partial P_z}{\partial z}\right) + \frac{\zeta_y}{\overline{\rho}}\left(\frac{\partial u}{\partial x}\frac{\partial\theta_e}{\partial x} + \frac{\partial w}{\partial x}\frac{\partial\theta_e}{\partial z}\right), \tag{10.6a}$$

$$PZ2 = \frac{1}{\overline{\rho}}\frac{\partial\theta_e}{\partial x}\frac{\partial B}{\partial x}, \tag{10.6b}$$

$$PZ3 = -\frac{L_f}{c_p}\frac{\zeta_y}{\overline{\rho}}\frac{\partial}{\partial x}\left(\frac{P_{18}\theta_e}{T}\right), \tag{10.6c}$$

$$PZ4 = -\frac{1}{c_p}\frac{\zeta_y}{\overline{\rho}}\frac{\partial}{\partial x}\left(\frac{Q_R\theta_e}{T}\right). \tag{10.6d}$$

Term $PZ1$ is related to dynamic and thermodynamic processes including the zonal and vertical advection of P_z and the interaction between vorticity and the gradients of wind and equivalent potential temperature. Term $PZ2$ is associated with the buoyancy force. Term $PZ3$ is associated with cloud microphysical processes. Term $PZ4$ includes the solar and IR radiative forcing.

Model domain mean and mass integration of (10.6) yields

$$\frac{\partial\,[P_z]}{\partial t} = [PZ1] + [PZ2] + [PZ3] + [PZ4]\,. \tag{10.7}$$

The domain mean and mass integration of the first term in (10.6a) is much smaller than the other terms in (10.6) and is thus excluded in (10.7).

To examine the processes responsible for the variation of $[P_z]$, the four terms in (10.7) and tendency of $[P_z]$ are calculated. $[PZ3]$ follows the tendency of $[P_z]$ closely in both phase and amplitude with a correlation coefficient of 0.97 and a RMS difference of 1.2×10^{-6} Ks^{-2}, which is much smaller than the standard deviation of the tendency of $[P_z]$ (3.7×10^{-6} Ks^{-2}). $[PZ2]$ is positive and has moderate amplitudes, whereas $[PZ1]$ is negative. Their correlation coefficients with the tendency of $[P_z]$ are smaller than 0.2, and their RMS differences with the tendency of $[P_z]$ are similar to the standard deviation of the tendency of $[P_z]$. $[PZ4]$ is negligible. Thus,

$$\frac{\partial [P_z]}{\partial t} \approx -\frac{L_f}{c_p}\left[\frac{\zeta_y}{\overline{\rho}}\frac{\partial}{\partial x}\left(\frac{P_{18}\theta_e}{T}\right)\right]. \tag{10.7a}$$

The dominance of $[PZ3]$ in the variation of $[P_z]$ indicates that the variation of the vertical component of the 2D ***CVV*** is controlled by the interaction between the vorticity and zonal gradient of cloud heating. (10.5a) shows that the cloud heating comes from the ice microphysical processes including vapor deposition, evaporation of liquid water from the surface of graupel, accretion of cloud water by snow, and melting of snow and graupel. This implies that P_{18} could be zero in water clouds. Thus, the variation of the vertical component of the 2D ***CVV*** is associated with the variation of the ice hydrometeor mixing ratio through the interaction between the dynamics and ice microphysics.

Gao et al. (2007) used the data from a 3D cloud-resolving model simulation during the TRMM Kwajalein Experiment (KWAJEX) to analyze 3D ***CVV*** (10.1a) and their relations to tropical oceanic convection as well as moist potential vorticity (PV). The model is forced by zonally uniform vertical velocity, zonal wind, and thermal and moisture advection based on 6-hourly KWAJEX observations. Kwajalein island is located at 8.44°N, 167.43°E and has the area of 15 km^2, which is the largest island in Kwajalein Atoll (2200 km^2). The model is integrated from 0600 LST 7 August 1999 to 0600 LST 12 August 1999 (a total of 5 days). Shie et al. (2003) conducted 3D simulations and compared them with radar observations during three active periods in KWAJEX. The simulation and observation show reasonable agreement in rainfall, apparent heat and moisture sources. Thus, hourly model domain mean data are used in the following discussions.

To compare ***CVV*** with *PV*, the *PV* is calculated. The *PV* can be expressed by

$$PV = \frac{\boldsymbol{\xi}\cdot\nabla\theta_e}{\overline{\rho}} = PV_1 + PV_2 + PV_3, \tag{10.8}$$

where $PV_1 = \zeta_x\partial\theta_e/\overline{\rho}\partial x$, $PV_2 = \zeta_y\partial\theta_e/\overline{\rho}\partial y$, $PV_3 = \zeta_z\partial\theta_e/\overline{\rho}\partial z$.

Figure 10.2a shows the time series of *IWP*+*LWP*, $[C_x]$, $[C_y]$, $[C_z]$, and $[PV]$. $[C_z]$ closely follows evolution of *IWP*+*LWP* whereas $[C_x]$ and $[C_y]$ do not. The linear correlation coefficient between $[C_z]$ and *IWP*+*LWP* is 0.97, whereas those between $[C_x]$ and *IWP*+*LWP*, and $[C_y]$ and *IWP*+*LWP* are 0.29 and 0.48, respectively. During the 5-day integration period, there are 7–8 convective events. A Student's *t*-test on the significance of the correlation coefficients is further conducted using 5 degrees of freedom, and a critical correlation coefficient at the 5% significant level is 0.75. Thus, the relation between *IWP*+*LWP* and $[C_z]$ is statically significant. $[PV]$ basically follows the evolution of *IWP*+*LWP* with the linear correlation coefficient of 0.73, which is marginally statistically significant. From (10.8), $[PV]$ is determined

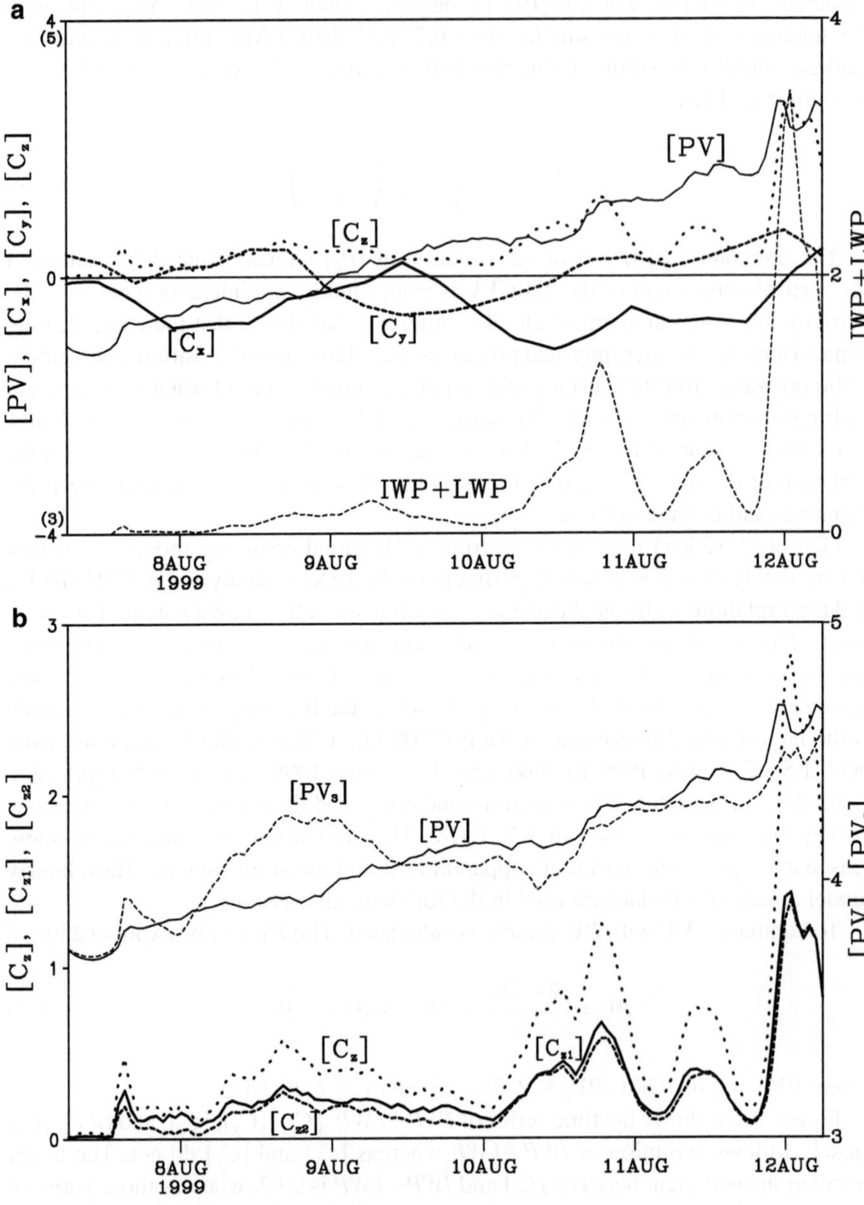

Fig. 10.2 Time series of (**a**) *IWP+LWP* (*light dashed*), [*PV*] (*light solid*), [C_x] (*dark solid*), [C_y] (*dark dashed*), and [C_z] (*dotted*), and (**b**) [C_z] (*dotted*), [C_{z1}] (*dark solid*), [C_{z2}] (*dark dashed*), [*PV*] (*light solid*), and [PV_3] (*light dashed*) in the 5-day 3D cloud-resolving model simulation during KWAJEX. Units are in mm for $IWP + LWP$, 10^{-3} K s^{-1} for [*PV*], [PV_3], [C_z], [C_{z1}], and [C_{z2}], and 10^{-1} K s^{-1} for [C_x] and [C_y]. The plotting scales in (**a**) are 3–5 × 10^{-3} K s^{-1} for [*PV*] and [PV_3], −4 to 4 × 10^{-3} K s^{-1} for [C_z], and −4 to 4 × 10^{-1} K s^{-1} for [C_x] and [C_y] (After Gao et al. 2007)

by three components. $[PV_3]$ have the similar magnitude to $[PV]$ (Fig. 10.2b), whereas two other components (not shown) are about one order of magnitudes smaller than $[PV_3]$. Thus, $[PV]$ is determined by the covariance between vertical component of vorticity (horizontal rotational circulations) and vertical gradient of moist potential gradient (stability parameter). Note that although the magnitude of $[PV_1]$ is small, the correlation coefficient between $[PV_1]$ and *IWP+LWP* is 0.54, which is slightly smaller than that between $[PV_3]$ and $IWP + LWP$ (0.62). Since *PV* is a scalar, the role of horizontal component of vorticity in convective development cannot be analyzed. $[C_z]$ is determined by the covariance between horizontal components of vorticity (the secondary circulation in horizontal-vertical framework that is directly associated with the dynamic aspect of convection and is described by the zonal and meridional components of relative vorticity) and horizontal gradients of equivalent potential temperature (that is directly associated with the thermodynamic aspect of the convection). The larger correlation between convection and vertical component of ***CVV*** indicates that the secondary circulation more directly represent convection. The comparison between the *PV* and ***CVV*** shows that the vector analysis can be used to study the roles of horizontal and vertical components of vorticity in convective development separately.

$[C_z]$ can be broken down into $[C_{z1}]$ and $[C_{z2}]$, where $C_{z1} = \zeta_x \partial\theta_e/\overline{\rho}\partial y$, and $C_{z2} = -\zeta_y \partial\theta_e/\overline{\rho}\partial x$. Only $[C_{z2}]$ appears in the 2D framework. Gao et al. (2004) in their 2D study of ***CVV*** during TOGA COARE showed the high linear correlation coefficient between $[C_{z2}]$ and *IWP+LWP*. To examine the contribution of $[C_{z1}]$ and $[C_{z2}]$ to the lag correlation between *IWP+LWP* and $[C_z]$, the weighted lag correlation coefficients between *IWP+LWP* and $[C_{z1}]$/$[C_{z2}]$ are calculated. The two correlation coefficients at lag hour 0 show same values, which indicate that $[C_{z1}]$ and $[C_{z2}]$ equally contribute to the correlation coefficient between cloud hydrometeors and vertical component of ***CVV***. Thus, vertical component of ***CVV*** can be a controlling parameter for studying tropical oceanic convection, regardless of the dimension of data (2D or 3D).

10.2 Moist Vorticity Vector

There is a plenty of water vapor over the tropics. Thus, the vapor gradient may be much more important than the temperature gradient. P_z can be broken into the two parts:

$$P_z = P_{z1} + P_{z2}. \tag{10.8a}$$

where

$$P_{z1} = -\frac{\zeta_y}{\overline{\rho}}\frac{\partial\theta}{\partial x}\exp\left(\frac{L_v q_v}{c_p T}\right), \tag{10.8b}$$

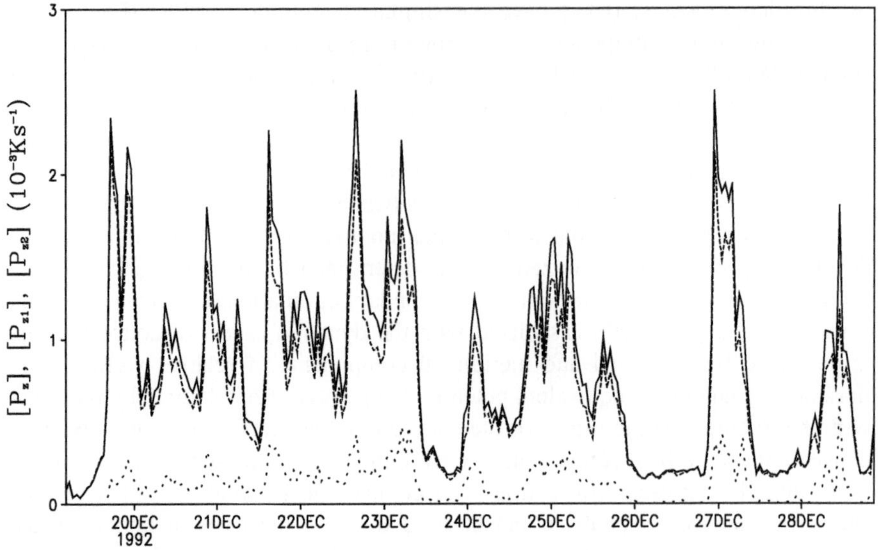

Fig. 10.3 Time series of $[P_z]$ (*solid line*), $[P_{z1}]$ (*dotted line*), and $[P_{z2}]$ (*dashed line*) in the 10-day 2D cloud-resolving model simulation during TOGA COARE. Units are in 10^{-3} K s^{-1} (After Gao et al. 2005)

$$P_{z2} = -\frac{\zeta_y}{\overline{\rho}}\theta\frac{\partial}{\partial x}\exp\left(\frac{L_v q_v}{c_p T}\right). \tag{10.8c}$$

P_{z1} is associated with the horizontal gradient of the potential temperature, and P_{z2} is related to the horizontal gradient of the specific humidity. Figure 10.3 shows time series of domain averaged and mass-integrated P_z, P_{z1}, and P_{z2} ($[P_z]$, $[P_{z1}]$, and $[P_{z2}]$). $[P_z]$ is largely determined by $[P_{z2}]$, implying the important role of the moisture in equivalent potential temperature gradient in the tropical deep convective regime.

The moist vorticity vector (***MVV***) in the 2D *x*–*z* frame is defined as

$$\mathbf{MVV} = \frac{\boldsymbol{\xi} \times \nabla q_v}{\overline{\rho}} = \frac{\zeta_y}{\overline{\rho}}\left(\frac{\partial q_v}{\partial z}\mathbf{i} - \frac{\partial q_v}{\partial x}\mathbf{k}\right). \tag{10.9}$$

MVV has zonal ($M_x = \zeta_y \partial q_v / \overline{\rho}\partial z$) and vertical ($M_z = -\zeta_y \partial q_v / \overline{\rho}\partial x$) components in the 2D *x*–*z* frame.

Figure 10.4 shows the time series of $[M_x]$, $[M_z]$, and *IWP*+*LWP*, respectively. The time evolution of $[M_z]$ is in phase with that of *IWP*+*LWP* as indicated by their correlation coefficient of 0.78. However, the correlation coefficient between $[M_x]$ and *IWP*+*LWP* is only 0.32. Thus, the tendency of the vertical component of 2D ***MVV*** is closely associated with the variation of tropical convection and its tendency equation will be derived next.

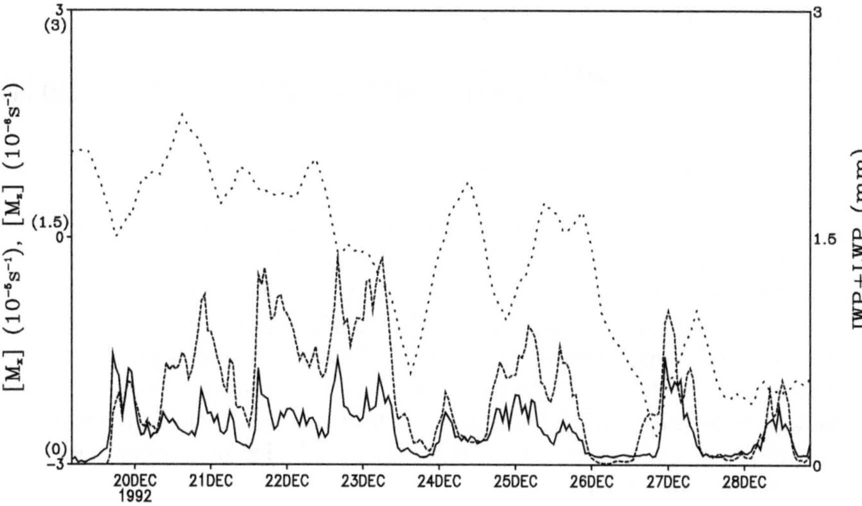

Fig. 10.4 Time series of $[M_x]$ (10^{-5}s^{-1}, *dotted line*), $[M_z]$ (10^{-6}s^{-1}, *solid line*), and *IWP+LWP* (mm, *dashed line*) in the 10-day 2D cloud-resolving model simulation during TOGA COARE. The plotting scales of $[M_x]$, $[M_z]$, and $IWP+LWP$ are -3 to $3\times10^{-5}\text{s}^{-1}$, 0–$3\times10^{-6}\text{s}^{-1}$, and 0–3 mm, respectively (After Gao et al. 2005)

Taking $-\dfrac{\partial q_v}{\partial x}\times(7.3)-\dfrac{\zeta_y}{\overline{\rho}}\dfrac{\partial(7.4b)}{\partial x}$, the tendency equation for the vertical component of the 2D ***MVV*** (M_z) can be expressed by

$$\frac{\partial M_z}{\partial t} = MZ1 + MZ2 + MZ3, \tag{10.10}$$

where

$$MZ1 = \left(-u\frac{\partial M_z}{\partial x} - w\frac{\partial M_z}{\partial z}\right) + \frac{\zeta_y}{\overline{\rho}}\left(\frac{\partial u}{\partial x}\frac{\partial q_v}{\partial x} + \frac{\partial w}{\partial x}\frac{\partial q_v}{\partial z}\right), \tag{10.10a}$$

$$MZ2 = \frac{1}{\overline{\rho}}\frac{\partial q_v}{\partial x}\frac{\partial B}{\partial x}, \tag{10.10b}$$

$$MZ3 = \frac{\zeta_y}{\overline{\rho}}\frac{\partial}{\partial x}\left(S_{qv}\right). \tag{10.10c}$$

Term $MZ1$ is related to dynamic and thermodynamic processes including the zonal and vertical advection of M_z and the interaction between vorticity and the gradients of wind and specific humidity. Term $MZ2$ is associated with the buoyancy force. Term $MZ3$ is associated with cloud microphysical processes including condensation and deposition.

Model domain mean and mass integration of (10.10) yields

$$\frac{\partial [M_z]}{\partial t} = [MZ1] + [MZ2] + [MZ3]. \tag{10.11}$$

The domain mean and mass integration of the first term in (10.10a) is much smaller than the other terms in (10.10) and is thus excluded in (10.11).

To examine the processes responsible for the variation of $[M_z]$, the three terms in (10.11) and the tendency of $[M_z]$ are calculated. $[MZ1]$ is always positive with the magnitudes of up to 3×10^{-9} s^{-2} whereas $[MZ3]$ varies between -5 and 5×10^{-9} s^{-2}. $[MZ3]$ follows the tendency of $[M_z]$ roughly with a correlation coefficient of 0.8 and a RMS difference of 1.05×10^{-9} s^{-2}, which is similar to the standard deviation of the tendency of $[M_z]$ (1.06×10^{-9} s^{-2}). A positive tendency for $[M_z]$ is initiated by positive $[MZ3]$; $[MZ1]$ contributes later. $[MZ2]$ has negligibly small magnitudes compared to $[MZ3]$ and $[MZ1]$. Thus, $[MZ3]$ has major contributions to the tendency of $[M_z]$, and $[MZ1]$ has the significant modifications. Since the vapor condensation and deposition rates are important parts of S_{qv}, the variation of the vertical component of the 2D ***MVV*** is associated with the total hydrometeor mixing ratio through the interaction between the dynamics and the water/ice microphysics.

The 3D ***MVV*** can be expressed by

$$\mathbf{MVV} = \frac{\vec{\xi} \times \nabla q_v}{\overline{\rho}} = M_x\mathbf{i} + M_y\mathbf{j} + M_z\mathbf{k}, \tag{10.12}$$

where $M_x = \zeta_y \partial q_v / \overline{\rho} \partial z - \zeta_z \partial q_v / \overline{\rho} \partial y$, $M_y = \zeta_z \partial q_v / \overline{\rho} \partial x - \zeta_x \partial q_v / \overline{\rho} \partial z$, $M_z = \zeta_x \partial q_v / \overline{\rho} \partial y - \zeta_y \partial q_v / \overline{\rho} \partial x$.

Since moisture gradient is much important than temperature gradient in the tropical deep convective regime, it determines the variation of equivalent potential temperature gradient. Thus, $(\boldsymbol{\xi} \cdot \nabla q_v) / \overline{\rho}$ is defined as the moist vorticity (*MV*). To compare ***MVV*** with *MV*, the *MV* is calculated. The *MV* can be expressed by

$$MV = \frac{\boldsymbol{\xi} \cdot \nabla q_v}{\overline{\rho}} = MV_1 + MV_2 + MV_3, \tag{10.13}$$

where $MV_1 = \zeta_x \partial q_v / \overline{\rho} \partial x$, $MV_2 = \zeta_y \partial q_v / \overline{\rho} \partial y$, $MV_3 = \zeta_z \partial q_v / \overline{\rho} \partial z$.

Figure 10.5a shows the time series of *IWP*+*LWP*, $[M_x]$, $[M_y]$, $[M_z]$, and $[MV]$. Vertical component of ***MVV*** intimately follows the variation of *IWP*+*LWP* with the correlation coefficient of 0.95. $[MV]$ and *IWP* + *LWP* are correlated with the coefficient of 0.75. $[MV_3]$ have the similar magnitude to $[MV]$ (Fig. 10.5b), whereas two other components (not shown) are about one order of magnitudes smaller than $[MV_3]$. Horizontal components of ***MVV*** ($[M_x]$ and $[M_y]$) are loosely correlated with cloud hydrometeors for only -0.26 and 0.17, respectively, which are not statistically significant. $[M_z]$ can be broken down into $[M_{z1}]$ and $[M_{z2}]$, where $M_{z1} = \zeta_x \partial q_v / \overline{\rho} \partial y$, and $M_{z2} = -\zeta_y \partial q_v / \overline{\rho} \partial x$. Only $[M_{z2}]$ appears in the 2D framework and has the high linear correlation coefficient with *IWP*+*LWP*. In the 3D framework, $[M_{z1}]$ is slightly

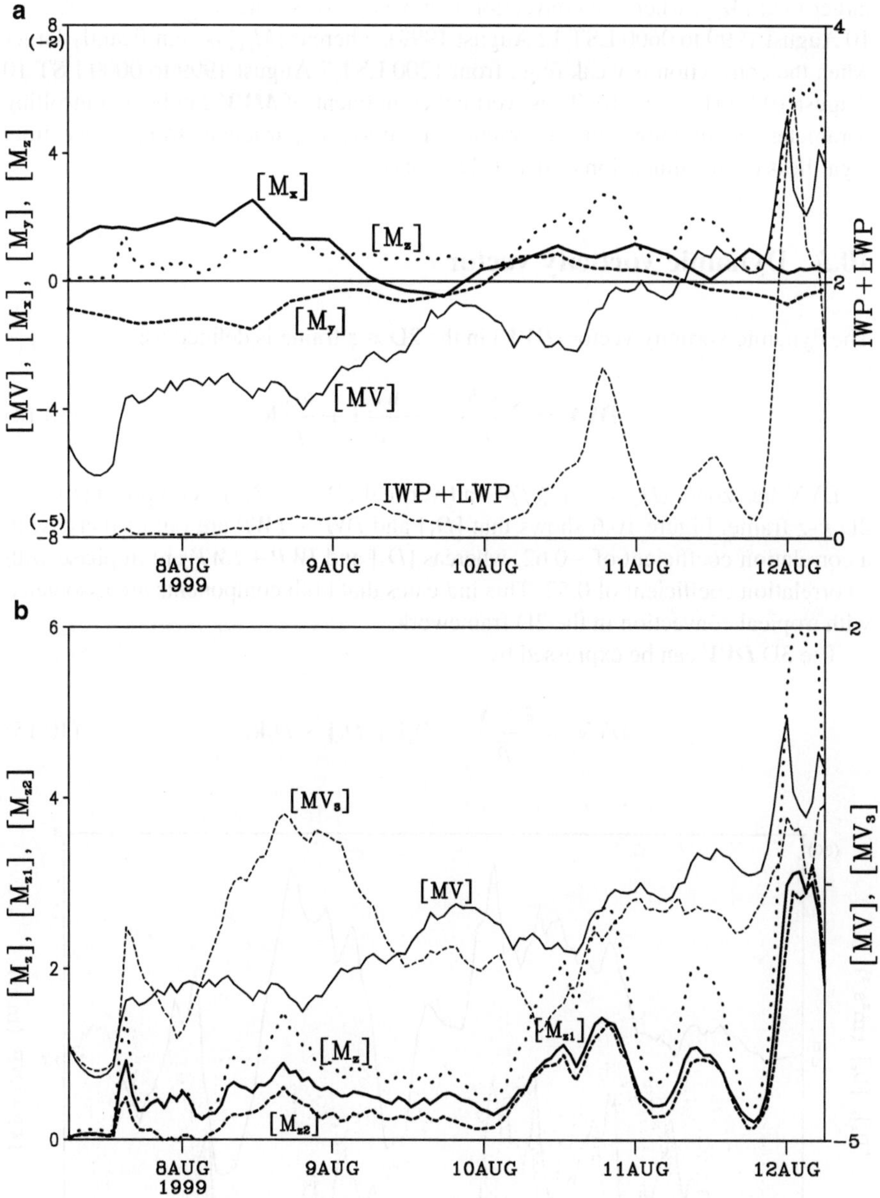

Fig. 10.5 Time series of (**a**) *IWP+LWP*, $[MV]$, $[M_x]$, $[M_y]$, $[M_z]$, and (**b**) $[M_z]$, $[M_{z1}]$, $[M_{z2}]$, $[MV]$, $[MV_3]$ in the 5-day 3D cloud-resolving model simulation during KWAJEX. Units are mm for $IWP + LWP$, 10^{-7} s^{-1} for $[MV]$, $[MV_3]$, $[M_z]$, $[M_{z1}]$, $[M_{z2}]$, and 10^{-5} s^{-1} for $[M_x]$, $[M_y]$. The plotting scales in (**a**) are -5 to -2×10^{-7} s^{-1} for $[MV]$, -8 to 8×10^{-7} s^{-1} for $[M_z]$, and -8 to 8×10^{-5} s^{-1} for $[M_x]$ and $[M_y]$ (After Gao et al. 2007)

larger than $[M_{z2}]$ when the convection is moderate or strong (e.g., from 0000 LST 10 August 1999 to 0600 LST 12 August 1999), whereas $[M_{z1}]$ is significantly larger when the convection is weak (e.g., from 1200 LST 7 August 1999 to 0000 LST 10 August 1999) (Fig. 10.5b). Thus, vertical component of ***MVV*** can be a controlling parameter for studying tropical oceanic convection, in particular strong convection, regardless of the dimension of data (2D or 3D).

10.3 Dynamic Vorticity Vector

The dynamic vorticity vector (***DVV***) in the 2D *x–z* frame is defined as:

$$\mathbf{DVV} = \frac{\boldsymbol{\xi} \times \mathbf{V}}{\rho} = -\frac{w\zeta_y}{\rho}\mathbf{i} + \frac{u\zeta_y}{\rho}\mathbf{k}. \quad (10.14)$$

DVV has zonal ($D_x = -w\zeta_y/\overline{\rho}$) and vertical ($D_z = u\zeta_y/\overline{\rho}$) components in the 2D *x–z* frame. Figure 10.6 shows that $[D_x]$ and $IWP + LWP$ are out of phase with a correlation coefficient of –0.62, whereas $[D_z]$ and $IWP + LWP$ are in phase with a correlation coefficient of 0.52. This indicates that both components are associated with tropical convection in the 2D framework.

The 3D ***DVV*** can be expressed by

$$\mathbf{DVV} = \frac{\boldsymbol{\xi} \times \mathbf{V}}{\overline{\rho}} = D_x\mathbf{i} + D_y\mathbf{j} + D_z\mathbf{k}, \quad (10.15)$$

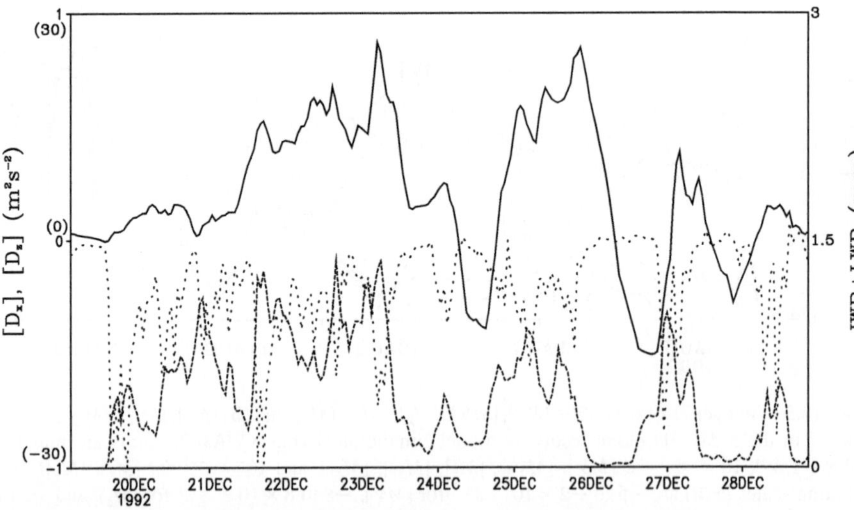

Fig. 10.6 Time series of $[D_x]$ (m^2s^{-2}, *dotted line*), $[D_z]$ (m^2s^{-2}, *solid line*) and $IWP + LWP$ (mm, *dashed line*) in the 10-day 2D cloud-resolving model simulation during TOGA COARE. The plotting scales of $[D_x]$, $[D_z]$ and $IWP + LWP$ are -1 to $1\ m^2s^{-2}$, -30 to $30\ m^2s^{-2}$, and 0–3 mm, respectively (After Gao et al. 2005)

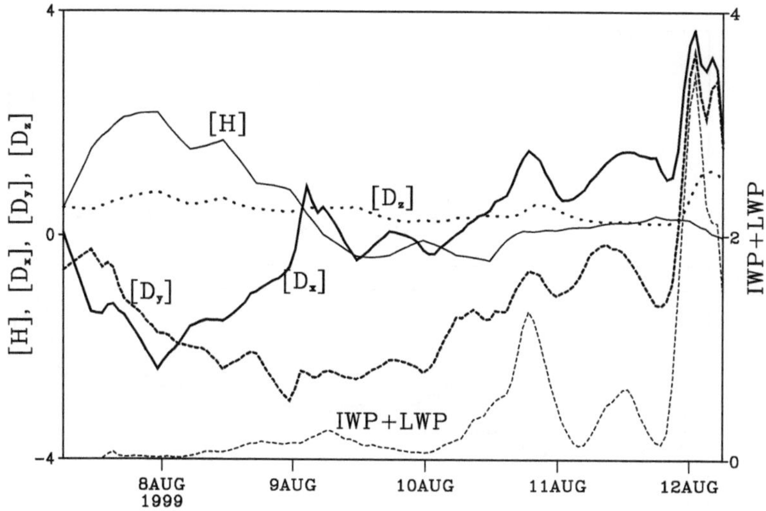

Fig. 10.7 Time series of *IWP + LWP* (*light dashed*), [H] (*light solid*), [D_x] (*dark solid*), [D_y] (*dark dashed*), and [D_z] (*dotted*) in the 5-day 3D cloud-resolving model simulation during KWAJEX. Units are in mm for *IWP + LWP*, 10^2 m^2 s^{-2} for [H], m^2 s^{-2} for [D_x] and [D_y], and 10^2 m^2 s^{-2} for [D_z] (After Gao 2007)

where $D_x = D_{x1} + D_{x2}$, $D_y = D_{y1} + D_{y2}$, $D_z = D_{z1} + D_{z2}$; $D_{x1} = \zeta_y w/\overline{\rho}$, $D_{x2} = -\zeta_z v/\overline{\rho}$, $D_{y1} = \zeta_z u/\overline{\rho}$, $D_{y2} = -\zeta_x w/\overline{\rho}$, $D_{z1} = \zeta_x v/\overline{\rho}$, and $D_{z2} = -\zeta_y u/\overline{\rho}$.

To compare ***DVV*** with helicity (H), H is calculated. H can be expressed by

$$H = \frac{\boldsymbol{\xi} \cdot \mathbf{V}}{\overline{\rho}} = H_1 + H_2 + H_3, \tag{10.16}$$

where $H_1 = \zeta_x u/\overline{\rho}$, $H_2 = \zeta_y v/\overline{\rho}$, $H_3 = \zeta_z w/\overline{\rho}$.

Figure 10.7 shows that [D_x] and [D_y] follow evolution of *IWP+LWP* more closely than [D_z] does. The linear correlation coefficients between [D_x] and IWP+LWP and [D_y] and *IWP+LWP* are 0.77 and 0.81, respectively, whereas the linear correlation coefficient between [D_z] and *IWP+LWP* is 0.45. The linear relations between [D_x] and *IWP+LWP* and [D_y] and *IWP+LWP* are statistically significant, whereas the linear relation between [D_z] and *IWP+LWP* is not. Further analysis shows that [D_x] and [D_y] are determined by [D_{x2}] and [D_{y1}], respectively (Fig. 10.8b, c), whereas [D_z] is determined by both [D_{z1}] and [D_{z2}] after the two-day integration (Fig. 10.8d).

The 3D ***DVV*** is different from the 2D ***DVV*** in the two ways. Firstly, the zonal component of the 2D ***DVV*** is out of phase with the convection whereas that of the 3D ***DVV*** is in phase with the convection. This is due to that fact that the zonal component of the 3D ***DVV*** is determined by [D_{x2}], which is excluded from the zonal component of the 2D ***DVV***. Secondly, the linear correlation coefficient between the vertical component of the 3D ***DVV*** and the convection is significantly smaller than

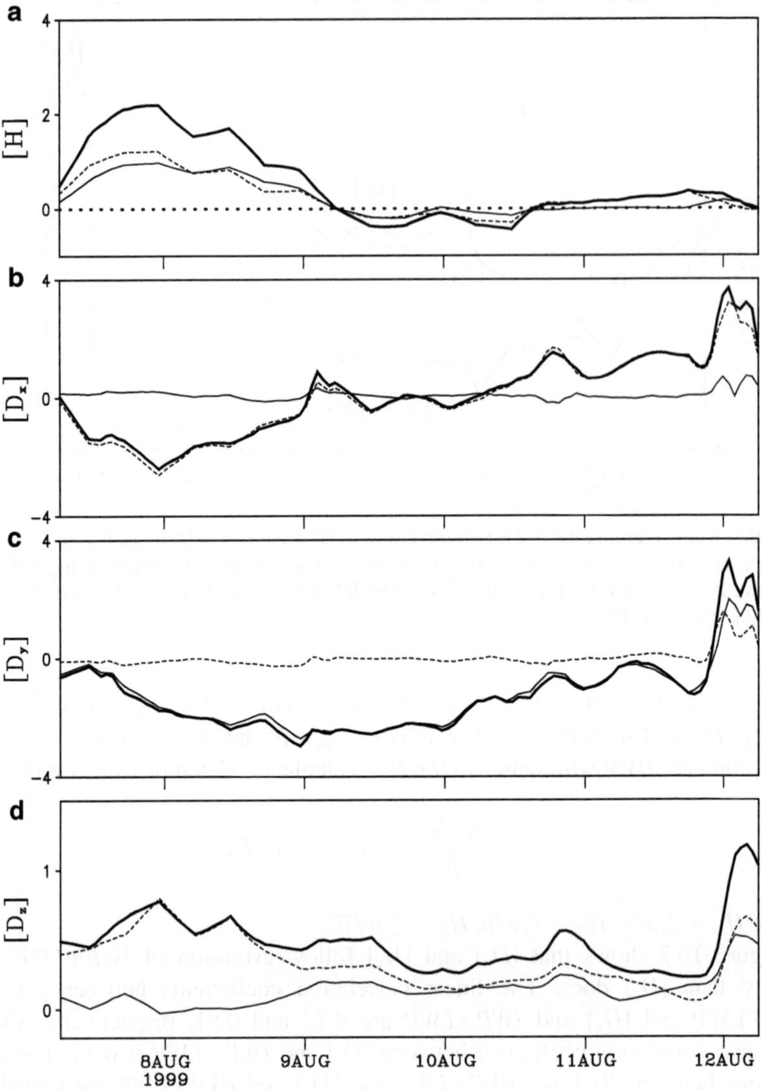

Fig. 10.8 Time series of (**a**) $[H]$ (*dark solid*), $[H_1]$ (*light solid*), $[H_2]$ (*dashed*), $[H_3]$ (*dot*), (**b**) $[D_x]$ (*dark solid*), $[D_{x1}]$ (*light solid*), $[D_{x2}]$ (*dashed*), (**c**) $[D_y]$ (*dark solid*), $[D_{y1}]$ (*light solid*), $[D_{y2}]$ (*dashed*), and (**d**) $[D_z]$ (*dark solid*), $[D_{z1}]$ (*light solid*), $[D_{z2}]$ (*dashed*) in the 5-day 3D cloud-resolving model simulation during KWAJEX. Units are in 10^2 m^2 s^{-2} for $[H]$, m^2 s^{-2} for $[D_x]$ and $[D_y]$, and 10^2 m^2 s^{-2} for $[D_z]$ (After Gao 2007)

those between the horizontal components and the convection, whereas the linear correlation coefficients between the vertical component of the 2D ***DVV*** and the convection and the horizontal component of the 2D ***DVV*** and the convection have similar magnitudes.

$[H]$ and *IWP*+*LWP* are loosely out of phase with the linear correlation coefficient of −0.28, which is not statistically significant. From (10.16), $[H]$ is determined by three components. $[H_1]$ and $[H_2]$ have similar magnitudes, which are much larger than that of $[H_3]$ (Fig. 10.8a). Thus, $[H]$ is determined by the covariance between horizontal components of vorticity and horizontal wind. The results indicate that the linear correlations between the 3D ***DVV*** and tropical oceanic convection are better than those between the helicity and tropical oceanic convection.

Figure 10.7 shows that during the strong upward motion at midnight of 11 August 1999, ***DVV*** has a large fluctuation correspondingly whereas $[H]$ does not. Due to the conservation property of equivalent potential temperature, the particles are restricted to move on the isentropic surface during an adiabatic process. During the development of convective systems, all particles move vertically on the isentropic surface that is nearly perpendicular to the earth surface, whereas 3D vorticity vector develops horizontally, which directly leads to a small $[H]$ and a large ***DVV***. Thus, ***DVV*** better captures strong convective signals than $[H]$ does.

When the wind is separated into rotational and divergent components, ***DVV*** contains both interactions between divergence wind and vorticity and rotational wind and vorticity, whereas $[H]$ only includes the interaction between divergence wind and vorticity. The large signals of ***DVV*** with strong upward motions during the convective development suggest the importance of the interaction between rotational wind and vorticity in the strong convective events.

References

Bennetts DA, Hoskins BJ (1979) Conditional symmetric instability- a possible explanation for frontal rainbands. Quart J Roy Meteor Soc 105:945–962

Betchov R (1961) Semi-isotropic turbulence and helicoidal flows. Phys Fuilds 4:925–926

Cao Z, Cho H (1995) Generation of moist vorticity in extratropical cyclones. J Atmos Sci 52:3263–3281

Cho H, Cao Z (1998) Generation of moist vorticity in extratropical cyclones. Part II: sensitivity to moisture distribution. J Atmos Sci 55:595–610

Danielsen EF, Hipskind RS (1980) Stratospheric-tropospheric exchange at polar latitudes in summer. J Geophys Res 85:393–400

Droegemeier KK, Lazarus SM (1993) The influence of helicity on numerically simulated convective storms. Mon Wea Rev 121:2005–2029

Emanuel KA (1979) Inertial instability and mesoscale convective systems. Part I: linear theory of inertial instability in rotating viscous fluids. J Atmos Sci 36:2425–2449

Ertel H (1942) Ein neuer hydrodynamischer wirbelsatz. Meteorology Zeitschr Braunschweigs 6:277–281

Gao S (2007) A three-dimensional dynamic vorticity vector associated with tropical oceanic convection. J Geophys Res. doi:10.1029/2007JD008247, (c) American Geophysical Union. Reprinted with permission

Gao S, Lei T, Zhou Y (2002) Moist potential vorticity anomaly with heat and mass forcings in torrential rain system. Chin Phys Lett 19:878–880

Gao S, Ping F, Li X, Tao WK (2004) A convective vorticity vector associated with tropical convection: A 2D cloud-resolving modeling study. J Geophys Res 109. doi:10.1029/2004JD004807, (c) American Geophysical Union. Reprinted with permission

Gao S, Cui X, Zhou Y, Li X, Tao WK (2005) A modeling study of moist and dynamic vorticity vectors associated with 2D tropical convection. J Geophys Res 110. doi:10.1029/2004JD005675, (c) American Geophysical Union. Reprinted with permission

Gao S, Li X, Tao WK, Shie CL, Lang S (2007) Convective and Moist Vorticity Vectors Associated with Three-Dimensional Tropical Oceanic Convection During KWAJEX, J Geophys Res 112. doi:10.1029/2006JD007179, (c) American Geophysical Union. Reprinted with permission

Hoskins BJ, Berrisford P (1988) A potential vorticity perspective of the storm of 15–16 October 1987. Weahter 43:122–129

Li X, Sui CH, Lau KM (2002) Dominant cloud microphysical processes in a tropical oceanic convective system: a 2-D cloud resolving modeling study. Mon Wea Rev 130:2481–2491

Lilly DK (1986) The structure, energetics and propagation of rotating convective storms. Part II: helicity and storm stabilization. J Atmos Sci 43:126–140

Montgomery MT, Farrell BF (1993) Tropical cyclone formation. J Atmos Sci 50:285–310

Shie CL, Tao WK, Simpson J (2003) Simulated KWAJEX convective systems using a 2D and 3D cloud resolving model and their comparisons with radar observations. 31st conference on radar meteorology, Seattle, Washington, DC, 6–12 August 2003

Tan Z, Wu R (1994) Helicity dynamics of atmospheric flow. Adv Atmos Sci 11:175–188

Thorpe AJ (1985) Diagnosis of balanced vortex structure using potential vorticity. J Atmos Sci 42:397–406

Xu Q (1992) Formation and evolution of frontal rainbands and geostrophic potential vorticity anomalies. J Atmos Sci 49:629–648

Chapter 11
Diurnal Variations of Tropical Oceanic Convection

Abstract In this chapter, diurnal variation of tropical rainfall is analyzed by calculating the diurnal composite of water vapor and heat budgets. The theoretical analysis of condensation and deposition shows that diurnal variation of temperature is primarily responsible for that of condensation and deposition. Diurnal variations of convective and stratiform rainfall are also examined based on the analysis of surface rainfall budget.

Keywords Diurnal variation • Condensation • Deposition • Diurnal composite

The diurnal variation of tropical oceanic convection is one of the most important components in tropical variability and plays a crucial role in regulating tropical hydrological and energy cycles. The dominant diurnal signal is the nocturnal peak in precipitation that occurs in the early morning. The second rainfall peak appears in the early afternoon. The nocturnal rainfall maximum may be primarily caused by the IR cooling or radiational difference between cloudy regions and clear-sky regions. The afternoon rainfall maximum may result from the increase of *SST*. The cloud-resolving model simulations could provide a powerful tool in enhancing the understanding of physical mechanisms associated with the formation of two rainfall peaks through the sensitivity experiments and associated budget analysis of heat, water vapor, and surface rainfall. In this chapter, a four-decade history of observational and numerical studies of diurnal rainfall variations is briefly reviewed. The nocturnal rainfall peak is discussed by analyzing domain-mean cloud-resolving model simulation data during TOGA COARE. The nocturnal and afternoon rainfall peaks are examined by analyzing grid data from a coupled ocean–cloud-resolving atmosphere model simulation. The diurnal variations of convective and stratiform rainfall are analyzed using the equilibrium data from cloud-resolving model simulations.

© Springer International Publishing Switzerland 2016

X. Li, S. Gao, *Cloud-Resolving Modeling of Convective Processes*,
Springer Atmospheric Sciences, DOI 10.1007/978-3-319-26360-1_11

11.1 Introduction

Kraus (1963) analyzed precipitation data collected from nine weather ships, showed nocturnal rainfall peaks, and suggested that solar heating and IR cooling tend to suppress convection during daytime and enhance convection during nighttime, respectively. The solar radiative heating reduces rainfall by evaporating cloud drops and lowering vapor condensation rate. Gray and Jacobson (1977) found that heavy rainfall is 2–3 times greater in the morning than in the late afternoon and evening and argued that the nocturnal rainfall peaks are enhanced by the secondary circulation forced by the radiational differences between cloudy regions and clear-sky regions. The cloud-radiative forcing causes upward motion and convection during nighttime through the low-level convergence. Randall et al. (1991) carried out sensitivity experiments using a general circulation model and showed that with the absence of the radiative effects of clouds, the phase of simulated diurnal rainfall cycle is not changed but the amplitude is much weakened. Xu and Randall (1995) conducted the experiments with the cloud-resolving model and showed the diurnal variation regardless of whether radiation is interactive or non-interactive, indicating the mechanism proposed by Gray and Jacobson plays a secondary role in the diurnal processes. Tao et al. (1996) carried out a series of cloud-resolving model simulations to study cloud–radiation interaction mechanisms and emphasized that the increase of surface precipitation by IR cooling is due to the increased relative humidity. The surface rainfall is not sensitive to the cloud-top cooling and cloud-base warming and differential cooling between clear-sky and cloudy regions. Liu and Moncrieff (1998) from their cloud-resolving model simulations found that the simulated diurnal variation is primarily due to the direct interaction between radiation and convection, and the mechanism proposed by Gray and Jacobson (1977) is a secondary factor.

Sui et al. (1997) conducted the diurnal analysis using the observational data from TOGA COARE. The data is first categorized into the disturbed and undisturbed periods by calculating the standard deviation of brightness temperature measured by the GMS operated by the Japanese Meteorological Agency. Over the disturbed periods, total surface rain rates well as convective and stratiform rain rates reach the maxima at 0300 LST. Fractional coverage for stratiform clouds has a maximum at 0300 LST whereas fractional coverage for convective clouds does not show a significant diurnal variation. Diurnal variation of the rain rate histogram shows the evolution of nocturnal rainfall has a growing phase from 2200 to 0300 LST, when a wide range of convection (rain rate is higher than 0.5 mm h^{-1}) becomes enhanced with most occurrences within 0.5–5 mm h^{-1}. The nocturnal rainfall is associated with anomalous ascending motion in the layer between 500 and 200 mb at 0400 LST. Over the undisturbed periods, the surface rain rate is very small, but shows a maximum from 1200 to 1800 LST. The diurnal variation of rain rate histogram shows that the evolution of afternoon rainfall has a growing phase from 1200 to 1800 LST, when most occurrences of rain rates are within 0.2–0.4 mm h^{-1}. The afternoon rainfall peak is associated with the maximum *SST* after the solar radiation flux reaches the maximum. Based on the observational analysis, Sui et al. suggested that the nocturnal rainfall peak is related to the destabilization by radiative cooling during nighttime, and the falling temperature increases the available precipitable water that could be easier for the surface precipitation.

Sui et al. (1998) conducted the cloud-resolving model simulations to test their nocturnal rainfall mechanism. An experiment with the imposed large-scale ascending motion and a time-invariant *SST* generates a positive rainfall anomaly in the night and a negative rainfall anomaly in the day. The maximum simulated rain rate occurs around 0200 LST. Two additional experiments are carried out: one experiment with a zero-imposed vertical velocity and a time-invariant *SST* and the other experiment with the cloud–radiation interaction suppressed. All three experiments show a dominant nocturnal rainfall maximum. The results imply that cloud–radiation interaction does not play a crucial role in the formation of the nocturnal rainfall peak. The common feature in all the experiments is the falling temperature induced by the nocturnal IR cooling. Thus, the numerical experiments support the suggestion by Sui et al. (1997) that the nocturnal rainfall peak is related to more (less) available precipitable water in the night (day) due to the diurnal cooling/heating cycle. They also conducted the experiment with zero-imposed vertical velocity and a zonally uniform, diurnally varied *SST* and found that the simulated diurnal variations still have a nocturnal rainfall maximum but with a weaker magnitude and a secondary rainfall peak in the afternoon. This indicates that the maximum *SST* in the afternoon induces the unstable atmosphere that eventually leads to the rainfall peak.

11.2 Diurnal Variation of Model Domain-Mean Surface Rainfall

Diurnal variations of tropical oceanic precipitation are analyzed with the surface rainfall equation (2.5) using hourly model domain-mean simulation data in COARE. Figure 11.1 shows diurnal composites of P_s and four contributing components (Q_{WVT}, Q_{WVF}, Q_{WVE}, and Q_{CM}). The maximum surface rain rates (~0.4 mm h^{-1}) occur early morning and afternoon. Q_{WVE} has a constant of roughly 0.2 mm h^{-1}. Q_{CM} has fluctuations around 0 with amplitudes of 0.1 mm h^{-1} or less. Q_{WVT} and Q_{WVF} show significant diurnal cycles. Q_{WVF} has its maximum (~0.2 mm h^{-1}) at hour 16, decreases during nighttime, and reaches its minimum (~0 mm h^{-1}) at hour 10. Since the *SST* does not show any strong diurnal signals (Fig. 4.1c), a forced rainfall peak in the afternoon is associated with the imposed large-scale vertical velocity. Q_{WVT} has its maximum (~0.2 mm h^{-1}) at hours 5–7 and decreases to its minimum (~−0.1 mm h^{-1}) at hour 14. Maximum Q_{WVT} occurs at hours 5–7 when Q_{WVF} reaches its minimum. Positive Q_{WVT} signifies a local atmospheric drying. Thus, the rainfall maximum in the early morning is in phase with the local atmospheric drying.

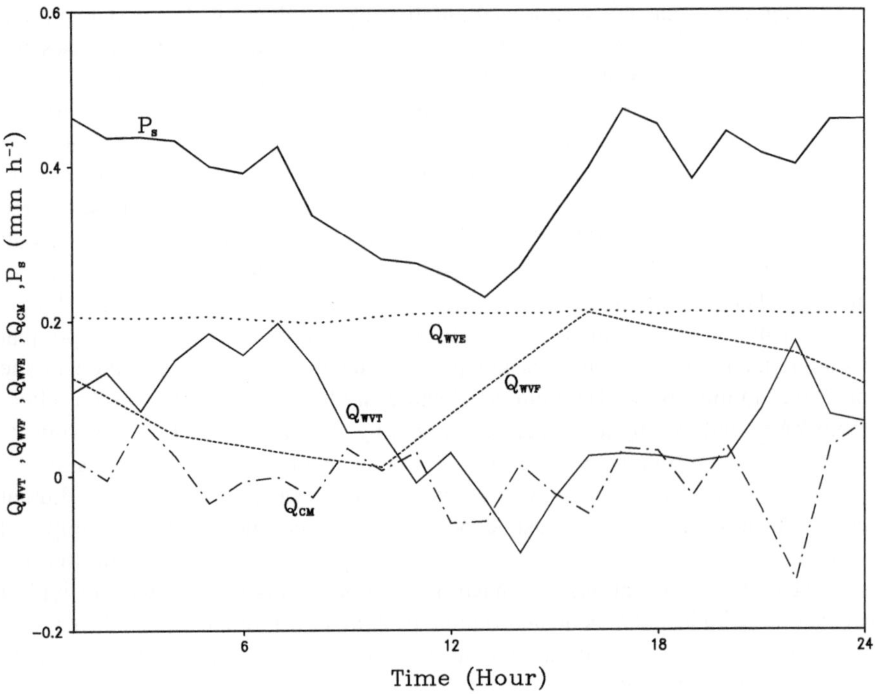

Fig. 11.1 Diurnal composites of model domain-mean Q_{WVT} *(light solid)*, Q_{WVF} *(dashed)*, Q_{WVE} *(dot)*, Q_{CM} *(dashed dot)*, and P_s *(dark solid)* simulated in COARE (Unit is mm h^{-1})

While the vapor and cloud budgets basically represent the redistributions of the vapor and cloud hydrometeors, the thermal forcing may act to trigger the variations of vapor, clouds, and surface rainfall. Thus, the diurnal composite of model domain and mass-weighted mean thermal budget is analyzed to explain the nocturnal rainfall peak. The diurnal composite of the mean thermal budget (3.3) is shown in Fig. 11.2, and the diurnal composite of *CAPE* for the reversible moist adiabatic process (2.11 and 2.12b) is also calculated and shown in Fig. 11.3. After the solar heating reaches its maximum at noon, the solar heating starts to decrease its magnitude. The net radiation becomes cooling around hour 15 when IR cooling becomes dominant. The net radiative cooling reaches its maximum rate after hour 18 when solar radiation vanishes. The advective cooling (Q_{HF}) associated with the ascending motion in the afternoon also acts to lower local temperature. The similar cooling rates from Q_{HF} and Q_{RAD} occur from hour 18 to hour 8. Thus, the radiative and advective cooling lowers local temperature (positive Q_{HT}) from noon to midnight. The temperature decrease eventually halts the increase of *CAPE* around hour 20 (Fig. 11.3) after sunset. Continuous radiative and advective cooling causes a significant release of *CAPE*, a clear signal for destabilizing the atmosphere. The *CAPE* is transferred to energize nocturnal convective development. The *CAPE* reaches minimum around hour 8.

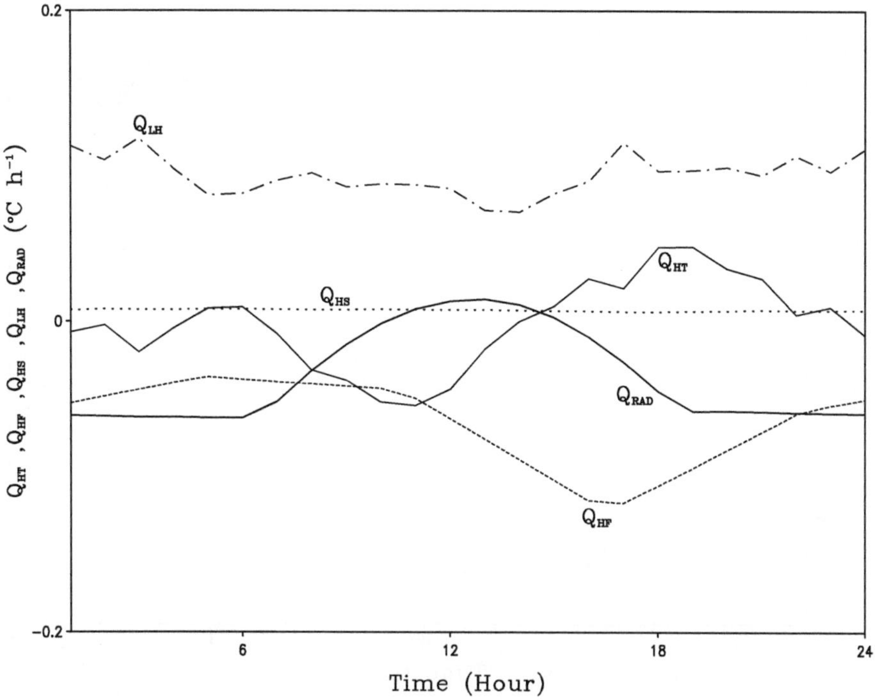

Fig. 11.2 Diurnal composites of model domain-mean Q_{HT} (*light solid*), Q_{HF} (*dashed*), Q_{HS} (*dot*), Q_{LH} (*dashed dot*), and Q_{RAD} (*dark solid*) simulated in COARE (Unit is °C h^{-1})

To understand the peaks of Q_{WVT} and Q_{WVF}, model domain-mean PW budget (3.1a) can be also written as

$$Q_{WVT} + Q_{WVF} + Q_{WVE} = Q_{WVOUT} - Q_{WVIN}. \tag{11.1}$$

Here, Q_{WVOUT} ($=[P_{CND} + \Sigma P_{DEP}]$) is a vapor sink, whereas Q_{WVIN} ($=[P_{REVP}+P_{MLTS} + P_{MLTG}]$) is a vapor source.

Diurnal composites of Q_{WVOUT} and Q_{WVIN} (Fig. 11.4) display that the diurnal amplitude of Q_{WVOUT} (0.4 mm h^{-1}) is more than twice as large as that of Q_{WVIN} (<0.2 mm h^{-1}), although both show similar diurnal signals to P_s. Thus, the diurnal variation of Q_{WVOUT} is consistent with the diurnal variation of $Q_{WVT} + Q_{WVF} + Q_{WVE}$, which is largely contributed by Q_{WVT} and Q_{WVF} as indicated in Fig. 11.1. The afternoon peak of Q_{WVOUT} occurs at hour 17, one hour later after Q_{WVF} reaches its maximum at hour 16 (Fig. 11.1). This suggests that the afternoon Q_{WVF} maximum associated with the imposed large-scale upward motion whose maximum (3 cm s^{-1}) appears around 340 mb (not shown) is responsible for the afternoon Q_{WVOUT} peak through large vapor condensation and deposition rates. The large magnitudes of Q_{WVOUT} at hours 5–7 account for the early morning Q_{WVT} peak. Thus, a favorable coupling vapor–temperature (moist and cool) condition has

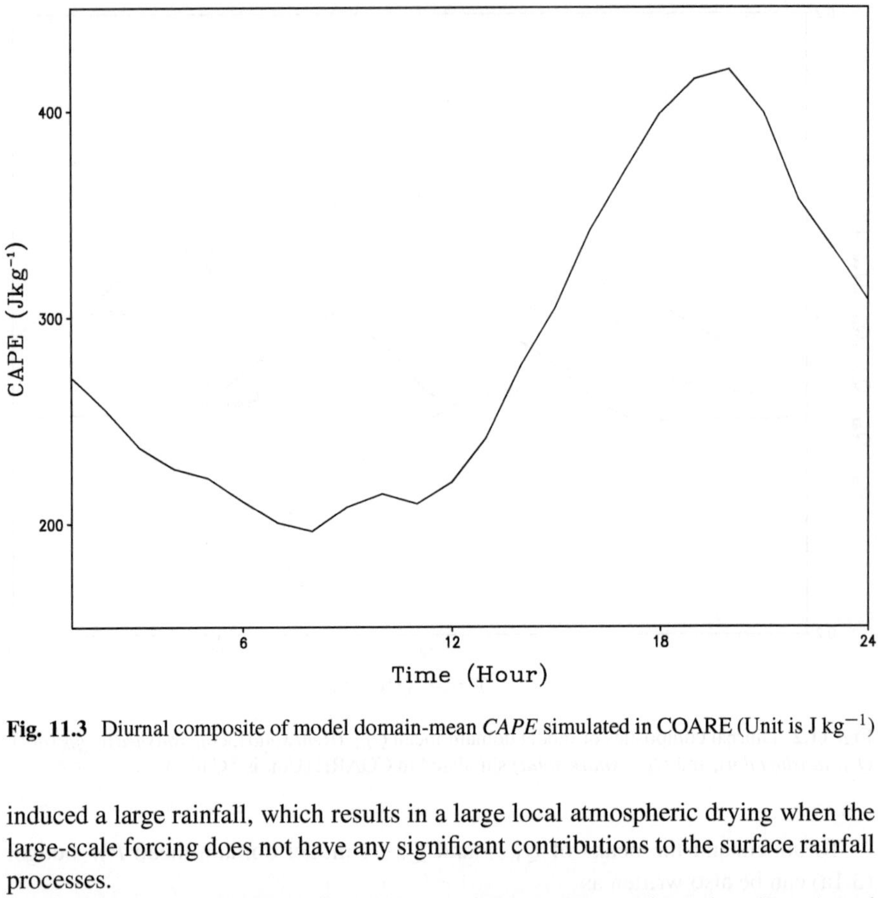

Fig. 11.3 Diurnal composite of model domain-mean *CAPE* simulated in COARE (Unit is J kg^{-1})

induced a large rainfall, which results in a large local atmospheric drying when the large-scale forcing does not have any significant contributions to the surface rainfall processes.

Li (2004) conducted a scale analysis of the vapor condensation (P_{CND}) and deposition (P_{DEP}) rates to explain the maxima in the condensation and deposition rates in the morning. Following Tao et al. (1989), the sum of P_{CND} and (P_{DEP}) from (1.16) and (1.26) can be expressed by

$$P_{CND} + P_{DEP} = \frac{q_v - (q_{ws} + q_{is})}{\Delta t}, \tag{11.2}$$

since

$$\frac{A_1 q_c q_{ws} + A_2 q_i q_{is}}{q_c + q_i} \frac{L_v (T - T_{oo}) + L_s (T_o - T)}{c_p (T_o - T_{oo})} \ll 1.$$

Each variable can be decomposed into a daily mean (m) and a diurnal anomaly (d). $T = T_m + T_d$ and $q_v = q_{vm} + q_{vd}$. The diurnal anomaly of $P_{CND} + P_{DEP}$ then becomes

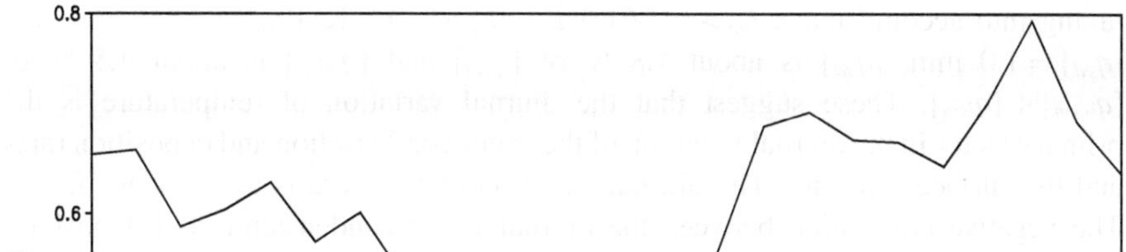

Fig. 11.4 Diurnal composites of model domain-mean Q_{WVOUT} (*solid*) and Q_{WVIN} (*dashed*) simulated in COARE (Unit is mm h^{-1})

$$(P_{CND} + P_{DEP})_d = \frac{q_{vd} - q_{wisd}}{\Delta t}, \tag{11.3}$$

where

$$q_{wisd} = -\frac{E_1 T_d}{(T_m - F_1)^2} q_{wsm} - \frac{E_2 T_d}{(T_m - F_2)^2} q_{ism}. \tag{11.3a}$$

Note that the relations $1/(1+x) = 1 - x$ and $e^x = 1 + x$ for $x << 1$ are used in the derivation of (11.3). (11.3) indicates that the diurnal variation of vapor condensation and deposition is determined by the diurnal variation of temperature and specific humidity. The diurnal variation of temperature and specific humidity is therefore further analyzed.

The linear correlation coefficients between the diurnal anomalies of vapor condensation and deposition rates and mass-weighted mean temperature and between the diurnal anomalies of vapor condensation and deposition rates and *PW* are −0.68 and 0.52, respectively. The correlation coefficient for 24 samples at the 99 % confidence level is 0.5. Thus, the temperature correlation is well above, but the moisture correlation is only marginally above the 99 % confidence level. Furthermore, taking into account that $< T_m > = 261.5$ K, $<T_d> = 0.5$ K, $[q_{vm}] = 54.7$ mm, and $[q_{vd}] = 1.0$ mm, $[q_{vd}]$ is about 1.8 % of $[q_{vm}]$ and $[q_{wisd}]$ is about 4.5 % of $[q_{wsm}] + [q_{ism}]$. These suggest that the diurnal variation of temperature is the primary factor in the diurnal variation of the vapor condensation and deposition rates and the surface rain rate. The diurnal variation of moisture is a secondary factor. The negative correlation between the diurnal vapor condensation and deposition anomalies and the mass-weighted mean temperature indicates that colder temperatures cause lower saturated mixing ratios making it easier for water vapor to be condensed and deposited into precipitation. Therefore, nocturnal radiative cooling leads to colder air temperatures that make it easier for clouds to develop and hence rainfall.

11.3 Diurnal Analysis with Grid Simulation Data from a Coupled Model

Gao et al. (2006) analyzed the grid data from the 2D coupled ocean–cloud-resolving atmosphere model simulation with the imposed COARE forcing by categorizing the data into two groups: the amplitude of diurnal *SST* variation is smaller than 0.6 °C in case W whereas it is larger than 0.6 °C in case S. The amplitudes of diurnal *SST* composites are 0.8 °C in case S and 0.3 °C in case W (Fig. 11.5b in left panel), respectively. The *SST* peaks occur at hour 16 in both cases. The diurnal variations of surface rain rates in the two cases are virtually out of phase (Fig. 11.5a in left panel). In case W, the large rain rate occurs in the first half of the day, with the maximum of 0.67 mm h^{-1} at hours 6–10. The rain rate reaches the minimum of 0.31 mm h^{-1} at hour 16. In case S, the surface rain rate decreases in the early morning and reaches the minimum of 0.17 mm h^{-1} at hours 10–11. It increases to its peak of 0.54 mm h^{-1} at hour 16. The rainfall peaks around hours 6–10 in case W are virtually different from nocturnal rainfall peaks around hours 2–3 in the previous studies (also see Sect. 8.1). However, the rainfall peaks appear at hours 2–3 when the diurnal variations of rainfall in case S are included in the mean variations, which is consistent with nocturnal rainfall peak in the observed area-mean surface rain rates.

To examine the diurnal variations of clouds, *LWP* and *IWP* are analyzed. Like the surface rain rates in case W versus in case S, the diurnal variations of both *LWP* and *IWP* in the two cases are out of phase (Fig. 11.5a in right panel). The *LWP* and *IWP* have the similar magnitudes. The *IWP* is slightly larger than the *LWP* in case W whereas the *IWP* is slightly smaller than the *LWP* in case S. The differences between the *IWP* and *LWP* become larger when the surface rain rates hit the maximum in both cases. To examine the cloud source, the diurnal composites of the P_{CND} and ΣP_{DEP} are plotted (Fig. 11.5b in right panel). Like the *LWP* and *IWP* in case W versus in case S, the diurnal variations of vapor condensation and deposition rates in the two cases are out of phase. The vapor condensation rates are much larger than

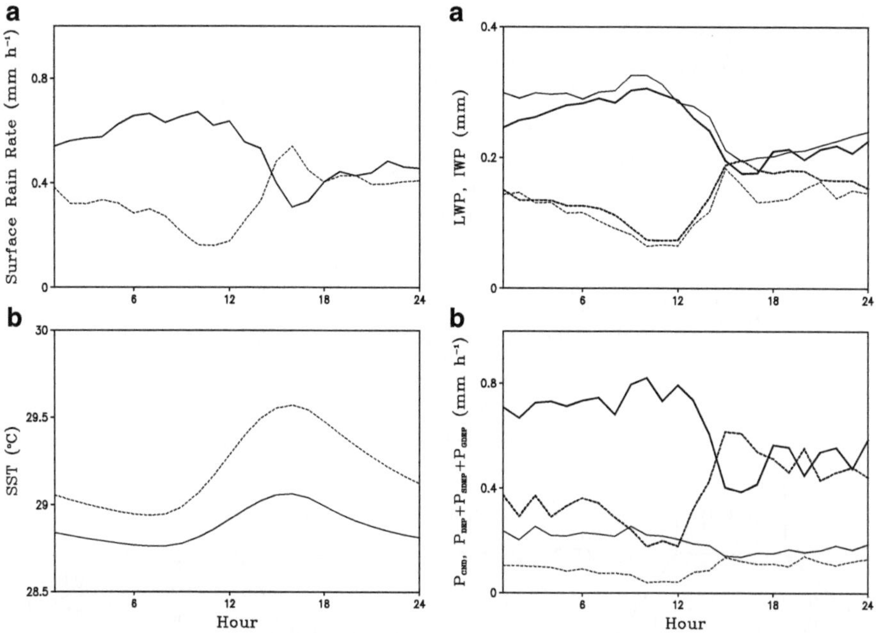

Fig. 11.5 Left panel: diurnal composites of (**a**) P_s (mm h^{-1}) and (**b**) *SST* (°C). *Right panel*: diurnal composites of (**a**) *LWP* (dark) and *IWP* (light) (mm) and (**b**) P_{CND} (dark) and sum of P_{DEP}, P_{SDEP}, and P_{GDEP} (light) (mmh^{-1}). *Solid* and *dashed lines* denote cases W and S in the coupled model simulation, respectively (After Gao et al. 2006)

the vapor deposition rates in both cases, in particular, in case W because the vapor deposition rates are almost constant. Thus, the major cloud sources come from the vapor condensation process. In case W, the P_{CND} varies from 0.4 to 0.8 mm h^{-1} whereas the ΣP_{DEP} is about 0.2 mmh^{-1}. The P_{CND} has a dramatic decrease from hour 12 to hour 15 when the *LWP* has a significant decrease. In case S, the P_{CND} varies from 0.2 to 0.65 mmh^{-1} whereas the ΣP_{DEP} is about 0.1 mmh^{-1}. The P_{CND} has a dramatic increase from hour 12 to hour 15 when the *LWP* increases its values twice. The maximum of the P_{CND} is 0.62 mm h^{-1}. Large differences in magnitudes between the vapor condensation and deposition rates and similarities of magnitudes between the *LWP* and *IWP* imply the important exchange processes between the water and ice clouds.

Cloud microphysics budgets in case W are calculated by averaging within hours 1–10 (Fig. 11.6a) whereas the budgets in case S are calculated by averaging within hours 16–20 (Fig. 11.6b) when the surface rain rates reach their peaks and the differences between the *IWP* and *LWP* are the largest. The *IWP* (0.3 mm) is 7 % larger than the *LWP* (0.28 mm) in case W whereas the *IWP* (0.14 mm) is 32 % smaller than the *LWP* (0.18 mm) in case S. In case W, vapor condensation rate ($[P_{CND}]$) is 0.73 mm h^{-1} and vapor deposition rates ($[\Sigma P_{DEP}]$) are 0.25 mm h^{-1}. Vapor deposition rates contribute to the cloud growth by 26 %. In case S, the vapor

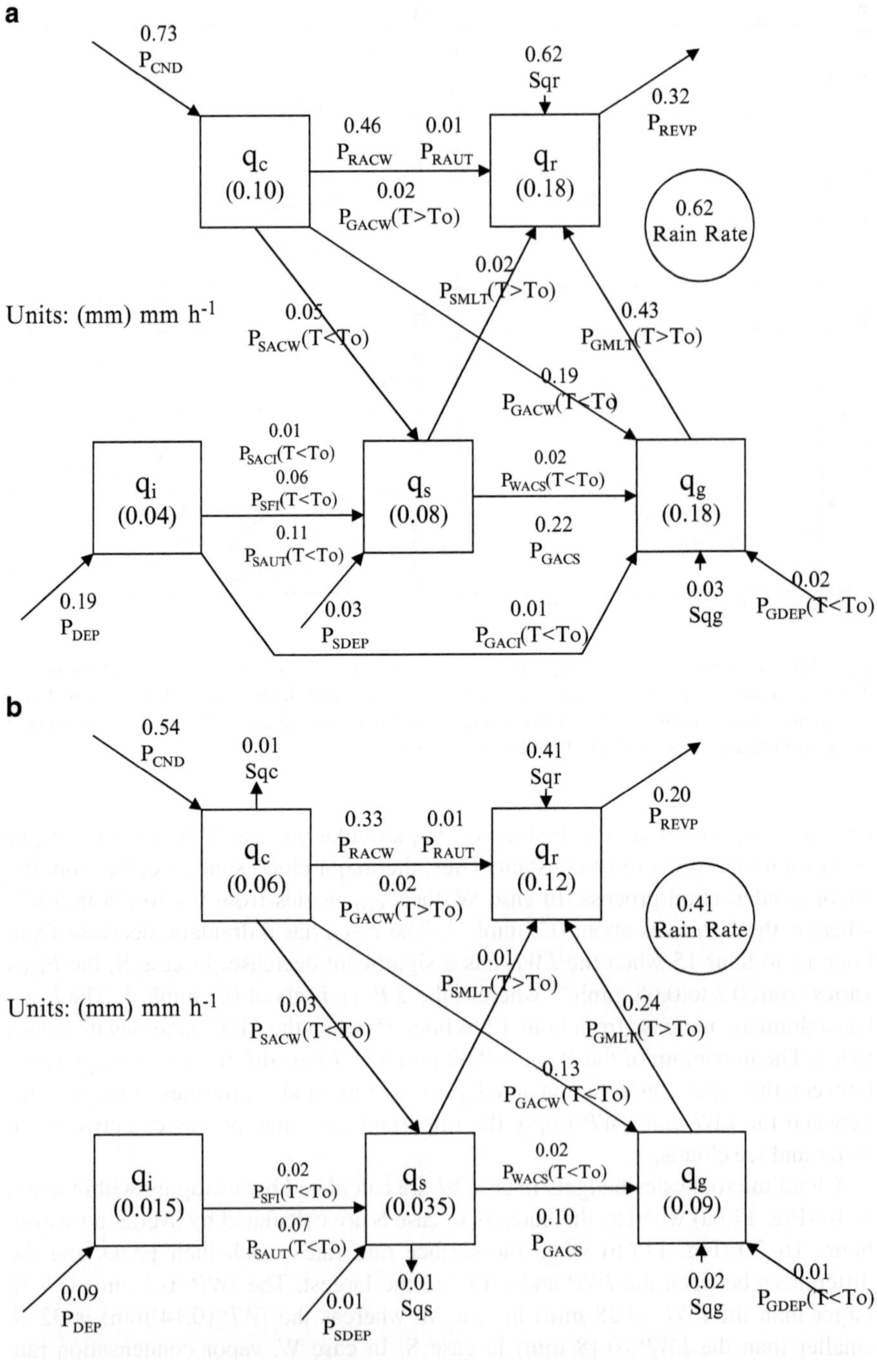

Fig. 11.6 Cloud microphysics budgets in (**a**) case W (averaged within hours 1–10) and (**b**) case S (averaged within hours 16–20) simulated in coupled model experiment. Units for cloud hydrometeors and conversions are mm and mmh^{-1}, respectively (After Gao et al. 2006)

condensation rate is 0.54 mmh^{-1} and the vapor deposition rates are 0.12 mm h^{-1}. The vapor deposition rates contribute to the cloud growth by 18 %. Although rates are different in the two cases, 67 % of $[P_{CND}]$ goes to the growth of rain mainly through $[P_{RACW}]$ and 30–33 % contributes to the growth of precipitation ice mainly through $[P_{GACW}]$. The major sources responsible for rainfall are $[P_{RACW}]$ and $[P_{GMLT}]$. Rainfall source comes from $[P_{GMLT}]$ by 48 % in case W and 41 % in case S, respectively. Thus, smaller *IWP* in case S results from smaller vapor deposition rates, which is in turn due to warmer air temperature by solar heating, in comparison with those in case W. Thus, vapor deposition rates are directly responsible for the growth of ice clouds.

11.4 Diurnal Variations of Convective and Stratiform Rainfall

Cui (2008) analyzed diurnal variations of convective and stratiform rainfall using hourly equilibrium simulation data from experiment SST29, in which the zero vertical velocity and constant *SST* of 29 °C are imposed in the model during the integration (also see Sect. 11.2). The convective rain rate is higher than the stratiform rain rate during early morning and evening. The diurnal variation of convective rain rate is mainly determined by that of vapor convergence over convective regions, whereas diurnal variation of stratiform rainfall is determined by that of local vapor and hydrometeor loss over raining stratiform regions.

The effects of diurnal variation of *SST* on diurnal variations of convective and stratiform rainfall can be further examined by comparing SST29 with SST29D1. A diurnally varied *SST* with the time mean of 29 °C and diurnal amplitude of 1 °C is imposed in SST29D1 (also see Sect. 11.3). Hourly equilibrium simulation data are used to make a diurnal composite of surface rainfall. The differences in diurnal variation of model domain-mean surface rain rates between SST29D1 and SST29 are mainly caused by those of convective surface rain rates (Fig. 11.7) since the stratiform rain rates in both experiments show similar diurnal variations. The differences in diurnal variation of vapor convergences (Q_{WVF}) between SST29D1 and SST29 mainly account for the differences in diurnal variation of convective rain rates (Fig. 11.7c) because the local vapor changes (Q_{WVT}) and surface evaporation rates (Q_{WVE}) have at least one order of magnitudes smaller than vapor convergences, and hydrometeor sources in both experiments display similar variations (Fig. 11.7).

Since higher *SST* induced by solar radiative heating significantly enhances convective rain rate for hour 15 to hour 20, the surface rainfall budgets averaged in this period in both SST29 and SST29D1 are analyzed. The model domain-mean calculations of surface rainfall equation in the late afternoon (Table 11.1) show that the surface rain rate is larger in SST29D1 (0.1322 mm h^{-1}) than in SST29 (0.1023 mm h^{-1}) largely because the local atmospheric moistening rate is smaller in SST29D1 (−0.0131 mm h^{-1}) than in SST29 (−0.0369 mm h^{-1}) and partly

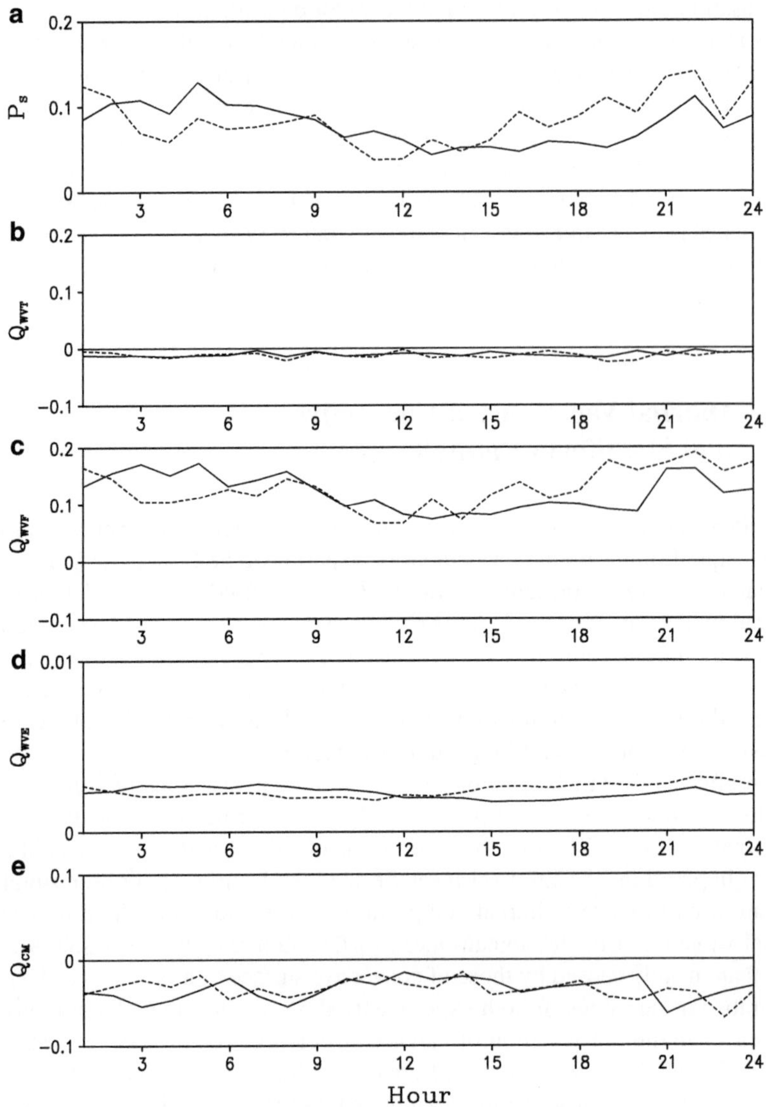

Fig. 11.7 Diurnal composites of (**a**) P_s, (**b**) Q_{WVT}, (**c**) Q_{WVF}, (**d**) Q_{WVE}, and (**e**) Q_{CM} over convective regions simulated in SST29 (*solid*) and SST29D1 (*dashed*) (Unit is mm h^{-1})

because the surface evaporation rate is larger in SST29D1 (0.1456 mm h^{-1}) than in SST29 (0.1341 mm h^{-1}). The partition calculations of surface rain rate in the late afternoon reveal that model domain-mean surface rain rate is larger in SST29D1 than in SST29 mainly because the convective surface rain rate is larger in SST29D1 (0.0862 mm h^{-1}) than in SST29 (0.0547 mm h^{-1}), which is mainly due to the

Table 11.1 Time means of fractional coverage, P_s, Q_{WVT}, Q_{WVF}, Q_{WVE}, and Q_{CM} over clear-sky regions, raining stratiform regions, convective regions, and non-raining stratiform regions and their sums (model domain means) averaged using diurnal composite data from hour 15 to hour 20 in (a) SST29 and (b) SST29D1

(a)	Clear-sky regions	Raining stratiform regions	Convective regions	Non-raining stratiform regions	Model domain mean
P_s	0.00	0.0476	0.0547	0.0	0.1023
Q_{WVT}	−0.0056	0.0178	−0.0115	−0.0375	−0.0369
Q_{WVF}	−0.0690	−0.0001	0.0929	−0.0235	0.0
Q_{WVE}	0.0744	0.0040	0.0019	0.0538	0.1341
Q_{CM}	0.0002	0.0258	−0.0286	0.0072	0.0046
(b)					
P_s	0.0	0.0459	0.0862	0.0	0.1322
Q_{WVT}	0.0059	0.0164	−0.0157	−0.0197	−0.0131
Q_{WVF}	−0.0845	−0.0062	0.1374	−0.0467	0.0
Q_{WVE}	0.0770	0.0048	0.0027	0.0611	0.1456
Q_{CM}	0.0016	0.0309	−0.0382	0.0054	−0.0003

Unit is mm h^{-1}

fact that the vapor convergence rate over convective regions is larger in SST29D1 (0.1374 mm h^{-1}) than in SST29 (0.0929 mm h^{-1}). Due to cyclic boundary condition in the model, the vapor convergence in convective regions is largely balanced by vapor divergence in rainfall-free regions. In SST29D1, the surface evaporation (0.1381 mm h^{-1}) is used to offset vapor divergence (−0.1312 mm h^{-1}) in rainfall-free regions. In SST29, the surface evaporation (0.1282 mm h^{-1}) offsets vapor divergence (−0.0925 mm h^{-1}) and enhances water vapor (−0.0431 mm h^{-1}) in rainfall-free regions. SST29D1 transports more water vapor from rainfall-free regions to convective regions and thus produces larger convective rainfall and model domain-mean surface rainfall than SST29 does.

To examine cloud-radiative effects on diurnal variations of tropical convective and stratiform rainfall, additional experiment SST29NCR (without cloud-radiative effects) is carried out (also see Sect. 11.4). The magnitude of diurnal variation of convective rain rate is larger in SST29NCR than in SST29 whereas the diurnal phases in the two experiments are similar (Fig. 11.8a). The diurnal variations of stratiform rain rate in the two experiments are similar in both magnitude and phase (Fig. 11.9a). This indicates that the magnitude of diurnal variation of convective rain rate is sensitive to the cloud-radiative effects whereas the phase of diurnal variation of convective rain rate and both magnitude and phase of diurnal variation of stratiform rain rate are not.

Over convective regions, the positive difference in rain rate for SST29NCR-SST29 is mainly contributed to by the positive difference in vapor convergence since the differences in local vapor and hydrometeor changes are negative and the difference in surface evaporation is small (Fig. 11.8). Over stratiform regions, the similar rain rates in the two experiments are due to the fact that the negative

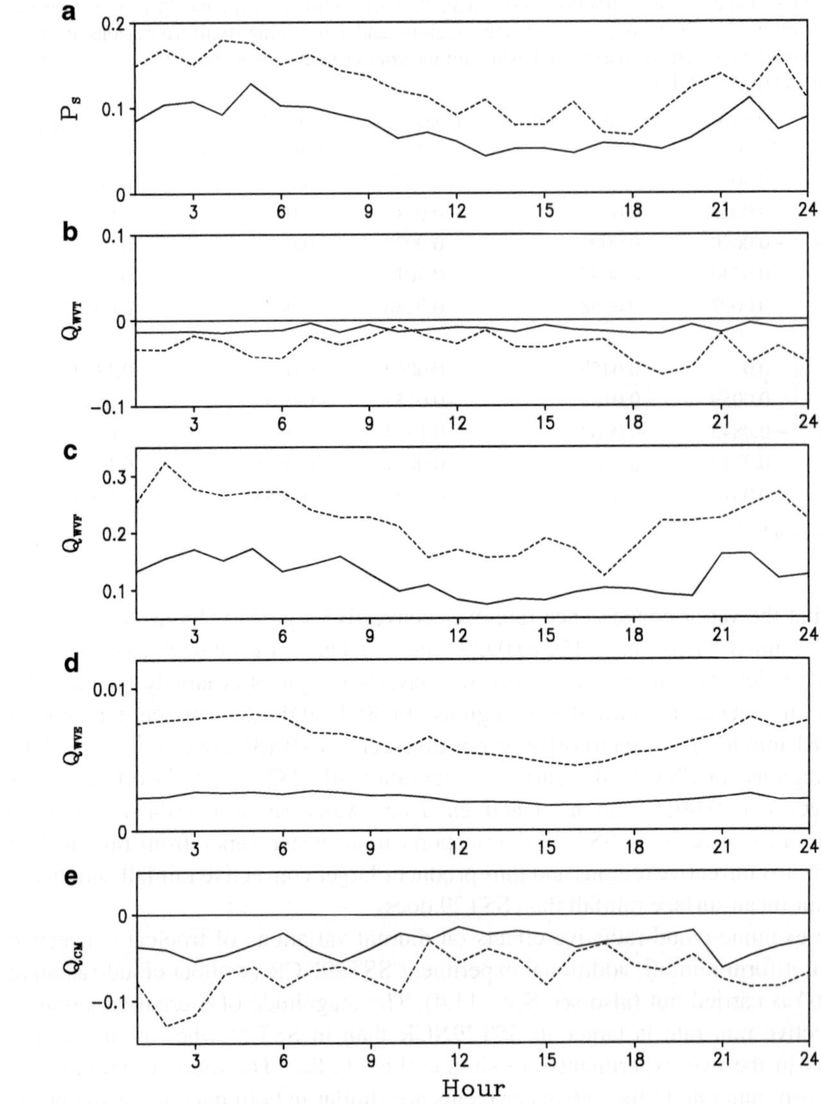

Fig. 11.8 Diurnal composites of (**a**) P_s, (**b**) Q_{WVT}, (**c**) Q_{WVF}, (**d**) Q_{WVE}, and (**e**) Q_{CM} over convective regions simulated in SST29 (*solid*) and SST29NCR (*dashed*) (The unit is mm h^{-1})

difference in vapor convergence for SST29NCR-SST29 is nearly balanced by the positive differences in local vapor and hydrometeor changes and surface evaporation (Fig. 11.9).

The positive difference in vapor convergence over convective regions (Fig. 11.8c) for SST29NCR-SST29 is balanced by the negative difference in vapor divergence over stratiform regions (Fig. 11.9c) and rainfall-free regions (Fig. 11.10b) since

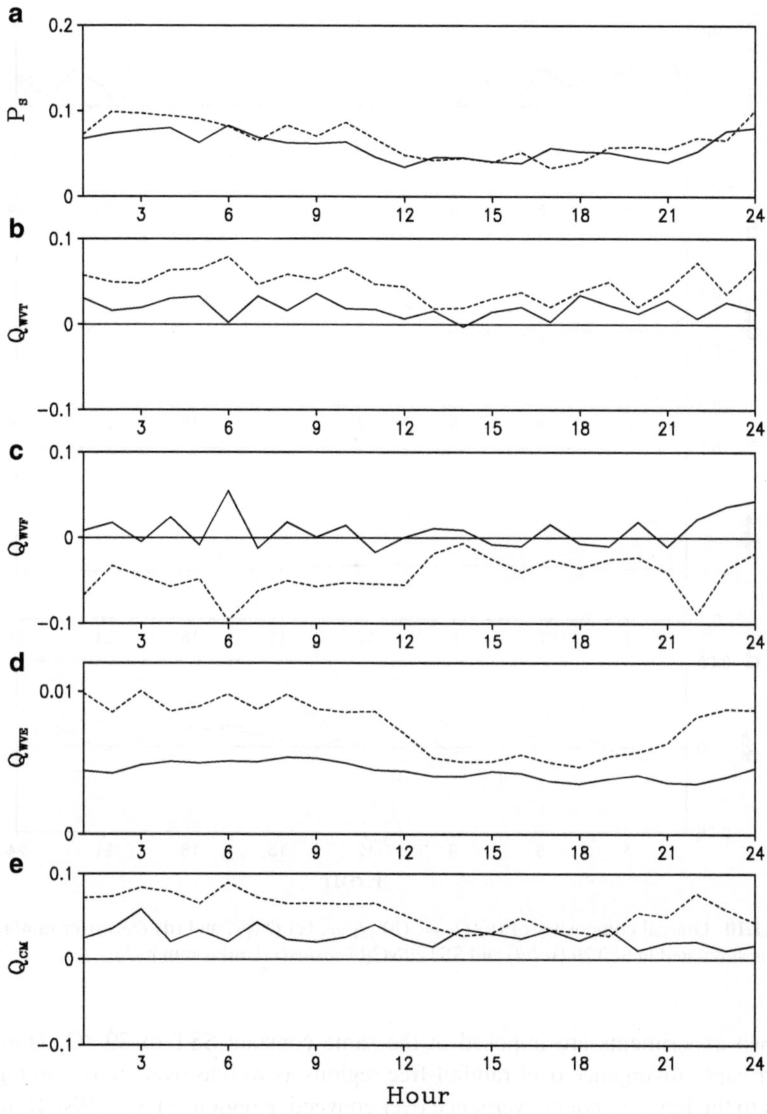

Fig. 11.9 As in Fig. 11.8 except over raining stratiform regions

the lateral boundary in the model is cyclic. Over rainfall-free regions, the negative difference in vapor divergence is mainly compensated by the positive difference in surface evaporation (Fig. 11.10c). The exclusion of cloud-radiative effects in the model produces cold and dry equilibrium atmosphere (see Sect. 11.4) than the inclusion does. The drier atmosphere in SST29NCR leads to larger surface evaporation and vapor divergence than the warmer atmosphere in SST29 does while

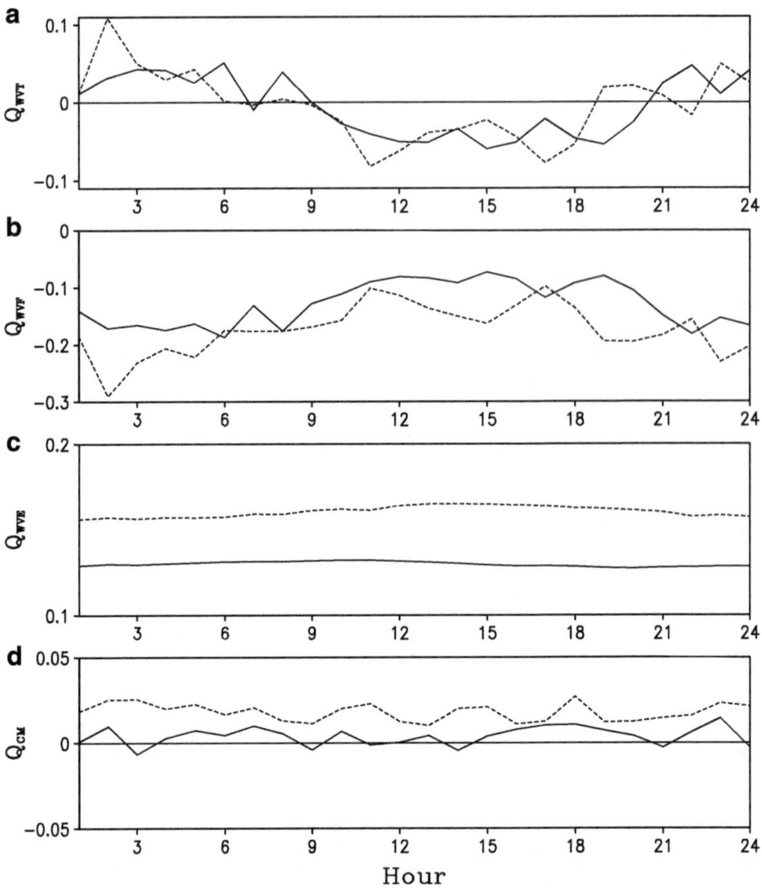

Fig. 11.10 Diurnal composites of (**a**) Q_{WVT}, (**b**) Q_{WVF}, (**c**) Q_{WVE}, and (**d**) Q_{CM} over rainfall-free regions simulated in SST29 (*solid*) and SST29NCR (*dashed*) (Unit is mm h^{-1})

the two experiments are imposed in the same constant SST of 29 °C. Thus, the larger vapor divergence over rainfall-free regions as well as over stratiform regions leads to the larger vapor convergence over convective regions in SST29NCR than in SST29.

References

Cui X (2008) A cloud-resolving modeling study of diurnal variations of tropical convective and stratiform rainfall. J Geophys Res 113. doi:10.1029/2007JD008990

Gao S, Ping F, Li X (2006) Cloud microphysical processes associated with the diurnal variations of tropical convection: a 2D cloud resolving modeling study. Meteorog Atmos Phys 91:9–16

Gray WM, Jacobson RW (1977) Diurnal variation of deep cumulus convection. Mon Weather Rev 105:1171–1188

Kraus EB (1963) The diurnal precipitation change over the sea. J Atmos Sci 20:546–551

Li X (2004) Cloud modeling in the tropical deep convective regime. In: Zhu X (ed) Observation, theory, and modeling of atmospheric variability. World Scientific, Hackensack

Liu C, Moncrieff MW (1998) A numerical study of the diurnal cycle of tropical oceanic convection. J Atmos Sci 55:2329–2344

Randall DA, DA Harshvardhan D (1991) Diurnal variability of the hydrologic cycle in a general circulation model. J Atmos Sci 48:40–62

Sui CH, Lau KM, Takayabu Y, Short D (1997) Diurnal variations in tropical oceanic cumulus ensemble during TOGA COARE. J Atmos Sci 54:639–655

Sui CH, Li X, Lau KM (1998) Radiative-convective processes in simulated diurnal variations of tropical oceanic convection. J Atmos Sci 55:2345–2359

Tao WK, Simpson J, McCumber M (1989) An ice-water saturation adjustment. Mon Weather Rev 117:231–235

Tao WK, Lang S, Simpson J, Sui CH, Ferrier BS, Chou MD (1996) Mechanisms of cloud-radiation interaction in the tropics and midlatitude. J Atmos Sci 53:2624–2651

Xu KM, Randall DA (1995) Impact of interactive radiative transfer on the macroscopic behavior of cumulus ensembles. Part II: mechanisms for cloud-radiation interactions. J Atmos Sci 52: 800–817

Chapter 12
Precipitation Efficiency

Abstract Precipitation efficiency is defined as the ratio of rain rate to rain source. In this chapter, precipitation efficiency is defined in surface rainfall, cloud, and rain budgets. The analysis demonstrates that the precipitation efficiency in cloud budget is larger than that in surface rainfall budget, but it is smaller than that in rain budget. Previous studies show precipitation efficiency could be negative or over 100 %. Precipitation efficiency ranges in 1–100 % when it is defined in budgets. The responses of precipitation efficiency to physical factors and the dependence of precipitation efficiency on rain type are discussed.

Keywords Precipitation efficiency • Surface rainfall budget • Cloud budget • Rain budget

Precipitation efficiency is an important physical parameter in vapor, cloud, and surface rainfall budgets. Although it has been intensively studied for more than five decades (e.g., Braham 1952), it remains a difficult and complex quantity for definition and estimation. Generally, precipitation efficiency is precipitation divided by sources associated with the precipitation in convective systems. There are two ways to define the precipitation efficiency. It can be defined in the point view of large-scale water vapor budgets, in which the water vapor convergence and surface evaporation are traditionally considered as major indirect precipitation sources. It also can be defined in the point view of cloud microphysical budget, in which vapor condensation and deposition are considered as the direct rainfall sources. Calculations of precipitation efficiency show that they could be more than 100 %. This indicates that some precipitation sources may not be included in the definitions. To fix it, new definitions of precipitation efficiency are introduced by Sui et al. (2007), in which the precipitation efficiency is less than or equal to 100 %. In this chapter, the definition of precipitation efficiency and the relation between the precipitation efficiency and other physical parameters will be addressed in Sect. 12.1 based on Li et al. (2002), Sui et al. (2005, 2007), Shen et al. (2013), and Zhou et al. (2014). The relationship between precipitation efficiency and physical factors will be discussed in Sect. 12.2 based on Zhou et al. (2014). The dependence of precipitation efficiency on rain type will be examined in Sect. 12.3 based on Gao and Li (2011).

© Springer International Publishing Switzerland 2016

X. Li, S. Gao, *Cloud-Resolving Modeling of Convective Processes*,
Springer Atmospheric Sciences, DOI 10.1007/978-3-319-26360-1_12

12.1 Definition of Precipitation Efficiency

For large-scale applications involving cumulus parameterization (e.g., Kuo 1965, 1974), the precipitation efficiency is defined as the ratio of the surface rain rate to the sum of the surface evaporation and the vertically integrated horizontal and vertical vapor advection (vapor convergence), which is referred to as large-scale precipitation efficiency (*LSPE*). *LSPE* can be written as

$$LSPE = \frac{P_s}{Q_{WVF} + Q_{WVE}}. \tag{12.1}$$

Braham (1952) proposed a vapor cycle in which the water vapor is transported into the air column by dynamic convergence and the vapor is condensed to liquid or frozen water. The condensational hydrometeors partially fall down to form the precipitation and partially leave in the atmosphere to form the clouds and defined the precipitation efficiency as the ratio of the surface rain rate and total moisture influx of the thunderstorm. For the thunderstorm with a typical life span of 25 min, Braham estimated that only 10 % of the water vapor transporting into the thunderstorms is eventually measured as surface rainfall. With the data from surface to 700 mb, Newton (1963, 1966) suggested that the precipitation efficiency is about 50 %. Foote and Fankhauser (1973) used the radar-determined rain rate and vapor convergence derived by aircraft data at 800 mb and showed that the precipitation efficiency is about 60 %. Auer and Marwitz (1968) defined the precipitation as the ratio of surface rain rate to the moisture flux through the cloud base and calculated the precipitation efficiency with the radar data averaged within 1 h and found that the precipitation efficiency associated with the thunderstorms that produce hail is about 55 %, whereas the efficiencies of the thunderstorms that do not produce any hail are nearly 100 % or more. Heymsfield and Schotz (1985) calculated the efficiency associated with a severe squall line over Oklahoma using the moisture budget in which rainfall is balanced by moisture influx, anvil outflow and entrainment-induced vapor loss and found that the precipitation efficiency is only 25–40 % and argued that the low precipitation efficiency may be due to large moisture losses in the upper troposphere. Ferrier et al. (1996) calculated precipitation efficiency with the cloud-resolving model simulation data and found that the precipitation efficiency is sensitive to the inflow depths and is 26–66 %. Dowell et al. (1996) showed that the precipitation efficiency of individual cells could vary considerably across a large convective system and suggested that the precipitation efficiency should be understood as a time average over the history of a precipitation-producing weather system. Dowell et al.'s estimate of precipitation efficiency is 44 %.

For smaller scale cloud-resolving models (e.g., Li et al. 1999), the precipitation efficiency is defined as the ratio of the surface rain rate to the sum of the vertically-integrated condensation and deposition rates. This is referred to as cloud-microphysics precipitation efficiency (*CMPE*). The *CMPE* is similar to the precipitation efficiency defined by Weisman and Klemp (1982), Lipps and Hemler (1986), Chong and Hauser (1989), and Tao et al. (2004). *CMPE* can be written as

$$CMPE = \frac{P_s}{Q_{WVOUT}}. \quad (12.2)$$

Lipps and Hemler (1986) defined the cloud efficiency as the total rain reaching the ground divided by the total positive cloud condensation occurring over a specific time interval and analyzed mass continuity of water substance in which the total cloud condensation is balanced by total cloud water evaporation, rain water evaporation, total rain at the ground, residual amounts of cloud water and rain water with the cloud-resolving model simulation data during the GATE period and found that the precipitation efficiency is 42 %. Chong and Hauser (1989) analyzed water budgets over convective and stratiform regions for a tropical squall line in West Africa using Doppler radar data. In the water budgets, surface precipitation is contributed by the mass of water condensed, the mass of water lost due to evaporation in cloud region, the mass of water lost due to evaporation into environment, and the mass of water transported by horizontal flow between convective and stratiform region. Chong and Hauser found that the precipitation efficiency is 47–57 % for the convective region and 45–57 % for the stratiform region, respectively. Ferrier et al. (1996) used the simulation data averaged over the last 3 h of each experiment and displayed the precipitation efficiency of 24–45 %. Tao et al. (2004) examine the relationship of precipitation efficiency with moisture and wind with cloud-resolving model simulation data during TOGA COARE, GATE, SCSMEX, and Atmospheric Radiation Measurement (ARM) and showed that the precipitation efficiency is 30–45 %, individual precipitation efficiency is insensitive to environmental moisture, and it decreases with increasing mid-tropospheric wind shear.

Li et al. (2002) calculated the simulation data in COARE and found that the *LSPE* can exceed 100 % for strong convection. This suggests that the surface rain rate could be larger than the total moisture convergence, which is contrary to the assumption in Kuo's scheme (1965, 1974) that a small portion of the surface evaporation and moisture convergence (say 5 %) is used to moisten the atmosphere.

To examine the *LSPE* and local vapor and cloud change, the surface rainfall equation (2.5) is divided by the vapor source ($Q_{WVF} + Q_{WVE}$) to yield

$$LSPE - \frac{Q_{WVT}}{Q_{WVF} + Q_{WVE}} - \frac{Q_{CM}}{Q_{WVF} + Q_{WVE}} = 1. \quad (12.3)$$

The role of cloud hydrometeors can be ignored in the surface rainfall processes by assuming that $[\overline{S}_{qv}] \cong P_s$. Thus, (12.3) becomes

$$LSPE - \frac{Q_{WVT}}{Q_{WVF} + Q_{WVE}} = 1. \quad (12.4)$$

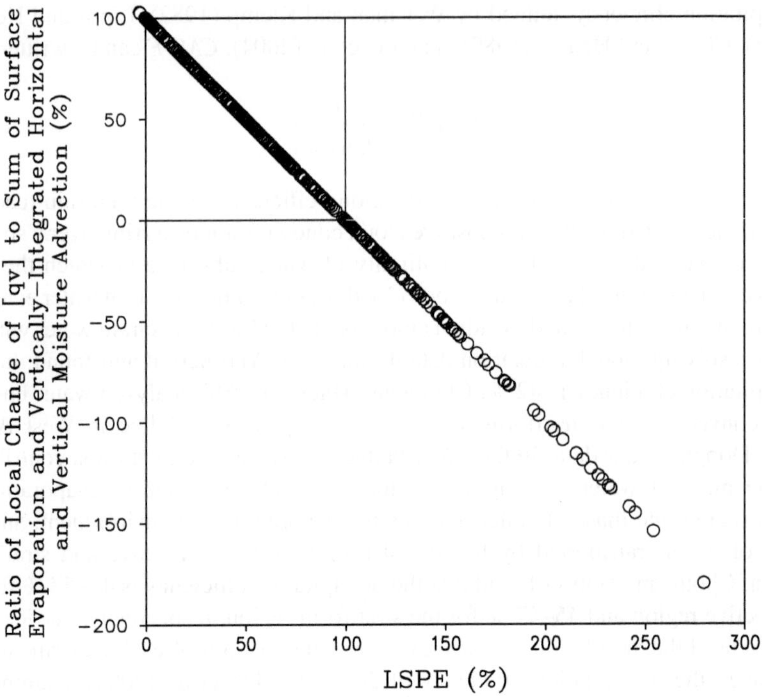

Fig. 12.1 Ratio of local *PW* change to the sum of moisture convergence and surface evaporation versus *LSPE* calculated using hourly zonal-mean simulation data from COARE. Unit is % (After Li et al. 2002)

The ratio of the local *PW* change ($-Q_{WVT}$) to the moisture sources ($Q_{WVF} + Q_{WVE}$) and *LSPE* is shown in Fig. 12.1, and – Q_{WVT} and P_s is shown in Fig. 12.2. The upper left square in Figs. 12.1 and 12.2 indicates that the light rain regime where the atmosphere is always moistened by the moisture convergence and surface evaporation. Figures 12.1 and 12.2 show that in the heavy rain regime (outside the upper left square) up to 60 % of the moisture source for precipitation came from the drying processes of the environmental atmosphere. Newton (1963, 1966), Foote and Fankhauser (1973), Gamache and Houze (1983), and Ferrier et al. (1996) also showed that the precipitation efficiency defined by Braham (1952) could be larger than 100 %. The cumulus parameterization schemes (e.g., Kuo 1965, 1974) moisten atmosphere by consuming 5 % of the moisture source, which eliminates the drying processes. Figure 12.2 shows that the drying processes in the rainfall production are important. Thus, the cumulus parameterization scheme should allow the precipitation rate to be larger than the rate of the sum of the moisture convergence and surface evaporation in some circumstances in order to avoid moisture bias.

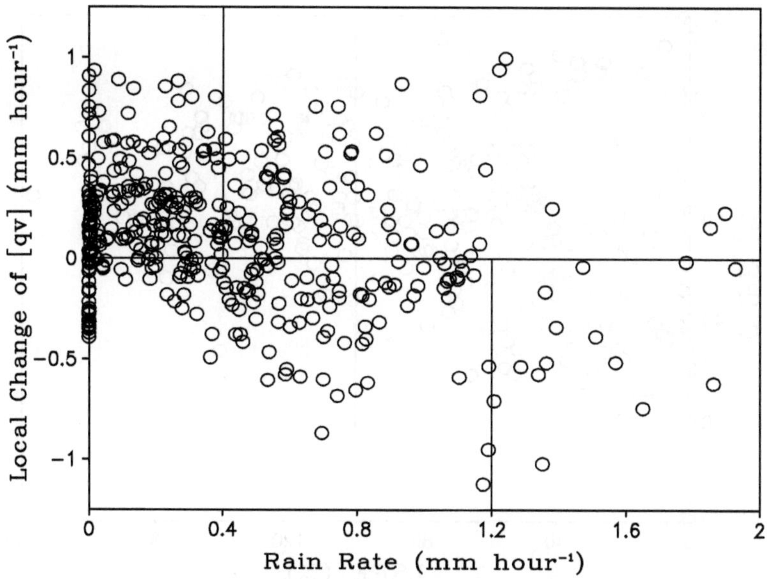

Fig. 12.2 Local *PW* change versus surface rain rate calculated using hourly zonal-mean simulation data from COARE. Unit is mm h^{-1} (After Li et al. 2002)

The regression relation between Q_{WVOUT} and $Q_{WVF} + Q_{WVE}$ can be expressed by

$$Q_{WVOUT} = 0.17 + 0.84\,(Q_{WVF} + Q_{WVE})\,. \tag{12.5}$$

Thus, *CMPE* and *LSPE* have the following statistic relation:

$$CMPE = 1.19\,LSPE. \tag{12.6}$$

(12.6) indicates that *CMPE* is larger than *LSPE*. Figure 12.3 shows that *LSPE* is about 80 % of *CMPE* when *LSPE* is smaller than 60 %, and the *CMPE* is a constant of about 70 % when *LSPE* is larger than 60 %. Figure 12.3 also indicates that (12.6) is valid only when both *CMPE* and *LSPE* are smaller than 100 %.

Since large-scale forcing is imposed in vapor budget whereas lateral boundary condition causes zero zonal-mean hydrometeor convergence, the zonal-mean calculations show that *CMPE* is smaller than 100 % whereas *LSPE* could be larger than 100 % (Fig. 12.3). Sui et al. (2005b) calculated *CMPE* and *LSPE* using data averaged over 96, 48, and 24 km from the grid data in COARE and found that the statistical relation

$$Q_{WVOUT} = Q_{WVF} + Q_{WVE} \tag{12.7}$$

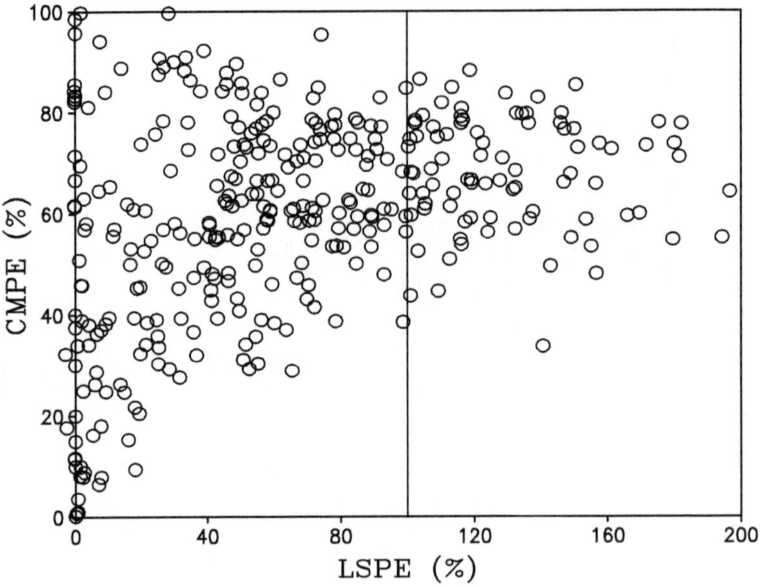

Fig. 12.3 *CMPE* versus *LSPE* calculated using hourly zonal-mean simulation data from COARE. Unit is % (After Li et al. 2002)

is a valid approximation, in particular, for the 24-km averages. Thus, *CMPE* and *LSPE* is statistically equivalent as indicated by Fig. 12.4.

The *CMPE* provides a basis for identifying the relevant processes determining precipitation efficiency.

$$CMPE = 1 - \frac{[Q_{WVIN}]}{[Q_{WVOUT}]} + \frac{[CONV_C]}{[Q_{WVOUT}]}. \quad (12.8)$$

Here $[CONV_c] = CONV_{IWP} + CONV_{LWP}$ [see (6.3a) and (6.3b)]. (12.8) indicates that the positive values of $[CONV_c]$ make *CMPE* larger through the advection of clouds into the region of interest, while negative values of $[CONV_c]$ make *CMPE* smaller. Figure 12.5 shows that the larger *CMPE* is associated with hydrometeor convergence, whereas smaller *CMPE* is associated with hydrometeor divergence. *CMPE* could be even larger than one if $[CONV_c] > [Q_{WVIN}]$. Figure 12.5 displays many cases of $CMPE > 100$ %. The above results indicate a significant effect of hydrometeor convergence on surface precipitation processes.

Since the hydrometeor convergence is expected to be a function of the strength of convection, the dependence of *CMPE* on rainfall rate is examined using the data averaged in three grid area sizes of 96, 48, and 24 km in COARE. Figure 12.6 shows that *CMPE* values spread over a wide range and become even larger than 100 % in light-rain conditions (surface rainfall rates are smaller than 5 mm h^{-1} for 96- and 48-km averages and smaller than 10 mm h^{-1} for 24-km average). But the *CMPE* tends to converge to a threshold value with increasing surface rainfall rate.

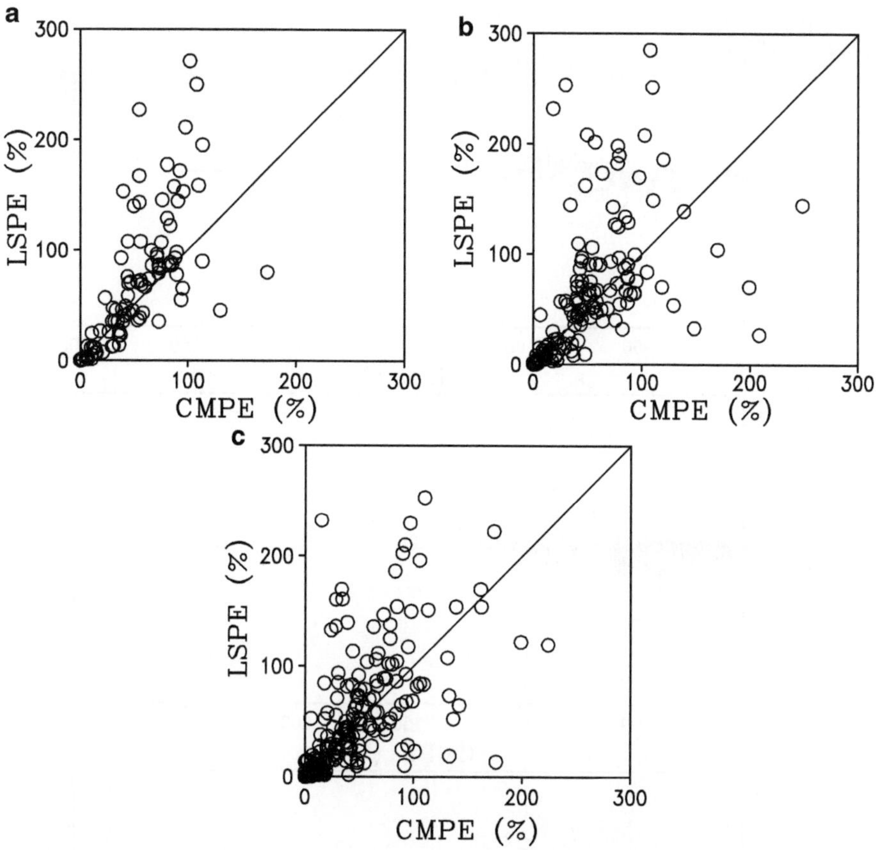

Fig. 12.4 *LSPE* versus *CMPE* averaged within (**a**) 96 km, (**b**) 48 km, and (**c**) 24 km in COARE. Unit is %. The *diagonal lines* denote *CMPE* = *LSPE* (After Sui et al. 2005)

From above discussions, both *LSPE* and *CMPE* could be greater than 100 % and *LSPE* could be negative, which is not physically meaningful. This could be due to exclusion of some sources associated with precipitation or inclusion of precipitation sink in the definition of precipitation efficiency. For example, local atmospheric drying and local hydrometeor loss could be major contributors to surface rain rate during the weakening and dissipating stages of tropical oceanic convection while vapor divergence occurs (see Table 7.1). Local atmospheric drying and local hydrometeor loss could make *LSPE* larger than 100 % since they are not included as the rainfall sources in the definition of *LSPE* [see (12.1)]. Vapor divergence could make *LSPE* negative since it is included in the definition of *LSPE* as the precipitation source. Thus, Sui et al. (2007) introduced new definitions for *LSPE* and *CMPE*. The new *LSPE* is defined as

$$LSPE = \frac{P_s}{RSWVCB}, \tag{12.9}$$

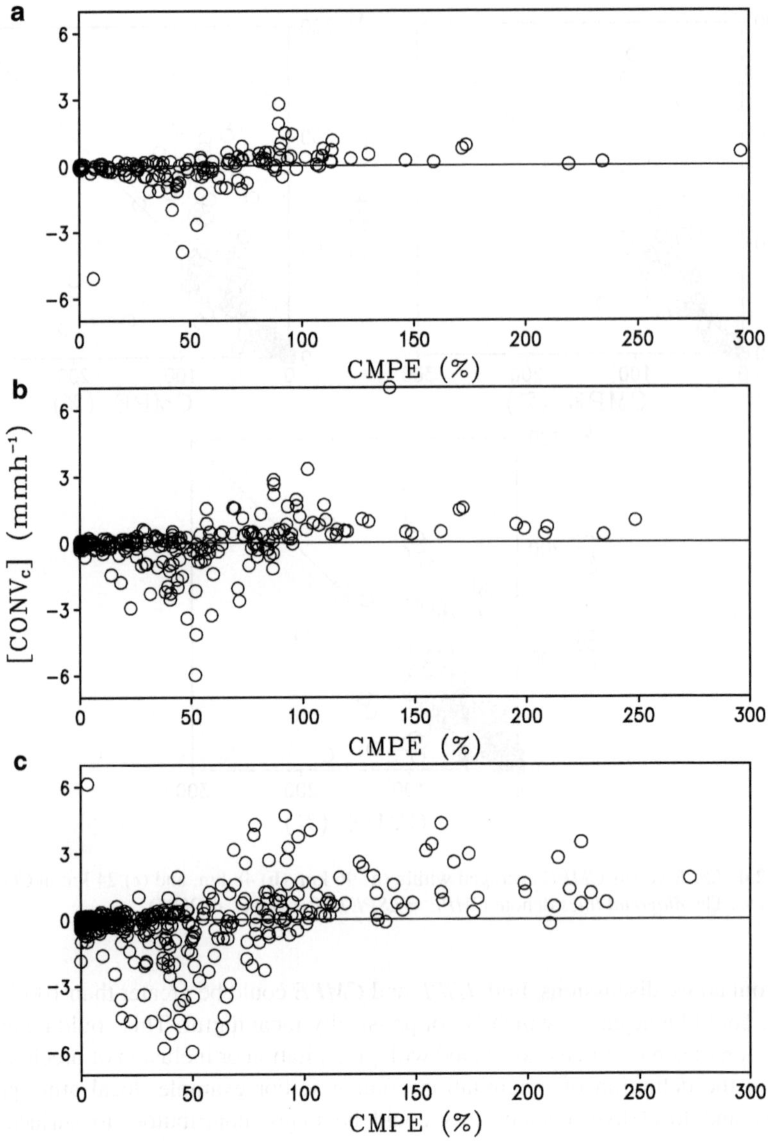

Fig. 12.5 [$CONV_C$] (mm h^{-1}) versus *CMPE* (%) averaged within (**a**) 96 km, (**b**) 48 km, and (**c**) 24 km in COARE (After Sui et al. 2005)

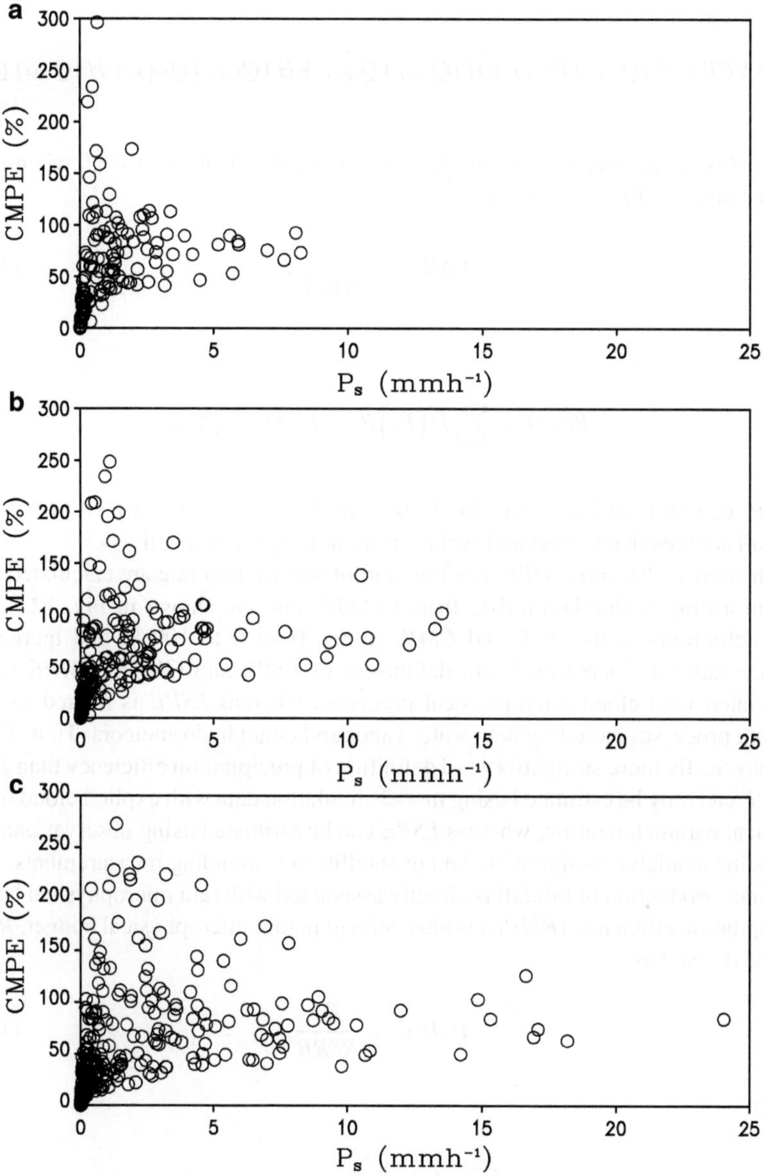

Fig. 12.6 *CMPE* (%) versus P_s (mm h^{-1}) averaged within (**a**) 96 km, (**b**) 48 km, and (**c**) 24 km in COARE (After Sui et al. 2005)

where

$$RSWVCB = H(Q_{WVT})Q_{WVT} + H(Q_{WVF})Q_{WVF} + H(Q_{WVE})Q_{WVE} + H(Q_{CM})Q_{CM}. \tag{12.9a}$$

Here, H is the Heaviside function, $H(F)=1$ when $F>0$, and $H(F)=0$ when $F \le 0$.

The new *CMPE* is defined as

$$CMPE = \frac{P_s}{RSCB}, \tag{12.10}$$

where

$$RSCB = \sum_{I=1}^{7} H\left(P_I\right)P_I + H(Q_{CM})Q_{CM}, \tag{12.10a}$$

and P_I can be found in (2.6a). In (12.9) and (12.10), *RSCB* and *RSWVCB* are the rainfall sources from cloud and surface rainfall budgets, respectively.

The new *LSPE* and *CMPE* as a function of surface rain rate are calculated using 96-km averaged simulation data from COARE and are shown in Fig. 12.7. With new definitions, both *LSPE* and *CMPE* range from 0 to 100 % and increase as surface rain rate increases. From definitions of *LSPE* and *CMPE*, *CMPE* is only associated with cloud microphysical processes whereas *LSPE* is related to water cycling processes including both water vapor and cloud hydrometeors. Thus, *CMPE* is a physically more straightforward definition of precipitation efficiency than *LSPE*. *CMPE* can only be estimated using model simulation data with explicit cloud microphysical parameterization, whereas *LSPE* can be estimated using observational data including available assimilation data of satellite and sounding measurements.

Since production of rainfall is directly associated with rain microphysical budget, precipitation efficiency (*RMPE*) is also defined in rain microphysical budget. *RMPE* can be defined as

$$RMPE = \frac{P_S}{RSRB}, \tag{12.11}$$

where

$$RSRB = \sum_{I=1}^{12} H\left(RP_I\right)RP_I + H(Q_{RM})Q_{RM}. \tag{12.11a}$$

RP_I and Q_{RM} can be found in (2.7a) and (2.7b), respectively. *RSRB* is smaller than *RSCB*, and both are much smaller than *RSWVCB* (Fig. 12.8a). Thus, *LSPE* is much smaller than *RMPE* and *CMPE*, while *RMPE* is larger than *CMPE* (Fig. 12.8b), i.e.,

$$RMPE > CMPE > LSPE. \tag{12.12}$$

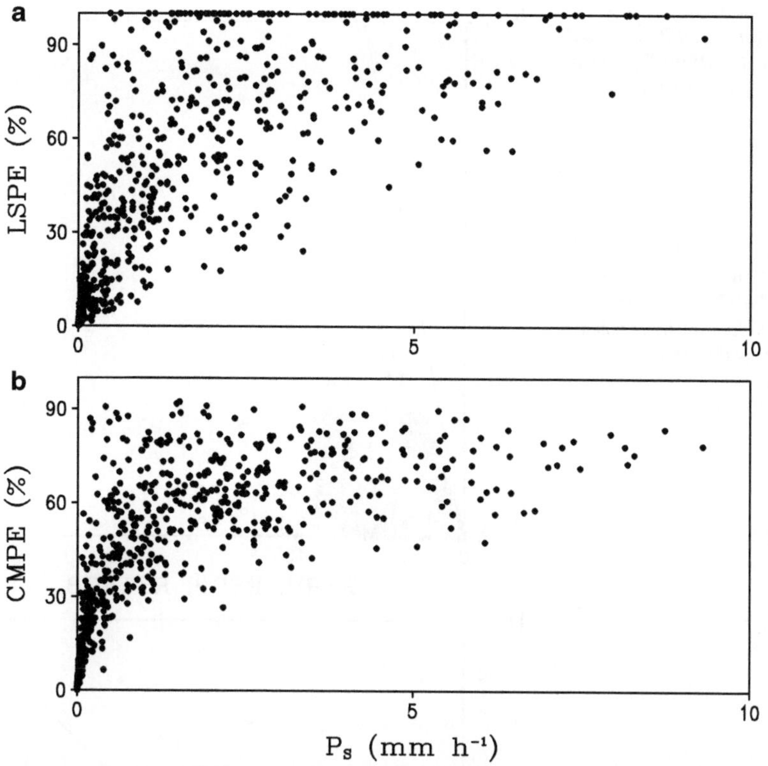

Fig. 12.7 (**a**) *LSPE* (%) vs P_s (mm h^{-1}) and (**b**) *CMPE* vs P_s using hourly 96-km averaged data from COARE. *LSPE* and *CMPE* are defined in (12.9) and (12.10), respectively (After Sui et al. 2007)

From cloud microphysical budget, (12.10a) becomes

$$RSCB = RSRB + RSCWCISGB, \tag{12.13}$$

where

$$\begin{aligned} RSCWCISGB = & \sum_{I=1}^{9} H\left(CWP_I\right) CWP_I + H\left(Q_{CWM}\right) Q_{CWM} \\ & + \sum_{I=1}^{9} H\left(CIP_I\right) CIP_I + H\left(Q_{CIM}\right) Q_{CIM} \\ & + \sum_{I=1}^{15} H\left(SP_I\right) SP_I + H\left(Q_{SM}\right) Q_{SM} \\ & + \sum_{I=1}^{14} H\left(GP_I\right) GP_I + H\left(Q_{GM}\right) Q_{GM}, \end{aligned} \tag{12.14}$$

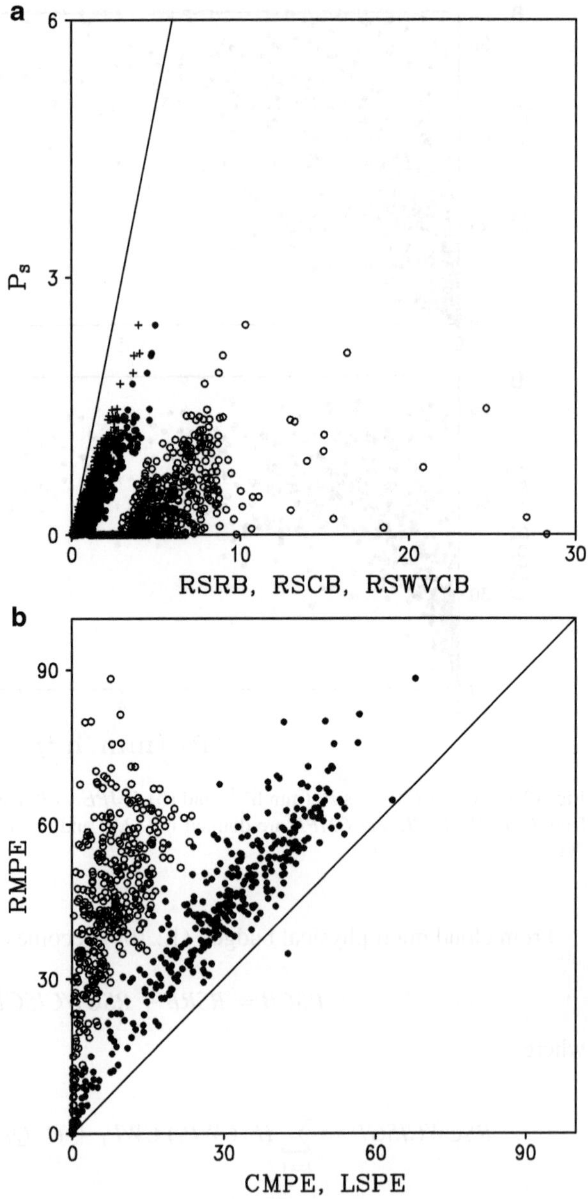

Fig. 12.8 (**a**) P_S versus *RSRB* (*cross*), *RSCB* (*closed circle*), and *RSWVCB* (*open circle*), and (**b**) *RMPE* versus *CMPE* (*closed circle*) and *LSPE* (*open circle*). Units are mm h^{-1} for P_S, *RSRB*, *RSCB*, and *RSWVCB*, and % for *RMPE*, *CMPE*, and *LSPE*. *Straight line* in (**a**) denote $P_S = RSRB$, $P_S = RSCB$, and $P_S = RSWVCB$, whereas *straight line* in (**b**) denotes $RMPE = CMPE$ and $RMPE = LSPE$ (After Zhou et al. 2014)

$$CWP_I = \Big(- P_{SACW}, -P_{RAUT}, -P_{RACW}, -P_{SFW}\,(T < T_o)\,, -P_{GACW}, P_{CND}, \\ -P_{IHOM}\,(T < T_{oo})\,, P_{IMLT}\,(T > T_o)\,, -P_{IDW}\,(T_{oo} < T < T_o) \Big), \tag{12.14a}$$

$$CIP_I = \Big(- P_{SAUT}\,(T < T_o)\,, -P_{SACI}\,(T < T_o)\,, -P_{RACI}\,(T < T_o)\,, \\ -P_{SFI}\,(T < T_o)\,, -P_{GACI}\,(T < T_o)\,, P_{IHOM}\,(T < T_{oo})\,, \\ -P_{IMLT}\,(T > T_o)\,, P_{IDW}\,(T_{oo} < T < T_o, P_{DEP}) \Big), \tag{12.14b}$$

$$SP_I = \Big(P_{SAUT}\,(T < T_o)\,, P_{SACI}\,(T < T_o)\,, P_{SACW}\,(T < T_o)\,, \\ P_{SFW}\Big(T < T_o, P_{SFI}\,(T < T_o)\,, P_{RACI}\,(T < T_o)\,, \\ -P_{RACS}\,(T > T_o)\,, -P_{GACS}, -P_{SMLT}\,(T > T_o)\,, \\ -P_{RACS}\,(T < T_o)\,, P_{SACR}\,(T < T_o)\,, \\ P_{SDEP}\,(T < T_o)\,, -P_{MLTS}\,(T > T_o)\,, \\ P_{IACR}\,(T < T_o)\,, -P_{WACS}\,(T < T_o)\Big), \tag{12.14c}$$

$$GP_I = \Big(P_{RACI}\,(T < T_o)\,, P_{GACI}\,(T < T_o)\,, \\ P_{GACW}\,(T < T_o)\,, P_{SACW}\,(T < T_o)\,, P_{GACS}, \\ P_{IACR}\,(T < T_o)\,, P_{GACR}\,(T < T_o)\,, P_{RACS}\,(T < T_o)\,, P_{GFR}\,(T < T_o)\,, \\ P_{WACS}\,(T < T_o)\,, -P_{GMLT}\,(T > T_o)\,, P_{GDEP}\,(T < T_o)\,, -P_{MLTG}\,(T > T_o)\,, \\ P_{SACR}\,(T < T_o)\Big), \tag{12.14d}$$

$$Q_{CWM} = -\frac{\partial\,[q_c]}{\partial t} - \left[u\frac{\partial q_c}{\partial x}\right] - \left[w\frac{\partial q_c}{\partial z}\right], \tag{12.14e}$$

$$Q_{CIM} = -\frac{\partial\,[q_i]}{\partial t} - \left[u\frac{\partial q_i}{\partial x}\right] - \left[w\frac{\partial q_i}{\partial z}\right], \tag{12.14f}$$

$$Q_{SM} = -\frac{\partial\,[q_s]}{\partial t} - \left[u\frac{\partial q_s}{\partial x}\right] - \left[w\frac{\partial q_s}{\partial z}\right], \tag{12.14g}$$

$$Q_{GM} = -\frac{\partial\,[q_g]}{\partial t} - \left[u\frac{\partial q_g}{\partial x}\right] - \left[w\frac{\partial q_g}{\partial z}\right]. \tag{12.14h}$$

Here, CWP_I, CIP_I, SP_I, and GP_I denote rainfall source/sink terms from cloud water, cloud ice, snow, and graupel microphysical processes, respectively, and P_I denotes rainfall source/sink terms from cloud microphysical processes shown in Table 1.2.

Since it consists of positive values only, *RSCWCISGB* always is a rainfall source because rainfall is generally associated with cloud water and ice microphysical processes. Thus, *RSCB* is larger than *RSRB* as indicated in Fig. 12.8a.

From (12.10) and (12.12), we have

$$\frac{RMPE}{CMPE} = \frac{RSCB}{RSRB} > 0. \tag{12.15}$$

(12.9) and (12.10) show a common item $[H(Q_{CM})Q_{CM}]$. The three water vapor terms [the second to fourth term in (12.9a)) are cloud source. Thus, we have the following relation:

$$H\left(Q_{WVF}\right)Q_{WVF} + H\left(Q_{WVE}\right)Q_{WVE} + H\left(Q_{CM}\right)Q_{CM} \geq \sum_{I=1}^{7} H\left(P_I\right)P_I \tag{12.16}$$

$$RSWVCB > RSCB. \tag{12.17}$$

$$\frac{CMPE}{LSPE} = \frac{RSWVCB}{RSCB} > 0. \tag{12.18}$$

(12.15) and (12.18) lead to (12.12).

12.2 The Relationship Between Precipitation Efficiency and Physical Factors

The precipitation is associated with various dynamic and thermodynamic factors. Thus, the relationship between precipitation efficiency and the physical factors will be examined in this study. Mass-weighted mean temperature, sea surface temperature (*SST*), and precipitable water (*PW*) averaged over model domain denote a vertical mean state of environmental thermal and vapor water conditions. Convective available potential energy (*CAPE*), calculated for reversible moist adiabatic process (2.13) over model domain, measures environmental stratification. Model domain mean water vapor convergence (Q_{WVF}) is associated with upward motions imposed in the model during the model integration. Vertical wind shear affects precipitation through conversion between model domain mean kinetic energy and perturbation kinetic energy (Wang et al 2009; Shen et al. 2011), which is calculated by the difference in imposed zonal winds between 10.2 and 2.7 km in this study. Cloud ratio (*CR*), defined as the ratio of ice water path (the sum of mass integrations of mixing ratio of cloud ice, snow, and graupel) to liquid water path (the sum of mass integrations of cloud water and rain) by Sui and Li (2005), gauges relative importance of ice and water clouds. *RMPE* does not show any relationship with mass-weighted mean temperature, *PW*, and *SST* (Fig. 12.9a–c). This implies that *RMPE* has weak relationship to the mean state of environmental conditions. *RMPE* decreases as *CAPE*, Q_{WVF}, and *CR* increase (Figs. 12.9d, e and g). *RMPE* increases as the negative vertical wind shear weakens (Fig. 12.9f). Since *RMPE* is slightly larger than *CMPE*, and their linear correlation coefficient is 0.92, the

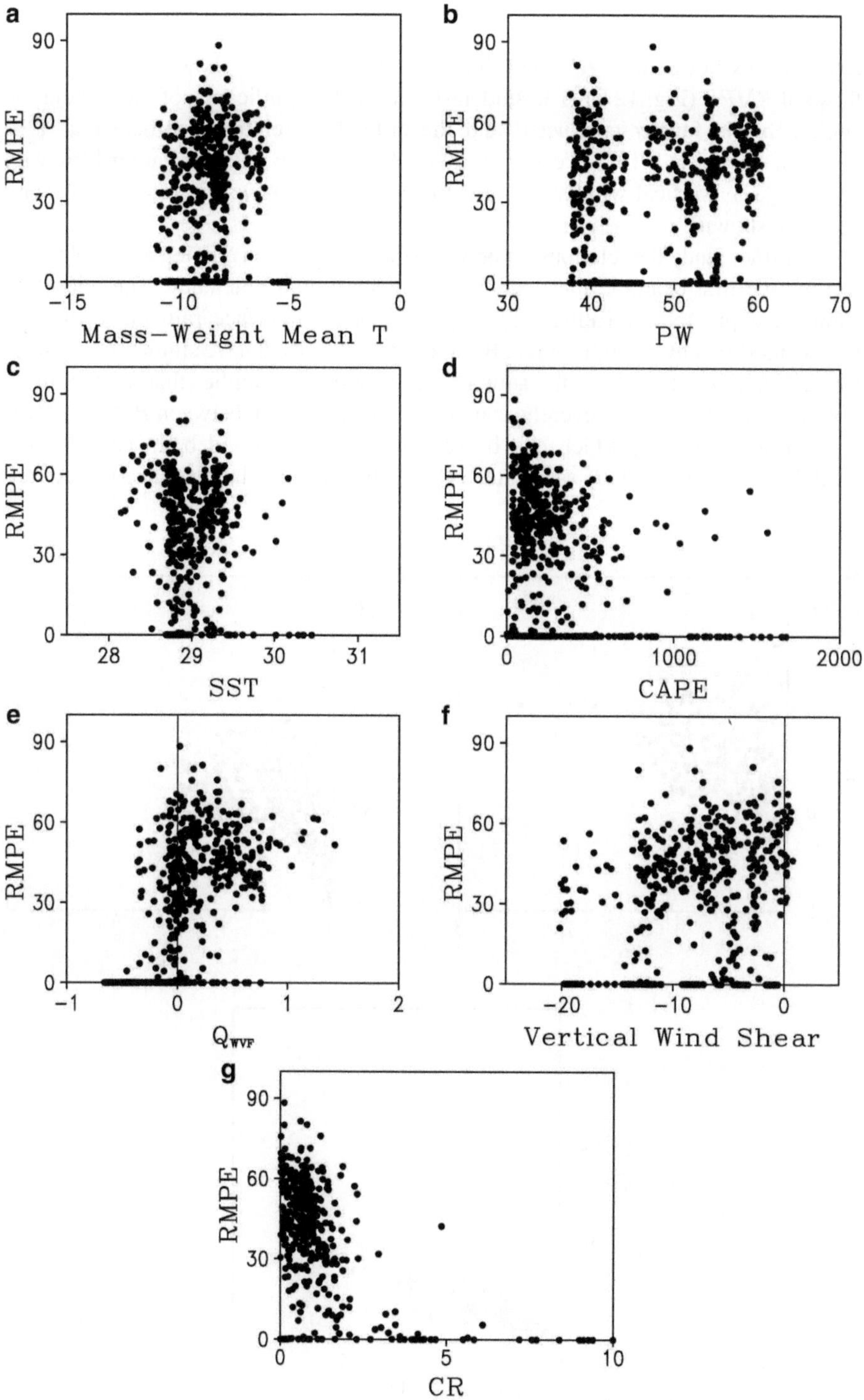

Fig. 12.9 *RMPE* (%) versus (**a**) mass-weighted mean temperature (°C), (**b**) precipitable water (*PW*; mm), (**c**) SST (°C), (**d**) *CAPE* (Jkg^{-1}), (**e**) model domain mean water vapor convergence (Q_{WVF}; mm h^{-1}), (**f**) vertical wind shear (m s^{-1}), and (**g**) cloud ratio (CR) (After Zhou et al. 2014)

relationships between *CMPE* and the physical factors (not shown) are similar to those of *RMPE* (Fig. 12.9). The student-t test on the significance of the correlation coefficients is further conducted and the critical correlation coefficient at 1 % significant level is 0.13. Like *RMPE* and *CMPE*, *LSPE* shows relationships with *CAPE*, Q_{WVF}, vertical wind shear, and *CR* but does not have any relationship with *SST* (not shown).

To further study the relationship between the precipitation efficiency and physical factors, we first examine the definition of precipitation efficiency. The definition shows that precipitation efficiency is proportional to surface rain rate, which is largely modified by rainfall source. Both *RMPE* and *CMPE* have similar magnitudes for surface rain rate and rainfall source (Fig. 12.8a). This implies that the modification from rainfall source severely scatters the linear relation between *RMPE*/*CMPE* and surface rain rate, which can be seen in Fig. 12.10a and b. In contrast, the rainfall source (*RSWVCB*) associated with *LSPE* is much larger than surface rain

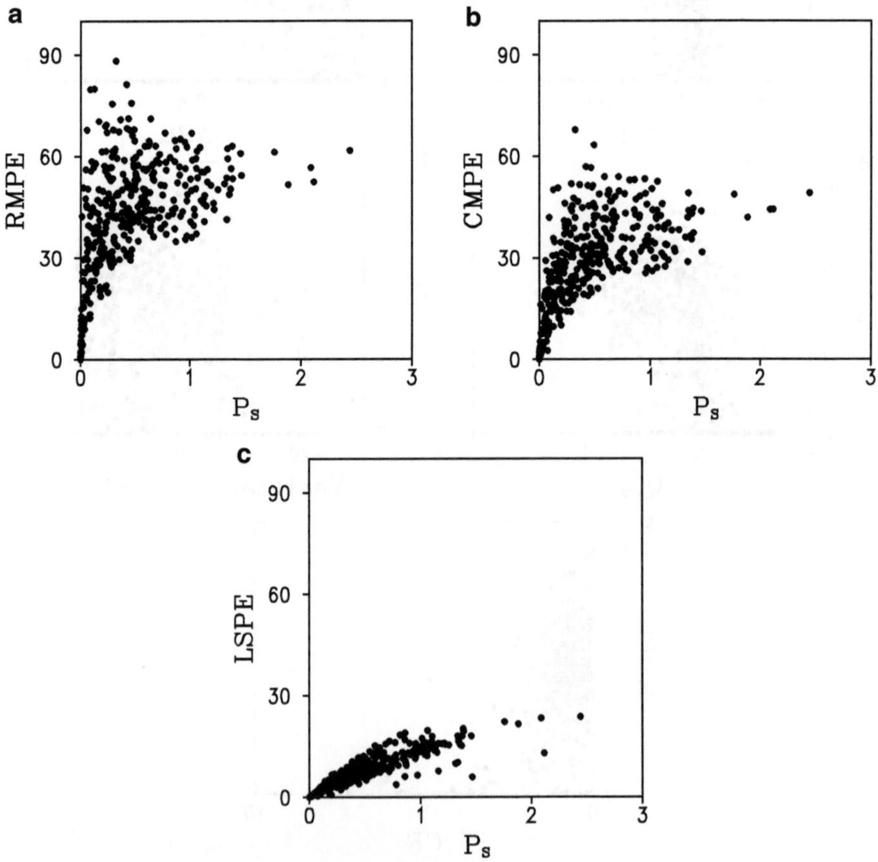

Fig. 12.10 (**a**) *RMPE* versus P_S, (**b**) *CMPE* versus P_S, and (**c**) *LSPE* versus P_S. Units are % for *RMPE*, *CMPE*, and *LSPE* and mm h^{-1} for P_S (After Zhou et al. 2014)

rate (Fig. 12.8a), which leads to a linear correlation between *LSPE* and surface rain rate with a coefficient of 0.92 (Fig. 12.10c). All precipitation efficiencies generally increase as surface rain rate increases.

Rain rate decreases as *CAPE* increases (Fig. 12.11d) and *CR* increases from 1 to 5 (Fig. 12.11g). The rainfall is a result of the release of *CAPE*. As the rainfall weakens, *CAPE* starts to build up. Such relationship between rainfall and *CAPE* forms a basic scientific promise that parameterizes cumulus and associated rainfall for large-scale circulation models. The increase of *CR* from 1 to 5 favors the development of stratiform clouds. Convective rainfall increases when *CR* increases from 0 to 1. *RMPE* and *CMPE* decrease with increasing *CR* from 0 to 1 (Fig. 12.9g), indicating strong effects of rainfall source on modification of linear relation between precipitation efficiencies and rain rate in rain and cloud microphysical budgets.

Rain rate increases as model domain mean water vapor convergence increases (Fig. 12.11e) and negative vertical wind shear weakens (Fig. 12.11f). Large-scale vertical velocity is a key forcing imposed in cloud model simulation. Thus, the mean water vapor convergence associated with imposed upward motions is a key process for production of precipitation as indicated by the similar magnitudes of Q_{WVF} and P_S in Fig. 12.11e. The rainfall source associated with *LSPE* (Fig. 12.8a) is much larger than the mean water vapor convergence due to a large cancellation between the rainfall source and rainfall sink dominated by water vapor divergence associated with downward motions in a cyclic later boundary condition furnished in the cloud model used in this study.

Rain rate is associated with the secondary circulation, whose intensity is measured by perturbation kinetic energy (K'). Thus, rain rate and perturbation kinetic energy are highly correlated with the coefficient of 0.59 (Fig. 12.12). Vertical wind shear may affect rainfall by changing conversion between model domain mean kinetic energy and perturbation kinetic energy [see (2.12 and 2.12a)]. The conversion between the mean ($\overline{K}$) and perturbation (K') kinetic energy ($C_u\left(\overline{K}, K'\right)$) is a term in the budget of perturbation kinetic energy. The tendency of the perturbation kinetic energy ($\frac{\partial K'}{\partial t}$) is negatively correlated with $C_u\left(\overline{K}, K'\right)$ and their linear correlation coefficient is −0.26. $C_u\left(\overline{K}, K'\right)$ is positively correlated with the vertical wind shear and their linear correlation coefficient is 0.14, which marginally exceeds critical correlation coefficient. Thus, the relationship between rain rate and vertical wind shear cannot be physically explained.

12.3 The Dependence of Precipitation Efficiency on Rain Type

Since the rainfall sources only count the positive values of rainfall processes, calculating the area-averaged grid-scale data first and then taking positive terms (referred as to the mean data calculation hereafter) may be significantly smaller than accumulating positive rainfall source terms at each model grid first and then

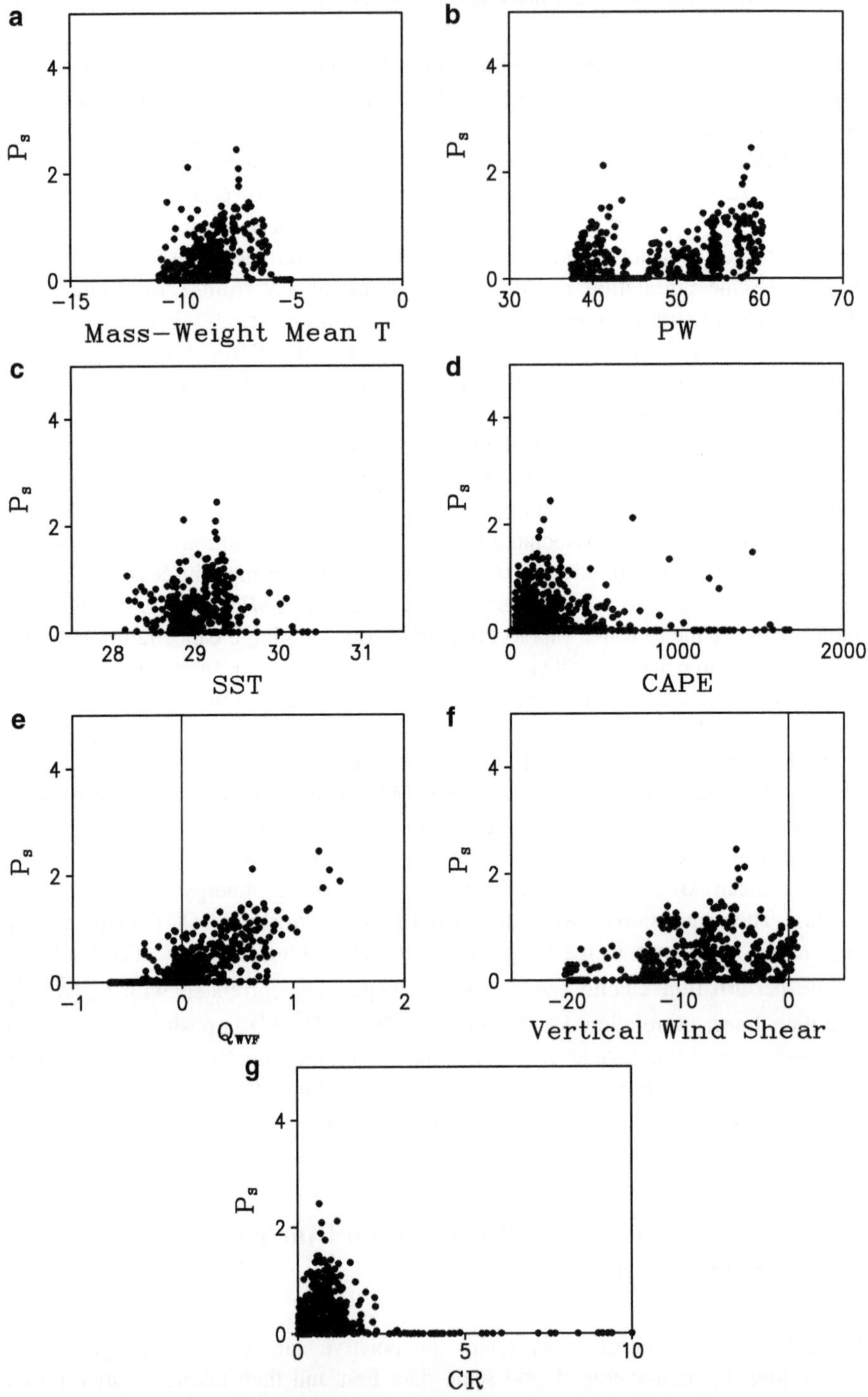

Fig. 12.11 P_S (mm h^{-1}) versus (**a**) mass-weighted mean temperature (°C), (**b**) precipitable water (*PW*; mm), (**c**) SST (°C), (**d**) *CAPE* (Jkg^{-1}), (**e**) model domain mean water vapor convergence (Q_{WVF}; mm h^{-1}), (**f**) vertical wind shear (m s^{-1}), and (**g**) cloud ratio (CR) (After Zhou et al. 2014)

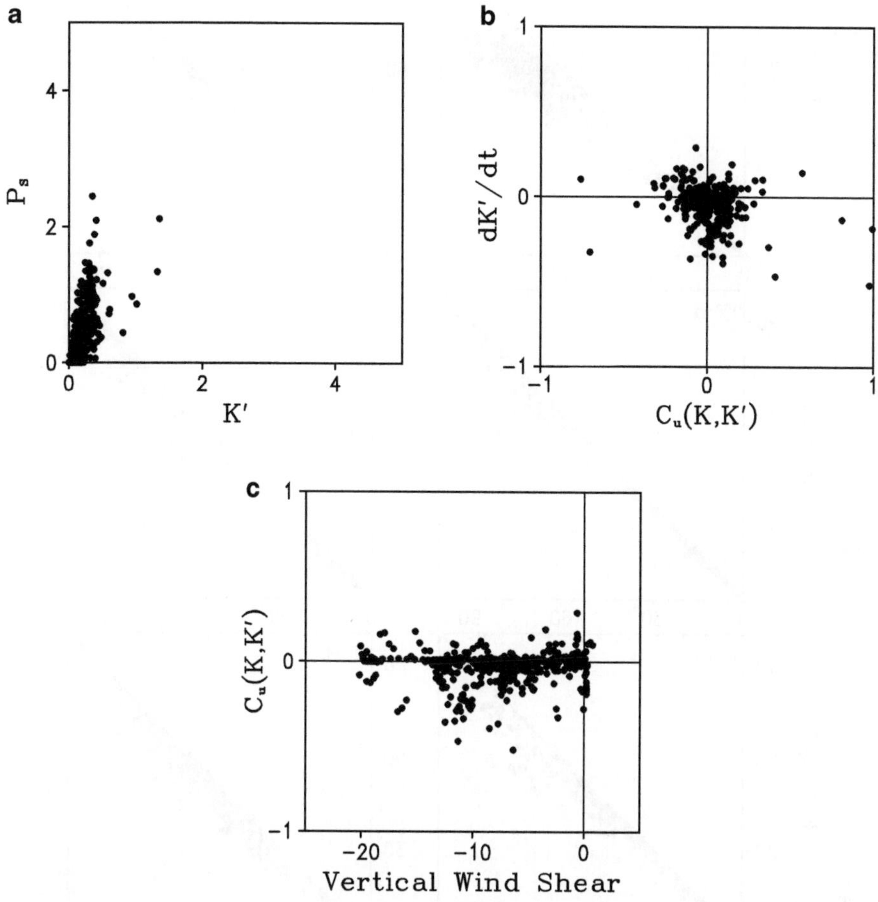

Fig. 12.12 (**a**) P_S (mm h^{-1}) versus K′ (10^{11} J) (**b**) $\frac{\partial K'}{\partial t}$ (10^7 Js^{-1}) versus $C_u\left(\overline{K}, K'\right)$ (10^7 Js^{-1}), and (**c**) $C_u\left(\overline{K}, K'\right)$ versus vertical wind shear (ms^{-1}) (After Zhou et al. 2014)

make an average (referred as to grid-scale data calculation hereafter) because the rainfall sinks (the negative values of rainfall processes) offset the rainfall sources in the mean data calculation. The former and the latter are respectively denoted by subscripts "M" and "P" in this study. The comparison between the two calculations shows the spatial-scale dependence of precipitation efficiency on rain type. The rainfall is separated into eight types based on surface rainfall equation (see Table 6.1). tfm is not discussed in this section because its rain rate is negligibly small.

The two calculations for *RMPE* are compared in Fig. 12.13. $RMPE_M$ is higher than $RMPE_P$ in the "mean" calculation for the entire rainfall area. Their RMS difference is 11.9 % (Table 12.1), which is smaller than the standard deviation of $RMPE_P$ (17.6 %, Table 12.2). The RMS differences are greatly reduced when *RMPE* is calculated for each rain type. The RMS is the smallest for TFM whereas they are

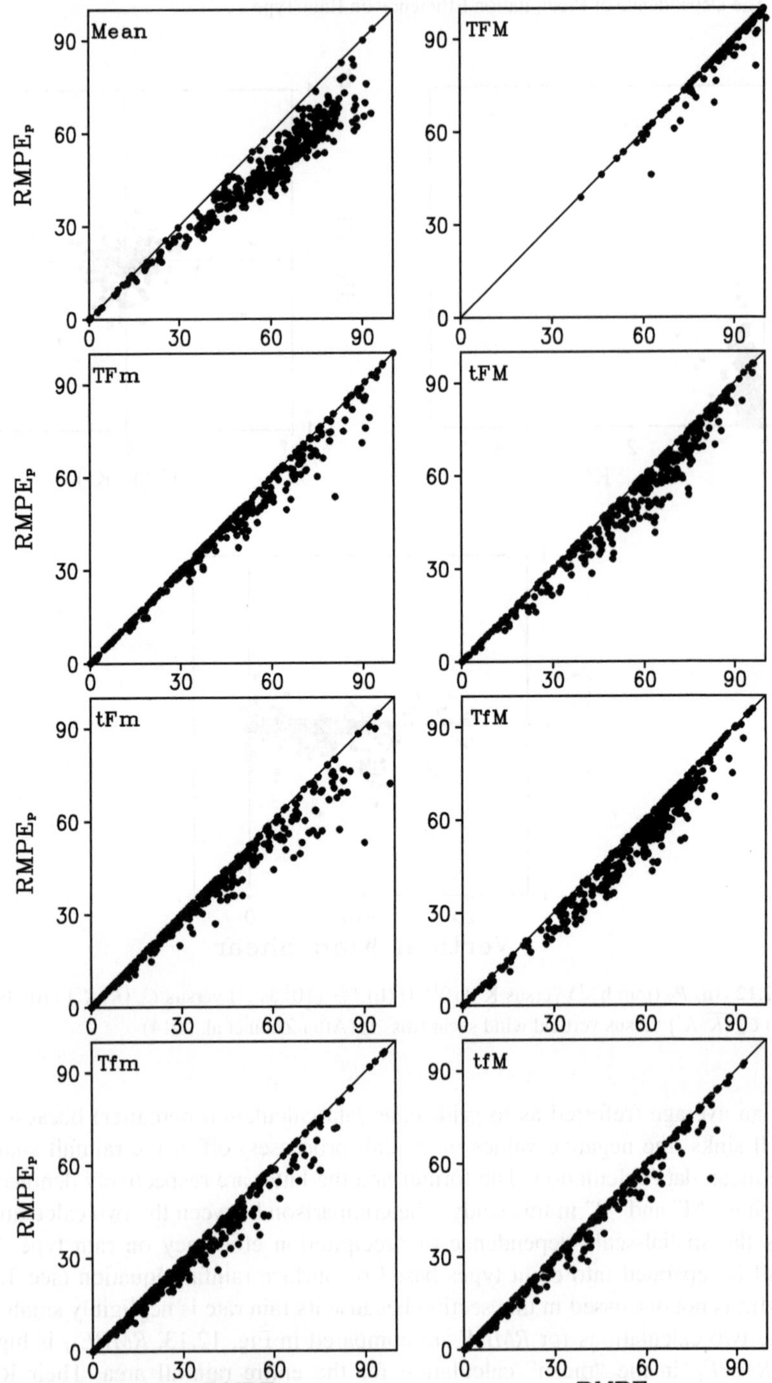

Fig. 12.13 $RMPE_P$ versus $RMPE_M$. $RMPE_M$ is calculated from area-mean data over rainfall regions (Mean) and regions of seven rain types, whereas $RMPE_P$ is calculated from the accumulations of rainfall sources from each model grid over rainfall regions and regions of seven rain types. *Diagonal lines* denote $RMPE_P = RMPE_M$ (After Gao and Li 2011)

Table 12.1 RMS differences (%) between $RMPE_P$ and $RMPE_M$ [RMS(RM)], $CMPE_P$ and $CMPE_M$ [RMS(CM)], $RMPE_P$ and $CMPE_M$ [RMS(RM,CM)], and $RMPE_P$ and $LSPE$ [RMS(RM,LS)] calculated over rainfall regions (Mean) and regions of seven rainfall types

	RMS(RM)	RMS(CM)	RMS(RM,CM)	RMS(RM,LS)
Mean	11.9	22.9	11.9	31.0
TFM	2.8	0.1	4.5	19.2
TFm	4.0	0.1	17.6	16.0
tFM	5.8	7.0	13.3	34.8
tFm	5.4	0.1	19.2	28.1
TfM	5.6	6.6	14.3	39.3
Tfm	4.2	0.2	15.0	25.2
tfM	3.4	1.0	16.7	17.2

After Gao and Li (2011)

Table 12.2 Standard deviations (%) of $RMPE_P$, $RMPE_M$, $CMPE_P$, $CMPE_M$, and $LSPE$ calculated over rainfall regions (Mean) and regions of seven rainfall types

	$RMPE_P$	$RMPE_M$	$CMPE_P$	$CMPE_M$	$LSPE$
Mean	17.6	21.3	15.4	21.3	32.5
TFM	12.0	11.4	10.7	10.7	0.0
TFm	23.9	25.1	19.6	19.6	21.7
tFM	24.0	24.9	25.2	27.5	16.4
tFm	21.0	23.2	14.5	14.5	9.3
TfM	21.7	22.1	19.3	22.3	10.1
Tfm	21.0	22.0	15.7	15.7	10.6
tfM	24.0	24.7	17.9	18.3	18.8

After Gao and Li (2011)

over 5 % for tFM, tFm, and TfM (Table 12.1); there is much less scatter in TFM compared to the other rain types. This indicates that the RMS difference calculated over the whole rainfall area results primarily from tFM, tFm, and TfM.

The RMS difference of *CMPE* (22.9 %) is double that for *RMPE* in the calculations over the entire rain area; *CMPE* has a larger scatter than *RMPE* (Table 12.1 and Figs. 12.13 and 12.14). Like *RMPE*, $CMPE_M$ is higher than $CMPE_P$. Unlike *RMPE*, Table 12.2 shows that the RMS difference for CMPE is larger than the standard deviation of $CMPE_P$ (15.4 %) and $CMPE_M$ (21.3 %). This suggests that the calculation of *CMPE* is more spatial-scale dependent than the calculation of *RMPE*. There are virtually no differences between the mean and grid-scale data calculations of *CMPE* ($CMPE_P = CMPE_M$) for TFM, TFm, tFm, and Tfm (Fig. 12.14), indicating spatial-scale independence for the calculations of *CMPE* for these rain types. Thus, the RMS difference of *CMPE* calculated over rainfall area comes mainly from tFM and TfM as indicated by their RMS differences in Table 12.1. The calculation for each rain type reveals that the RMS difference of *CMPE* is well below the standard deviation of $CMPE_P$ and $CMPE_M$ (Tables 12.1 and 12.2).

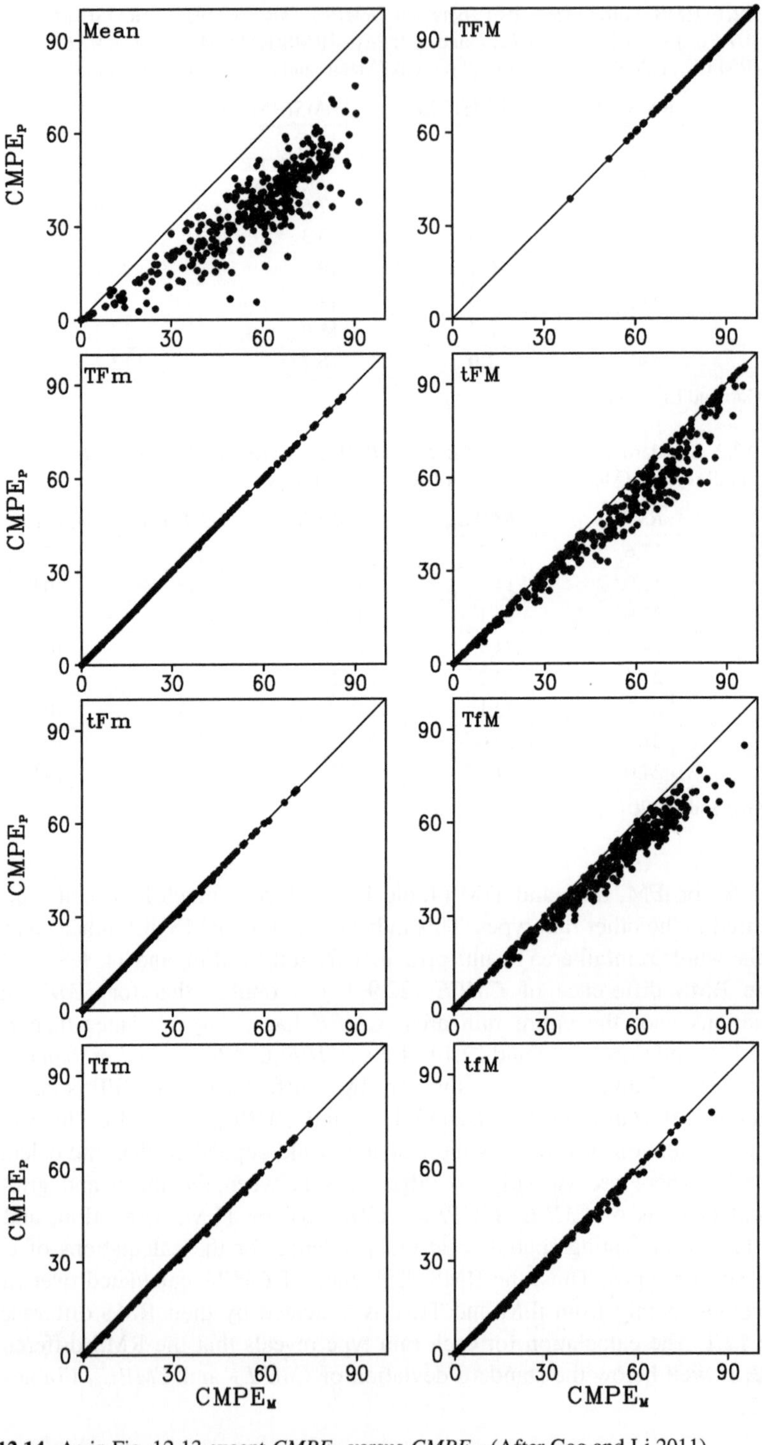

Fig. 12.14 As in Fig. 12.13 except $CMPE_P$ versus $CMPE_M$ (After Gao and Li 2011)

When $CMPE_M$ is compared to the reference precipitation efficiency ($RMPE_P$) over the entire rainy area, Fig 12.15 shows that $CMPE_M$ is generally higher than $RMPE_P$. Their RMS difference is 11.9 % (Table 12.1), which is smaller than the standard deviation of $RMPE_P$ (17.6 %, Table 12.2). This implies that $CMPE_M$ calculated over the entire rainfall area can be used to represent the reference precipitation efficiency. TFM shows much less scatter than the other rain types do because only fewer than 30 hourly samples of $CMPE_M$ deviate from $RMPE_P$ (Fig. 12.15). The RMS difference in TFM (4.5 %) is much smaller than the standard deviations of $RMPE_P$ (12.0 %) and $CMPE_M$ (10.7 %). The other rain types have RMS differences of at least 13.0 %. $CMPE_M$ is generally lower than $RMPE_P$ in TFm, tFm, Tfm, and tfM. In tFM and TfM, $CMPE_M$ and $RMPE_P$ are similar in many hourly samples, but $CMPE_M$ is much lower than $RMPE_P$ in some hourly samples, which makes their RMS differences similar to those of TFm, tFm, Tfm, and tfM. Although the RMS differences for the six rain types are well below the standard deviations of $RMPE_P$ (over 21.0 %), they can be similar to the standard deviation of $CMPE_M$ (15.7 %) for Tfm or even be larger than the standard deviation of $CMPE_M$ (14.5 %) for tFm. These six rain types have similar contributions to the RMS differences in the calculations over the entire rainfall area.

Figure 12.16 reveals that *LSPE* is generally higher than the reference precipitation efficiency ($RMPE_P$) for the total rainfall area. *LSPE* can be up to 100 % whereas $RMPE_P$ is less than 100 %. Their RMS difference of 31.0 % is larger than the standard deviation of $RMPE_P$ (17.6 %), but it is marginally smaller than the standard deviation of *LSPE* (32.5 %). Because all of the rainfall processes contribute to surface rainfall in TFM, *LSPE* is always 100 %. Meanwhile, $RMPE_P$ in TFM is generally lower than 100 %, which leads to a large RMS difference (19.2 %). The RMS difference in TFM is well above the standard deviations of $RMPE_P$ (12.0 %) and *LSPE* (0 %). *LSPE* is generally lower than $RMPE_P$ in the six other rain types and their RMS differences are over 16.0 % with the largest RMS difference up to 39.3 % in TfM. The RMS differences in tFM, tFm, TfM, and Tfm are larger than the standard deviations of $RMPE_P$ and *LSPE*. The RMS differences are smaller than the standard deviations of $RMPE_P$ and are marginally smaller than the standard deviations of *LSPE* in TFm and tfM. Thus, *LSPE* often cannot be used to estimate *RMPE* over the rainfall area and for each rain type.

The time-mean calculations over rainfall area in Table 12.3 reveal that $CMPE_M$ is similar to $RMPE_M$ but is higher than the reference precipitation efficiency ($RMPE_P$). *LSPE* is much higher than all other time-mean estimates of precipitation efficiency. All time-mean estimates of precipitation efficiency in TFM are higher than those in the six other rain types. In TFM, LSPE is 100 % and other estimates are about 86.0 %. All time-mean estimates of *LSPE* in the six other rain types are well below those of $RMPE_P$. The time-mean estimates of $CMPE_M$ are similar to those of $RMPE_P$ in tFM and TfM, but they are significantly lower than those of $RMPE_P$ in TFm, tFm, Tfm, and tfM.

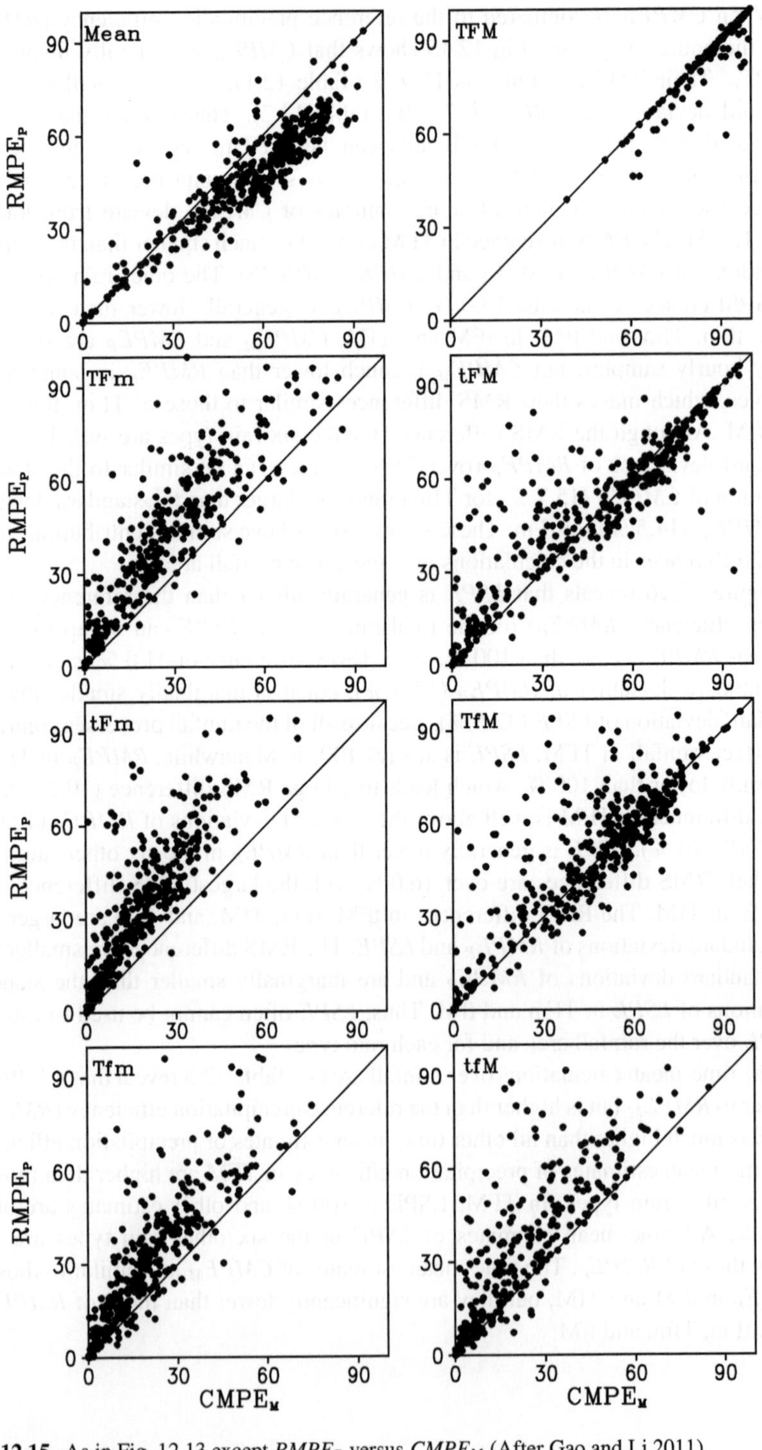

Fig. 12.15 As in Fig. 12.13 except $RMPE_P$ versus $CMPE_M$ (After Gao and Li 2011)

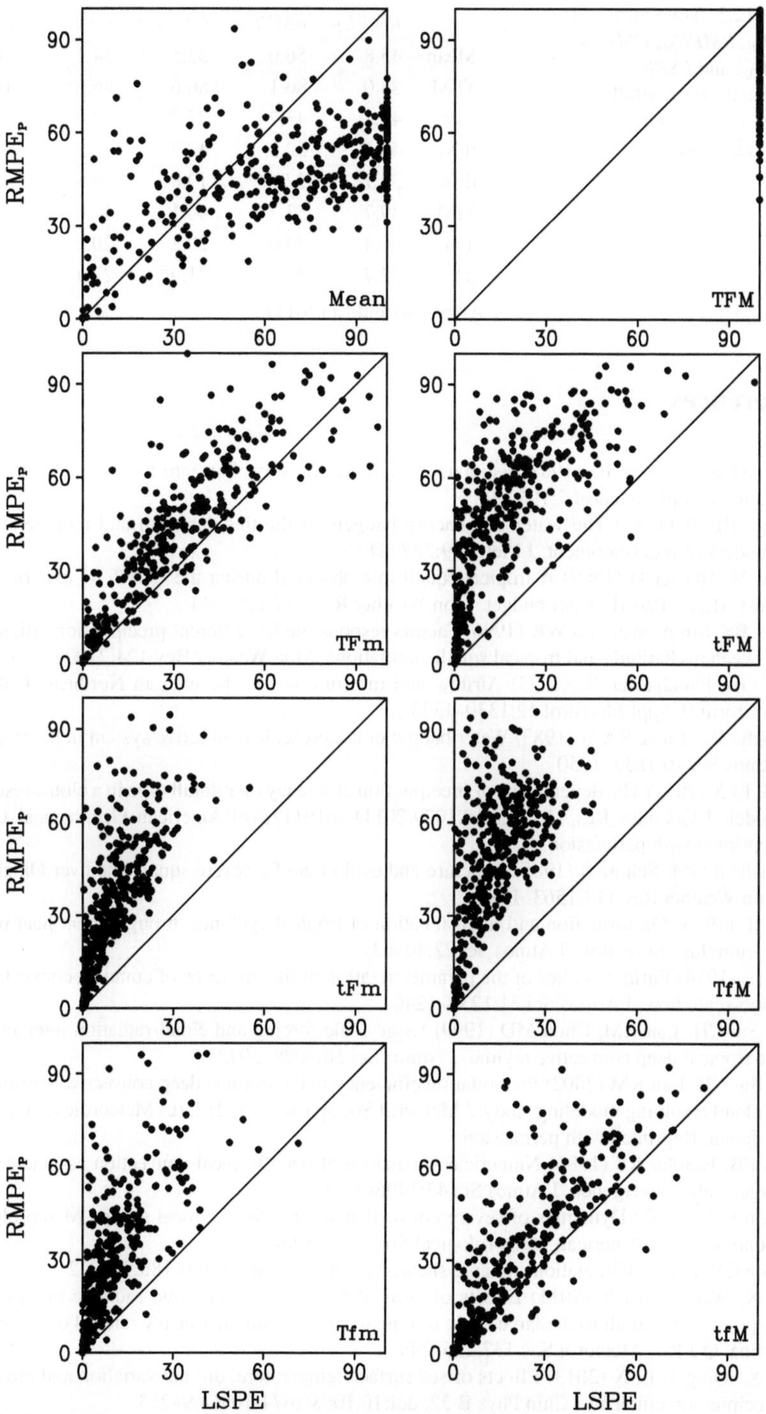

Fig. 12.16 As in Fig. 12.13 except $RMPE_P$ versus $LSPE$ (After Gao and Li 2011)

Table 12.3 Time means of $RMPE_P$, $RMPE_M$, $CMPE_P$, $CMPE_M$, and $LSPE$ calculated over rainfall regions (Mean) and regions of seven rain types

	$RMPE_P$	$RMPE_M$	$CMPE_P$	$CMPE_M$	$LSPE$
Mean	45.8	56.0	33.5	54.1	64.5
TFM	85.0	86.1	86.6	86.6	100.0
TFm	41.1	43.2	27.7	27.7	30.6
tFM	48.6	52.5	40.7	45.5	18.2
tFm	34.7	37.6	19.8	19.8	11.1
TfM	49.7	53.7	40.1	45.0	14.7
Tfm	30.4	33.0	20.2	20.2	10.5
tfM	32.7	34.5	21.9	22.4	22.7

After Gao and Li (2011)

References

Auer AH Jr, Marwitz JD (1968) Estimates of air and moisture flux into hailstorms on the High Plains. J Appl Meteorol 7:196–198

Braham RR Jr (1952) The water and energy budgets of the thunderstorm and their relation to thunderstorm development. J Meteor 9:227–242

Chong M, Hauser D (1989) A tropical squall line observed during the CORT 81 experiment in West Africa. Part II: water budget. Mon Weather Rev 117:728–744

Ferrier BS, Simpson J, Tao WK (1996) Factors responsible for different precipitation efficiencies between midlatitude and tropical squall simulations. Mon Weather Rev 124:2100–2125

Foote GB, Fankhauser JC (1973) Airflow and moisture budget beneath an Northeast Colorado hailstorm. J Appl Meteorol 12:1330–1353

Gamache JF, Houze RA Jr (1983) Water budget of a mesoscale convective system in the tropics. J Atmos Sci 40:1835–1850

Gao S, Li X (2011) The dependence of precipitation efficiency on rainfall type in a cloud-resolving model. J Geophys Res 116. doi:10.1029/2011JD016117, (c) American Geophysical Union. Reprinted with permission

Heymsfield GM, Schotz S (1985) Structure and evolution of a severe squall line over Oklahoma. Mon Weather Rev 113:1563–1589

Kuo HL (1965) On formation and intensification of tropical cyclones through latent heat release by cumulus convection. J Atmos Sci 22:40–63

Kuo HL (1974) Further studies of the parameterization of the influence of cumulus convection on large-scale flow. J Atmos Sci 31:1232–1240

Li X, Sui CH, Lau KM, Chou MD (1999) Large-scale forcing and cloud-radiation interaction in the tropical deep convective regime. J Atmos Sci 56:3028–3042

Li X, Sui CH, Lau KM (2002) Precipitation efficiency in the tropical deep convective regime: a 2-D cloud resolving modeling study. J Meteorol Soc Jpn 80:205–212, (c) Meteorological Society of Japan. Reprinted with permission

Lipps FB, Hemler RS (1986) Numerical simulation of deep tropical convection associated with large-scale convergence. J Atmos Sci 43:1796–1816

Newton CW (1963) Dynamics of severe convective storms. Severe local storms Meteorological monograph 27. American Meteorological Society, Boston

Newton CW (1966) Circulations in large sheared cumulonimbus. Tellus 18:699–712

Shen X, Wang Y, Li X (2011) Effects of vertical wind shear and cloud radiatve processes on responses of rainfall to the large-scale forcing during pre-summer heavy rainfall over southern China. Q J Roy Meteorol Soc 137:236–249

Shen X., Qing T, Li X (2013) Effects of sea surface temperature, diurnal variation and clouds on precipitation efficiency. Chin Phys B 22. doi:10.1088/1674-1056/094213

Sui CH, Li X (2005) A tendency of cloud ratio associated with the development of tropical water and ice clouds. Terr Atmos Oceanic Sci 16:19–434

Sui CH, Li X, Yang MJ, Huang HL (2005) Estimation of oceanic precipitation efficiency in cloud models. J Atmos Sci 62:4358–4370, (c) American Meteorological Society. Reprinted with permission

Sui CH, Li X, Yang MJ (2007) On the definition of precipitation efficiency. J Atmos Sci 64:4506–4513

Tao WK, Johnson D, Shie CL, Simpson J (2004) The atmospheric energy budget and large-scale precipitation efficiency of convective systems during TOGA COARE, GATE, SCSMEX, and ARM: cloud-resolving model simulations. J Atmos Sci 61:2405–2423

Wang D, Li X, Tao WK, Wang Y (2009) Effects of vertical wind shear on convective development during a landfall of severe tropical storm Bilis (2006). Atmos Res 94:270–275

Weisman ML, Klemp JB (1982) The dependence of numerically simulated convective storms on vertical wind shear and buoyancy. Mon Weather Rev 110:504–520

Zhou Y, Gao S, Li X (2014) Precipitation efficiency and its relationship with physical factors. Chin Phys B 23. doi:10.1088/1674-1056/23/6/064210, (c) Chinese Physical Union. Reprinted with permission

Chapter 13
Air-Sea Coupling

Abstract In this chapter, small-scale air-sea coupling is studied using coupled ocean-cloud resolving atmosphere model. The differences in mixed-layer temperature, salinity, and depth between 1D and 2D ocean models coupled with cloud-resolving atmosphere model are examined through the analysis of mixed layer heat and salinity budgets. The relationships between rain rate and sea surface temperature and between surface evaporation flux and sea surface temperature over cloudy area are analyzed to study the role of air-sea coupling in regulating surface rainfall processes. The relative importance of surface evaporation versus atmospheric moisture advection and condensation is examined at different time scales.

Keywords Air-sea coupling • Coupled ocean-cloud-resolving atmosphere model • Mixed layer • Temperature • Salinity

Air-sea interaction is one of the important processes that affect both atmospheric and oceanic variability. The atmosphere may affect *SST* and upper temperature stratification through changing ocean surface heat fluxes. The ocean may affect the atmospheric convection and associated thermodynamic distributions through changing the boundary layer stability. In this chapter, the brief history of coupled modeling studies of air-sea interaction processes is reviewed. The 2D coupled ocean-cloud resolving atmosphere model and its applications to study effects of small scale fluctuations associated with atmospheric convection and precipitation on spatial distribution of ocean mixed-layer temperature and salinity and role of air-sea coupling in the surface rainfall process are discussed based on Li et al. (2000) and Gao et al. (2006) in this chapter.

© Springer International Publishing Switzerland 2016

X. Li, S. Gao, *Cloud-Resolving Modeling of Convective Processes*,
Springer Atmospheric Sciences, DOI 10.1007/978-3-319-26360-1_13

13.1 Introduction

Precipitation and the associated stratification of salinity affect *SST* by changing the mixed layer depth (e.g., Miller 1976; Li et al. 1998) and the upper ocean thermal structure by forming a barrier layer between the halocline and the thermocline (e.g., Godfrey and Lindstrom 1989; Lukas and Lindstrom 1991; Vialard and Delecluse 1998a, b). Miller (1976) found precipitation could induce shallow mixed layers in numerical simulations. Since the effect of heating/cooling is inversely related to the mixed-layer depth, shallow mixed layers cause large temperature changes than do deep mixed layers with the same thermal forcing. Cooper (1988) found that salinity effects could account for as much as a 0.5 °C temperature bias and a 0.1 m s^{-1} velocity bias near the surface after 110 days of integration over the Indian Ocean. Murtugudde and Busalacchi (1998) found that the differences in annual mean *SST* between simulations with and without salinity and climatological precipitation could be as much as 0.5 °C, indicating the inclusion of salinity effects is necessary to simulate realistic climatic systems. Yang et al. (1999) also found that in the western Pacific warm pool, *SST* would be 0.6 °C lower if there were no salinity effect associated with precipitation.

Sui et al. (1997) employed a mixed layer model to study the role of vertical solar absorption profile in the diurnal ocean temperature simulations and their impacts in the intraseasonal variability. Due to the asymmetric diurnal variation for shoaling and deepening of the mixed layer, the cumulative effects of diurnal mixing cycles are essential to maintain a stable upper-ocean thermal stratification and to simulate a realistic evolution of mixed layer and temperature at the intraseasonal time. Further sensitivity tests of mixed layer to diurnal cycles indicate that the inclusion of diurnal convective-radiative processes in the atmosphere-ocean systems in the coupled models affects the capability of simulating intraseasonal variability.

Li et al. (1998) further included the salinity in the ocean mixed-layer model to examine the impacts of the precipitation and associated upper-ocean salinity stratification in the ocean mixed layer. The inclusion of salinity and precipitation-induced fresh water flux in the simulation shows much deep mixing occurs when the rainfall appears during nighttime since the fresh water flux induces a much shallower mixed layer with a large deepening rate. The salinity contributes more to the density stratification than does the temperature, which takes care of upper-ocean stability. The inclusion of the salinity stratification could cause the entrainment of warmer water into the ocean mixed layer since the salinity stratification takes care of the upper-ocean stability, whereas the exclusion of salinity in the simulation only shows entrainment of cold water into the ocean mixed layer since the thermal stratification accounts for the upper-ocean stability. Since the Kraus-Tuner mixing parameterization scheme (Niiler and Kraus 1977) requires both thermal and saline stratifications to determine the mixed-layer depth, decoupled salinity experiments are further conducted to examine the effect of thermal stratification on saline structure. The experiments reveal that a large fresh water input could cause a large difference (0.2 PSU) in the salinity between the experiments with and without the thermal stratification. The simulations indicate that the inclusion of precipitation-induced fresh water flux and salinity stratification improve the simulation of thermal evolution in the ocean mixed layer.

The simulation of the mixed-layer temperature relies largely on the vertical solar absorption profile. The simulated amplitudes of the mixed-layer temperature (<1 °C) are significantly smaller than the observed amplitudes (1–3 °C), implying that the mixed layer in the simulation does not absorb enough solar heat. Sui et al. (1998) calculated mixed-layer heat budget with the observed *SST* data to retrieve the amount of heat absorbed in the observed mixed layer to maintain the observed amplitudes of the *SST* and found that more than 39 % of the net surface solar irradiance is absorbed within the first 0.45 m, which is higher than previous estimates. The vertical solar absorption profile is then modified and the simulation with the modified solar profile yields more realistic amplitudes of the *SST* at both diurnal and intraseasonal time scales.

13.2 Development of a Cloud-Resolving Air-Sea Coupling System

Li et al. (2000) developed a 2D coupled ocean-cloud resolving atmosphere model to study small-scale air-sea coupling processes (see Sect. 1.1). The simulation from 0400 LST 18 December to 0400 LST 25 December 1992 during TOGA COARE shows the small-scale structures in the mixed layer associated with surface rainfall (Fig. 13.1). When the effects of fresh water flux and salinity were included in the coupled model, differences in the model domain mean mixed-layer temperature and salinity between 1D and 2D experiments were about 0.4 °C and 0.3 PSU, respectively (Fig. 13.2). The mean salinity difference was larger than the mean temperature difference in terms of their contributions to the mean density difference. In the 2D experiment, the surface heat flux showed a significant diurnal signal with the dominance of downward solar radiation during daytime and upward flux (IR radiative, sensible, and latent heat fluxes) during nighttime at each grid, although the amplitude was affected by precipitation. Thus, there was a strong thermal correlation between grids (Fig. 13.3a). Narrow cloudy areas were surrounded by broad cloud-free areas. The mean precipitation could occur, whereas the precipitation might not occur during most of the integration period. Thus, there is very low correlation between model domain mean and grid value of the fresh water fluxes (Fig. 13.3b). Since the rain rates have significant spatial variations, the fresh water flux has much larger spatial fluctuations than the saline entrainment does. Therefore, the fresh

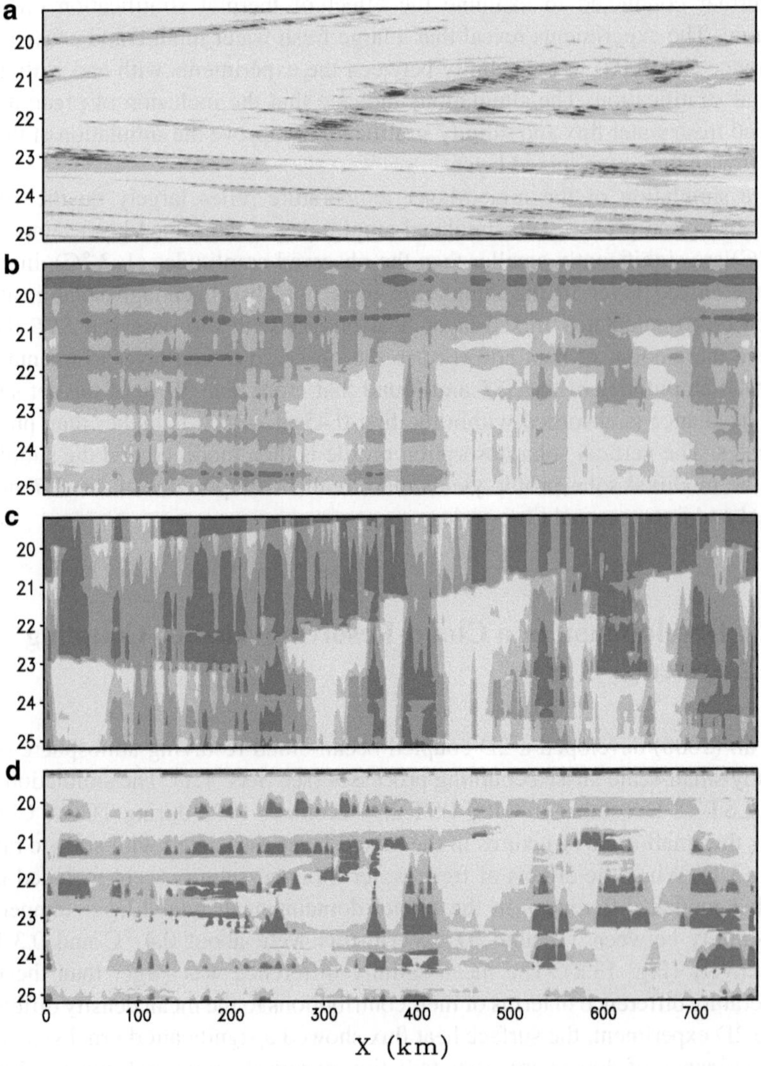

Fig. 13.1 Temporal and zonal distribution of (**a**) P_s, (**b**) T_m, (**c**) S_m, and (**d**) h_m. *Light shadings* denote 0–10 mm h^{-1} for P_s, 28.75–29.25 °C for T_m, 33.75–34 PSU for S_m, and 2–10 m for h_m, respectively; *medium shadings* denote 5–10 mm h^{-1} for P_s, 29.25–29.75 °C for T_m, 34–34.25 PSU for S_m, and 10–20 m for h_m, respectively; *dark shadings* denote >10 mm h^{-1} for P_s, >29.75 °C for T_m, >34,25 PSU for S_m, and >20 m for h_m, respectively (After Li et al. (2000)

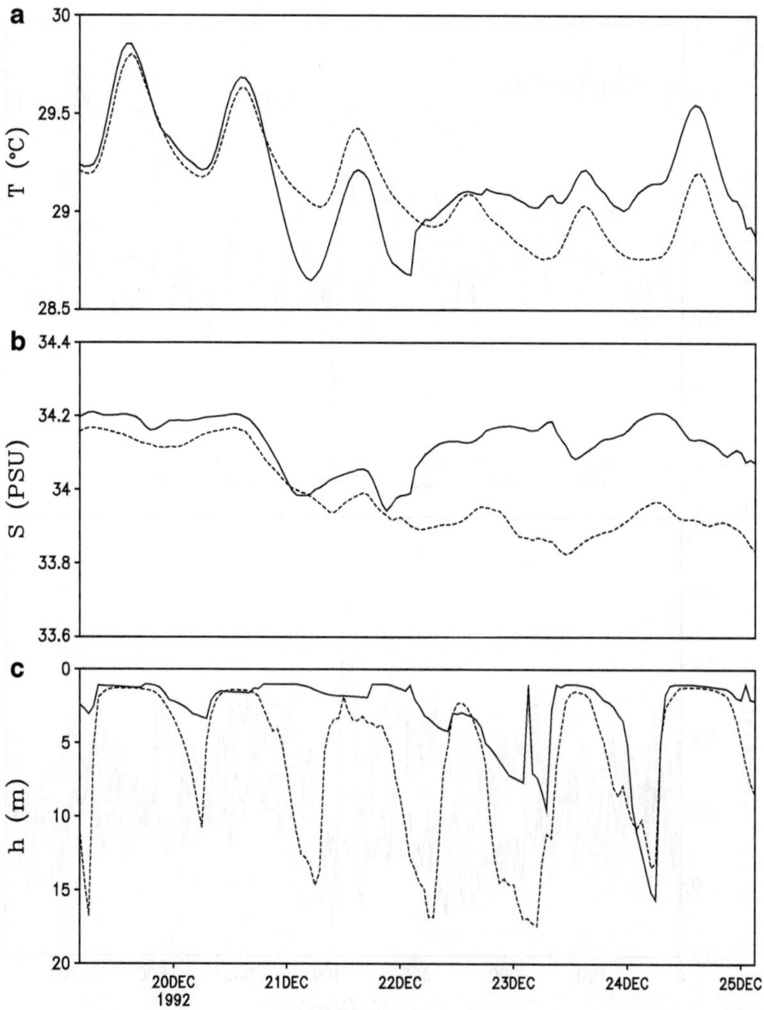

Fig. 13.2 Temporal evolution of (**a**) model domain mean mixed-layer temperature (°C), (**b**) salinity (PSU), and (**c**) depth (m) simulated in coupled model experiments with 1D (*solid*) and 2D (*dashed*) ocean mixed-layer models, respectively (After Li et al. 2000)

water flux determines large spatial salinity fluctuations, which contributes to large mean salinity difference between the 1D ocean model experiment and the 2D ocean model experiment.

The mean mixed layer temperature in both experiments shows a similarity in the first two days of model integrations (Fig. 13.2a). The mixed-layer temperature simulated by the 1D ocean model experiment becomes 1 °C lower than that

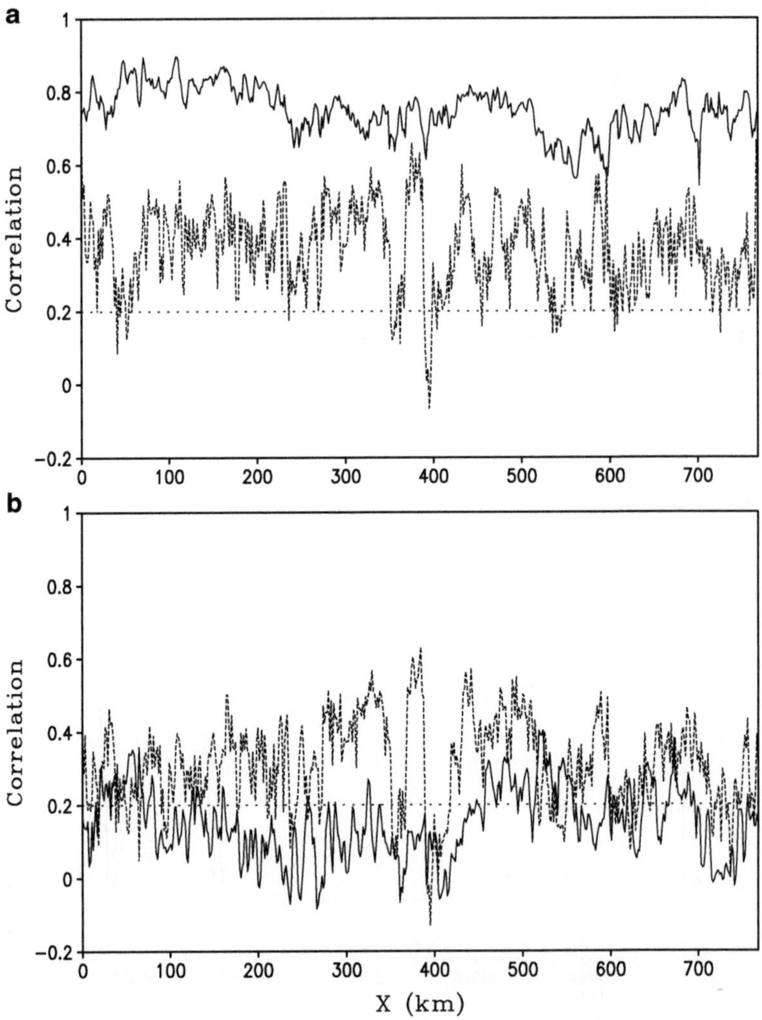

Fig. 13.3 Zonal distributions of correlation coefficients between grid-point value and model domain mean for (**a**) thermal entrainment (TE), and surface thermal forcing (STF), (**b**) saline entrainment (SE), and fresh water forcing (FWF) in the 2D coupled model simulation. *Dashed* and *solid lines* denote entrainment and forcing respectively. The correlation coefficient curves above upper dotted straight line exceed 95 % confidence level (After Li et al. 2000)

simulated by the 2D ocean model experiment from later night of 20 December 1992 and it is higher after the afternoon of 22 December. The mean heat budgets averaged from 2000 LST 20 December to 0400 LST 21 December show that the thermal forcing in the 1D ocean model experiment (−2.38 °C day^{-1}) is larger than that in the 2D ocean model experiment (−1.14 °C day^{-1}) due to the fact that the mixed layer depth in the 1D ocean model experiment (1.1 m) is much smaller than that in the 2D ocean model experiment (7.8 m). As a result, the mixed-layer temperature in the 1D ocean model experiment (−2.07 °C day^{-1}) decreases at a higher rate than that in the 2D ocean model experiment (−0.89 °C day^{-1}) does. The mean heat budgets averaged from 1400 LST to midnight of 22 December reveal that the thermal entrainment (0.61 °C day^{-1}) is nearly balanced by the thermal forcing (−0.54 °C day^{-1}) in the 1D ocean model experiment, whereas the thermal forcing (−0.92 °C day^{-1}) cannot be compensated by the thermal entrainment (0.38 °C day^{-1}) in the 2D ocean model experiment. Thus, the mixed layer shows a slight warming in the 1D ocean model experiment, whereas the mixed layer has a cooling in the 2D ocean model experiment.

The mean mixed layer salinity in both experiments shows a similarity in the first three days of model integrations (Fig. 13.2b). After 22 December, the salinity in the 2D ocean model experiment becomes 0.3 PSU lower than that in the 1D ocean model experiment. To explain this salinity difference, the domain mean salinity budgets averaged from 0100 LST to 1100 LST 22 December are analyzed. In the 1D ocean model experiment, the saline entrainment (0.33 PSU day^{-1}) overcomes fresh water forcing (−0.16 PSU day^{-1}) to enhance the ocean mixed-layer salinity at the rate of 0.17 PSU day^{-1}. In the 2D ocean model experiment, the saline entrainment (0.17 PSU day^{-1}) and salinity advection (0.16 PSU day^{-1}) are nearly balanced by fresh water forcing (−0.34 PSU day^{-1}) so that the domain mean salinity is a constant during this period.

13.3 Role of Air-Sea Coupling in Surface Rainfall Process

The ocean may affect the formation and development of clouds by *SST* through changing ocean surface fluxes, including surface evaporation flux. One of the most important air-sea interaction processes is intensification of convection with increasing *SST* from 26.5 to 29.5 °C over the tropical deep convective regime as shown in various analyses of monthly-mean data (e.g., Gadgil et al. 1984; Graham and Barnett 1987; Gutzler and Wood 1990; Zhang 1993; Waliser and Graham 1993; Webster 1994; Waliser 1996; Lau et al. 1997). Gao et al. (2006) analyzed hourly 50-day 2D coupled model simulation during TOGA COARE and calculated cloud-weighted P_s and cloud-weighted *SST* by averaging P_s and *SST* over cloudy area. P_s increases with increasing *SST* from 28 to 30 °C (Fig. 13.4), which is qualitatively consistent with the results from previous studies. The linear correlation coefficient calculated with 1161 samples is 0.18, which is above the 1 % significance level (0.081). However, the variance between cloud-weighted P_s and cloud-weighted *SST*

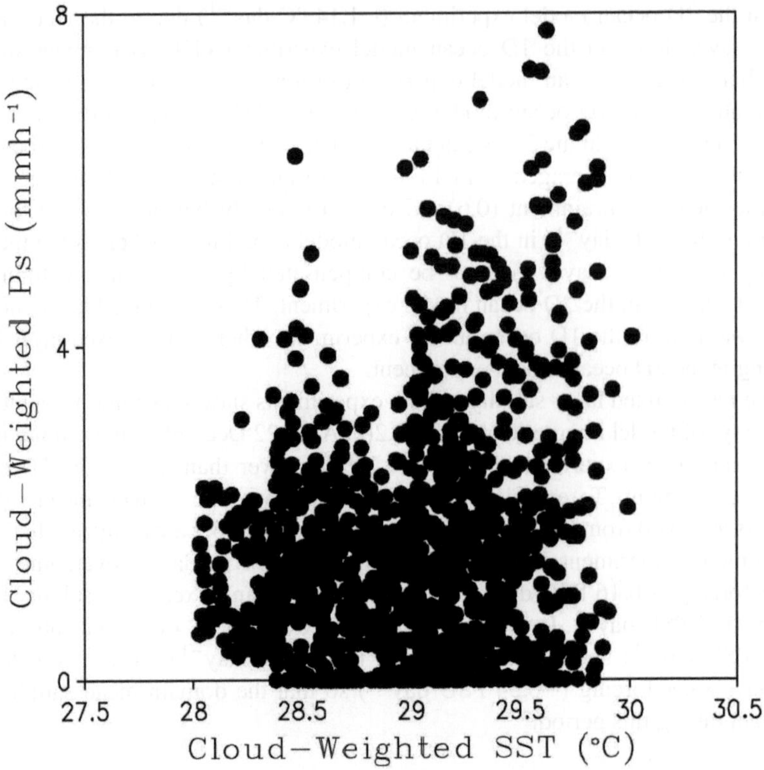

Fig. 13.4 Cloud-weighted P_s (mm h^{-1}) versus cloud-weighted *SST* (°C) (After Gao et al. 2006)

only is 3.2 %, which implies that the other 96.8 % of the variation in rainfall is explained by variations in other physical processes. Surface rainfall equation (2.4) shows contribution of surface evaporation to surface rain rate. Thus, *SST* affects atmospheric precipitation through changing surface evaporation (Q_{WVE}). Cloud-weighted Q_{WVE} decreases with increasing *SST* (Fig. 13.5) as indicated by their linear correlation coefficient of −0.35. Only 12.3 % of the variation in surface evaporation flux is explained by the variation in *SST*, whereas most (87.7 %) of the variation in surface evaporation flux is accounted for by variations in the other processes in the hourly data analysis. Thus, surface evaporation flux associated with *SST* has negligible impacts in the variation of atmospheric water vapor and surface rainfall over tropical cloudy regions. The negative correlation coefficient also indicates that the warm *SST* cannot produce a large surface evaporation flux to moisten the atmosphere over cloudy regions at a short timescale.

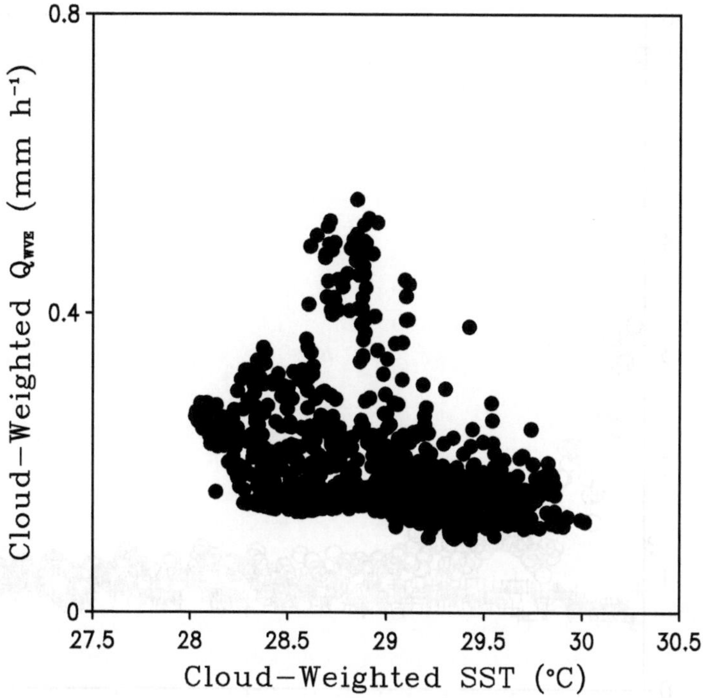

Fig. 13.5 Cloud-weighted Q_{WVE} versus cloud-weighted *SST* (°C) (After Gao et al. 2006)

To examine the time-dependent importance of surface evaporation flux in atmospheric *PW* budget and ocean mixed-layer thermal budget over cloudy regions, magnitudes of surface evaporation flux and sum of condensation and atmospheric moisture convergence in *PW* budget and magnitudes of surface thermal forcing and the sum of ocean thermal entrainment rate and thermal advections are calculated using the data averaged over various time frames shown in Fig. 13.6. For the *PW* budget averaged over the time frames that are less than 2 days, maximum magnitudes of the sum of condensation and atmospheric moisture convergence are much larger (2–6 times) than maximum magnitudes of surface evaporation flux. For the *PW* budget averaged over the time frames that are more than 2 days, they are about the same. In the ocean mixed-layer thermal budget, surface thermal forcing and the sum of ocean thermal entrainment rate and thermal advections have similar maximum magnitudes regardless of the time frames in which the analyzed data are averaged. This indicates that the impacts of the upper ocean in atmospheric moisture through surface evaporation flux associated with *SST* over cloudy regions may be time-dependent.

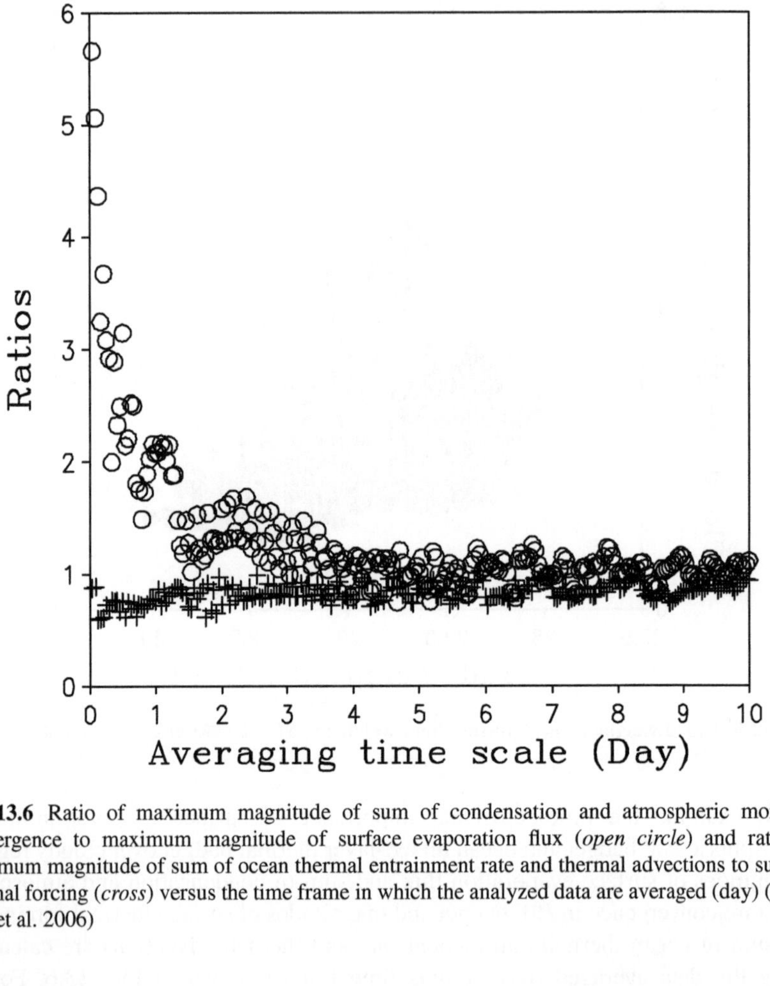

Fig. 13.6 Ratio of maximum magnitude of sum of condensation and atmospheric moisture convergence to maximum magnitude of surface evaporation flux (*open circle*) and ratio of maximum magnitude of sum of ocean thermal entrainment rate and thermal advections to surface thermal forcing (*cross*) versus the time frame in which the analyzed data are averaged (day) (After Gao et al. 2006)

References

Cooper NS (1988) The effect of salinity on tropical ocean models. J Phys Oceanogr 18:697–707

Gadgil S, Joseph PV, Joshi NV (1984) Ocean-atmospheric coupling over monsoon regions. Nature 312:141–143

Gao S, Ping F, Cui X, Li X (2006) Short timescale air-sea coupling in the tropical deep convective regime. Meteorog Atmos Phys 93:37–44

Godfrey JS, Lindstrom EJ (1989) The heat budget of the equatorial western Pacific surface mixed layer. J Geophys Res 94:8007–8017

Graham N, Barnett TP (1987) Sea surface temperature, surface wind divergence, and convection over tropical oceans. Science 238:657–659

Gutzler DS, Wood TM (1990) Structure of large-scale convective anomalies over the tropical oceans. J Clim 6:2049–2062

Lau KM, Wu HT, Bony S (1997) The role of large-scale atmospheric circulation in the relationship between tropical convection and sea surface temperature. J Clim 10:381–392

Li X, Sui CH, Adamec D, Lau KM (1998) Impacts of precipitation in the upper ocean in the western Pacific warm pool during TOGA COARE. J Geophys Res 103:5347–5359

Li X, Sui CH, Lau KM, Adamec D (2000) Effects of precipitation on ocean mixed-layer temperature and salinity as simulated in a 2-D coupled ocean-cloud resolving atmosphere model. J Meteor Soc Japan 78:647–659, (c) Meteorological Society of Japan. Reprint with permission

Lukas R, Lindstrom E (1991) The mixed layer of the western equatorial Pacific Ocean. J Geophys Res 96:3343–3457

Miller JR (1976) The salinity effect in a mixed-layer ocean model. J Phys Oceanogr 6:29–35

Murtugudde R, Busalacchi AJ (1998) Salinity effects in a tropical ocean model. J Geophys Res 103:3283–3300

Niiler PP, Kraus EB (1977) One-dimensional models. In: Kraus EB (ed) Modeling and prediction of the upper layers of the Ocean. Pergamon, New York

Sui CH, Li X, Lau KM, Adamec D (1997) Multi-scale air-sea interactions during TOGA COARE. Mon Weather Rev 125:448–462

Sui CH, Li X, Lau KM (1998) Selective absorption of solar radiation and upper ocean temperature in the equatorial western Pacific. J Geophys Res 103:10313–10321

Vialard J, Delecluse P (1998a) An OGCM study for the TOGA decade. Part I: role of salinity in the physics of the western Pacific fresh pool. J Phys Oceanogr 28:1071–1088

Vialard J, Delecluse P (1998b) An OGCM study for the TOGA decade. Part II: barrier layer formation and variability. J Phys Oceanogr 28:1089–1106

Waliser DE (1996) Formation and limiting mechanisms for very high sea surface temperature: linking the dynamics and thermodynamics. J Clim 9:161–188

Waliser DE, Graham NE (1993) Convective cloud systems and warm pool sea surface temperatures: coupled interactions and self-regulation. J Geophys Res 98:12881–12893

Webster PJ (1994) Role of hydrological processes in ocean–atmosphere interactions. Rev Geophys 32:427–476

Yang S, Lau KM, Schopf PS (1999) Sensitivity of the tropical Pacific Ocean to precipitation induced freshwater flux. Clim Dyn 15:737–750

Zhang C (1993) Large-scale variability of atmospheric deep convection in relation to sea surface temperature in the Tropics. J Clim 6:1898–1912

Chapter 14
Climate Equilibrium States

Abstract In this chapter, cloud-resolving model is integrated to reach climate equilibrium states to study effects of sea surface temperature and its diurnal variation, diurnal variation of radiation, and cloud-radiative and microphysical processes on climate equilibrium states. Heat and water vapor budgets are analyzed to highlight dominant physical processes that are responsible for these effects. The effects of zonal perturbations of sea surface temperature on climate equilibrium states are also examined.

Keywords Climate equilibrium states • Sea surface temperature • Diurnal variation • Radiative and microphysical processes • Heat budget • Water vapor budget • Zonal perturbation

Tropical climate equilibrium states are essentially determined by the nonlinear interactions of multiscale physical processes including the large-scale and cloud dynamics, cloud microphysics, radiative and surface processes, turbulence, and ocean mixing processes. The cloud-resolving modeling of convective–radiative equilibrium states provides ways to enhance the understanding of these controlling processes. The equilibrium studies with cloud-resolving models have been conducted for two decades since Nakajima and Matsuno first ran an equilibrium simulation with a 2D cloud-resolving model in 1988. In this chapter, a brief history of equilibrium studies with cloud-resolving modeling is reviewed. Effects of *SST*, diurnal variations, and cloud microphysical and radiative processes on tropical equilibrium states are discussed based on Gao et al. (2007), Ping et al. (2007), Cui and Gao (2008), and Gao (2008).

14.1 Introduction

Nakajima and Matsuno (1988) used a 2D cloud-resolving model with a constant radiative cooling profile and water microphysical schemes to study tropical cloud clusters. Their experiments with the 50-h integrations simulate tropical quasi-equilibrium states, which was the first successful experiment for equilibrium

© Springer International Publishing Switzerland 2016

X. Li, S. Gao, *Cloud-Resolving Modeling of Convective Processes*,
Springer Atmospheric Sciences, DOI 10.1007/978-3-319-26360-1_14

cloud-resolving model simulation. Nakajima and Matsuno found that individual clouds with a horizontal scale of 1 km and time scale of 1 h are embedded in cloud clusters with a horizontal scale of 100 km and time scale of 10 h, which is due to the fact that the formation of cold air at the foot of an individual cloud limits a small horizontal scale and a short life cycle for the individual cloud, whereas the density current associated with the cloud air pool triggers the formation of a new cloud along the edge of the pool extending a long horizontal scale and a long life cycle for cloud clusters. Their results also showed that the model cloud can be terminated by evaporative cooling. Islam et al. (1993) employed a 3D model to study the predictability of tropical mesoscale rainfall. The balance between the surface fluxes and constant radiative cooling is imposed in the model that leads to an equilibrium state after 50 h of the integration. Held and Hemler (1993) used a 2D model with interactive cloud-radiative forcing and water and ice microphysical schemes to study convective–radiative equilibrium in tropical moist convection and found that a quasi-biennial oscillation (QBO)-like variability with a period of 60 days is associated with the evolution of model domain-mean zonal wind, and the localization of tropical convection is a result of zero model domain-mean zonal wind.

Lau et al. (1993) and Sui et al. (1994) studied the tropical water and energy cycles and their roles in the tropical system by integrating the 2D cloud-resolving model to the climate equilibrium states. The model uses initial conditions from the 1956 Marshall Islands experiment in the central Pacific and is imposed with a time-invariant horizontally uniform large-scale vertical velocity and a constant *SST* of 28 °C, in which the simulated atmosphere is conditionally unstable below the freezing level and close to neutral above the freezing level. After the adjustment in about 20 days, the simulations reach the quasi-equilibrium states with the temperature of 258 K and *PW* of 51 mm. In the convective–radiative equilibrium conditions, the two thirds of surface rainfall come from the convective clouds while one third comes from the stratiform clouds. The *PW* budget shows that three fourths and one fourth of the total moisture supply are from the moisture advection associated with the imposed large-scale vertical velocity and the surface evaporation flux, respectively. The total moisture supply is completely converted into surface rainfall. The heat budget reveals that the cooling from the radiation and advection associated with the imposed large-scale vertical velocity is mainly balanced by the latent heat release associated with the precipitation processes. Grabowski et al. (1996) integrated the 2D cloud-resolving model for 24 days with similar initial conditions and found a quasi-equilibrium state with a temperature of 263 K and *PW* of 70 mm. Thus, tropical equilibrium states are warmer and more humid in Grabowski et al. (1996) than in Sui et al. (1994).

Robe and Emanuel (1996) used a 3D cloud-resolving model with a constant radiative cooling profile and water microphysical schemes to simulate statistical equilibrium. The simulations reach equilibrium states due to the balance between the net upward mass flux by moist convection and the net radiative cooling. The cloud mass flux increases with increasing radiative cooling. The mean updraft velocity is independent of the strength of the radiative forcing. Tompkins and Craig

(1998) used a 3D cloud-resolving model with interactive cloud-radiative forcing and water and ice microphysical schemes and also simulated a statistical equilibrium state. After 30 days of integration, they found that the adjustment time scale could be different; for instance, vertical mass flux adjusts to the equilibrium state much more quickly than thermodynamic variables do. The simulated convection is organized due to interaction between radiative, convective, and surface fluxes. Xu and Randall (1998) carried out the sensitive tests of quasi-equilibrium states to large-scale advective cooling and moistening and found that the time-varying large-scale forcing has no significant impact on the long-term behaviors while they affect short-term variations. Xu and Randall (1999) showed that simulated statistical equilibrium state is between the cold and dry regime in Sui et al. (1994) and the warm and humid regime in Grabowski et al. (1996) and found that the statistical equilibrium states are more sensitive to the transient large-scale forcing than to the magnitude of the forcing. Xu and Randall compared the model differences between Sui et al. and Grabowski et al. and found that Sui et al.'s model may be unable to maintain the initial wind profile reducing the surface wind speed and weakening surface evaporation. Tao et al. (1999) further compared the models by Sui et al. and Grabowski et al. and found that the mass-weighted relative humidity is 10 % higher in Grabowski et al. than in Sui et al., the *CAPE* in Grabowski et al. is also larger than that of Sui et al., and the microphysical parameterization schemes, grid sizes, and domains between their models are quite different. To explain the differences of equilibrium states between Sui et al. and Grabowski et al.'s experiments, sensitivity tests with the 2D cloud-resolving model are conducted by Tao et al. (1999). The equilibrium states are not sensitive to the initial conditions whereas they are sensitive to the minimum surface speed prescribed in the calculation of surface fluxes. Equilibrium thermodynamic states depend on the surface evaporation, where the surface wind plays a central role. The small surface evaporation associated with weak surface winds produces a cold and dry equilibrium state whereas the large evaporation associated with strong surface winds causes a warm and humid equilibrium state.

Randall et al. (1994) studied the deviations from the statistical equilibrium state in cloud-resolving model simulations, which could be due to an imbalance between a transient large-scale dynamic process such as waves and transient radiative processes associated with multi-time scale variations. Gao et al. (2006) analyzed the simulation data in COARE to study tropical heat and water vapor quasi-equilibrium and cycle. The data in the quasi-equilibrium states are averaged for two groups: surface rain rate is larger than mean value (0.29 mm h^{-1}) in group A and it is smaller in group B. As group-mean local atmospheric drying in group A (−0.09 mmh^{-1}) changes to local atmospheric moistening in group B (0.17 mmh^{-1}), group-mean local atmospheric warming (0.05 °Ch^{-1}) changes to local atmospheric cooling (−0.03 °Ch^{-1}), the surface rain rate decreases from 0.63 to 0.07 mmh^{-1}, the *CAPE* increases from 234 to 407 Jkg^{-1}, and the sum of surface evaporation and moisture convergence decreases from 0.48 to 0.21 mmh^{-1}. This implies that enhanced surface rainfall associated with the development of convection is a result of the release of more unstable energy and consumption of more water vapor. The energy and water vapor sources set a constraint for the tropical heat/water cycling. Convection develops with the enhanced rainfall as a result of the consumption of water vapor, and the release of *CAPE*, which causes the local atmospheric drying and warming. The convection and rainfall are suppressed until the atmosphere becomes more stable with small *CAPE* and low amount of water vapor. *CAPE* is generated and moisture is accumulated to rebuild a favorable environmental condition for the development of convection. Such cloud–environment interaction limits the deviation of thermodynamic state from its mean state and the life cycle of the clouds. Thus, clouds and associated precipitation serve as a regulator that ensures the approximate thermal and moisture balance and small variations of local water vapor and temperature.

Ramanathan and Collins (1991) conducted the observational study using NASA's Earth Radiation Budget Experiment (ERBE) and proposed that the thermostat effect is necessary to counteract the super-greenhouse warming and accounts for limiting *SST* over the western Pacific warm pool to a rather uniform distribution between 29 and 30 °C. Lau et al. (1994) further used the cloud-resolving model to investigate the cirrus-cloud thermostat effect for tropical *SST* by analyzing the net radiation flux at the top of the atmosphere and the net heat at the ocean–atmosphere interface. The model is integrated with the constant *SST*s of 28 and 30 °C and with and without the large-scale forcing, respectively. The net radiation flux at the top of the atmosphere is comprised of the net-absorbed solar radiation averaged over clear-sky regions, the IR radiation emitted by the ocean surface, atmospheric greenhouse effect, as well as IR and shortwave cloud forcing whereas the net heat flux at the ocean–atmosphere interface consists of solar and IR radiative and sensible and latent heat fluxes. The comparison of the experiments with the same large-scale forcing but different *SST*s shows that the largest changes in the components contributing to the net radiation flux at the top of the atmosphere are due to the emission of the surface IR radiation and greenhouse warming by the increase in water vapor which offset each other in a large part. The magnitude of the emission of the surface IR radiation is smaller than the greenhouse warming, suggesting no apparent "super-greenhouse" effect. The changes in IR and shortwave cloud forcing are small, which are insensitive to the changes in the *SST*s, because the change in the *SST* induces the change in low and mid-tropospheric clouds but does not have any impact on upper tropospheric clouds. The change in the net heat flux at ocean–atmosphere interface is mainly due to the change in the surface latent heat flux. The increase in the *SST* induces surface cooling by increasing surface evaporation. The increase in the *SST* produces 13 % increase in surface precipitation. The comparison of the experiments with the same *SST*s but different large-scale forcing (with and without the forcing) displays that the largest changes in the budget at the top of the atmosphere occur in the shortwave and IR cloud forcing, which cancel each other out in a large part. The experiment without the forcing undergoes a large reduction of greenhouse effect by decreasing the moisture. At the ocean–atmosphere interface, the largest change appears in the surface radiative flux due to the largest difference of clouds between the experiments with and without large-scale forcing.

14.2 Effects of *SST* on Equilibrium Climate

Surface boundary has an important impact on convective development as well as equilibrium climate. Chao (1961) investigated effects of surface heating on convective development by introducing a surface heat balance condition in which the surface solar heat flux transports upward by turbulence. His analytical model solution showed intensification of a vertical circulation associated with the surface forcing under an unstable stratification. Li et al. (1964) found a cloud with a size of 3 km in horizontal length and 2 km in depth corresponds to a surface heating perturbation of 0.4 °C and 2 km in horizontal length. In an oceanic coupled system, *SST* is affected by surface radiative and heat flux. *SST* in turn impacts the atmosphere through surface sensible heat and evaporation fluxes. In an atmospheric system alone, *SST* becomes an important forcing. Gao et al. (2007) studied effects of *SST* on tropical equilibrium states by conducting three experiments in a model framework of zero-imposed vertical velocity and vertically invariant zonal wind. Experiments SST27, SST29, and SST31 are imposed by time-invariant *SST*s of 27, 29, and 31 °C, respectively. A vertical wind shear has important impacts on convective development and configuration. For example, Chao and Chen (1964) in their linear analytical analysis showed that the vertical wind shear intensifies convective development when it is larger than 10^{-2} s^{-1}. The model produced unicell-type convection in a unidirectional wind shear (e.g., Dudhia et al. 1987; Tao et al. 1995) whereas the model generated multicell-type convection in a reversing wind shear (Tao et al. 2003). To emphasize the effects of *SST*, imposed zonal wind is vertically invariant.

Figure 14.1 shows time series of model domain and mass-weighted mean temperatures and model domain-mean *PW*. Temperature and *PW* reach equilibrium states in three experiments during the 40-day integrations. Model domain and mass-weighted mean temperatures averaged from 31 to 40 days are −6.2 °C in SST27, −3.0 °C in SST29, and 0.1 °C in SST31, respectively. Amounts of the mean *PW* averaged from day 31 to day 40 are 37.1 mm in SST27, 44.9 mm in SST29, and 52.5 mm in SST31, respectively. This indicates that cold *SST* produces cold and dry equilibrium states whereas warm *SST* generates warm and moist equilibrium states.

The temperature difference in condensational heating for SST27-SST29 is negative whereas the temperature differences in radiation and surface sensible heat flux are positive (Fig. 14.2a). Thus, the negative difference in condensational heating accounts for the negative temperature difference. The *PW* difference in condensation for SST27–SST29 is positive whereas the difference in surface evaporation flux is negative (Fig. 14.2b). Thus, the negative evaporation-induced difference is responsible for the negative *PW* difference. The analysis of heat and *PW* budgets reveals that the colder *SST* in SST27 pumps a smaller amount of water vapor from the ocean to the atmosphere than warmer SST in SST29 does, which leads to an atmospheric drying in SST27. Less water vapor in SST27 causes less condensation and associated heating than in SST29, which causes an atmospheric cooling in SST27.

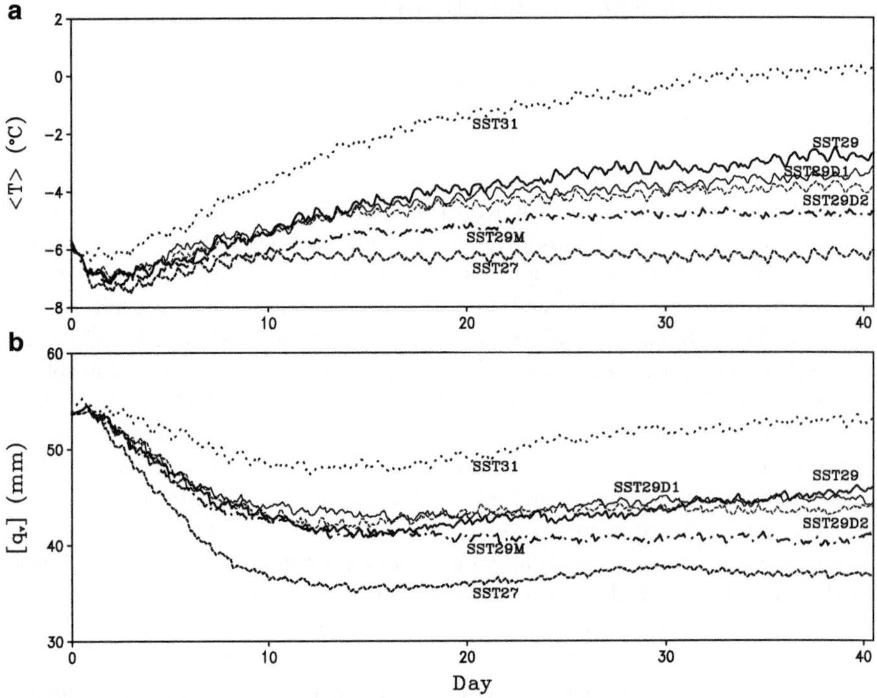

Fig. 14.1 Time series of (**a**) model domain and mass-weighted mean temperature (°C) and (**b**) model domain-mean *PW* (mm) in SST27 (*dark dashed*), SST29 (*dark solid*), SST31 (*dot*), SST29M (*dark dot dashed*), SST29D1 (*light solid*), and SST29D2 (*light dashed*) (After Gao et al. 2007)

Equilibrium cloud and rainfall properties in SST29 are calculated using simulation data from day 31 to 40 (Table 14.1a). Time and model domain-mean surface rain rate ($P_s = 0.130$ mm h^{-1}) is mainly from the mean surface evaporation rate ($Q_{WVE} = 0.134$ mm h^{-1}). Time and model domain-mean *LWP* (0.086 mm) is larger than time and model domain-mean *IWP* (0.60 mm). Fractional coverage is 50.1 % for non-raining stratiform regions, 3.2 % for convective regions, and 5.8 % for raining stratiform regions. Raining clouds (0.081 mm) have a larger *LWP* than non-raining clouds (0.005 mm) do whereas non-raining clouds (0.036 mm) have a larger *IWP* than do raining clouds (0.023 mm). Over clear-sky regions, the surface evaporation flux ($Q_{WVE} = 0.074$ mm h^{-1}) has a smaller magnitude than the vapor divergence ($Q_{WVF} = -0.088$ mm h^{-1}) has, which yields an atmospheric drying ($Q_{WVT} = 0.013$ mm h^{-1}). Over raining stratiform regions, all four processes contribute to surface rainfall. Among them, the cloud source ($Q_{CM} = 0.025$ mm h^{-1}) is the largest while the evaporation ($Q_{WVE} = 0.004$ mm h^{-1}) is smallest. Over convective regions, the vapor convergence ($Q_{WVF} = 0.116$ mm h^{-1}) is a main source for surface rainfall as well as a hydrometeor gain/convergence ($Q_{CM} = -0.033$ mm h^{-1}). Over non-raining stratiform regions, the vapor divergence

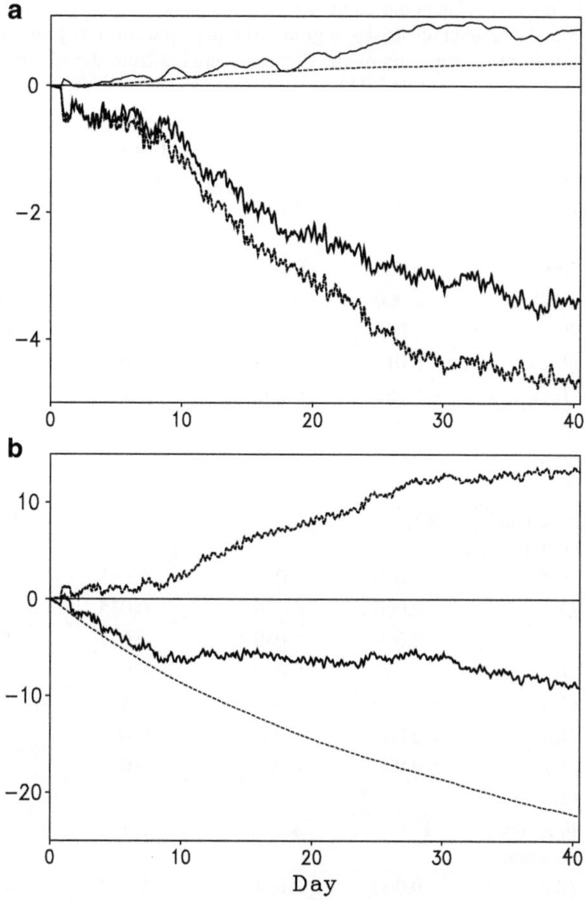

Fig. 14.2 Time series of (**a**) mass-weighted mean temperature (°C) difference (*dark solid*) for SST27-SST29 and differences caused by condensational heating (*dark dashed*), radiative heating (*light solid*), and surface sensible heat flux (*light dashed*) and (**b**) *PW* (mm) difference (*dark solid*) and the differences caused by condensation (*dark dashed*) and surface evaporation flux (*light dashed*) (After Gao et al. 2007)

($Q_{WVF} = -0.034$ mm h^{-1}) has a smaller magnitude than does the surface evaporation flux ($Q_{WVE} = 0.053$ mm h^{-1}), which leads to an atmospheric moistening ($Q_{WVT} = -0.021$ mm h^{-1}).

Equilibrium cloud and rainfall properties in SST27 (Table 14.1b) are compared with those in SST29 to examine the effects of SST on equilibrium cloud and rainfall processes. Time and model domain-mean surface rain rates are similar in SST27 and SST29 (0.130 mm h^{-1}) although the mean surface evaporation rate is smaller in SST27 (0.119 mm h^{-1}) than in SST29 (0.134 mm h^{-1}). Similar rain rates in both experiments are due to the local atmospheric drying in SST27 (0.012 mm h^{-1}). While the mean IWP is insensitive to the change of SST, the decrease in the imposed SST from 29 to 27 °C causes a decrease in the mean LWP from 0.084 to 0.074 mm. The decrease in the imposed SST from 29 to 27 °C leads to a decrease in the fractional coverage for convective regions from 3.2 to 2.9 %, in raining stratiform regions from 5.8 to 5.6 %, and in non-raining stratiform clouds from 50.1 to 45.8 %.

Table 14.1 Fractional cloud coverage (%), *IWP*, *LWP* (mm), P_s, Q_{WVT},Q_{WVF}, Q_{WVE}, and Q_{CM} (mm h^{-1}) over clear-sky regions, raining stratiform regions, convective regions, non-raining stratiform regions, and model domain averaged from days 31 to 40 in (a) SST29, (b) SST27, (c) SST29M, and (d) SST29D2

(a)	Clear-sky regions	Raining stratiform regions	Convective regions	Non-raining stratiform regions	Model Domain mean
Fractional coverage	40.7	5.8	3.2	50.1	100
IWP	0.000	0.018	0.005	0.036	0.060
LWP	0.000	0.041	0.040	0.005	0.086
P_s	0.000	0.056	0.074	0.000	0.130
Q_{WVT}	0.013	0.019	−0.011	−0.021	0.000
Q_{WVF}	−0.088	0.007	0.116	−0.034	0.000
Q_{WVE}	0.074	0.004	0.002	0.053	0.134
Q_{CM}	0.000	0.025	−0.033	0.002	−0.005
(b)					
Fractional coverage	45.5	5.6	2.9	45.8	100
IWP	0.000	0.022	0.005	0.033	0.061
LWP	0.000	0.036	0.035	0.004	0.074
P_s	0.000	0.059	0.071	0.000	0.130
Q_{WVT}	0.019	0.019	−0.009	−0.017	0.012
Q_{WVF}	−0.089	0.007	0.111	−0.029	0.000
Q_{WVE}	0.069	0.004	0.002	0.044	0.119
Q_{CM}	0.000	0.029	−0.034	0.002	−0.002
(c)					
Fractional coverage	47.8	4.7	3.1	44.2	100
IWP	0.000	0.021	0.007	0.031	0.059
LWP	0.000	0.036	0.043	0.005	0.084
P_s	0.000	0.065	0.103	0.000	0.168
Q_{WVT}	0.022	0.024	−0.013	−0.024	0.010
Q_{WVF}	−0.113	0.004	0.149	−0.039	0.000
Q_{WVE}	0.091	0.005	0.003	0.056	0.155
Q_{CM}	0.000	0.032	−0.037	0.007	0.003
(d)					
Fractional coverage	49.5	4.6	3.0	42.7	100
IWP	0.000	0.018	0.006	0.032	0.056
LWP	0.000	0.035	0.040	0.005	0.080
P_s	0.000	0.058	0.091	0.000	0.150
Q_{WVT}	0.020	0.021	−0.010	−0.020	0.011
Q_{WVF}	−0.103	0.001	0.134	−0.032	0.000
Q_{WVE}	0.082	0.004	0.003	0.048	0.137
Q_{CM}	0.000	0.032	−0.035	0.004	0.002

Although the mean Q_{WVT}, Q_{WVF}, and Q_{CM} are negligibly small, they may contribute to surface rainfall processes over specific regions. Area-weighted values (that are calculated by dividing the values in Table 14.1a, b with corresponding fractional cloud coverage in each region) are used for discussion in this paragraph. Over clear-sky regions, Q_{WVE} is 0.152 mm h^{-1} in SST27 and 0.182 mm h^{-1} in SST29. Q_{WVF} is −0.196 mm h^{-1} in SST27 and −0.213 mm h^{-1} in SST29. Q_{WVT} is 0.041 mm h^{-1} in SST27 and 0.032 mm h^{-1} in SST29. Both surface evaporation flux and the vapor divergence increase their magnitudes with increasing SST, whereas the local atmospheric drying rate decreases with increasing SST. Over raining stratiform regions, P_s is 1.054 mm h^{-1} in SST27 and 1.073 mm h^{-1} in SST29. The stratiform rain rate is insensitive to the variation of SST. Over convective regions, convective rain rate is larger in SST27 (2.448 mm h^{-1}) than in SST29 (2.313 mm h^{-1}) because the vapor convergence rate is larger in SST27 (3.828 mm h^{-1}) than in SST29 (3.625 mm h^{-1}). The vapor convergence is a dominant factor in convective rainfall. The vapor convergence rate and the surface rain rate increase as the *SST* decreases. Over non-raining stratiform regions, Q_{WVE} is 0.096 mm h^{-1} in SST27 and 0.106 mm h^{-1} in SST29. Q_{WVF} is −0.063 mm h^{-1} in SST27 and −0.068 mm h^{-1} in SST29. Q_{WVT} is −0.037 mm h^{-1} in SST27 and −0.042 mm h^{-1} in SST29. Q_{WVF}, Q_{WVE}, and Q_{WVT} increase their magnitudes with increasing SST. Thus, the SST-induced decrease in the surface evaporation flux suppresses the atmospheric moistening and vapor divergence.

14.3 Effects of Diurnal Variation on Equilibrium Climate

To examine effects of diurnal variation of solar radiative heating on tropical equilibrium states, the experiment SST29M with a fixed daily-mean cosine of solar zenith angle is conducted and compared with SST29 (Gao et al. 2007). Two experiments with diurnally varying SSTs are also conducted to study effects of diurnal variations of *SST* on tropical equilibrium states (Gao et al. 2007). Experiments SST29D1 and SST29D2 have diurnally varying SSTs with the mean of 29 °C and diurnal differences of 1 and 2 °C, respectively. The maximum and minimum SSTs occur at 1600LST and 0700 LST, respectively.

Model domain and mass-weighted mean temperatures in SST29M and SST29 have similar magnitudes and evolution in the first two days and their differences become larger later on (Fig. 14.1a). The zonal-mean equilibrium temperature averaged in the last 10 days of the integration in SST29M is −4.8 °C, which is 1.8 °C colder than the equilibrium temperature in SST29. Amounts of zonal-mean *PW* in SST29M and SST29 have similar magnitudes and evolution in the first 18 days. Later on, the *PW* in SST29M becomes smaller than it does in SST29 (Fig. 14.1b). Zonal-mean equilibrium *PW* averaged in the last 10 days of the integration in SST29M is 40.6 mm, which is 4.3 mm smaller than the equilibrium *PW* in SST29. These indicate that the simulation with a daily-mean solar zenith angle produces a colder and drier equilibrium state than does the simulation with a diurnally varied solar zenith angle.

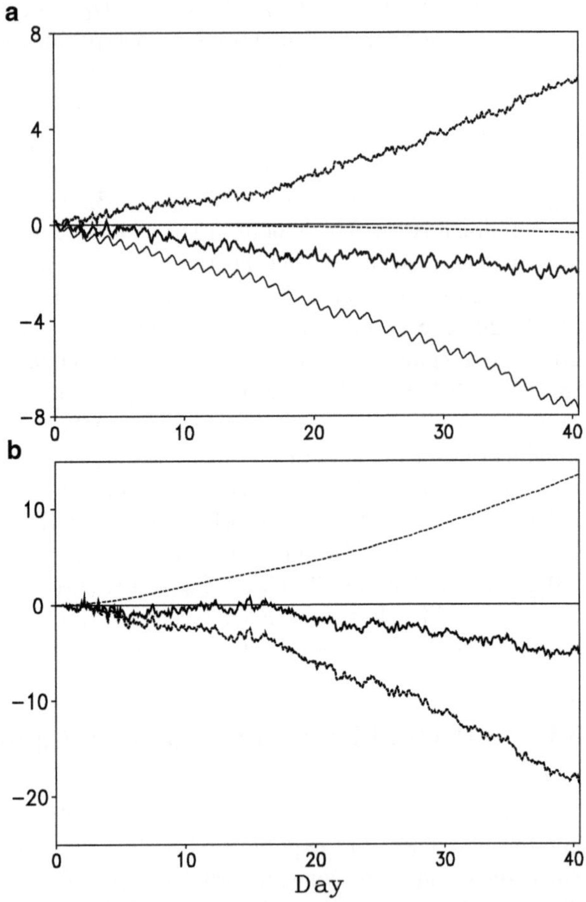

Fig. 14.3 As in Fig. 14.2 except for the difference for SST29M-SST29 (After Gao et al. 2007)

The difference in heat budgets between SST29M and SST29 shows that the temperature difference in condensational heating is positive whereas the temperature differences in radiation and surface sensible heat flux are negative (Fig. 14.3a). The negative difference in radiation has a much larger magnitude than the negative difference in surface sensible heat flux does. Thus, the negative radiation-induced difference leads to the negative temperature difference for SST29M-SST29. A separation analysis between solar heating and IR cooling reveals that the negative difference in solar heating accounts for the negative temperature difference since the difference in solar heating is negative whereas the difference in IR cooling is positive (Fig. 14.4a). The negative difference in solar heating for SST29M-SST29 is mainly contributed to by the negative difference in clear-sky regions in the first few days (Fig. 14.4c) and in cloudy regions later on (Fig. 14.5b). This indicates that the clouds enhance the negative difference in solar heating.

The difference in *PW* budgets for SST29M-SST29 shows that the negative *PW* difference is mainly determined by the negative difference in condensation since

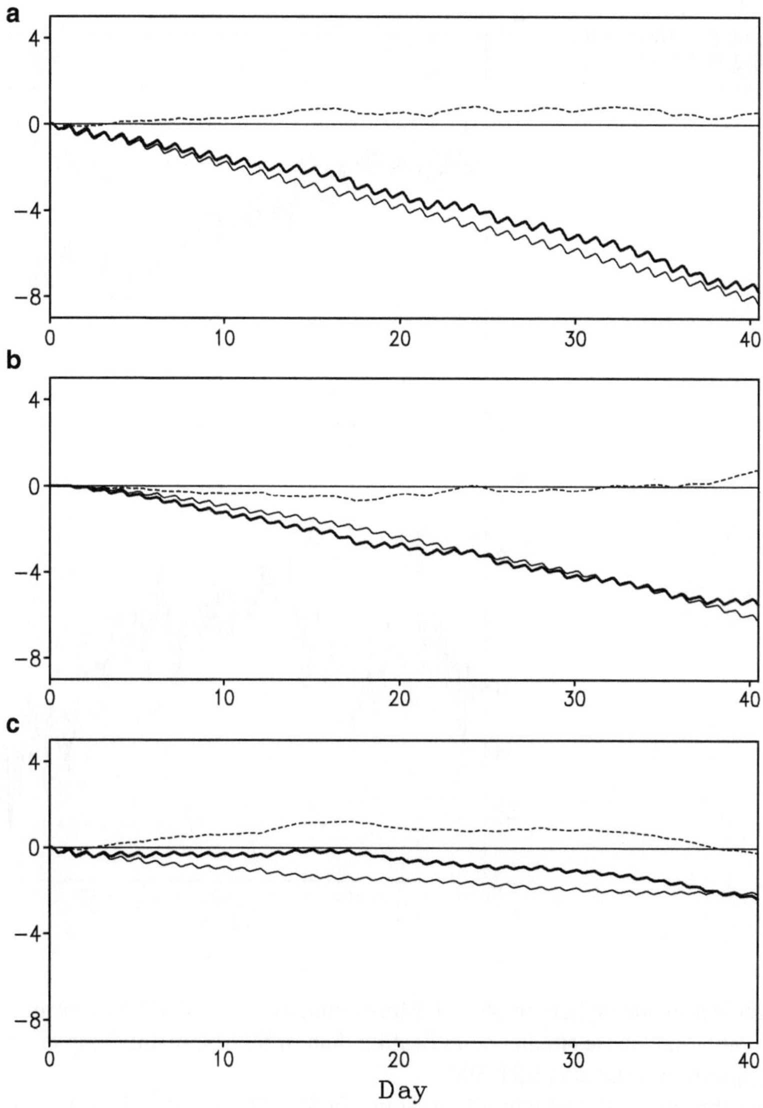

Fig. 14.4 Time series of mass-weighted mean temperature (°C) differences for SST29M-SST29 caused by radiation (*dark solid*), solar heating (*light solid*), and IR cooling (*light dashed*) in (**a**) horizontal domain mean, (**b**) cloudy regions, and (**c**) clear-sky regions (After Gao et al. 2007)

the difference in surface evaporation flux is positive whereas the difference in condensation is negative (Fig. 14.3b). This indicates that SST29M produces more condensation and consumes more water vapor than SST29 does.

Initially, SST29M produces smaller solar heating and colder temperature than SST29 does. The colder atmosphere in SST29M causes more condensation than

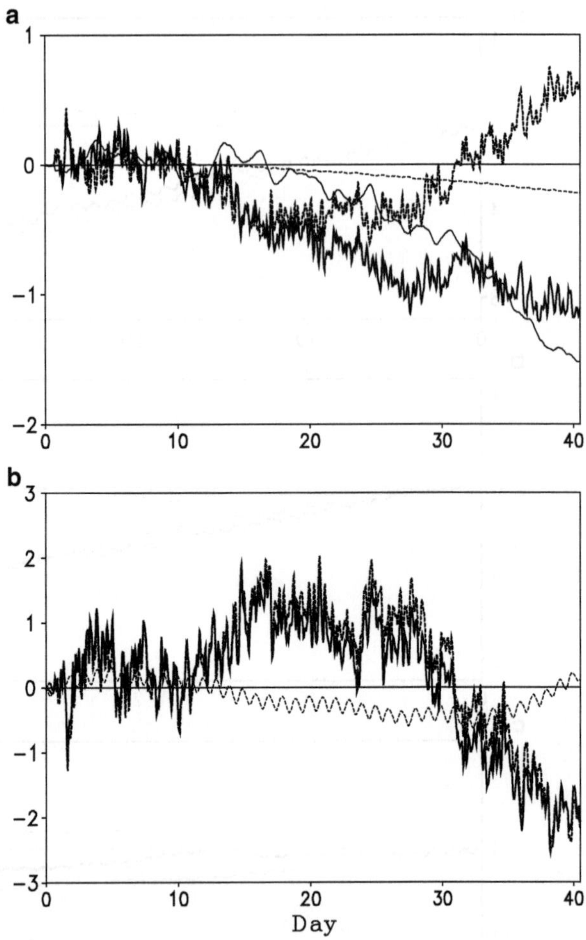

Fig. 14.5 As in Fig. 14.2 except for the difference for SST29D2-SST29 (After Gao et al. 2007)

does the warm atmosphere in SST29. More condensates in SST29M consume more water vapor and cause smaller solar heating than in SST29, maintaining colder and drier equilibrium states in SST29M.

Equilibrium cloud and rainfall properties in SST29M (Table 14.1c) is compared with those in SST29 to examine the effects of diurnal variation of solar zenith angle on equilibrium cloud and rainfall processes. The mean surface rain rate is larger in SST29M (0.168 mm h^{-1}) than in SST29 (0.130 mm h^{-1}) due to the fact that the mean surface evaporation rate is larger in SST29M (0.155 mm h^{-1}) than in SST29 (0.134 mm h^{-1}) and that the mean local atmospheric drying also contributes to surface rain rate in SST29M (0.010 mm h^{-1}). Thus, the exclusion of diurnal variation of solar zenith angle in the simulation enhances surface rain rate through enhanced surface evaporation fluxes and the atmospheric drying process. Both the mean *IWP* and *LWP* are not very sensitive to diurnal variation of solar zenith angle. While fractional coverage of convective regions is insensitive to the diurnal variation of solar zenith angle, SST29M produces a smaller fractional coverage of both raining and non-raining stratiform regions and larger fractional coverage of clear-sky regions than SST29 does. In convective regions, SST29M generates larger *IWP* and *LWP* than SST29 does. In raining stratiform regions, SST29M produces a larger *IWP* and a smaller *LWP* than SST29 does. In non-raining stratiform regions, *IWP* is 14 % smaller in SST29M than in SST29 although both experiments have same *LWP*.

SST29M produces larger mean surface rain rate than SST29 does mainly because SST29M (0.103 mm h^{-1}) has larger time-mean convective rain rate than SST29 (0.074 mm h^{-1}) does. The difference in time-mean stratiform rain rate for SST29M-SST29 also contributes to the difference in mean surface rain rate. The vapor convergence largely determines convective rain rate in both experiments. Thus, the exclusion of diurnal variation of solar zenith angle in the simulation enhances water vapor convergence in convective regions, convective rain rate, and thus model domain-mean surface rain rate. SST29M generates larger surface evaporation rate than SST29 does over clear-sky regions, while the surface evaporation rates in both experiments are similar over non-raining stratiform regions. Thus, SST29M has larger water vapor divergence than SST29 does over clear-sky regions. Therefore, the exclusion of diurnal variation of solar zenith angle in the simulation enhances surface evaporation over clear-sky regions and water vapor transport from clear-sky regions to convective regions and increases convective rainfall.

To examine the effects of diurnally varied *SST* on tropical equilibrium states, two additional experiments SST29D1 and SST29D2 are carried out. SST29D1 and SST29D2 show slightly warmer horizontal domain and mass-weighted mean temperatures in the first 10 days and colder temperatures after day 10 than SST29 does (Fig. 14.1a). Model domain-mean equilibrium temperatures averaged from day 31 to day 40 are −3.6 °C in SST29D1 and −3.9 °C in SST29D2, which are 0.6 and 0.9 °C colder than the equilibrium temperature in SST29, respectively. Amounts of model domain-mean *PW* in SST29D1 and SST29D2 show larger magnitudes than are shown in SST29 between day 10 and day 30, and they become smaller in the last 10 days of the integrations (Fig. 14.1b). Model domain-mean equilibrium *PW* amounts averaged in the last 10 days are 44.6 mm in SST29D1 and 43.6 mm in SST29D2, which are 0.3 mm and 1.3 mm smaller than the equilibrium *PW* in SST29, respectively.

The differences in *PW* budgets for SST29D2-SST29 reveal that SST29D2 enhances surface evaporation flux and condensation through a warm afternoon *SST* anomaly than SST29 does initially (Fig. 14.5b). The positive difference in condensation for SST29D2-SST29 induces a positive difference in IR cooling in the first 10 days (Fig. 14.6). The positive temperature difference is mainly caused by the positive differences in IR cooling and condensational heating (Fig. 14.5a). The positive difference in IR cooling comes from the positive difference in IR cooling in clear-sky regions (Fig. 14.6). From day 10 to day 20, the negative difference in condensational heating mainly accounts for the negative temperature difference. From day 20 to day 30, the negative differences in condensational

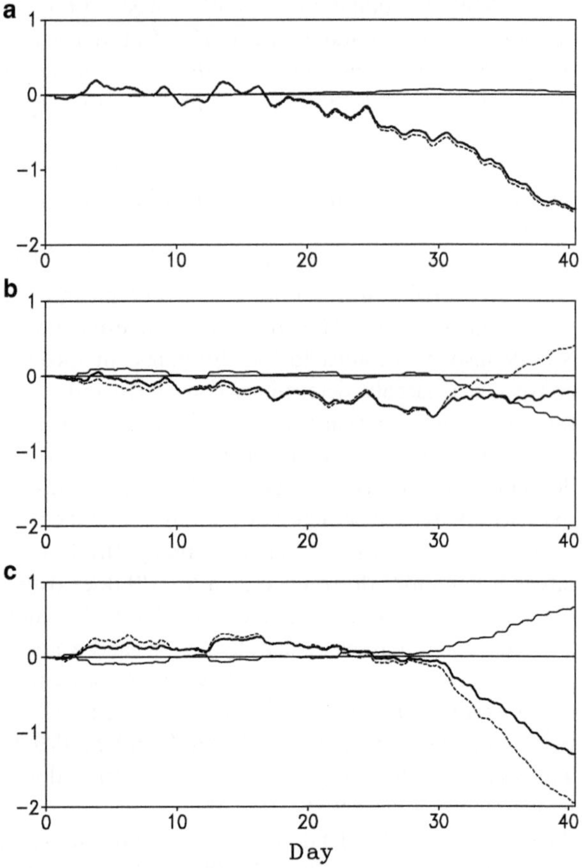

Fig. 14.6 As in Fig. 14.4 except for the difference for SST29D2-SST29 (After Gao et al. 2007)

heating and radiation are responsible for the negative temperature difference. The negative difference in radiation results from the negative difference in IR cooling in cloudy regions (Fig. 14.6b). In the last 10 days, the negative differences in IR cooling in clear-sky regions (Fig. 14.6c) as well as in solar heating in cloudy regions (Fig. 14.6b) keep the negative temperature difference since the difference in radiation remains negative, whereas the difference in condensational heating becomes positive (Fig. 14.5a).

The differences in *PW* budgets for SST29D2-SST29 show that the difference in condensation determines the *PW* difference because the positive difference in surface evaporation flux is small (Fig. 14.5b). The positive differences in condensation as well as in surface evaporation contribute to the positive *PW* difference in the first 10 days. The positive difference in condensation accounts for the positive *PW* difference from day 10 to day 30 since the difference in condensation is positive whereas the difference in surface evaporation is negative. The negative difference in condensation is mainly responsible for the negative *PW* difference in the last 10 days.

Initially, a warm afternoon *SST* anomaly in SST29D2 enhances surface evaporation and condensation, compared to SST29. More condensation induces less IR cooling in SST29D2 than in SST29. The warmer atmosphere in SST29D2 produces a smaller surface evaporation flux than is found in SST29 around day 10. Compared to SST29, the smaller surface evaporation flux in SST29D2 pumps less moisture into the atmosphere and causes less condensation and less moisture consumption from day 10 to day 30. Meanwhile, the smaller condensation in SST29D2 releases less latent heat and more IR cooling than in SST29, which leads to a colder temperature in SST29D2. In the last 10 days, a colder atmosphere in SST29D2 yields more condensation and more moisture consumption than in SST29, which results in a small amount of *PW* in SST29D2. More condensation in SST29D2 leads to more latent heat and larger IR cooling than in SST29. The larger IR cooling in SST29D2 maintains a colder atmosphere than in SST29.

Equilibrium cloud and rainfall properties in SST29D2 (Table 14.1d) are compared with those in SST29 to examine the effects of diurnal variation of *SST* on equilibrium cloud and rainfall processes. Time and model domain-mean surface rain rate is smaller in SST29D2 (0.150 mm h^{-1}) than in SST29 (0.130 mm h^{-1}) due to the fact that local atmospheric drying contributes to surface rain rate in SST29D2 (0.011 mm h^{-1}) and that the local hydrometeor loss occurs in SST29D2 (0.002 mm h^{-1}), whereas the local hydrometeor gain appears in SST29 (−0.005 mm h^{-1}). Thus, the inclusion of diurnal *SST* variation in the simulation enhances surface rain rate through the atmospheric drying and hydrometeor loss. Both time and model domain-mean *IWP* and *LWP* are 9 % smaller in SST29D2 than in SST29. While fractional coverage of convective regions is not very sensitive to the diurnal *SST* variation, SST29D2 produces smaller fractional coverage of stratiform regions and larger clear-sky regions than SST29 does. In convective regions, time-mean *IWP* and *LWP* are insensitive to the diurnal *SST* variation. Time-mean stratiform *LWP* contributes less to time and model domain-mean *LWP* in SST29D2 than in SST29 although time-mean stratiform *IWP* has same contributions to time and model domain-mean *IWP* in both experiments. SST29D2 produces larger time and model domain-mean surface rain rate than SST29 does mainly because SST29D2 (0.091 mm h^{-1}) has larger time-mean convective rain rate than SST29 (0.074 mm h^{-1}) does, which is due to the fact that the vapor convergence is larger in SST29D2 (0.134 mm h^{-1}) than in SST29 (0.116 mm h^{-1}). Thus, the inclusion of diurnal *SST* variation in the simulation enhances water vapor convergence in convective regions that intensifies convective rain rate and thus model domain-mean surface rain rate. SST29D2 generates larger surface evaporation rate and local atmospheric drying rate than SST29 does over clear-sky regions, which leads to larger water vapor divergence in SST29D2 than in SST29 over clear-sky regions. Therefore, the inclusion of diurnal *SST* variation in the simulation enhances surface evaporation over clear-sky regions and water vapor transport from clear-sky regions to convective regions that increases convective rainfall.

14.4 Cloud Microphysical and Radiative Effects on Equilibrium Climate

To study microphysical and radiative effects of tropical equilibrium states, two sensitivity experiments (SST29NIR and SST29NIM) are conducted by Ping et al. (2007). The control experiment SST29 includes ice microphysics that fully interacts with radiation. Experiment SST29NIR is identical to SST29 except that SST29NIR excludes ice-cloud-radiative effects by setting ice hydrometeor mixing ratio to zero in the calculation of radiation. The comparison between SST29NIR and SST29 shows the radiative effects of ice clouds on tropical equilibrium states. Experiment SST29NIM excludes ice-cloud variables and associated ice microphysical processes by setting ice hydrometeor mixing ratio to zero during the model integration. The comparison between SST29NIM with SST29NIR shows the microphysical effects of ice clouds on the tropical equilibrium states. The comparison between SST29NIM and SST29 reveals both microphysical and radiative effects of ice clouds on tropical equilibrium states.

Mean temperature and *PW* reach equilibrium states in three experiments during the 40-day integrations (Fig. 14.7). Model domain and mass-weighted mean temperatures averaged from 31 to 40 days are −7.4 °C in SST29NIR and −5.9 °C in SST29NIM, which are 4.4 and 2.9 °C colder than the equilibrium temperature in SST29, respectively. Amounts of model domain-mean *PW* averaged from day 31 to day 40 are 35.7 mm in SST29NIR and 41.4 mm in SST29NIM, which are 9.2 and 3.5 mm smaller than the equilibrium *PW* in SST29, respectively. The comparison between SST29NIR and SST29 shows that the exclusion of radiative effects of ice clouds in the simulation causes cold and dry equilibrium states. The comparison between SST29NIM and SST29NIR indicates that the exclusion of microphysical effects of ice clouds in the simulation induces warm and moist equilibrium states. The comparison between SST29NIM and SST29 reveals that the exclusion of ice microphysics produces a cold and dry tropical equilibrium state.

To understand the physical processes that are responsible for these similarities and differences, model domain-mean heat and *PW* budgets are calculated by averaging the simulation data in the first 30 days and are shown in Table 14.2. The negative difference in radiative heating is responsible for the negative difference in temperature tendency for SST29NIR-SST29 whereas the negative condensation difference accounts for the negative difference in *PW* tendency. The negative radiation difference for SST29NIR-SST29 is mainly determined by the negative difference in model domain-mean IR cooling (Table 14.2c). This indicates that the simulation that excludes the ice-radiative effects yields a larger IR cooling rate through allowing more outgoing IR radiation from the atmosphere than the simulation that includes the ice-radiative effects does.

To examine microphysical effects of ice clouds on equilibrium states, SST29NIM is compared with SST29NIR since the radiative effects of ice clouds are excluded in both experiments. The positive difference in radiative heating is responsible for the positive difference in temperature tendency for SST29NIM-SST29NIR whereas

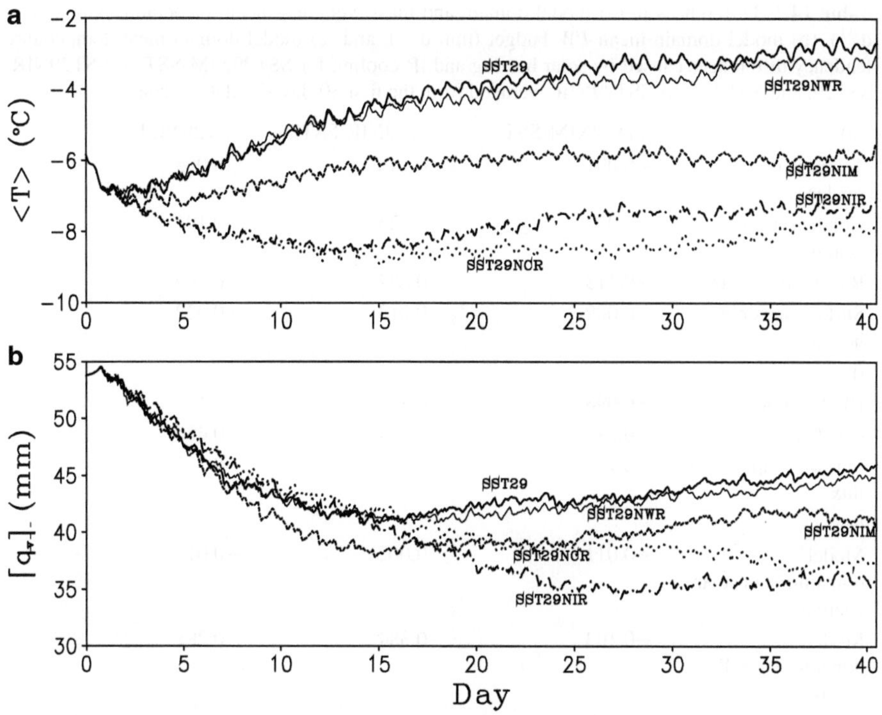

Fig. 14.7 Time series of (**a**) model domain and mass-weighted mean temperature (°C) and (**b**) model domain-mean *PW* (mm) in SST29 (*dark solid*), SST29NIM (*dark dashed*), SST29NIR (*dark dot dashed*), SST29NWR (*light solid*), and SST29NCR (*dot*) (After Ping et al. (2007) and Gao (2008)

the positive condensation difference accounts for the positive difference in *PW* tendency (Table 14.2). The positive difference in IR cooling explains the positive temperature tendency difference (Table 14.2c). The differences in temperature and *PW* budgets for SST29NIM-SST29 are similar to those for SST29NIR-SST29. Thus, the radiative effects of ice clouds on thermodynamic equilibrium states are dominant and the microphysical effects of ice clouds are secondary.

The exclusion of ice-radiative effects in SST29NIR increases *IWP* in raining stratiform and convective regions and thus model domain-mean *IWP* while it does not change model domain-mean *LWP* (Tables 14.1a and 14.3a). The exclusion of ice microphysical effects in SST29NIM mainly enhances *LWP* in non-raining stratiform regions and thus model domain-mean *LWP* (Table 14.3a, b). The exclusion of ice microphysical and radiative effects enhances *LWP* in non-raining stratiform and convective regions and thus model domain-mean *LWP* (Tables 14.1a and 14.3b). The exclusion of ice-radiative effects leads to a larger model domain-mean surface rain rate in SST29NIR than in SST29 (Tables 14.1a and 14.3a). SST29NIR produces a larger convective rain rate than SST29 does because SST29NIR has a larger vapor convergence than SST29 does. Due to the cyclic lateral boundary

Table 14.2 Differences in (a) model domain and mass-weighted mean temperature budget (°C d^{-1}), (b) model domain-mean *PW* budget (mm d^{-1}), and (c) model domain-mean temperature tendency differences caused by solar heating and IR cooling for SST29NIM-SST29, SST29NIR-SST29, and SST29NIM-SST29NIR averaged from the first 30-day simulation data

(a)	SST29NIM-SST29	SST29NIR-SST29	SST29NIM-SST29NIR
Temperature tendency	−0.091	−0.146	0.055
Condensational heating	0.032	0.223	−0.191
Radiative heating	−0.113	−0.382	0.269
Surface sensible heat flux	−0.009	0.013	−0.022
(b)			
PW tendency	−0.098	−0.212	0.114
Condensation	−0.283	−0.709	0.426
Surface evaporation flux	0.186	0.437	−0.251
(c)			
Model domain-mean solar heating	−0.011	0.002	−0.013
Model domain-mean IR cooling	−0.103	−0.384	0.281

After Ping et al. (2007)

condition, the larger vapor convergence in convective regions is caused by the larger vapor divergence in clear-sky regions in SST29NIR that is balanced by the larger surface evaporation in SST29NIR, compared to those in SST29. The larger surface evaporation in clear-sky regions is caused by drier air and larger clear-sky coverage in SST29NIR because of the same constant *SST* imposed in both experiments.

The exclusion of ice microphysical effects yields a smaller model domain-mean surface rain rate in SST29NIM than in SST29NIR (Table 14.3a, b). SST29NIM produces both smaller convective and stratiform rain rates than SST29NIR does because it has a smaller vapor convergence in convective regions and smaller local atmospheric drying and hydrometeor loss rates in raining stratiform regions. In clear-sky regions, vapor divergences in the two experiments are similar. In non-raining stratiform regions, the vapor divergence in SST29NIM is one order of magnitude smaller than that in SST29NIR because the surface evaporation rate is smaller in SST29NIM. Compared to SST29NIR, the smaller surface evaporation rate in SST29NIM is caused by more humid air and smaller coverage of non-raining stratiform regions. The exclusion of ice microphysical and radiative effects in simulation produces a slightly smaller model domain-mean surface rain rate in SST29NIM than in SST29 (Tables 14.1a and 14.3b), indicating that the ice microphysical effects and ice-radiative effects on surface rainfall processes are canceled out.

Table 14.3 As in Table 14.1 except for (a) SST29NIR and (b) SST29NIM

(a)	Clear-sky regions	Raining stratiform regions	Convective regions	Non-raining stratiform regions	Model domain mean
Fractional coverage	64.3	4.5	3.2	28.0	100
IWP	0.000	0.028	0.011	0.029	0.069
LWP	0.000	0.032	0.048	0.005	0.085
P_s	0.000	0.065	0.121	0.000	0.186
Q_{WVT}	0.001	0.030	−0.024	0.003	0.010
Q_{WVF}	−0.125	−0.018	0.197	−0.053	0.000
Q_{WVE}	0.122	0.007	0.005	0.044	0.178
Q_{CM}	0.002	0.046	−0.056	0.006	−0.003
(b)					
Fractional coverage	72.6	2.3	3.4	21.5	100
IWP	0.000	0.000	0.000	0.000	0.000
LWP	0.000	0.039	0.052	0.026	0.117
P_s	0.000	0.038	0.086	0.000	0.124
Q_{WVT}	0.007	0.009	−0.008	−0.006	0.002
Q_{WVF}	−0.126	0.018	0.111	−0.005	0.000
Q_{WVE}	0.117	0.003	0.003	0.024	0.147
Q_{CM}	0.002	0.008	−0.021	−0.012	−0.023

After Ping et al. (2007)

To examine radiative effects of water and ice clouds on tropical equilibrium states, two additional sensitivity experiments (SST29NWR and SST29NCR) are carried out by Gao (2008). SST29NWR and SST29NCR are identical to SST29 except that the radiative effects of water clouds and clouds (both water and ice clouds) are excluded by setting water and total hydrometeor mixing ratios to zero, respectively, in the cloud optical thickness and radiation calculations. Mean temperature and *PW* reach equilibrium states in SST29NWR and SST29NCR during the 40-day integrations (Fig. 14.7). Model domain and mass-weighted mean equilibrium temperatures averaged from 31 to 40 days are −8.1 °C in SST29NCR and −3.4 °C in SST29NWR. Model domain-mean amounts of equilibrium *PW* averaged from day 31 to day 40 are 37.5 mm in SST29NCR and 44.0 mm in SST29NWR. The magnitudes of temperature differences for SST29NWR-SST29 (−0.4 °C) and SST29NCR- SST29NIR (−0.7 °C) are much smaller than those for SST29NIR-SST29 (−4.4 °C) and SST29NCR-SST29NWR (−4.7 °C). The magnitudes of the *PW* differences for SST29NWR-SST29 (−0.9 mm) and SST29NCR-SST29NIR (1.8 mm) are also significantly smaller than those for SST29NIR-SST29 (−9.0 mm) and SST29NCR-SST29NWR (−6.5 mm). Thus, SST29 and SST29NWR have similar warm and humid equilibrium states whereas SST29NCR and SST29NIR have similar cold and dry equilibrium states. The similarities in model domain-mean temperature and *PW* between SST29 and SST29NWR and

Fig. 14.8 As in Fig. 14.2 except for the difference for SST29NCR-SST29 (After Gao 2008)

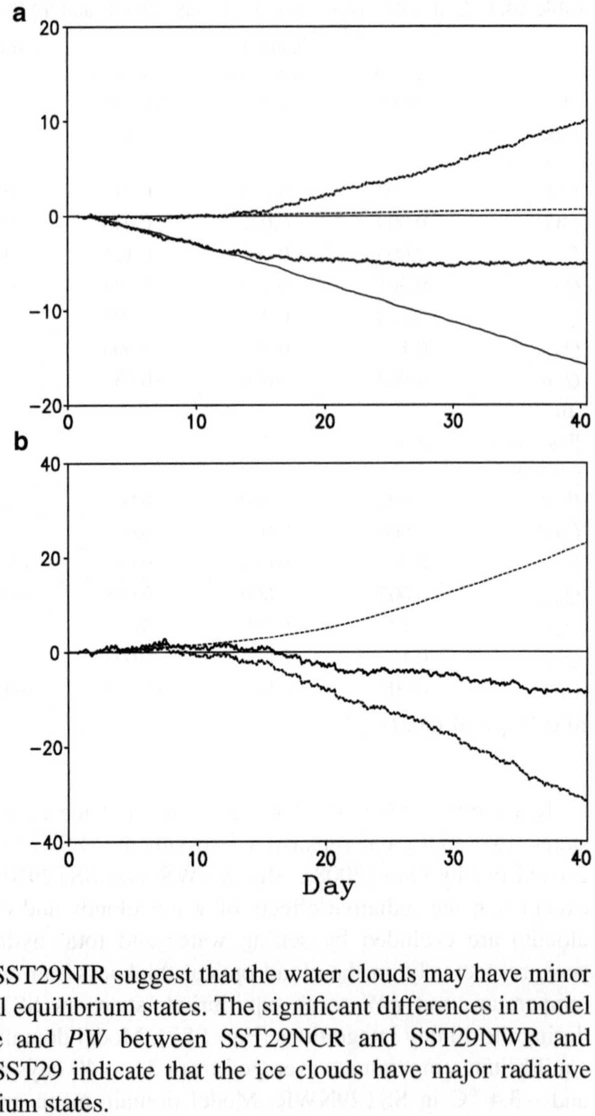

between SST29NCR and SST29NIR suggest that the water clouds may have minor radiative effects on tropical equilibrium states. The significant differences in model domain-mean temperature and *PW* between SST29NCR and SST29NWR and between SST29NIR and SST29 indicate that the ice clouds have major radiative effects on tropical equilibrium states.

The negative radiation-induced temperature difference for SST29NCR-SST29 mainly accounts for the negative temperature difference whereas the negative condensation-induced *PW* difference for SST29NCR-SST29 is mainly responsible for the negative *PW* difference (Fig. 14.8). The temporal–vertical distributions of model domain-mean solar radiation-induced temperature differences for SST29NCR-SST29 (Fig. 14.9a) show that the temperature difference is negative in the upper troposphere due to the fact that SST29NCR absorbs less solar radiation and allows more solar radiation penetrating into the mid- and lower troposphere than SST29 does. Thus, SST29NCR could have more solar radiation to absorb

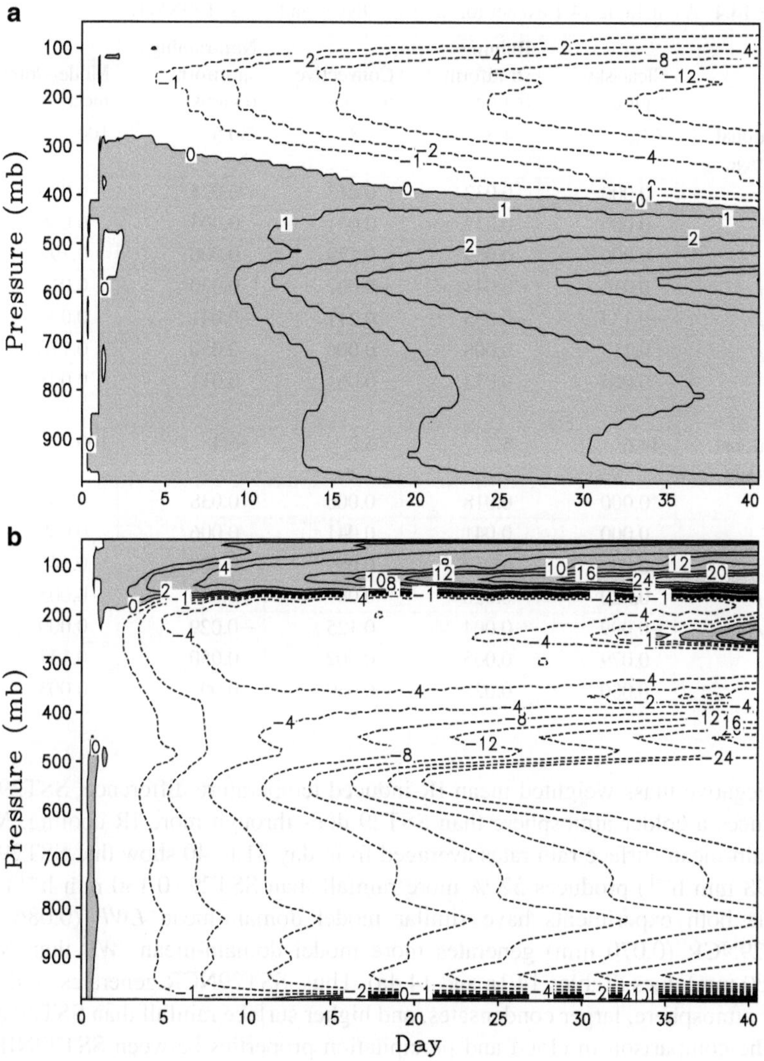

Fig. 14.9 Temporal and vertical distribution of model domain-mean temperature differences for SST29NCR-SST29 due to (**a**) solar and (**b**) IR radiation. Unit is °C. Positive differences are *shaded* (After Gao 2008)

in the mid- and lower troposphere than SST29 does. The negative solar radiation-induced temperature difference in the upper troposphere and the positive difference in the mid- and lower troposphere nearly cancel each other out in mass-weighted mean temperature calculations. Since ice clouds trap IR radiation, SST29NCR emits more IR radiation into space than SST29 does. Thus, the IR-induced temperature difference is negative in most of the troposphere (Fig. 14.9b), which causes

Table 14.4 As in Table 14.1 except for (a) SST29NCR and (b) SST29NWR

(a)	Clear-sky regions	Raining stratiform regions	Convective regions	Non-raining stratiform regions	Model domain mean
Fractional coverage	61.3	4.3	3.8	30.5	100
IWP	0.000	0.032	0.013	0.025	0.070
LWP	0.000	0.032	0.051	0.003	0.086
P_s	0.000	0.069	0.129	0.000	0.198
Q_{WVT}	0.016	0.043	−0.027	0.020	0.013
Q_{WVF}	−0.134	−0.035	0.211	−0.041	0.000
Q_{WVE}	0.113	0.008	0.006	0.050	0.176
Q_{CM}	0.005	0.053	−0.061	0.011	0.008
(b)					
Fractional coverage	44.6	5.7	3.2	46.4	100
IWP	0.000	0.018	0.006	0.038	0.062
LWP	0.000	0.041	0.041	0.006	0.088
P_s	0.000	0.057	0.089	0.000	0.146
Q_{WVT}	0.020	0.020	−0.009	−0.026	0.005
Q_{WVF}	−0.099	0.004	0.125	−0.029	0.000
Q_{WVE}	0.079	0.005	0.002	0.050	0.137
Q_{CM}	0.000	0.028	−0.030	0.004	0.003

the negative mass-weighted mean IR-induced temperature difference. SST29NCR produces a colder atmosphere than SST29 does through more IR cooling. Model domain-mean surface rain rates averaged from day 31 to 40 show that SST29NCR (0.198 mm h^{-1}) produces 52 % more rainfall than SST29 (0.130 mm h^{-1}) does. While both experiments have similar model domain-mean *LWP* (0.086 mm), SST29NCR (0.070 mm) generates more model domain-mean *IWP* than SST29 (0.060 mm) does (Tables 14.1a and 14.4a). Thus, SST29NCR generates colder and drier atmosphere, larger condensates, and higher surface rainfall than SST29 does.

The comparison in cloud and precipitation properties between SST29NIR and SSTNCR (Tables 14.3a and 14.4a) shows vapor convergence in both experiments determines convective rain rate, which contributes two thirds of time and model domain-mean surface rain rate. Cloud and rainfall properties in SST29NCR differ from those in SST29NIR in four ways. First, time and model domain-mean surface rain rate is larger in SST9NCR than in SST29NIR because the local hydrometeor loss occurs in SST29NCR whereas the local hydrometeor gain appears in ST29NIR. Second, fractional coverage of convective and non-raining stratiform regions is larger in SST29NCR than in SST29NIR whereas that of clear-sky and raining stratiform regions is smaller. Third, surface evaporation rate over non-raining stratiform regions is larger in SST29NCR than in SST29NIR whereas it is smaller over clear-sky regions. Fourth, over raining stratiform regions, vapor divergence rate in SST29NCR is twice as large as that in SST29NIR, which leads to a larger vapor convergence rate over convective regions in SST29NCR.

The comparison in cloud and rainfall properties between SST29NWR and SST29 (Tables 14.1a and 14.4b) shows that the time and model domain-mean surface rain rate is larger in SST29NWR than in SST29 due to the local vapor and hydrometeor loss in SST29NWR and the local hydrometeor gain in SST29. The difference in time and model domain-mean surface rain rate is from the difference in convective rain rate between the two experiments, which results from the difference in vapor convergence over convective regions. Compared to SST29, the larger vapor convergence over convective regions is caused by the larger vapor divergence over clear-sky regions in SST29NWR, which is balanced by the larger surface evaporation rate due to the larger fractional coverage of clear-sky regions in SST29NWR.

14.5 Effects of Zonal Perturbations of *SST* on Equilibrium States

The horizontal distribution of *SST* may be affected by clouds and precipitation. Since clouds and precipitation has a small spatial scale, *SST* may have large spatial perturbations in the presence of convective development. Cui and Gao (2008) conducted a series of sensitivity experiments of tropical equilibrium states to zonally perturbed SST. In their study, the zonal wave-like perturbations of *SST* with zonal wavenumbers 1, 2, 4, and 8 (the zonal scale of 768, 384, 192, and 96 km) are superimposed in zonal-mean *SST* of 29 °C in experiments SST29Z1, SST29Z2, SST29Z4, and SST29Z8, respectively. The amplitudes of zonal *SST* perturbations are 1 °C. Note that the model domain-mean *SST* is 29 °C in five experiments. The four perturbation experiments are compared to SST29.

During 40-day integrations, temperatures in all five experiments show quasi-equilibrium states with distinct differences (Fig. 14.10), whereas *PW* does not display distinct equilibrium differences in the five experiments (not shown). Compared to SST29 (−3.0 °C), SST29Z4 (−4.4 °C) and SST29Z8 (−3.7 °C) produce colder equilibrium states whereas SST29Z1 (−2.6 °C mm) and SST29Z2 (−2.3 °C) generate warmer equilibrium states. SST29Z8 and SST29Z4 stop their warming trends in days 4 and 8, respectively, causing cold equilibrium states while the other experiments continue their warming courses. To explain physical processes that are responsible for thermal equilibrium states, the differences in temperature budgets for SST29Z1-SST29, SST29Z2-SST29, SST29Z4-SST29, and SST29Z4-SST29 are calculated (Table 14.5). The analysis in days 4–7 shows that the positive difference in temperature tendency is mainly determined by the positive difference in radiative heating for SST29Z1-SST29. Fractional coverage for clear-sky regions is 56.9 % in SST29 with a uniform *SST* of 29 °C over the model domain. Fractional coverage for clear-sky regions in SST29Z1 is 49.3 % since half of the model domain in SST29Z1 has *SST* of colder than 29 °C. The positive difference in radiative heating

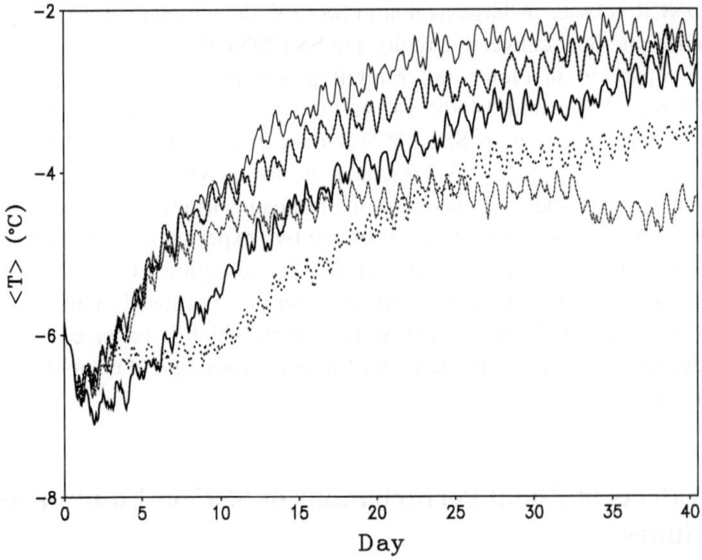

Fig. 14.10 Time series of model domain and mass-weighted mean temperature (°C) and (**b**) model domain-mean *PW* (mm) in SST29 (*dark solid*), SST29Z1 (*dark dashed*), SST29Z2 (*solid*), SST29Z4 (*dashed*), and SST29Z8 (*dot*) (After Cui and Gao 2008)

Table 14.5 Model domain and mass-weighted mean temperature budget differences (°C d^{-1}) for SST29Z1-SST29, SST29Z2-SST29, SST29Z4-SST29, and SST29Z8-SST29 based on the data averaged from days 4–7

	SST29Z1-SST29	SST29Z2-SST29	SST29Z4-SST29	SST29Z8-SST29
Temperature tendency	0.136	0.204	0.143	−0.154
Condensational heating	0.013	0.137	0.123	−0.121
Radiative heating	0.124	0.067	0.023	−0.030
Surface sensible heat flux	−0.002	0.000	−0.002	−0.002

After Cui and Gao (2008)

for SST29Z1-SST29 is determined by the positive difference in IR cooling over clear-sky regions (Table 14.6) since SST29Z1 has a smaller fractional coverage of clear-sky regions than SST29 does.

Table 14.5 reveals that the positive differences in temperature tendencies for SST29Z2-SST29 and SST29Z4-SST29 are caused by the positive differences in condensational heating during days 4–7. Although convection occupies similar sizes of areas in SST29 (43.1 %), SST29Z2 (45.8 %), and SST29Z4 (42.8 %), the convection in SST29Z2 and SST29Z4 is more organized than in SST29, explaining

Table 14.6 Temperature tendency differences in model domain-mean solar heating and IR cooling and corresponding contributions from clear-sky regions and cloudy regions for SST29Z1-SST29, SST29Z2-SST29, SST29Z4-SST29, and SST29Z8-SST29 based on the data averaged from days 4–7

	SST29Z1-SST29	SST29Z2-SST29	SST29Z4-SST29	SST29Z8-SST29
Model domain-mean solar heating	−0.015	−0.011	−0.011	0.000
Model domain-mean IR cooling	0.139	0.077	0.034	−0.030
Solar heating in clear-sky regions	−0.063	−0.029	−0.013	0.000
IR cooling in clear-sky regions	0.134	0.045	0.000	0.004
Solar heating in cloudy regions	0.048	0.018	0.002	0.000
IR cooling in cloudy regions	0.005	0.032	0.034	−0.034

Unit is in °C d^{-1}. After Cui and Gao (2008)

that more condensational heating occurs in SST29Z2 and SST29Z4 than in SST29. Meanwhile, the similar fractional coverage of clear-sky regions in SST29 (56.9 %), SST29Z2 (54.2 %), and SST29Z4 (57.2 %) significantly reduces the difference in IR cooling over clear-sky regions for SST29Z2-SST29 and SST29Z4-SST29 (Table 14.6), which causes smaller contributions to the positive differences in temperature tendencies for SST29Z2-SST29 and SST29Z4-SST29 than that for SST29Z1-SST29.

The negative difference in temperature tendency is mainly caused by the negative difference in condensational heating for SST29Z8-SST29 (Table 14.5). Since the distances between maximum SSTs are only 96 km in SST29Z8, convective systems interact with each other so that some convective systems are suppressed whereas other convective systems cannot grow freely due to the small areas of warm SST. As a result, the contribution from condensational heating to temperature tendency in SST29Z8 is smaller than in SST29 (Table 14.5). Again, similar fractional coverage of clear-sky regions in SST29 (56.9 %) and SST29Z8 (58.1 %) as well as cloudy regions in SST29 (43.1 %) and SST29Z8 (41.9 %) causes small differences in IR cooling over clear-sky regions and cloudy regions (Table 14.6), which account for small contributions from IR cooling to the positive difference in temperature tendency for SST29Z8-SST29.

The heat budgets averaged in days 4–7 are compared with those averaged in days 8–11 in SST29, SST29Z1, SST29Z2, and SST29Z4 (Table 14.7) to explain why SST29Z4 stops a warming course around day 8 that eventually leads to a

Table 14.7 Model domain and mass-weighted mean temperature budget difference (°C d^{-1}) between averages in days 8–11 and days 4–7 in SST29, SST29Z1, SST29Z2, and SST29Z4

	SST29	SST29Z1	SST29Z2	SST29Z4
Temperature tendency	0.067	−0.170	−0.174	−0.206
Condensational heating	−0.049	−0.170	−0.224	−0.194
Radiative heating	0.124	−0.002	0.048	−0.011
Surface sensible heat flux	−0.007	0.002	0.001	−0.002

After Cui and Gao (2008)

Table 14.8 Temperature tendency differences in model domain-mean solar heating and IR cooling and corresponding contributions from clear-sky regions and cloudy regions for SST29, SST29Z1, SST29Z2, and SST29Z4 between averages in days 8–11 and days 4–7

	SST29	SST29Z1	SST29Z2	SST29Z4
Model domain-mean solar heating	−0.017	0.001	−0.009	−0.006
Model domain-mean IR cooling	0.141	−0.002	0.058	−0.006
Solar heating in clear-sky regions	−0.080	−0.041	−0.076	−0.052
IR cooling in clear-sky regions	0.200	0.161	0.230	0.135
Solar heating in cloudy regions	0.063	0.042	0.067	0.046
IR cooling in cloudy regions	−0.059	−0.163	−0.172	−0.141

Unit is in °C d^{-1}. After Cui and Gao (2008)

cold equilibrium state (see Fig. 14.10). While SST29 has a positive difference in temperature tendency between averages in days 8–11 and days 4–7, SST29Z1, SST29Z2, and SST29Z4 show negative differences and SST29Z4 has the largest negative difference (Table 14.7). This shows that SST29Z4 turns out to have a colder equilibrium state than SST29 does. The positive difference in temperature tendency in SST29 is mainly determined by the positive difference in radiative heating. The negative differences in temperature tendency in SST29Z1, SST29Z2, and SST29Z4 are determined by the negative differences in condensational heating, and SST29Z2 has the largest negative difference. The positive difference in radiative heating offsets the negative difference in condensational heating in SST29Z2, thereby decreasing the magnitude of the negative difference in temperature tendency. The negative difference in radiative heating along with the negative difference in condensational heating in SST29Z4 serves to increase the magnitude of the negative difference in temperature tendency. The shrinking of zonal areas of tropical water clouds in SST29Z1, SST29Z2, and SST29Z4 reduces the vapor condensation and associated condensational heating in days 8–11, compared to those in days 4–7. Further analysis (Table 14.8) reveals that the positive differences in radiative heating in SST29 and SST29Z2 are caused by the positive difference in IR cooling over clear-sky regions. The small differences in radiative heating in SST29Z1 and SST29Z4 are from the cancelation in solar heating and IR cooling over clear-sky regions and cloudy regions, respectively. The positive differences in radiative heating in SST29 and SST29Z2 may be the results of significant reductions of clear-sky regions from days 4–7 (56.9 % in SST29 and 54.2 % in SST29Z2) to days 8–11 (48.3 % in SST29 and 44.8 % in SST29Z2).

References

Chao J (1961) A nonlinear analysis of development of thermal convection in a stratified atmosphere. Acta Meteor Sinica 31:191–204

Chao J, Chen L (1964) Effects of vertical wind shear on convective development. Acta Meteor Sinica 34:94–102

Cui X, Gao S (2008) Effects of zonal perturbations of sea surface temperature on tropical equilibrium states: a 2D cloud-resolving modeling study. Prog Nat Sci 18:413–419, (c) National Natural Science Foundation of China. Reprinted with permission

Dudhia J, Moncrieff MW, So DWK (1987) The two-dimensional dynamics of West African squall lines. Q J R Meteorol Soc 113:121–146

Gao S (2008) A cloud-resolving modeling study of cloud radiative effects on tropical equilibrium states. J Geophys Res 113. doi:10.1029/2007JD009177, (c) American Geophysical Union. Reprinted with permission

Gao S, Ping F, Li X (2006) Tropical heat/water vapor quasi-equilibrium and cycle as simulated in a 2D cloud resolving model. Atmos Res 79:15–29

Gao S, Zhou Y, Li X (2007) Effects of diurnal variations on tropical equilibrium states: a two-dimensional cloud-resolving modeling study. J Atmos Sci 64:656–664, (c) American Meteorological Society. Reprinted with permission

Grabowski WW, Moncrieff MW, Kiehl JT (1996) Long-term behaviour of precipitating tropical cloud systems: a numerical study. Q J R Meteorol Soc 122:1019–1042

Held IM, Hemler RS (1993) Radiative-convective equilibrium with explicit two-dimensional moist convection. J Atmos Sci 50:3909–3927

Islam S, Bras RL, Emanuel KA (1993) Predictability of mesoscale rainfall in the tropics. J Appl Meteorol 32:297–310

Lau KM, Sui CH, Tao WK (1993) A preliminary study of the tropical water cycle using the Goddard Cumulus Ensemble model. Bull Am Meteorol Soc 74:1313–1321

Lau KM, Sui CH, Chou MD, Tao WK (1994) An inquiry into the cirrus cloud thermostat effect for tropical sea surface temperature. Geophys Res Lett 21:1157–1160

Li X, Chao J, Hu Y (1964) A dynamic analysis of development of an anvil cumulonimbus cloud. Acta Meteor Sinica 34:225–232

Nakajima K, Matsuno T (1988) Numerical experiments concerning the origin of cloud clusters in the tropical atmosphere. J Meteor Soc Japan 66:309–329

Ping F, Luo Z, Li X (2007) Microphysical and radiative effects of ice clouds on tropical equilibrium states: a two-dimensional cloud-resolving modeling study. Mon Wea Rev 135:2794–2802, (c) American Meteorological Society. Reprinted with permission

Ramanathan V, Collins W (1991) Thermodynamic regulation of ocean warming by cirrus clouds deduced from observations of the 1987 El Nino. Nature 351:27–32

Randall DA, Hu Q, Xu KM, Krueger SK (1994) Radiative-convective disequilibrium. Atmos Res 31:315–327

Robe FR, Emanuel KA (1996) Moist convective scaling: some inferences from three-dimensional cloud ensemble simulations. J Atmos Sci 53:3265–3275

Sui CH, Lau KM, Tao WK, Simpson J (1994) The tropical water and energy cycles in a cumulus ensemble model. Part I: equilibrium climate. J Atmos Sci 51:711–728

Tao WK, Scala J, Ferrier B, Simpson J (1995) The effects of melting processes on the development of a tropical and a midlatitude squall line. J Atmos Sci 52:1934–1948

Tao WK, Simpson J, Sui CH, Shie CL, Zhou B, Lau KM, Moncrieff MW (1999) Equilibrium states simulated by cloud-resolving models. J Atmos Sci 56:3128–3139

Tao WK, Shie CL, Simpson J, Braun S, Johnson RH, Ciesielski PE (2003) Convective systems over the South China Sea: cloud-resolving model simulations. J Atmos Sci 60:2929–2956

Tompkins AM, Craig GC (1998) Radiative-convective equilibrium in a three-dimensional cloud-ensemble model. Q J R Meteorol Soc 124:2073–2097

Xu KM, Randall DA (1998) Influence of large-scale advective cooling and moistening effects on the quasi-equilibrium behavior of explicitly simulated cumulus ensembles. J Atmos Sci 55:896–909

Xu KM, Randall DA (1999) A sensitivity study of radiative-convective equilibrium in the tropics with a convection-resolving model. J Atmos Sci 56:3385–3399

Chapter 15
Remote Sensing Applications

Abstract In this chapter, vertical profiles of temperature, specific humidity, and mixing ratio of cloud hydrometeors from cloud-resolving models are applied to radiative transfer model to simulate microwave radiance responses to cloud hydrometeors. The sensitivity experiments of radiances to cloud hydrometeors are used to remove cloud-induced contamination from cloudy radiances. The comparison in radiance between simulations and observations is conducted to evaluate the effects of accretion of snow by graupel on radiance. The cloud microphysical budgets from the experiments with and without the accretion processes are also compared to analyze the effects of the accretion on cloud budget.

Keywords Radiance • Microwave channel • Cloud contamination • Accretion of snow by graupel

Microwave radiances are sensitive to vertical profiles of temperature, water vapor, and cloud hydrometeors. Thus, satellite microwave measurements are important sources that provide atmospheric thermodynamic profiles, cloud properties, and surface parameters, in particular over vast ocean, where conventional observations are not available. Cloud-resolving model simulations can provide vertical profiles of temperature, water vapor, and cloud hydrometeors to radiative transfer models, which can simulate radiances. The comparison between the simulated radiances and satellite-measured radiances can evaluate how well numerical models produce cloud properties. The sensitivity tests of radiances to cloud hydrometeors can be used to eliminate cloud contamination from cloudy radiances. Thus, responses of radiance to cloud hydrometeors, correction of cloud contamination on cloudy radiances, and comparison between observed and simulated radiances are discussed in this chapter.

15.1 Introduction

The microwave instruments were first flown aboard US satellites in 1972. In 1978, the NOAA launched the Microwave Sounding Unit (MSU) aboard the polar-orbiting satellites. Recently, the NOAA launched the Advanced MSU (AMSU), which

© Springer International Publishing Switzerland 2016

X. Li, S. Gao, *Cloud-Resolving Modeling of Convective Processes*,
Springer Atmospheric Sciences, DOI 10.1007/978-3-319-26360-1_15

contains twelve channels within the 50–60 GHz oxygen bands (AMSU-A channels 3–14), three channels around the 183 GHz water vapor line (AMSU-B channels 18–20), and five window channels (AMSU-A channels 1–2 and 15 and AMSU-B channels 16–17). The satellite microwave measurements are used to retrieve temperature (Spencer et al. 1990), water vapor (Grody 1980), ocean wind (Wentz et al. 1986), cloud liquid water (Weng and Grody 1994), ice cloud parameters (Weng and Grody 2000), precipitation (Ferraro 1997), snow cover (Grody 1991), and sea ice concentration (Grody et al. 2000). The retrieval of temperature and moisture is largely contaminated by the clouds. The cloud contamination on the AMSU measurements is dependent on the vertical profiles of cloud hydrometeors that are usually not available from observations. Thus, the correction of cloud contamination on AMSU measurements resorts to the radiative transfer simulation with reasonable vertical cloud structures from the cloud-resolving model.

The sensitivity of the microwave radiances to clouds has been investigated by many researchers in the remote sensing community. In particular, the relation between the microwave radiances and surface rain rate has been intensively studied in the past two decades. Wu and Weinman (1984) computed microwave radiances as a function of rain rate and cloud hydrometeors and found that the lower-frequency radiances are sensitive to liquid precipitation at low altitudes whereas the high-frequency radiances are more sensitive to the ice hydrometeor at the cloud top. Mugnai and Smith (1988) used a microwave radiative transfer model with cloud inputs from the cloud model of Hall (1980) to investigate the impact of time-dependent cloud microphysical structure on the transfer to space of passive microwave radiation at frequencies from 19.35 GHz to 231 GHz. They demonstrated that cloud water can have a major impact on the upwelling microwave radiation originating from both the surface and a rain layer placed below cloud base. They showed the different roles of absorption, emission, and scattering in determining the radiances at the top of the atmosphere in different stages of cloud development. Smith and Mugnai (1988) further showed that the relation between the microwave brightness temperature and surface rain rate is a function of cloud water processes and microwave frequencies. Smith and Mugnai (1989) found the important impacts of large ice particles on the microwave radiances. Yeh et al. (1990) used a microwave radiative transfer model to simulate the upwelling brightness temperatures at frequencies from 18 to 183 GHz based on the input hydrometeor profile derived from the ground-based radar reflectivities and found that the upwelling brightness temperatures at both 92 and 181 GHz are most sensitive to the cloud structures in the upper portion of storms. Adler et al. (1991) employed a 3D cloud–microwave radiative transfer model to study the relation between upwelling radiances and cloud properties including rain rate of a tropical oceanic squall line and showed that the relation between microwave brightness temperature and rain rate is a function of cloud structure and content as well as development stages of convection. Smith et al. (1992) investigated precipitation retrieval from the responses of microwave radiances at the top of the atmosphere to clouds based on the microwave radiative transfer simulations and found that the relation between the microwave radiance and surface rain rate is largely affected by the ice clouds. Mugnai et al. (1993)

developed a hybrid statistical–physical rainfall algorithm to produce liquid-ice profile information and surface rain rate. Muller et al. (1994) simulated the effects of water vapor and cloud properties on AMSU moisture channel measurements and found that the water vapor, cloud water, and ice have impacts on microwave radiances near 183 GHz.

Schaerer and Wilheit (1979) demonstrated theoretically a concept for obtaining water vapor profiles by using stronger resonance at 183 GHz and derived a weighting function at 183 GHZ. The retrievals of water vapor profiles have been intensively studied by Rosenkranz et al. (1982), Kakar (1983), and Kakar and Lambrightsen (1984). Wang et al. (1983) and Lambrightsen and Kakar (1985) retrieved the water vapor profiles from 183 GHz measurements made from the NASA WRB-57 aircraft using a nonlinear physical relaxation and a linear statistical retrieval algorithm, respectively. The abovementioned retrievals of water vapor profiles were limited to clear conditions. Isaacs and Deblonde (1987) derived a linear statistical retrieval algorithm that included clouds and found some shortcomings related to the inclusion of clouds. Wilheit (1990) developed an algorithm for retrieving water vapor profiles in clear and cloudy atmospheres from 183 GHz radiometric measurements based on numerical simulations.

15.2 AMSU Responses to Cloud Hydrometeors

Li and Weng (2001) use a new, fast, and accurate microwave radiative transfer model (Liu and Weng 2002) including scattering and polarization to conduct the radiance simulations with the cloud-resolving model outputs from experiment COARE (see Sect. 3.1) to study the AMSU-B responses to various cloud parameters. The model is applied to compute the radiances at AMSU-B frequencies and bandwidths (see Table 15.1) based on thermodynamic profiles and cloud information derived from the cloud-resolving model simulation on 1400 LST 21 December 1992, when a strong convection occurred. Figure 15.1 shows horizontal distributions of brightness temperatures and total transmittance for AMSU-B frequencies at nadir, cloud contents, surface rain rate, *PW*, and surface wind. Strong convection occurs within 500–570 km, where surface rain rate has a maximum of 70 mm hour^{-1}. The

Table 15.1 AMSU-B channel characteristics

Channel number	Center frequency (GHz)	No. of passbands	Bandwidth per passband (MHz)	NEΔT (K)
16	89(0.9)	2	1000	0.37
17	150(0.9)	2	1000	0.84
18	183.31(1)	2	500	1.06
19	183.31(3)	2	1000	0.70
29	183.31(7)	2	2000	0.60

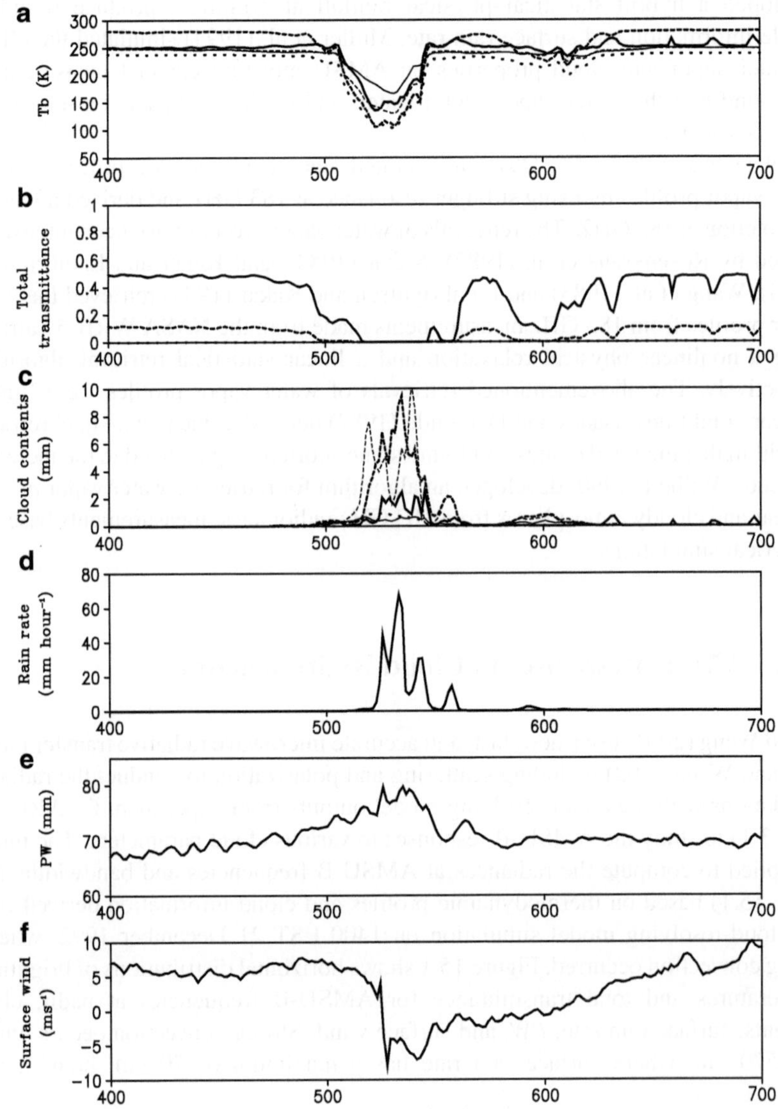

Fig. 15.1 Horizontal distributions of (**a**) brightness temperatures and (**b**) total transmittance for AMSU-B frequencies (*dark solid*, channel 16; dark *dashed*, channel 17; *dot*, channel 18; *light solid*, channel 19; *light dashed*, channel 20), (**c**) cloud hydrometeors (*dark solid*, cloud water; *dark dashed*, raindrop; *dot*, cloud ice; *light solid*, snow; *light dashed*, graupel), (**d**) surface rain rate, (**e**) *PW*, and (**f**) surface zonal wind simulated in experiment COARE on 1400 LST 21 December 1992 (After Li and Weng 2001)

convection is associated with large *PW*, large horizontal wind shear, and large magnitudes of cloud hydrometeors, in particular raindrop and graupel. The convection causes a large drop of brightness temperatures for all AMSU-B frequencies, which are as low as 100 K. The convection also causes zero total transmittance at channels 16 and 17, whereas total transmittance is zero at channels 18–20 because of strong water vapor absorption.

Simulated radiances at AMSU frequencies with both ice and water hydrometeors are compared with those simulated with ice hydrometeors or with water hydrometeors to examine the AMSU responses to cloud hydrometeors (Fig. 15.2). Simulated radiances at 23.8, 31.4, and 50.3 GHz are only sensitive to water hydrometeors. Both ice and water hydrometeors have impacts on the variations of radiances simulated at 52.8, 53.596, 54.4, 54.94, and 55.5 GHz. Only ice hydrometeors are responsible for the variations of radiances simulated at AMSU-B frequencies. The simulated radiances at 23.8 and 31.4 GHz increase with increasing *LWP* (Fig. 15.3a) whereas the simulated radiances at 52.8, 53.596$\pm$0.115, 89$\pm$0.9, 150$\pm$0.9, 183.31$\pm$3, and $\pm$7 GHz decrease with increasing *IWP* (Fig. 15.3b–d). The simulated radiances at 52.8, 53.596$\pm$0.115, 89$\pm$0.9, 150$\pm$0.9, 183.31$\pm$3, and $\pm$7 GHz are sensitive to the height whereas the simulated radiances at 23.8 and 31.4 GHz are insensitive to the height.

15.3 Correction of Cloud Contamination on AMSU Measurements

Radiances in cloudy and rainy conditions are affected by the clouds. The retrieval of atmospheric thermodynamic profiles requires correction of cloud contamination on radiances. Figure 15.4 shows the differences of brightness temperatures simulated with clouds minus those simulated without clouds at 52.8 and 53.596 GHz versus *LWP* and *IWP*. The comparison between the two figures clearly shows that the relation between the T_b difference and *IWP* is much better than that between the T_b difference and *LWP*. This suggests that the cloud contamination on 52.8 and 53.596 GHz should be corrected by *IWP* information. This is a typical case where the modeling information can serve as a guide for remote sensing.

Large differences between the simulated radiances with and without clouds exist at most of AMSU channels as shown in Fig. 15.5 (left panel). After corrections, the RMS differences between clear-sky and predicted clear-sky brightness temperatures over cloudy regions are smaller than 1 K, and the linear correlation coefficients are larger than 0.7 that exceed the 1 % confidence level (Fig. 15.5 in right panel). The contamination-corrected AMSU radiances are used to retrieve temperature and moisture profiles over cloudy regions. Figure 15.6 shows the retrieved specific humidity versus true specific humidity at selected vertical levels, where the RMS differences are smaller than 0.5 gkg^{-1} and the correlation coefficients are larger than 0.7.

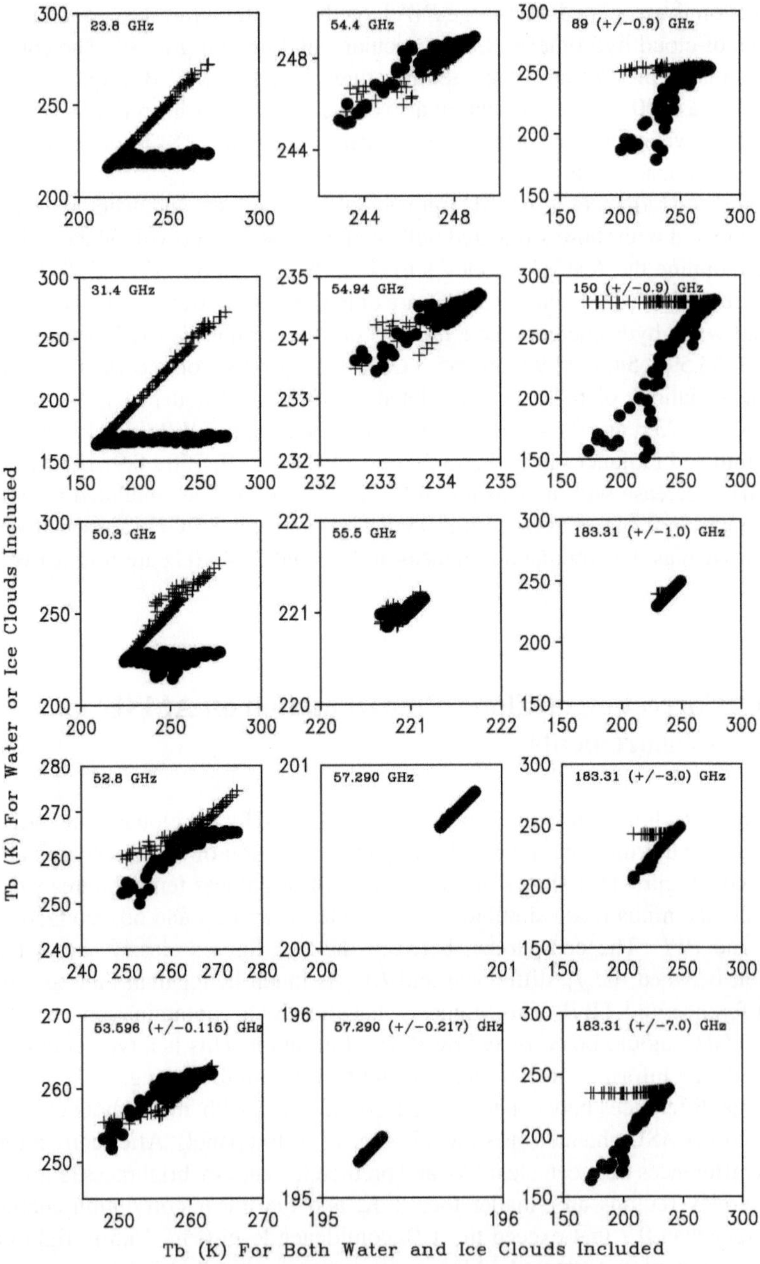

Fig. 15.2 Brightness temperatures (*K*) simulated with water hydrometeor (*cross*) or with ice hydrometeors (*close circle*) versus those simulated with both water and ice hydrometeors. The cloud data are from experiment COARE

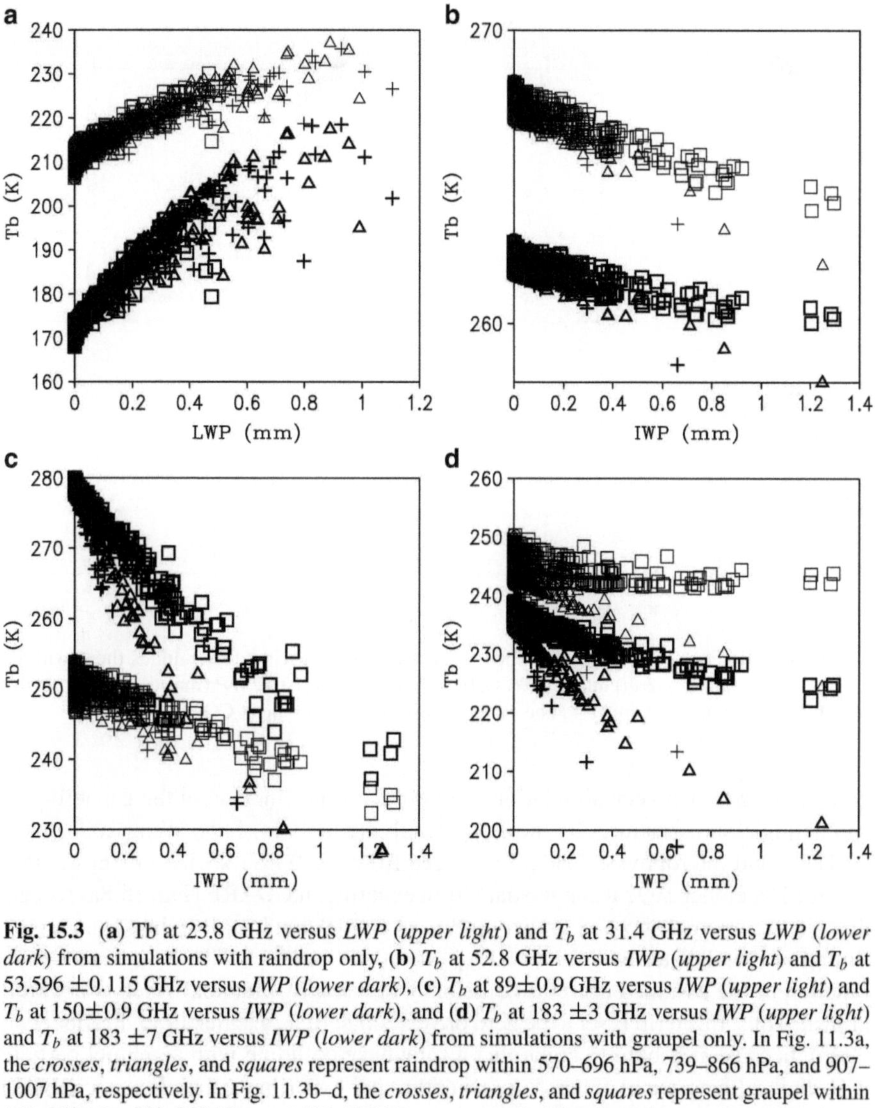

Fig. 15.3 (**a**) Tb at 23.8 GHz versus *LWP* (*upper light*) and T_b at 31.4 GHz versus *LWP* (*lower dark*) from simulations with raindrop only, (**b**) T_b at 52.8 GHz versus *IWP* (*upper light*) and T_b at 53.596 ±0.115 GHz versus *IWP* (*lower dark*), (**c**) T_b at 89±0.9 GHz versus *IWP* (*upper light*) and T_b at 150±0.9 GHz versus *IWP* (*lower dark*), and (**d**) T_b at 183 ±3 GHz versus *IWP* (*upper light*) and T_b at 183 ±7 GHz versus *IWP* (*lower dark*) from simulations with graupel only. In Fig. 11.3a, the *crosses*, *triangles*, and *squares* represent raindrop within 570–696 hPa, 739–866 hPa, and 907–1007 hPa, respectively. In Fig. 11.3b–d, the *crosses*, *triangles*, and *squares* represent graupel within 175–253 hPa, 282–380 hPa, and 416–530 hPa, respectively. The cloud data are from experiment COARE

15.4 Comparison Studies Between Simulated and Observed Radiances

The comparison between simulated and observed radiances can be used to evaluate cloud simulations. Water and ice hydrometeor-sensitive radiances are used to assess the water and ice cloud simulations. As an example, Fig. 15.7 shows the

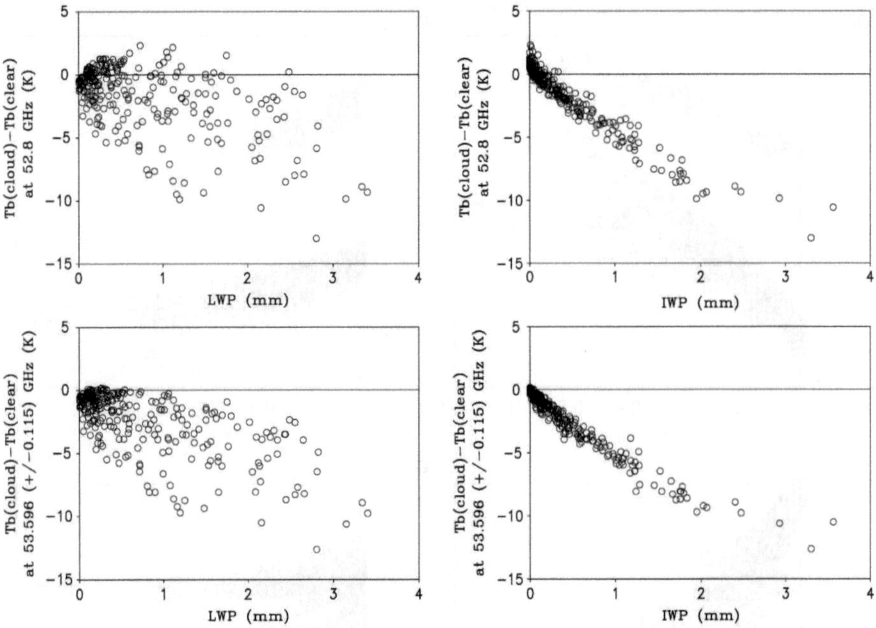

Fig. 15.4 Difference of brightness temperatures (*K*) simulated with clouds minus those without clouds at 52.8 (*upper panel*) and 53.596 GHz (*lower panel*) versus *LWP* (mm) in the *left panels* and *IWP* (mm) in the *right panels* (The cloud data are from experiment COARE)

good agreement between simulations and observations, indicating the capability of simulating cloud structures by the cloud-resolving model (Li and Weng 2003).

The cloud microphysics budget averaged from 2200 LST 19 December to 1000 LST 21 December 1992 using the data from experiment COARE (Fig. 15.8a) reveals that the accretion of snow by graupel ($[P_{GACS}] = 0.19$ mm h^{-1}) is as important as the riming of cloud water by graupel ($[P_{GACW}] = 0.19$ mm h^{-1}). The P_{GACS} is a strong function of the assumed accretion efficiency of snow by graupel, which is not well known. Thus, this term is set to be zero or very small (e.g., Ziegler 1985; Ferrier et al. 1995), because it is hard to argue that a snowflake colliding with a graupel particle will stick to the graupel. Therefore, an additional experiment with P_{GACS} excluded is carried out (Fig. 15.8b). The vapor condensation and deposition rates (P_{CND} and P_{DEP}) in the two experiments are similar. The *LWP* in the two experiments are about the same, whereas the *LWP* in the experiment without P_{GACS} (0.39 mm) is much larger than that in the experiment with P_{GACS} (0.21 mm). Compared to the experiment with P_{GACS}, the mixing ratio of graupel decreases significantly, whereas the mixing ratio of snow increases significantly and becomes the major component

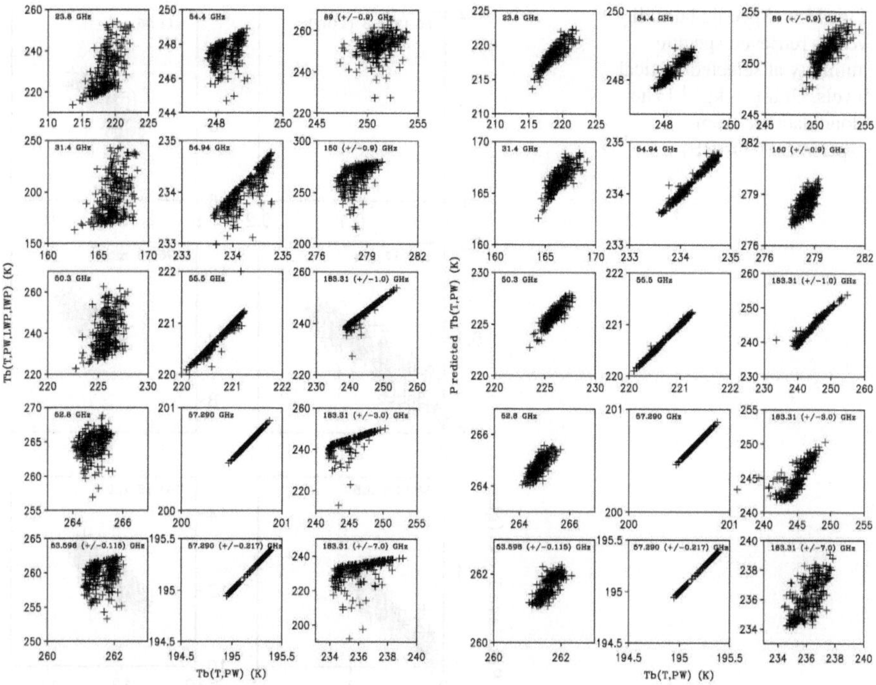

Fig. 15.5 Brightness temperatures (*K*) simulated with clouds versus those simulated without clouds at AMSU channel over cloudy regions before the correction of cloud contamination in the *left panels* and clear-sky brightness temperatures versus predicted clear-sky brightness temperatures over cloudy regions after the correction in the *right panels* (The cloud data are from experiment COARE)

of ice hydrometeors in the experiment without P_{GACS}. In the experiment without P_{GACS}, the conversion rate from snow to rain through the accretion of snow by rain ($[P_{RACS}] = 0.2$ mm h^{-1}) and the melting of snow ($[P_{SMLT}] = 0.1$ mm h^{-1}) is larger than the conversion from graupel to rain through the melting of graupel ($[P_{GMLT}] = 0.15$ mm h^{-1}) in the rain budget.

Figure 15.9 shows the pairs of brightness temperatures in observations and simulations without P_{GACS}. The variations of radiances in the experiment without P_{GACS} are closer to the observations than those in the experiment with P_{GACS} (Fig. 15.7). The signals are very significant in AMSU channels 16 and 17 because of significant changes of ice clouds. The comparison between simulations and observations suggests that the P_{GACS} suppress the development of precipitation ice unrealistically.

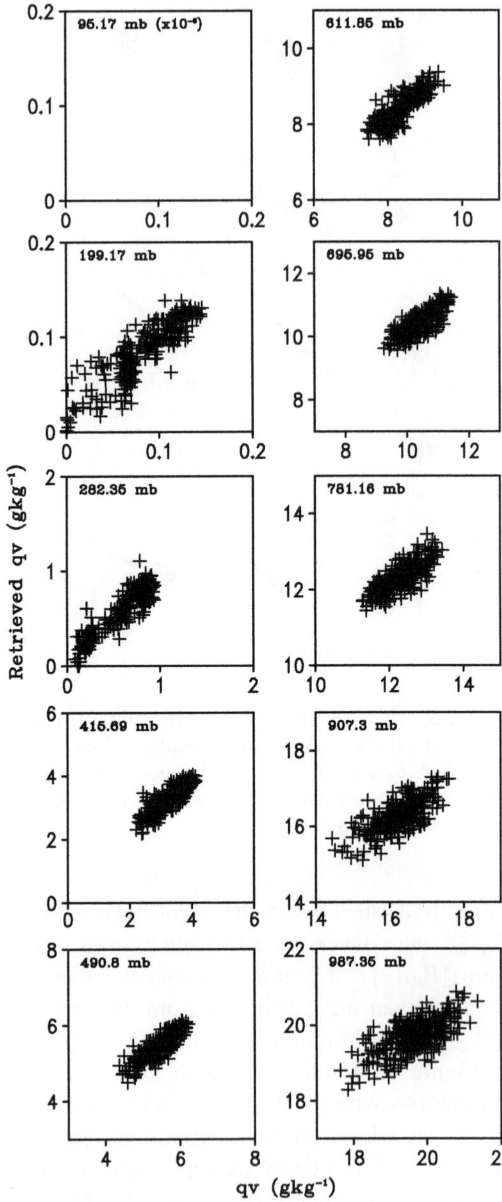

Fig. 15.6 Specific humidity versus retrieved specific humidity at selected vertical levels. Unit is g kg^{-1} (The cloud data are from experiment COARE)

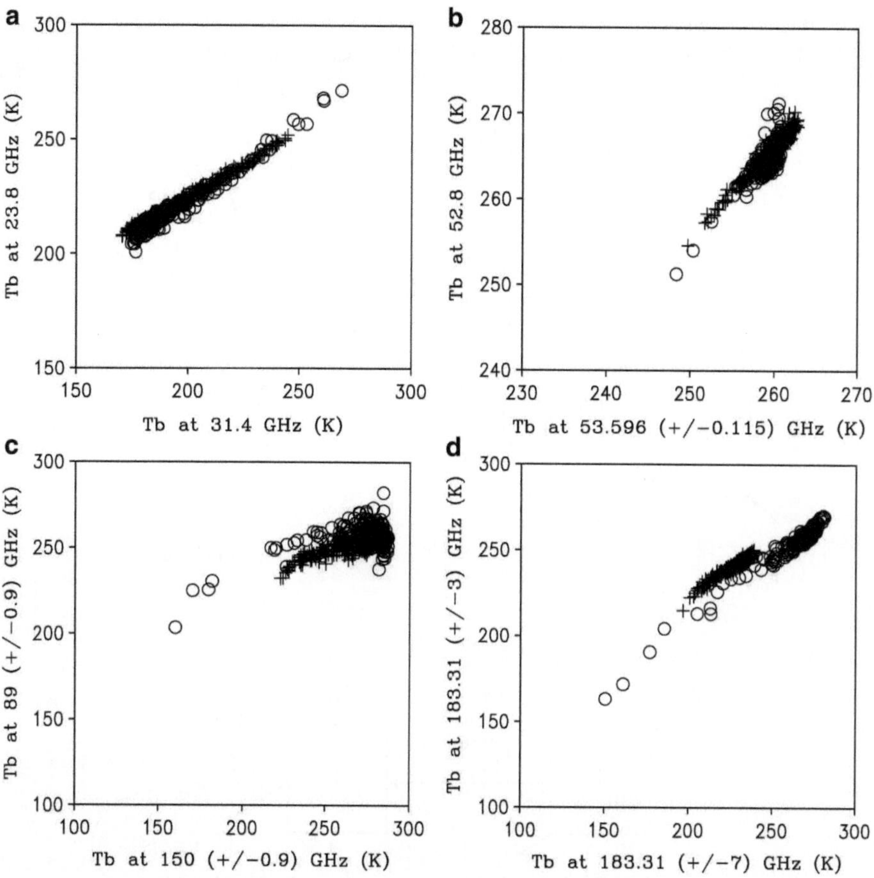

Fig. 15.7 (**a**) Brightness temperature T_b at AMSU-A channel 1 (23.8 GHz) versus channel 2 (31.4 GHz), (**b**) T_b at AMSU-A channel 4 (52.8 GHz) versus channel 5 [53.596 (+/−0.115) GHz], (**c**) T_b at AMSU-B channel 16 [89 (+/−0.9) GHz] versus channel 17 [150 (+/−0.9) GHz], and (**d**) T_b at AMSU-B channel 19 [183.31 (+/−3) GHz] versus channel 20 [183.31 (+/−7) GHz]. The dots represent the observations over (10°S–10°N, 140°E–180°) from NOAA-15 and 16 satellites in selected days of year 2001, whereas the crosses denote the simulations from microwave radiative transfer model developed by Liu and Weng (2002). The simulations use cloud information in COARE from 2200 LT 19 December to 1000 LT 21 December 1992. The simulation data are averaged in 32 grid points (48 km, mean), which is similar to the horizontal resolution of AMSU observation at nadir. All data are obtained in cloudy condition in which vertically integrated cloud water is larger than 0.1 mm (After Li and Weng 2003. The cloud data are from experiment COARE)

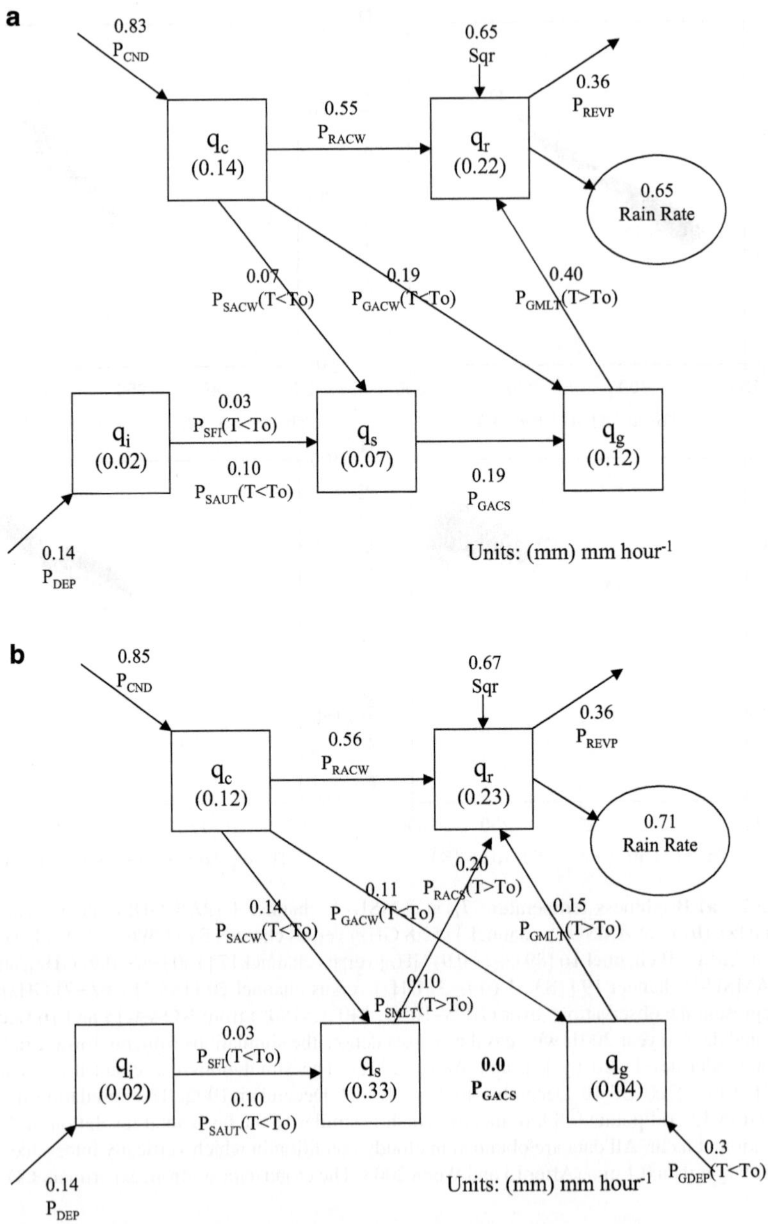

Fig. 15.8 36-h and model domain-mean cloud microphysics budgets averaged from 2200 LST 19 December to 1000 LST 21 December 1992 using the data from experiments (**a**) with P_{GACS} and (**b**) without P_{GACS} during TOGA COARE (After Li and Weng 2003)

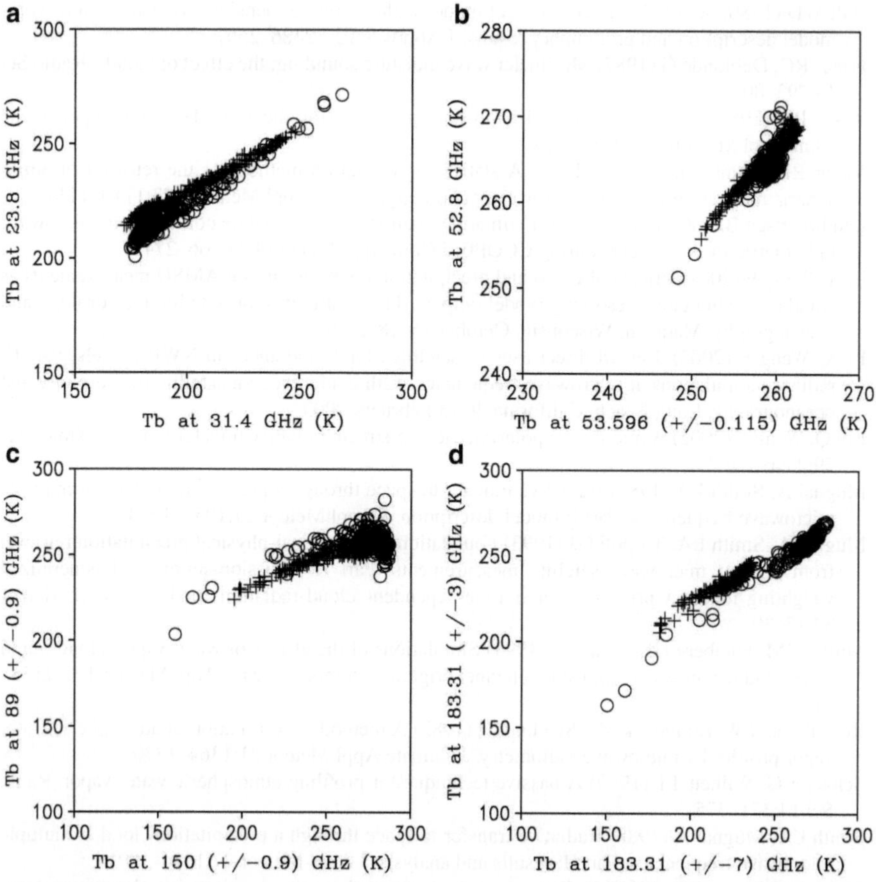

Fig. 15.9 Same as in Fig. 11.7 except for the experiment without P_{GACS} (After Li and Weng 2003)

References

Adler RF, Yeh HYM, Prasad N, Tao WK, Simpson J (1991) Microwave simulations of a tropical rainfall system with a three-dimensional cloud model. J Appl Meteorol 30:924–953

Ferraro R (1997) Special sensor microwave imager derived global rainfall estimates for climatological applications. J Geophys Res 102:16715–16735

Ferrier BS, Tao WK, Simpson J (1995) A double-moment multiple-phase four-class bulk ice scheme. Part II: simulations of convective storms in different large-scale environments and comparisons with other bulk parameterizations. J Atmos Sci 52:1001–1033

Grody N (1980) Atmospheric water content over the tropical Pacific derived from the Nimbus-6 scanning microwave spectrometer. J Appl Meteorol 19:986–996

Grody N (1991) Classification of snow cover and precipitation using the special sensor microwave imager. J Geophys Res 96:7423–7435

Grody N, Weng F, Ferraro R (2000) Application of AMSU for obtaining hydrological parameters. In: Pampaloni P, Paloscia S (eds) Microwave radiometery and remote sensing of Earth's surface and atmosphere. VSP, Zeist, Netherlands, pp 339–351

Hall WD (1980) A detailed microphysical model within a two-dimensional dynamic framework: model description and preliminary results. J Atmos Sci 37:2486–2507
Isaacs RG, Deblonde G (1987) Millimeter wave moisture sounding: the effect of clouds. Radio Sci 14:793–802
Kakar RK (1983) Retrieval of clear sky moisture profiles using the 183 GHz water vapor line. J Clim Appl Meteorol 22:1282–1289
Kakar RK, Lambrightsen BH (1984) A statistical correlation method for the retrieval of atmospheric moisture profiles by microwave radiometry. J Clim Appl Meteorol 23:1110–1114
Lambrightsen BH, Kakar RK (1985) Estimation of atmospheric moisture content from microwave radiometric measurements during CCOPE. J Clim Appl Meteorol 24:266–274
Li X, Weng F (2001) Effects of cloud and precipitation microphysics on AMSU measurements as simulated using cloud resolving model outputs. 11th conference on satellite meteorology and oceanography, Madison, Wisconsin, October 15–18, 2001
Li X, Weng F (2003) Toward direct uses of satellite cloudy radiances in NWP models. Part II: radiance simulations at microwave frequencies. 12th conference on satellite meteorology and oceanography, Long Beach, California, 9–13 February 2003
Liu Q, Weng F (2002) A microwave polarimetric two-stream radiative transfer model. J Atmos Sci 59:2396–2402
Mugnai A, Smith EA (1988) Radiative transfer to space through a precipitation cloud at multiple microwave frequencies. Part I: model description. J Appl Meteor 27:1055–1073
Mugnai A, Smith EA, Tropoli GJ (1993) Foundation for statistical-physical precipitation retrieval from passive microwave satellite measurements. Part II: emission-source and generalized weighting-function properties of a time-dependent cloud-radiation model. J Appl Meteor 32:17–39
Muller BM, Fuelberg HE, Xiang X (1994) Simulations of the effects of water vapor, cloud liquid water, and ice on AMSU moisture channel brightness temperatures. J Appl Meteorol 33:1133–1154
Rosenkranz PW, Komichak MJ, Staelin DH (1982) A method for estimation of atmospheric water vapor profiles by microwave radiometry. J Climate Appl Meteor 21:1364–1370
Schaerer G, Wilheit TT (1979) A passive technique for profiling atmospheric water vapor. Radio Sci 14:371–375
Smith EA, Mugnai A (1988) Radiative transfer to space through a precipitation cloud at multiple microwave frequencies. Part II: results and analysis. J Appl Meteor 27:1074–1091
Smith EA, Mugnai A (1989) Radiative transfer to space through a precipitation cloud at multiple microwave frequencies. Part III: influence of large ice particles. J Meteor Soc Japan 67:739–755
Smith EA, Mugnai A, Cooper HJ, Tropoli GJ, Xiang X (1992) Foundation for statistical-physical precipitation retrieval from passive microwave satellite measurements. Part I: brightness-temperature properties of a time-dependent cloud-radiation model. J Appl Meteor 31:506–531
Spencer R, Christy J, Grody N (1990) Global atmospheric temperature monitoring with satellite microwave measurements: method and results 1979–1984. J Clim 3:1111–1128
Wang JR, King JL, Wilheit TT, Szejwach G, Gesell LH, Nieman RA, Niver DS, Krupp BM, Gagliano JA (1983) Profiling atmospheric water vapor by microwave radiometry. J Clim Appl Meteorol 22:779–788
Weng F, Grody N (1994) Retrieval of cloud liquid water using the Special Sensor Microwave Imager (SSM/I). J Geophys Res 99:25535–25551
Weng F, Grody N (2000) Retrieval of ice cloud parameters using a microwave imaging radiometer. J Atmos Sci 57:1069–1081
Wentz F, Mattox L, Peteherych S (1986) New algorithms for microwave measurements of ocean winds application to SEASAT and the Special Sensor Microwave Imager. J Geophys Res 91:2289–2307
Wilheit TT (1990) An algorithm for retrieving water vapor profiles in clear and cloudy atmospheres from 183 GHz radiometric measurements: simulation studies. J Appl Meteorol 29:508–515

Wu R, Weinman JA (1984) Microwave radiances from precipitating clouds containing aspherical ice, combined phase, and liquid hydrometeors. J Geophys Res 89:7170–7178

Yeh HYM, Prasad N, Mack RA, Adler RF (1990) Aircraft microwave observations and simulations of deep convection from 18 to 183 GHz. Part II: model results. J Atmos Oceanic Technol 7:377–391

Ziegler CL (1985) Retrieval of thermal and microphysical variables in observed convective storms. Part 1: model development and preliminary testing. J Atmos Sci 42:1487–1509

Chapter 16
Precipitation Predictability

Abstract In this chapter, precipitation predictability is studied through the analysis of sensitivity of precipitation simulations to uncertainties of radiation calculations. The effects of large-scale forcing on precipitation predictability is examined by conducting sensitivity experiments with a weak forcing from tropical rainfall event and with a strong forcing from pre-summer torrential rainfall event. The ratio of the root-mean-squared difference between sensitivity and control experiments to the standard deviation of the control experiment is calculated to evaluate precipitation predictability. The ratios calculated from temporal and spatial average data are compared to identify the effects of large-scale forcing on precipitation predictability.

Keywords Precipitation predictability • Root-mean-squared difference • Standard deviation • Large-scale forcing

Governmental decision-making and public planning for severe weathers and associated disasters rely largely on accurate quantitative weather forecasts. Atmospheric predictability is associated with nonlinear interactions between momentum, thermodynamics, cloud microphysics, and radiation (e.g., Lorenz 1969; Ehrendorfer 1997; Palmer 2000; Hohenegger and Schar 2007; Wandishin et al. 2008). Such predictability has been intensively debated at various scales from synoptic scales to mesoscales and to cloud-resolving scales. Recently, the predictability of precipitation or quantitative precipitation forecast has been paid much attention by modeling scientists (e.g., Richard et al. 2003; Li et al. 2006) and operational forecasters. Accurate quantitative precipitation prediction relies on numerical modeling, which is much affected by model setup, quality of initial conditions, and calculations of physics including radiation and cloud microphysics parameterizations (e.g., Grabowski et al. 1998; Donner et al. 1999; Guichard et al. 2000; Xu et al. 2002; Petch and Gray 2001; Petch et al. 2002, 2008; Khairoutdinov and Randall 2003; Petch 2004, 2006; Cheng and Xu 2006; Li et al. 2006; Phillips and Donner 2006; Keil et al. 2008; Gao and Li 2008, 2009). Li et al. (2006) and Li and Shen (2010) conducted two-dimensional cloud-resolving model experiments

© Springer International Publishing Switzerland 2016

X. Li, S. Gao, *Cloud-Resolving Modeling of Convective Processes*,
Springer Atmospheric Sciences, DOI 10.1007/978-3-319-26360-1_16

to study the sensitivity of precipitation simulations to the uncertainty of initial conditions and associated vertical structure in the experiments with weak large-scale forcing. They found that the precipitation modeling is extremely sensitive to initial conditions through the uncertainty in calculation of vapor condensation and deposition rates. Physics-based condensation and deposition calculation requires quality temperature and water vapor information. Any tiny uncertainty from temperature and water vapor data could lead to significant errors in the calculation of condensation and deposition, which causes a large uncertainty in the estimate of precipitation.

Radiative heating calculation may contain uncertainty. The uncertainty may come from the uncertainty of temperature and water vapor over both clear-sky and cloudy regions and uncertainty of the calculation of cloud optical properties over cloudy regions. The analysis of satellite-retrieved data shows that the error of retrieved air temperature could be up to 1 K (e.g., Aires et al. 2002; Susskind et al. 2003) and the error of precipitable water could be as large as 3 mm (e.g., Grody et al. 2001). Ice crystals with various nonspherical shapes may cause large uncertainty in the radiative heating calculation through the calculation of cloud optical thickness (e.g., Fu et al. 1999; Yang et al. 2000). The schemes that parameterize radiation also contain uncertainty due to assumptions and approximation. Thus, the sensitivity of precipitation simulations to the uncertainty of radiation calculation is discussed in this chapter based on Ran and Li (2014).

A series of perturbation experiments are conducted using a two-dimensional cloud-resolving model and compared with the control experiment. Since the sensitivity of precipitation modeling depends largely on water vapor convergence associated with large-scale vertical velocity (e.g., Gao and Li 2009; Li and Shen 2010), effects of large-scale forcing on sensitivity of precipitation modeling are examined through the comparison between two experiments with different imposed large-scale forcing data sets. One set of experiments uses the large-scale forcing for tropical rainfall, and the other set of experiments uses the large-scale forcing for pre-summer heavy rainfall over southern China. The maximum large-scale upward motion of pre-summer rainfall event ($\sim$20 cm s^{-1}) is one order of magnitude larger than that of tropical rainfall event ($\sim$2 cm s^{-1}) (see Figs. 9.1 and 16.1). Thus, pre-summer and tropical rainfall events are referred to strong (SF) and weak (WF) forcing cases, respectively.

In WF, the development of a weak convective system is associated with a weak westerly wind burst over the equatorial area. the large-scale forcing is calculated for about 8 days from 1100 Local Standard Time (LST) 18–1700 LST 26 April 2003 over 150–160°E, EQ during the weak tropical rainfall event. The model domain in WF is completely covered by ocean. The daily mean sea surface temperature retrieved from the Tropical Rainfall Measuring Mission (TRMM) Microwave Imager (TMI) radiometer with a 10.7 GHz channel (Wentz et al. 2000) is imposed in WF. The large-scale forcing in SF is identical to that in the *P* experiments of Chap. 9. Figures 16.1 and 9.1 reveal that SF has a larger variation of upward motions than WF does.

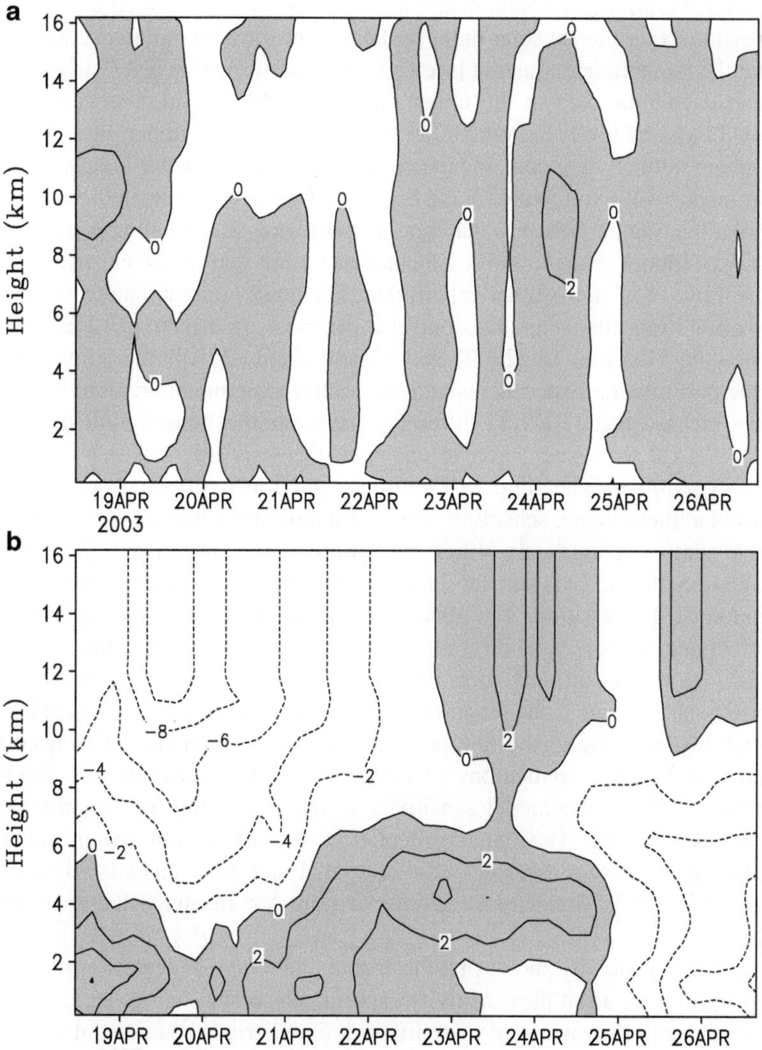

Fig. 16.1 Time–height distributions of (**a**) vertical velocity (cm s^{-1}) and (**b**) zonal wind (m s^{-1}) imposed in the model in WF. Upward motion in (**a**) and westerly wind in (**b**) are shaded (After Ran and Li 2014)

In addition to the control experiments in the two cases, the 12 perturbation experiments are conducted for each case. The perturbation experiments are identical to the control experiment, but perturbed in radiative heating term ($Q_R/\pi c_p$) in the equation of potential temperature after initial 6-h model spin-up. The radiative heating term consists of components of solar heating and infrared cooling. In the perturbation experiments, solar radiative heating or infrared radiative cooling terms of potential temperature equation [see (2.1a)] are perturbed by $\pm$ 1 % at each model grid and each time step in the upper (above 500 hPa), mid- (between 700 hPa and 500 hPa), and lower (below 700 hPa) troposphere. The upper, mid-, and lower troposphere defined here contains ice clouds, both ice and water clouds, and water clouds, respectively, following Li and Shen (2010). For example, in the experiments with solar heating perturbed in the upper troposphere, solar heating is perturbed by 1 % at each model grid above 500 hPa and each time step in one experiment and is perturbed by -1 % in the other experiment. The hourly output data from these two perturbation experiments and the control experiment are used to plot Fig. 16.3a.

Following Gao and Li (2009) and Li and Shen (2010), the simulation data from the perturbation experiments and the control experiment are used to calculate the root-mean-squared (RMS) difference between the perturbation experiments and the control experiment and to calculate the standard deviation (SD) of the control experiment. The ratio (RS ratio) of the RMS difference to SD is then calculated to measure the sensitivity of precipitation simulation to the uncertainty of radiative heating calculations. When the RS ratio is smaller than 1, the magnitude of simulation difference is smaller than that of variability, suggesting the reasonable simulation of the variation. The uncertainty of radiative heating calculations does not affect the model capability for simulating the variation. Thus, the model sensitivity to the uncertainty of radiative heating calculations is weak. When the RS ratio is larger than 1, the magnitude of simulation difference is larger than that of variability, suggesting the failure of the variation simulation. The uncertainty of radiative heating calculations changes the model capability for simulating the variation. Thus, the model sensitivity to the uncertainty of radiative heating calculations is strong. The standard deviations of surface rain rate in the control experiments of the two cases decrease as spatial and time scales for data average increase (Fig. 16.2). Standard deviations of rain rate are much larger in SF than in WF.

Figure 16.3 generally shows that the spatial and temporal scales for the RS ratio of 1 are similar for all of the sensitivity experiments in WF. In Fig. 16.3, the surface rain rate data are first averaged over different time periods and zonal distances and the resulting data are then used to calculate the RMS and SD to obtain the RS. Thus, the definitions of the spatial and temporal scales of RS are the time periods and zonal distances in which data are averaged for computing the RS. When solar radiation calculation is perturbed in the upper troposphere, the RS ratio of surface rain rate is generally smaller than 1 for the zonal and time scales of larger than 500 km and 6 h, respectively (Fig. 16.3a). When solar radiation calculation is perturbed in the mid-troposphere, the time scale for the RS ratio of less than 1 is longer than 3 h (Fig. 16.3b). When solar radiation calculation is perturbed in the lower troposphere, RS ratio could be larger than 1 for the zonal scale of about 600 km and time scales of about 8 h and 20 h (Fig. 16.3c). The zonal and time scales for the RS ratio of less than 1 are larger than 470 km and longer than 4 h when infrared radiation calculation

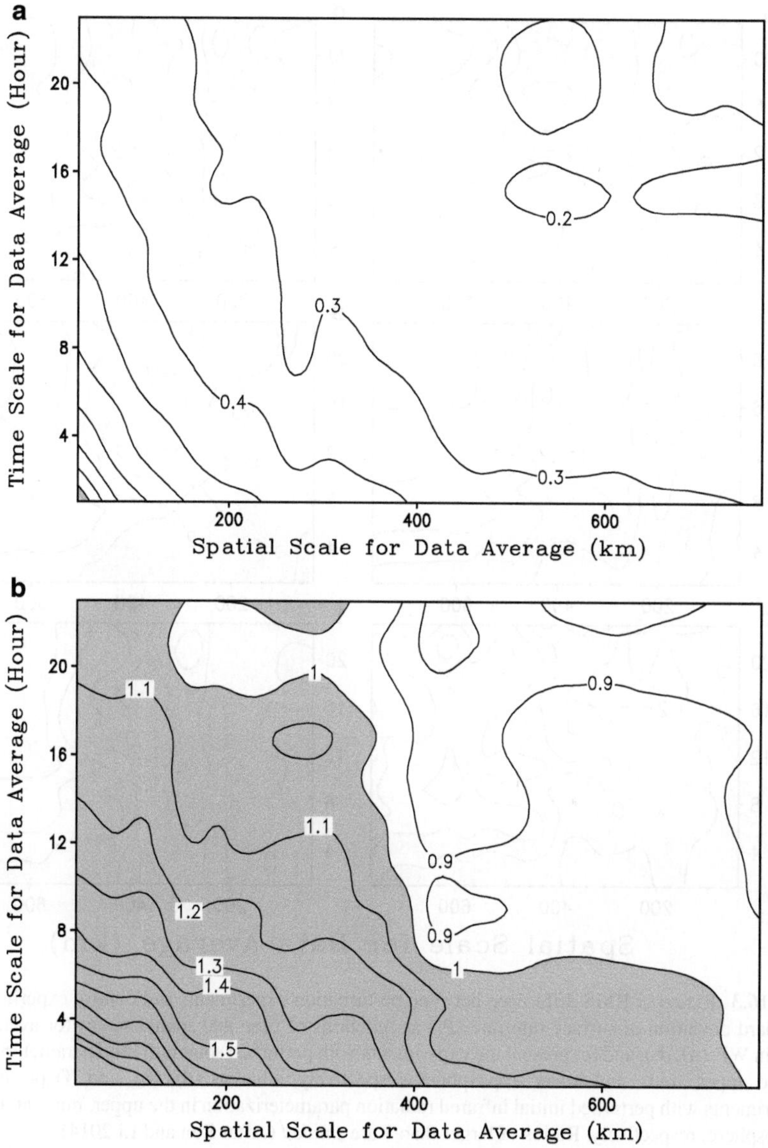

Fig. 16.2 Standard deviations of surface rain rate (P_s) in the control experiments as functions of time and spatial scales for average of data in (**a**) WF and (**b**) SF. Unit is mm h^{-1}. Standard deviations of larger than 1 mm h^{-1} are *shaded* (After Ran and Li 2014)

is perturbed in the lower troposphere whereas they increase to 530 km and 6 h when infrared radiation calculation is perturbed in the upper troposphere (Fig. 16.3d–f). Thus, precipitation simulation is more sensitive to the uncertainty of solar radiation calculation in the lower troposphere than in the mid- and upper troposphere, whereas

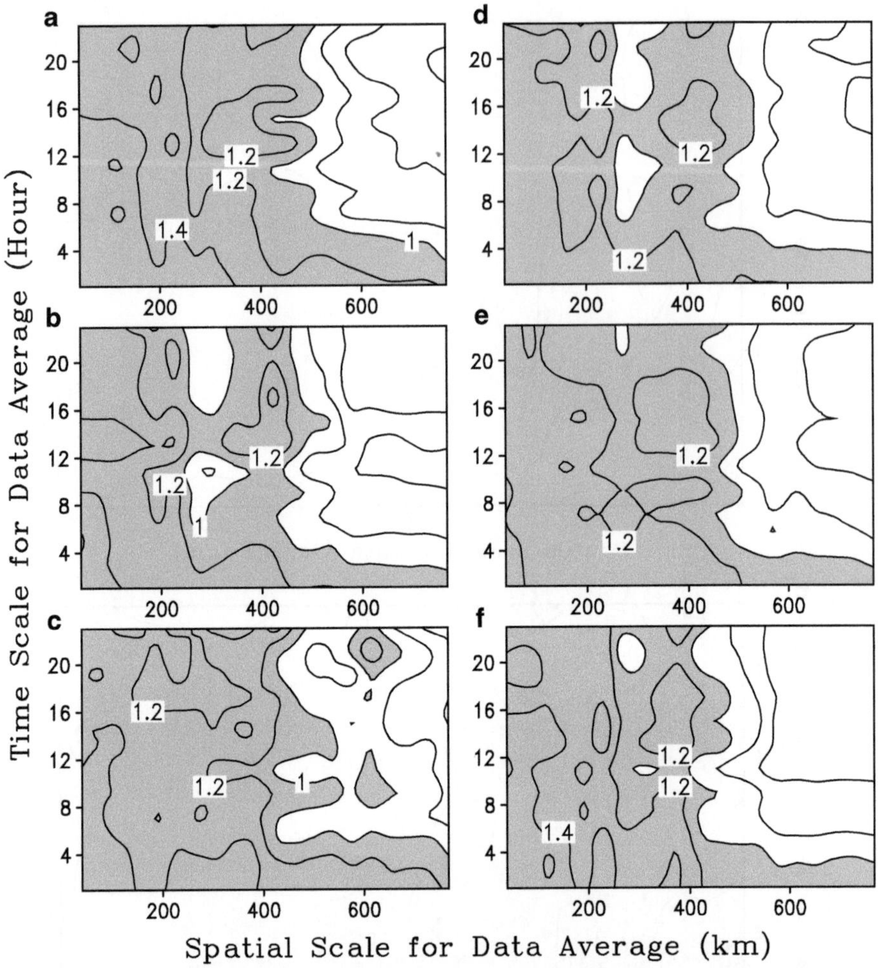

Fig. 16.3 Ratios of RMS difference between perturbation experiments and control experiment to standard deviation of surface rain rate (P_s) as functions of time and spatial scales for average of data in WF. (**a**), (**b**), and (**c**) present the experiments with perturbed solar radiation parameterization in the upper, mid-, and lower troposphere, respectively, whereas (**d**), (**e**), and (**f**) present the experiments with perturbed initial infrared radiation parameterization in the upper, mid-, and lower troposphere, respectively. Ratios of larger than 1 are *shaded* (After Ran and Li 2014)

it is less sensitive to the uncertainty of infrared radiation calculation in the lower troposphere than in the mid- and upper troposphere.

When large-scale forcing becomes stronger in SF, the zonal and time scales for the RS ratio of less than 1 generally are larger than 200 km and longer than 6 h, respectively, regardless of which vertical layers solar radiation calculations are perturbed in (Fig. 16.4a–c). When infrared radiation calculation in the upper

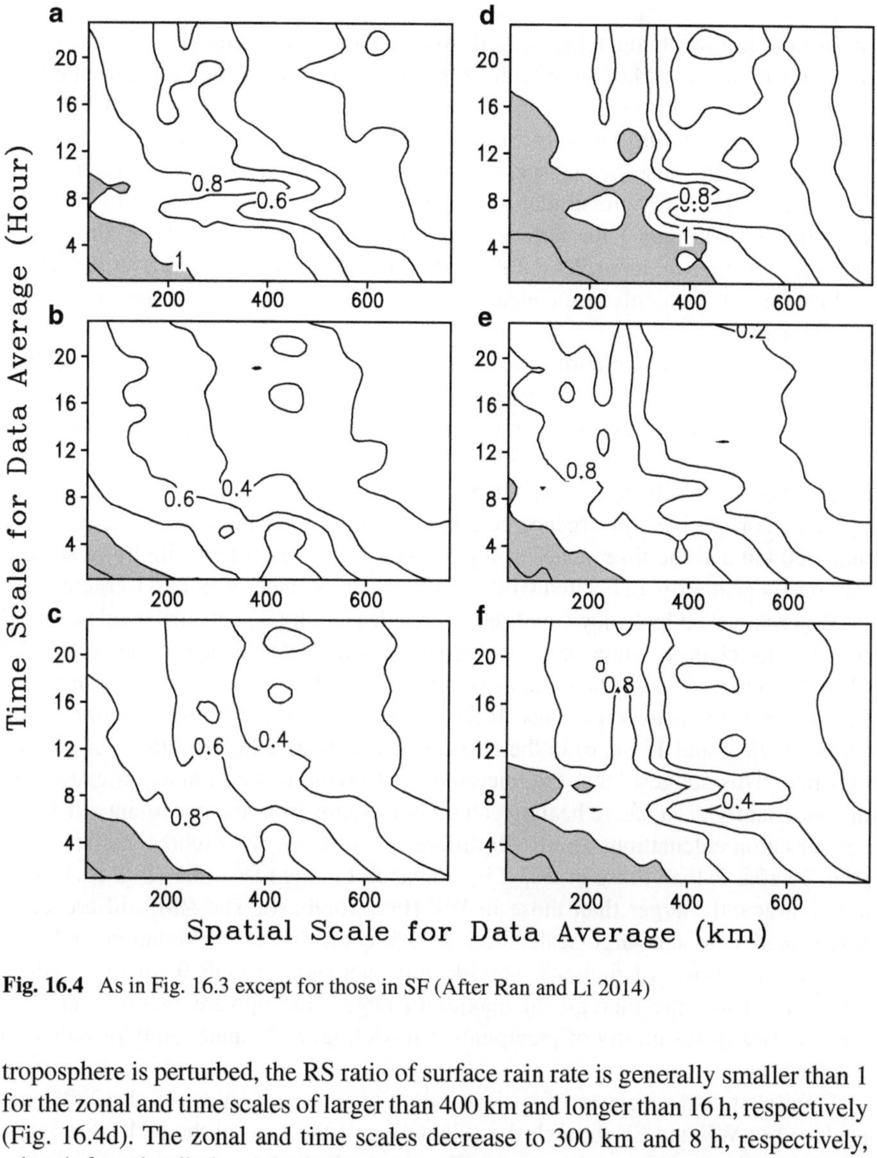

Fig. 16.4 As in Fig. 16.3 except for those in SF (After Ran and Li 2014)

troposphere is perturbed, the RS ratio of surface rain rate is generally smaller than 1 for the zonal and time scales of larger than 400 km and longer than 16 h, respectively (Fig. 16.4d). The zonal and time scales decrease to 300 km and 8 h, respectively, when infrared radiation calculation is perturbed in the mid-troposphere (Fig. 16.4e). When infrared radiation calculation is perturbed in the lower troposphere, the zonal and time scales are 370 km and 11 h, respectively (Fig. 16.4f). Thus, precipitation simulation is more sensitive to the uncertainty of infrared radiation calculation in the upper troposphere than in the lower troposphere. The comparison between Figs. 16.3 and 16.4 shows the zonal and time scales of surface rain rate for the RS ratio of 1 are larger in WF than in SF.

The surface rain rate is associated with both water vapor and cloud processes in surface rainfall budget [see (2.4)]. When radiation calculations in the lower troposphere are perturbed in WF, the zonal scales of local water vapor change for the RS ratio of less than 1 are smaller than those of surface rain rate while their time scales are similar (Fig. 16.5). The time scales of water vapor convergence are shorter than those of surface rain rate, but the zonal scales of water vapor convergence generally are similar to or smaller than those of surface rain rate. The RS ratios are less than 1 for surface evaporation and larger than 1 for hydrometeor change/convergence, regardless of the zonal and time scales for data average. Thus, surface rainfall is mainly associated with water vapor convergence and local water vapor change.

When radiation calculations are perturbed in the lower troposphere in SF, the zonal and time scales of water vapor convergence for the RS ratio of less than 1 are similar to those of surface rain rate, while they are much smaller than those of local water vapor change (Fig. 16.6). Like those in WF, The RS ratios of surface evaporation generally are smaller than 1. Unlike those in WF, the RS ratios for hydrometeor change/convergence could be less than 1 for the zonal scale of larger than 600 km and the time scales of longer than 4 h. Thus, the sensitivity of surface rain rate is primarily associated with the sensitivity of water vapor convergence.

Water vapor budget shows that the net condensation (S_{qv}) is attributable to local water vapor change, water vapor convergence, and surface evaporation [see (2.3); (2.4), (2.6) and (2.6a)]. The patterns of RS ratios of P_{CND} as functions of zonal and time scales for data average (Figs. 16.7b, d and 16.8b, d) are similar to those of Sqv (Figs. 16.7a, c and 16.8a, c) in the two cases, which are similar to those of surface rain rate. This suggests that the uncertainty of precipitation simulation caused by the uncertainty of radiative heating calculations stems from the uncertainty of vapor condensation calculation. The RMS differences in P_{CND} are slightly smaller in SF (Fig. 16.10a, c) than those in WF (Fig. 16.9a, c), but the SDs in SF (Fig. 16.10b, d) are significantly larger than those in WF (Fig. 16.9b, d). The RMS difference is less sensitive to the large-scale forcing, while the increased variation in large-scale forcing (upward motions) significantly increases rainfall fluctuation with a large SD. Thus, the increase in maximum large-scale upward motions leads to the decrease in sensitivity of precipitation modeling to the uncertainty of radiation calculation.

Finally, statistical errors of surface rain rate proposed by Li et al. (2006) are analyzed in WF and SF through the calculation of the ratio of the RMS difference of rain rate to the temporally and zonally averaged value. Figure 16.11 reveals that the spatial and temporal scales for the statistical error of 100 % are similar spatial and temporal averaging for all of the sensitivity experiments in WF, in which the statistical errors are smaller than 100 % for the spatial scale of data average of over 250 km and the time scale of over 2 h (Fig. 16.11). The spatial and temporal scales for the statistical error of 100 % decreases up to 50 km and 1 h in SF, respectively (Fig. 16.12). The spatial and temporal scales are larger when infrared radiation calculation is perturbed than when solar radiation calculation is perturbed.

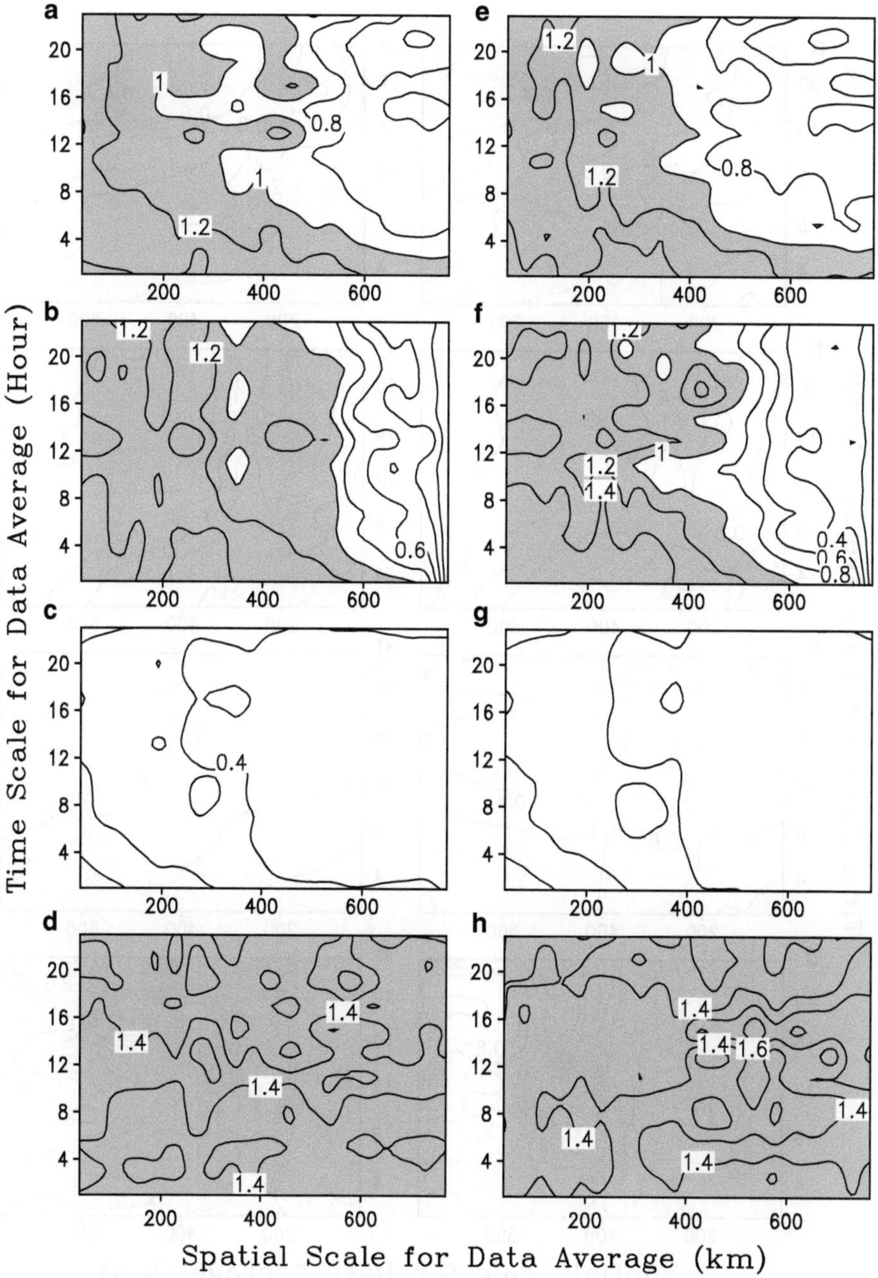

Fig. 16.5 Ratios of RMS differences between perturbation experiments and control experiment to standard deviations of water vapor tendency (Q_{WVT}) in (**a**) and (**e**), water vapor convergence (Q_{WVF}) in (**b**) and (**f**), surface evaporation (Q_{WVE}) in (**c**) and (**g**), and cloud source/sink (Q_{CM}) in (**d**) and (**h**) as functions of time and spatial scales for average of data in WF. *Left* and *right* panels, respectively, present the experiments with perturbed solar and infrared radiation parameterizations in the lower troposphere. Ratios of larger than 1 are *shaded* (After Ran and Li 2014)

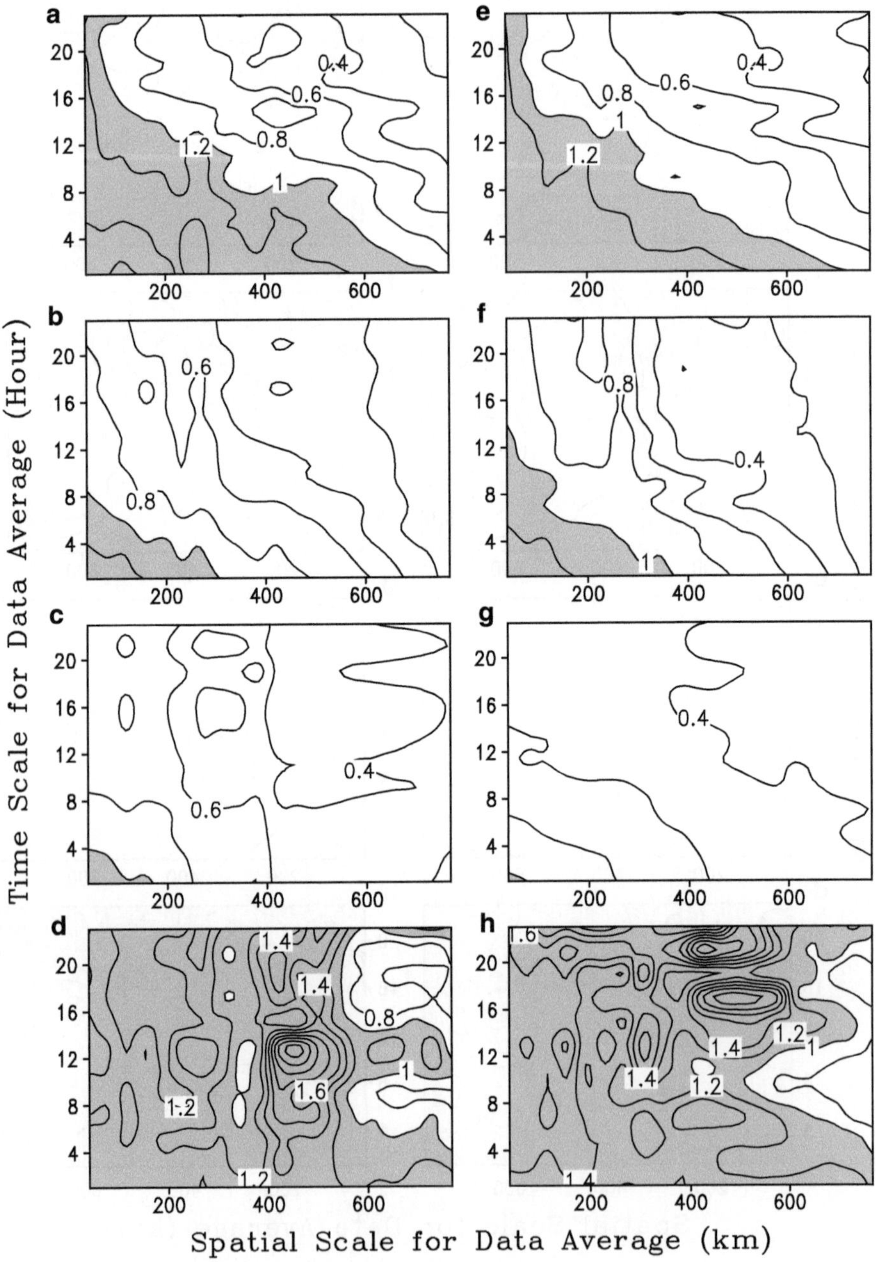

Fig. 16.6 As in Fig. 16.5 except for those in SF (After Ran and Li 2014)

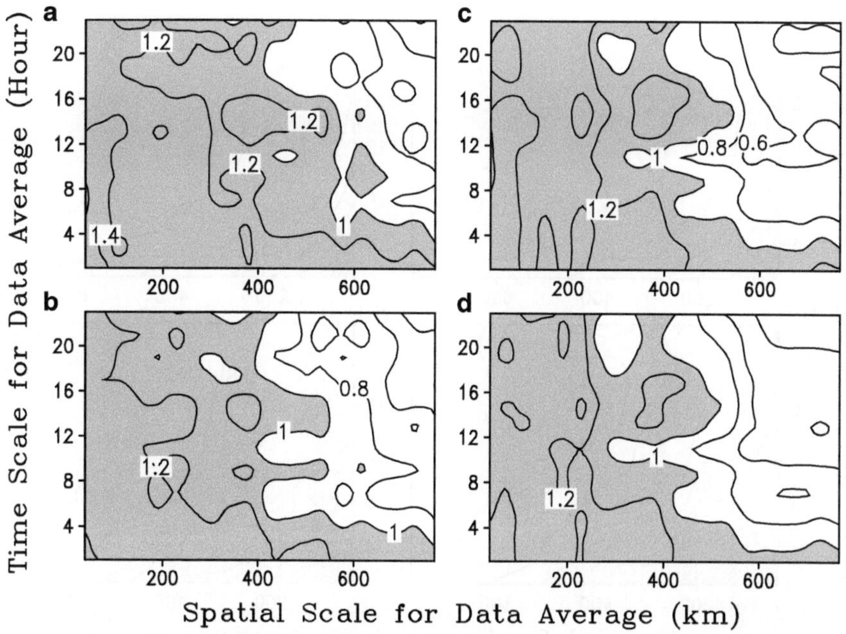

Fig. 16.7 Ratios of RMS differences between perturbation experiments and control experiment to standard deviations of water vapor sink (S_{qv}) in (**a**) and (**c**) and vapor condensation (P_{CND}) in (**b**) and (**d**) as functions of time and spatial scales for average of data in WF. *Left* and *right* panels, respectively, present the experiments with perturbed initial solar and infrared radiation parameterizations in the lower troposphere. Ratios of larger than 1 are *shaded* (After Ran and Li 2014)

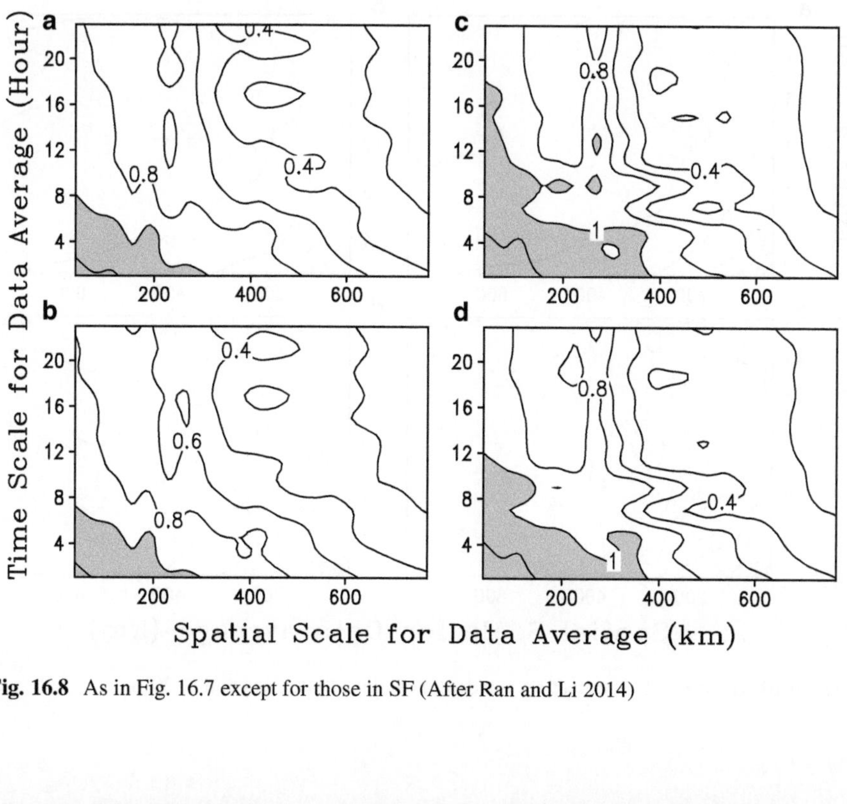

Fig. 16.8 As in Fig. 16.7 except for those in SF (After Ran and Li 2014)

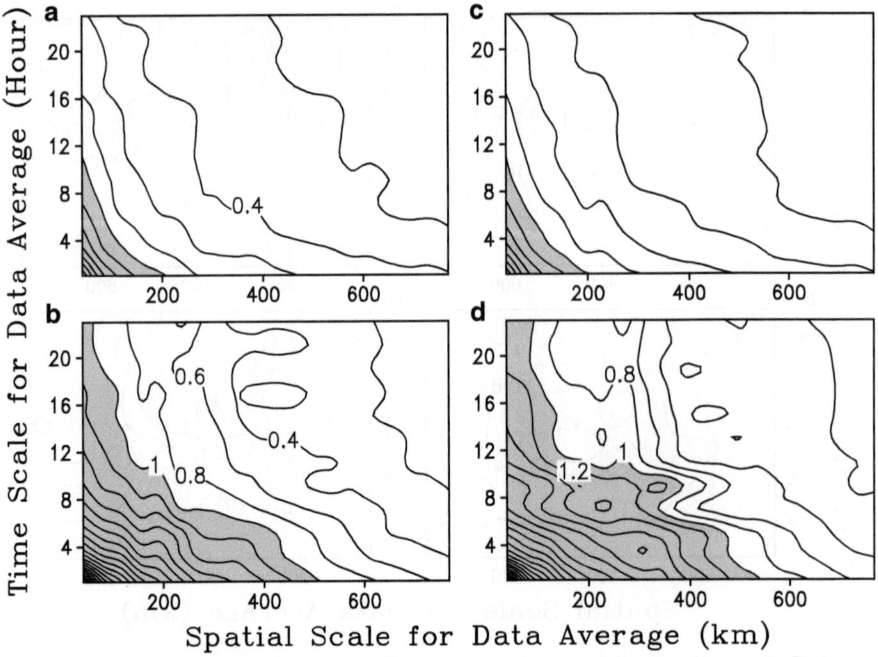

Fig. 16.9 RMS differences in P_{CND} (mm h^{-1}) between perturbation experiments as functions of time and spatial scales for average of data in WF (**a, c**) and SF (**b, d**). (**a**) and (**b**) present the experiments with perturbed solar radiation parameterization in the lower troposphere whereas (**c**) and (**d**) present the experiments with perturbed infrared radiation parameterization in the lower troposphere. RMS differences of larger than 1 mm h^{-1} are *shaded* (After Ran and Li 2014)

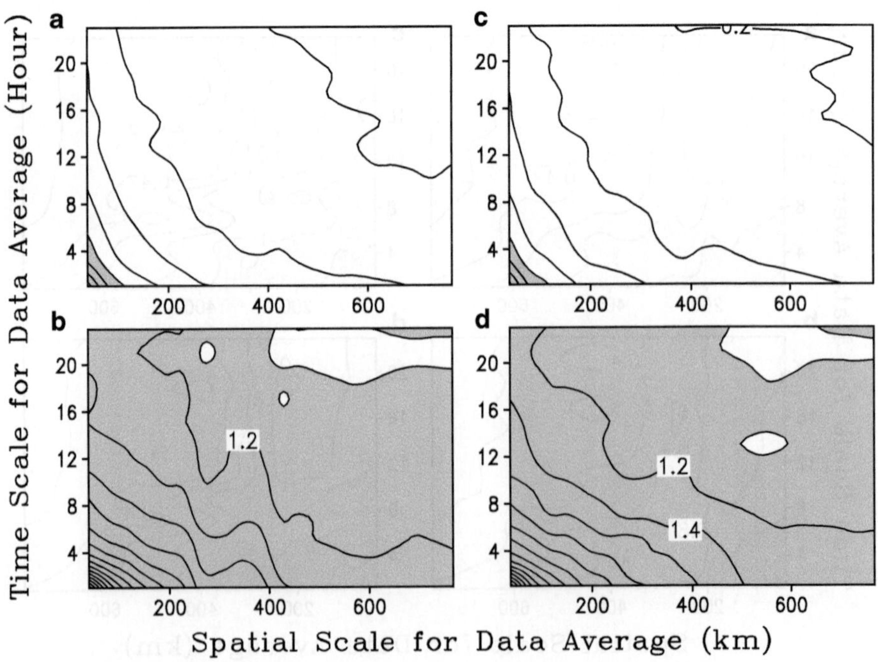

Fig. 16.10 As in Fig. 16.9 except for standard deviations (After Ran and Li 2014)

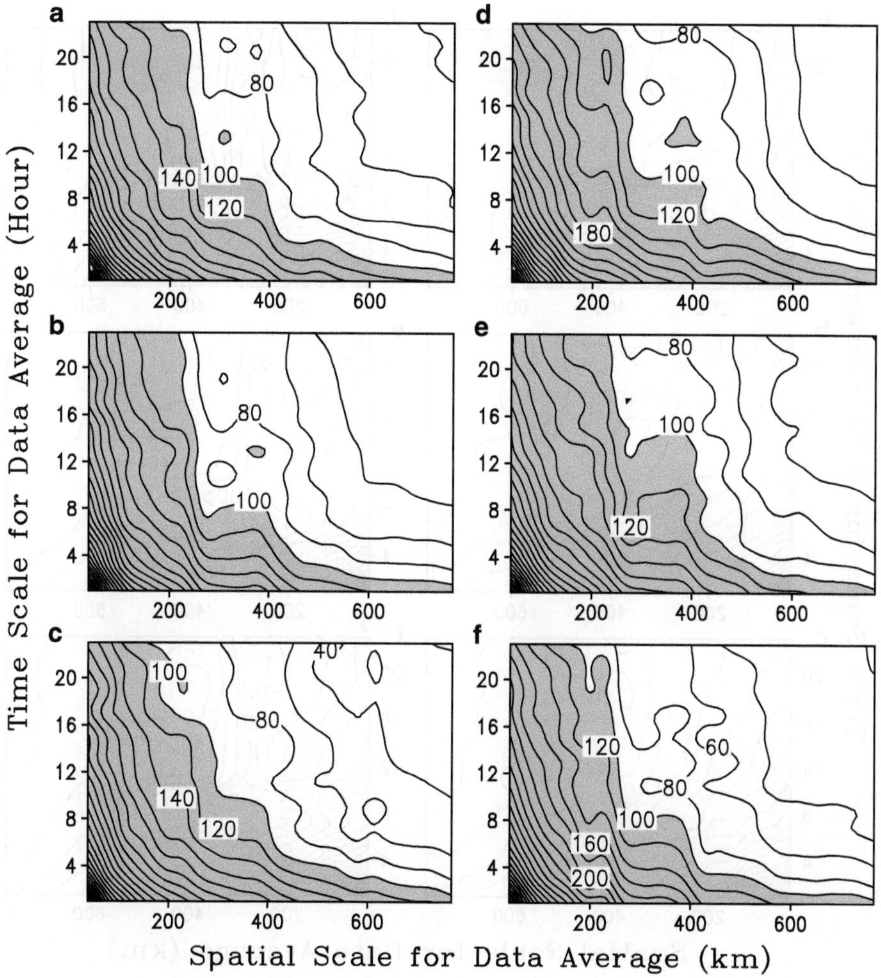

Fig. 16.11 Statistical errors (%) of surface rain rate as functions of time and spatial scales for average of data in WF. (**a**), (**b**), and (**c**) present the experiments with perturbed solar radiation parameterization in the upper, mid-, and lower troposphere whereas (**d**), (**e**), and (**f**) present the experiments with perturbed infrared radiation parameterization in the upper, mid-, and lower troposphere. Statistical errors of larger than 100 % are *shaded*

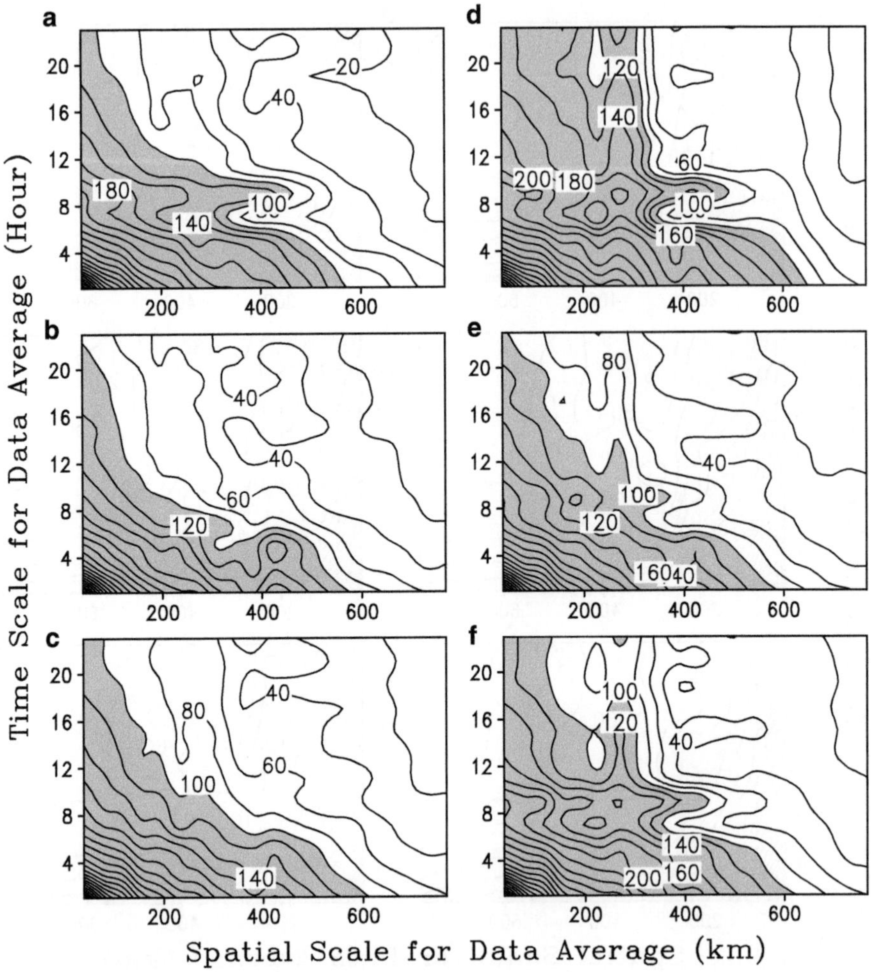

Fig. 16.12 As in Fig. 16.11 except for those in SF

References

Aires FW, Rossow B, Scott NA, Chedin A (2002) Remote sensing from the infrared atmospheric sounding interferometer instrument, 2, simultaneous retrieval of temperature, water vapor, and ozone atmospheric profiles. J Geophys Res 107. doi:10.1029/2001JD001591

Cheng A, Xu KM (2006) Simulation of shallow cumuli and their transition to deep convective clouds by cloud-resolving models with different their-order turbulence closures. Q J R Meteorol Soc 132:359–382

Donner LJ, Semen CJ, Hemler RS (1999) Three-dimensional cloud-system modeling of GATE convection. J Atmos Sci 56:1885–1912

Ehrendorfer M (1997) Predicting the uncertainty of numerical weather forecasts: a review. Meteor Z 6:147–183

Fu Q, Sun WB, Yang P (1999) Modeling of scattering and absorption cirrus ice particles at thermal infrared wavelengths. J Atmos Sci 56:2937–2947
Gao S, Li X (2008) Impacts of initial conditions on cloud-resolving simulations. Adv Atmos Sci 25:737–747
Gao S, Li X (2009) Dependence of the accuracy of precipitation and cloud simulation on time and spatial scales. Adv Atmos Adv 26:1108–1114
Grabowski WW, Wu X, Moncrieff MW, Hall WD (1998) Cloud-resolving model of tropical cloud systems during Phase III of GATE. Part II: effects of resolution and the third spatial dimension. J Atmos Sci 55:3264–3282
Grody N, Zhao J, Ferraro R, Weng F, Boers R (2001) Determination of precipitable water and cloud liquid water over oceans from the NOAA 15 advanced microwave sounding unit. J Geophys Res 106:2943–2953
Guichard F, Redelsperger JL, Lafore JP (2000) Cloud resolving simulations of convective activity during TOGA-COARE: sensitivity to external sources of uncertainties. Q J R Meteorol Soc 126:3067–3095
Hohenegger C, Schar C (2007) Atmospheric predictability at synoptic versus cloud-resolving scales. Bull Am Meteorol Soc 88:1783–1793
Keil C, Ropnack A, Craig GC, Schumann U (2008) Sensitivity of quantitative precipitation forecast to height dependent changes in humidity. Geophys Res Lett 35. doi:10.1029/2008GL033657
Khairoutdinov MF, Randall DA (2003) Cloud resolving modeling of the ARM summer 1997 IOP: model formulation, results, uncertainties, and sensitivities. J Atmos Sci 60:607–625
Li X, Shen X (2010) Sensitivity of cloud-resolving precipitation simulations to uncertainty of vertical structures of initial conditions. Q J R Meteorol Soc 136:201–212
Li X, Zhang S, Zhang DL (2006) Thermodynamic, cloud microphysics and rainfall responses to initial moisture perturbations in the tropical deep convective regime. J Geophys Res 111. doi:10.1029/2005JD006968
Lorenz EN (1969) Predictability of a low which possesses many scales of motion. Tellus 21:289–307
Palmer TN (2000) Predicting uncertainty in forecasts of weather and climate. Rep Prog Phys 63:71–116
Petch JC (2004) The predictability of deep convection in cloud-resolving simulations over land. Q J R Meteorol Soc 130:3173–3187
Petch JC (2006) Sensitivity studies of developing convection in a cloud-resolving model. Q J R Meteorol Soc 132:345–358
Petch JC, Gray MEB (2001) Sensitivity studies using a cloud-resolving model simulation of the tropical west Pacific. Q J R Meteorol Soc 127:2287–2306
Petch JC, Brown AR, Gray MEB (2002) The impact of horizontal resolution on the simulations of convective development over land. Q J R Meteorol Soc 128:2031–2044
Petch JC, Blossey PN, Bretherton CS (2008) Differences in the lower troposphere in two- and three-dimensional cloud-resolving model simulations of deep convection. Q J R Meteorol Soc 134:1941–1946
Phillips VT, Donner LJ (2006) Cloud microphysics, radiation and vertical velocities in two- and three-dimensional simulations of deep convection. Q J R Meteorol Soc 132:3011–3033
Ran L, Li X (2014) Sensitivity of cloud-resolving precipitation simulations to uncertainty of radiation calculation: effects of large-scale forcing. Q J R Meteorol Soc 140:838–845, (c) the Royal Meteorological Society. Reprinted with permission
Richard E, Cosma S, Tabary P, Pinty JP, Hagen M (2003) High-resolution numerical simulations of the convective system observed in the Lago Maggiore are on 17 September 1999 (MAP IOP2a). Q J R Meteorol Soc 129:543–563
Susskind J, Barnet CD, Blaisdell JM (2003) Retrieval of atmospheric and surface parameters from AIR/AMSU/HSB data in the presence of clouds. IEEE Trans Geosci Remote Sens 41:390–409
Wandishin MS, Stensrud DJ, Mullen SL, Wicker LJ (2008) On the predictability of mesoscale convective systems: two-dimensional simulations. Weather Forecast 23:773–785

Wentz FJ, Gentemann C, Smith D, Chelton D (2000) Satellite measurements of sea surface temperature through clouds. Science 288:847–850

Xu KM, Cederwall RT, Donner LJ, Grabowski WW, Guichard F, Johnson DE, Khairoutdinov M, Krueger SK, Petch JC, Randall DA, Seman CJ, Tao WK, Wang D, Xie SC, Yio JJ, Zhang MH (2002) An intercomparison of cloud resolving models with the atmospheric radiation measurement summer 1997 intensive observation period data. Q J R Meteorol Soc 128:593–624

Yang P, Liou KN, Wyser K, Mitchell D (2000) Parameterization of the scattering and absorption properties of individual crystals. J Geophys Res 105:4699–4718

Chapter 17
Modeling of Depositional Growth of Ice Crystal

Abstract The schemes that parameterize the depositional growth of ice crystal developed two decades ago use the assumption that ice crystal spectral density is independent from the size of ice crystal. Recently, a linear relation between the number and mass of ice crystal in the small range of ice crystal radius is assumed to develop new schemes. Four schemes are tested in a series of sensitivity experiments of pre-summer torrential rainfall event. In this chapter, four schemes are briefly discussed. The 5-day and model domain mean rain rate from sensitivity experiments are compared to the observation. The differences in cloud microphysical budget and vertical profiles of heat budget between sensitivity experiments are analyzed to highlight key physical processes for the improvement of precipitation simulation.

Keywords Depositional growth of ice crystal • Cloud microphysical budget • Heat budget • Ice crystal spectral density • Number and mass of ice crystal • Size of ice crystal

Ice clouds are an important ingredient of precipitation systems and play crucial roles in producing precipitation since the melting of precipitation ice is one of the major rainfall sources. They also affect heat balance by releasing latent heat during the formation of ice hydrometeor and by changing radiation through the prevention of solar radiation from reaching the earth surface and reduction of infrared radiation going out to the space. Thus, the accurate numerical modeling of clouds and precipitation relies on the improvement of the presentation of ice clouds and associated ice microphysical processes. Among various ice microphysical processes, Wegener-Bergeron-Findeisen (Wegener 1911; Bergeron 1935; Findeisen 1938) process is one of the important microphysical processes in determining cloud ice. Because the saturation water vapor pressure over water is higher than that over ice, ice crystals grow by vapor deposition at the expense of the supercooled droplets in mixed-phase clouds (Wegener 1911). The growth of ice crystals eventually leads to the formation of precipitating particles (Bergeron 1935). Some supercooled droplets freeze due to heterogeneous nucleation, while others evaporate with the resulting water vapor being deposited onto activated ice nuclei and ice crystals (Findeisen 1938). The depositional growth of snow from cloud ice (P_{SFI}) and the depositional growth of cloud ice from cloud water (P_{IDW}) are two important microphysical processes in

© Springer International Publishing Switzerland 2016

X. Li, S. Gao, *Cloud-Resolving Modeling of Convective Processes*,
Springer Atmospheric Sciences, DOI 10.1007/978-3-319-26360-1_17

modeling the Wegener-Bergeron-Findeisen process. Hsie et al. (1980) used a natural ice nucleus and the 50 μm radius of ice crystal in the calculations of P_{IDW} and P_{SFI}, respectively, which result in small amount of cloud ice. Based on the analysis of in-situ aircraft observations, Krueger et al. (1995) replaced a natural ice nucleus with an averaged mass of ice crystal in the calculation of P_{IDW} and increased the radius of ice crystal to 100 μm in the calculation of P_{SFI}. Li et al. (1999) used the Krueger schemes in the modeling of tropical rainfall and found a significant increase in cloud ice and a moderate increase in snow. The number of ice crystal may change with the size of ice crystal, but Hsie's and Krueger's schemes assume the independence of ice crystal spectral density from the size of ice crystal. Recently, Zeng et al. (2008) proposed a linear relation between the number and mass of ice crystal in the small range of ice crystal radius (less than 50 μm) to develop new schemes for P_{SFI} and P_{IDW} and revealed significant enhancement in cloud ice. Shen et al. (2014) modified Zeng's scheme in 40–50 μm. In this chapter, the effects of depositional growth of ice crystal on rainfall and clouds are discussed based on Shen et al. (2014) through the analysis of sensitivity experiments of pre-summer torrential rainfall case using various schemes of depositional growth of ice crystal.

Hsie et al. (1980) used a natural ice nucleus with a mass (m_n) of 1.05×10^{-15} g in the calculation of P_{IDW}, which can be expressed by

$$P_{IDW} = \frac{n_0 e^{\frac{1}{2}|T-T_o|}}{10^3 \rho} b_1 (m_n)^{b_2}, \tag{17.1}$$

where $n_0 = 10^{-8}$ m^{-3}; $T_0 = 0$ °C; ρ is the air density, which only is a function of height; b_1 and b_2 are the positive temperature-dependent coefficients tabulated by Koenig (1971). Hsie et al. (1980) calculated the timescale for the growth of small ice crystal from the radius of 40 μm to the radius of 50 μm in P_{SFI}, which can be written by

$$P_{SFI} = \frac{q_i}{\Delta t_1}, \tag{17.2}$$

where q_i is the mixing ratio of cloud ice; $\Delta t_1 \left[= \left(m_{I50}{}^{1-b_2} - m_{I40}{}^{1-b_2}\right)/b_1(1-b_2)\right]$ is the timescale needed for a crystal to grow from radius r_{I40} to radius r_{I50}; m_{I50} ($=4.8 \times 10^{-10}$ kg) is the mass of an ice crystal r_{I50} ($=50$ μm), and m_{I40} ($=2.46 \times 10^{-10}$ kg) is the mass of an ice crystal r_{I40} ($=40$ μm). Experiment H80 uses (17.1) and (17.2) to calculate the depositional growth of cloud ice.

Krueger et al. (1995) believed that the mass of a natural ice nucleus is much smaller than that of a typical ice crystal and replaced it with the averaged mass of an ice crystal in the calculation of P_{IDW}. Thus, P_{IDW} becomes

$$P_{IDW} = \frac{n_0 e^{\frac{1}{2}|T-T_o|}}{10^3 \rho} b_1 \left(\frac{\rho q_i}{n_0 e^{\frac{1}{2}|T-T_o|}}\right)^{b_2}. \tag{17.3}$$

The in-situ aircraft observations show that mean effective sizes of ice crystal in ice clouds vary from 24 μm to 124 μm (e.g., Krueger et al. 1995), which leads to more realistic estimate of timescale for the growth of ice crystal from 40 to 100 μm. Thus, Krueger et al. (1995) also used (17.2) for P_{SFI} except that $\Delta t_1 = \left({m_{I100}}^{1-b_2} - {m_{I40}}^{1-b_2}\right)/b_1\,(1-b_2)$, where $m_{I100}\left(= 3.84 \times 10^{-9} kg\right)$ is the mass and of an ice crystal r_{I100} (=10^2 μm). Experiment K95 is identical to H80 except that K95 uses those modified schemes of P_{SFI} and P_{IDW} to calculate the depositional growth of cloud ice. Note that the large-scale forcing in K95 is identical to that in P of Chap. 9

Both Hsie and Krueger schemes assume that the mass of ice crystal is independent of the size of ice crystal. Zeng et al. (2008) proposed that the number of ice crystals varies linearly with the mass of ice crystal within the small range of size of ice crystal. Thus, P_{IDW} and P_{SFI} are respectively modified as

$$P_{IDW} = \frac{2}{(b_2+1)(b_2+2)}\left[3b_2 q_i + (1-b_2)\, m_{I50}\mu\rho^{-1}N_i\right] b_1 m_{I50}^{b_2-1}, \tag{17.4}$$

$$P_{SFI} = \max\left[2b_1\left(3q_i - m_{I50}\mu\rho^{-1}N_i\right) m_{I50}^{b_2-1}, 0\right], \tag{17.5}$$

where $N_i = n_0 e^{\beta(T_0-T)}$, β varies from 0.4 to 0.6, and n_o varies from 10^{-9} cm^{-3} to 10^{-6} cm^{-3} (Fletcher 1962); μ(=1.2) is the ice particle enhancement factor due to a riming-splintering mechanism (Hallet and Mossop 1974). Three experiments (Z08L, Z08ML, and Z08MH) use (4) and (5) to calculate the depositional growth of cloud ice but they use difference parameters (β and n_o) in N_i. $\beta = 0.4$ and $n_o = 10^{-9}$ cm^{-3} for low ice crystal concentration in Z08L, $\beta = 0.4$ and $n_o = 10^{-9}$ cm^{-3} for moderately low ice crystal concentration in Z08ML, and $\beta = 0.55$ and $n_o = 10^{-7}$ cm^{-3} for moderately high ice crystal concentration in Z08MH [see Table 1 for category of ice crystal concentration in Zeng et al. (2011)].

Note that Zeng's scheme calculates depositional growth of cloud ice using the radius of ice crystal from 0 to 50 μm. Hsie et al. (1980) and Krueger et al. (1995) used the mass of the ice crystal with the radius of 40 μm as the base mass in the calculation of depositional growth of cloud ice. Thus, P_{IDW} and P_{SFI} proposed by Zeng et al. (2008) may be modified by calculating depositional growth of cloud ice using the ice radius from 40 to 50 μm (Shen et al. 2014):

$$\begin{aligned} P_{SFI} &= \int_0^{m_{I50}} mn(m)\frac{dm}{dt}\,dm - \int_0^{m_{I40}} mn(m)\frac{dm}{dt}\,dm \\ &= 2b_1\left(3f_1 q_i - f_2 m_{I50}\mu\rho^{-1}N_i\right) m_{I50}^{b_2-1}, \end{aligned} \tag{17.6}$$

$$\begin{aligned} P_{IDW} &= \int_0^{m_{I50}} n(m)\frac{dm}{dt}dm - \int_0^{m_{I40}} n(m)\frac{dm}{dt}dm \\ &= \frac{2}{(b_2+1)(b_2+2)}\left[3b_2 f_1 q_i + (1-b_2) f_2 m_{I50}\mu\rho^{-1}N_i\right] b_1 m_{I50}^{b_2-1}. \end{aligned} \tag{17.7}$$

Here, $f_1 = 1-(1/2)^{b_2-1}$, $f_2 = 1-(1/2)^{b_2}$. In the derivation of (17.6) and (17.7), the following relations are used.

$$\frac{dm}{dt} = b_1 m^{b2}, \tag{17.8}$$

$$n(m)\,dm = (C_1 - C_2 m)\,dm, \tag{17.9}$$

$$m_{I40} = \frac{m_{I50}}{2}, \tag{17.10}$$

where

$$C_1 = \frac{6}{m_{I50}^2}\left[\frac{2}{3} m_{I50}\mu P^{-1} Ni - q_i\right], \tag{17.11a}$$

$$C_2 = \frac{12}{m_{I50}^2}\left[\frac{1}{2} m_{I50}\mu P^{-1} Ni - q_i\right]. \tag{17.11b}$$

Experiment Z08mh is identical to Z08MH but it uses (17.6) and (17.7) to calculate the depositional growth of cloud ice. Six experiments are summarized in Table 17.1.

P_{IDW} has similar magnitudes in six experiments but P_{SFI} ranges from 1.3×10^{-8} mm h^{-1} in H80 to 0.29 mm h^{-1} in Z08MH (Table 17.2). The modified schemes lead to the significant decreases in P_{IDW} and P_{SFI} from Z08MH to Z08mh. The vertical profiles of the total hydrometeor mixing ratio and mass-integrated mixing ratios of cloud water, rain, cloud ice, snow, and graupel averaged over 5 day and model domain are similar in H80, K95, Z08L, and Z08ML (Figs. 17.1 and 17.2). The total hydrometeor mixing ratios in the upper troposphere are much larger in Z08MH and Z08mh than those in the four other experiments. The mass-integrated mixing ratios of rain, cloud ice, snow, and graupel in Z08MH and cloud ice and snow in Z08mh are larger than those in the four other experiments. The modified scheme reduces maximum total hydrometeor mixing ratio from 1.2 gkg^{-1} at 10 km in Z08MH to 0.6 gkg^{-1} at 12 km in Z08mh and decreases mass-integrated mixing ratios of rain, cloud ice, snow, and graupel from Z08MH to Z08mh.

Table 17.1 A summary of experiments with parameterization schemes of depositional growth of cloud ice (P_{SFI} and P_{IDW})

Experiment	Hsie et al. (1980)	Krueger et al. (1995)	Zeng et al. (2008)	Modified Zeng scheme
H80	Yes	No	No	No
K95	No	Yes	No	No
Z08L	No	No	Yes: $\beta = 0.4$; $n_o = 10^{-9}$ cm^{-3}	No
Z08ML	No	No	Yes: $\beta = 0.5$; $n_o = 5 \times 10^{-9}$ cm^{-3}	No
Z08MH	No	No	Yes: $\beta = 0.55$; $n_o = 10^{-7}$ cm^{-3}	No
Z08mh	No	No	No	Yes: $\beta = 0.55$; $n_o = 10^{-7}$ cm^{-3}

β and n_o are parameters for formulation of ice crystal concentration from Fletcher (1962). L, ML, and MH (mh) denote low, moderately low, and moderately high ice crystal concentration in the scheme of Zeng et al. (2008), respectively. The difference between Z08MH and Z08mh is that Z08MH and Z08mh uses original and modified Zeng's schemes (After Shen et al. 2014)

Table 17.2 Mass-integrations of depositional growth of snow from cloud ice ($[P_{SFI}]$) and the growth of cloud ice by the deposition of cloud water ($[P_{IDW}]$) averaged for 5 days over model domain in six experiments

Experiment	$[P_{SFI}]$	$[P_{IDW}]$
H80	0.13	1.3×10^{-8}
K95	0.06	7.5×10^{-5}
Z08L	0.19	3.3×10^{-4}
Z08ML	0.19	1.1×10^{-3}
Z08MH	0.21	0.29
Z08mh	0.10	2.5×10^{-4}

Unit is mm h^{-1}

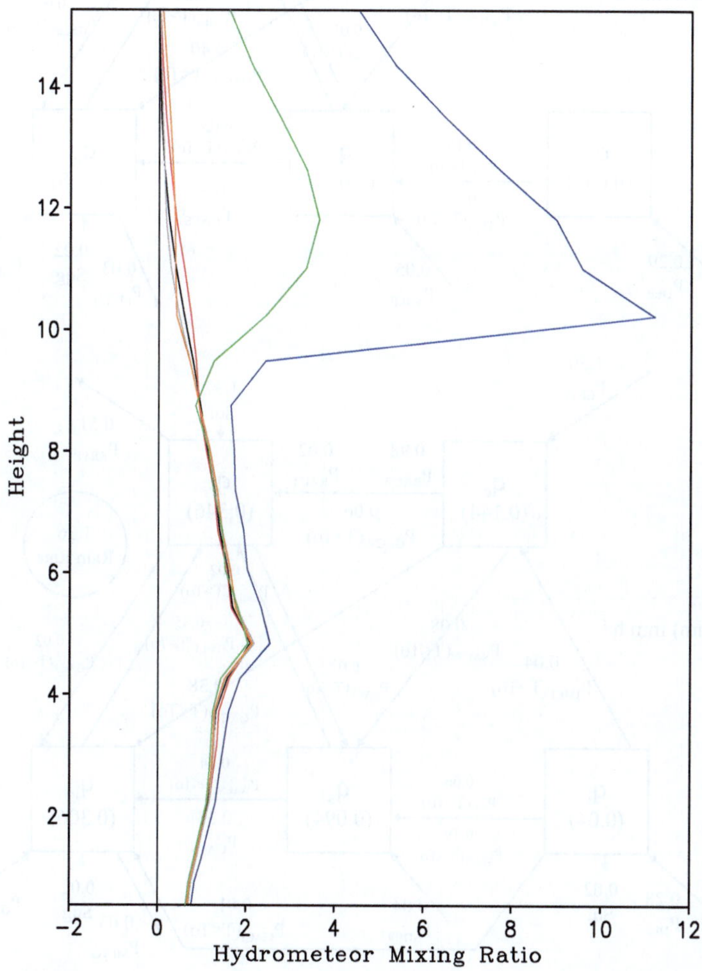

Fig. 17.1 Vertical profiles of hydrometeor mixing ratio (10^{-1} g kg^{-1}) averaged for 5 days over model domain in H80 (*black*), K95 (*red*), Z08L (*gray*), Z08ML (*orange*), Z08MH (*blue*), and Z08mh (*green*).

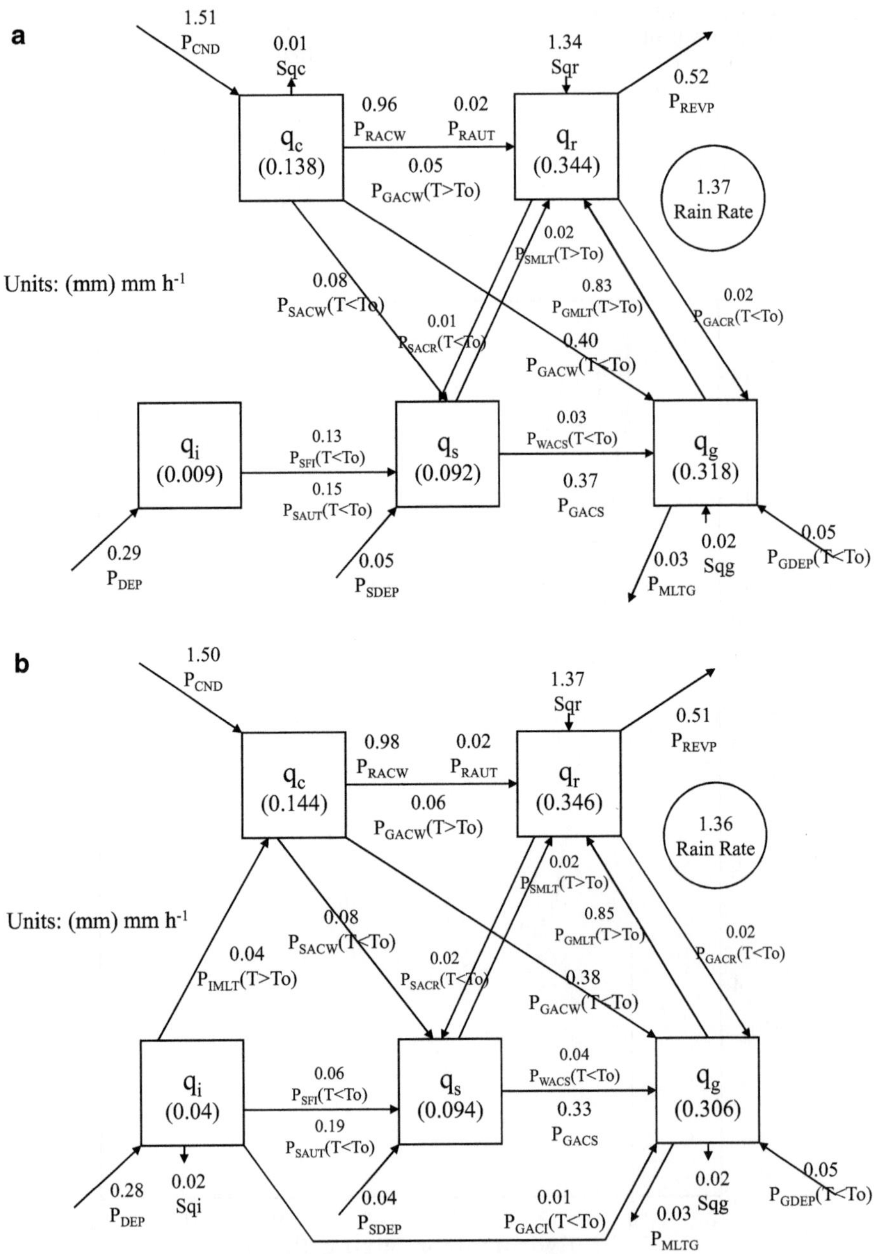

Fig. 17.2 Time mean cloud microphysics budgets in (**a**) H80, (**b**) K95, (**c**) Z08L, (**d**) Z08ML, (**e**) Z08MH, and (**f**) Z08mh. Units for cloud hydrometeors and conversions are mm and mm h^{-1}, respectively. Cloud microphysical conversion terms and their schemes can be found in Table 1.2. $T_0 = 0\ °C$

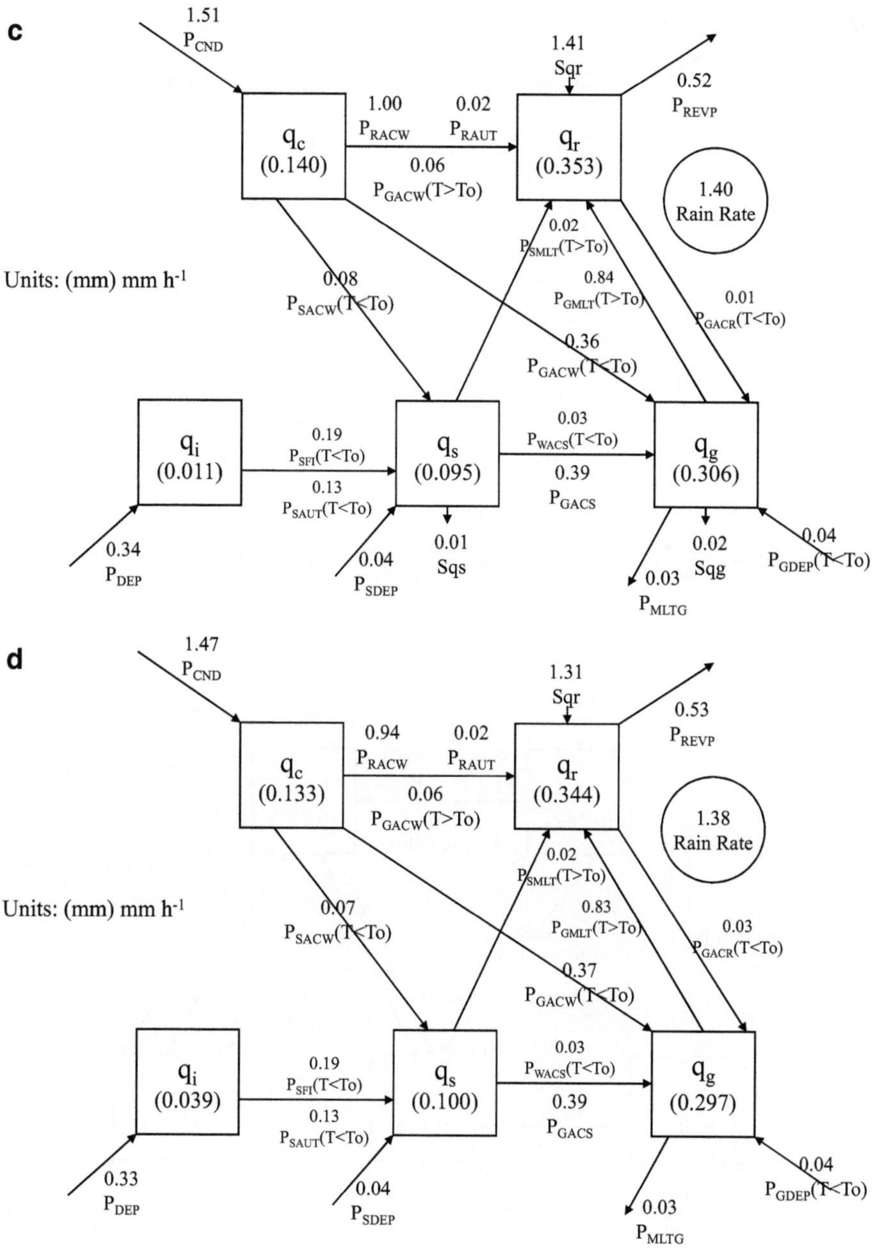

Fig. 17.2 (continued)

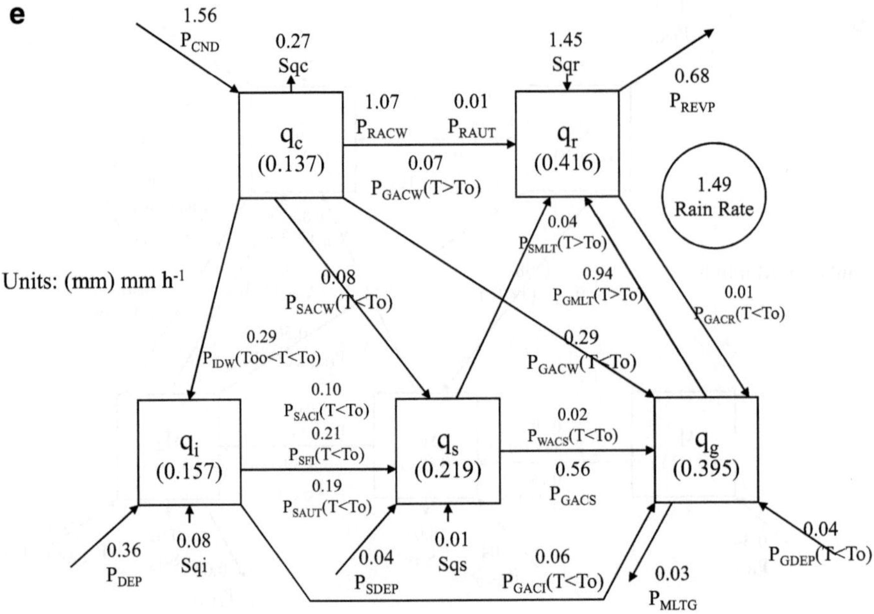

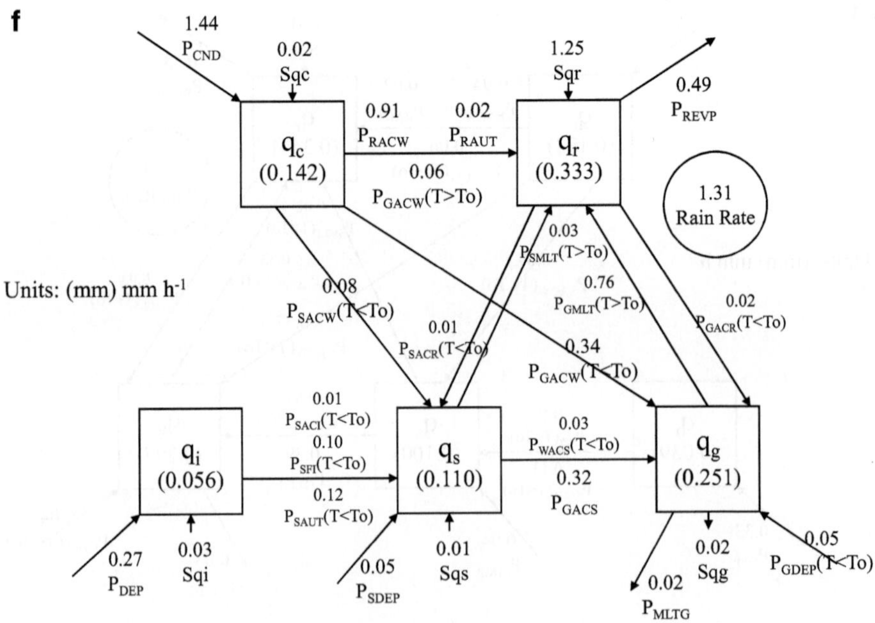

Fig. 17.2 (continued)

Table 17.3 (a) Cloud microphysical budgets (P_S, Q_{NC}, and Q_{CM}) averaged for 5 days over model domain in H80, K95K, Z08L, Z08ML, Z08MH, and Z08mh and (b) their differences for Z08mh -H80, Z08mh -K95K, Z08mh -Z08L, Z08mh -Z08ML, and Z08mh -Z08MH. The observed 5-day mean rain rate is 1.218 mm h^{-1}

(a)	H80	K95	Z08L	Z08ML	Z08MH	Z08mh
P_S	1.369	1.364	1.396	1.378	1.495	1.308
Q_{NC}	1.340	1.332	1.377	1.314	1.277	1.291
Q_{CM}	0.029	0.032	0.019	0.064	0.218	0.017

(b)	Z08mh -H80	Z08mh -K95	Z08mh -Z08L	Z08mh -Z08ML	Z08mh -Z08MH
P_S	−0.061	−0.056	−0.088	−0.070	−0.187
Q_{NC}	−0.049	−0.041	−0.086	−0.023	0.014
Q_{CM}	−0.012	−0.015	−0.002	−0.047	−0.201

Unit is mm h^{-1}

Table 17.4 (a) The net condensation (Q_{CM}) and its breakdown averaged for 5 days over model domain in Z08ML, Z08MH and Z08mh and (b) their differences for Z08mh -Z08ML and Z08mh -Z08MH

(a)	Z08ML	Z08MH	Z08mh
Q_{CM}	0.064	0.218	0.017
Q_{CMC}	−0.006	0.269	−0.016
Q_{CMR}	0.071	0.041	0.060
Q_{CMI}	−0.002	−0.082	−0.035
Q_{CMS}	0.001	−0.007	−0.009
Q_{CMG}	0.000	−0.003	0.017

(b)	Z08mh -Z08ML	Z08mh -Z08MH
Q_{CM}	−0.047	−0.201
Q_{CMC}	−0.010	−0.285
Q_{CMR}	−0.011	0.019
Q_{CMI}	−0.033	0.047
Q_{CMS}	−0.010	−0.002
Q_{CMG}	0.017	0.020

Unit is mm h^{-1}

Compared to observed mean rain rate (1.218 mm h^{-1}), H80, K95, Z08L, Z08ML, Z08MH, and Z08mh overestimate the mean rain rate by 12.4 %, 12.0 %, 14.6 %, 13.1 %, 22.7 %, and 7.4 %, respectively. To explain why the modified scheme improves the simulation of mean rain rate, the mean cloud microphysical budget is analyzed (Table 17.3). The modified scheme reduces the mean rain rate from Z08MH to Z08mh through the decrease in hydrometeor loss, while the vapor condensation (P_{CND}) barely changes. The decrease in hydrometeor loss corresponds to the change in cloud water from a significant loss in Z08MH to a slight gain in Z08mh (Table 17.4) through the severe reduction in P_{IDW} from Z08MH to Z08mh (Fig. 17.2e, f). The reduction in collection of cloud water by rain (P_{RACW}) from Z08MH to Z08mh corresponds mainly to the decrease in P_{CND}. The modified scheme also suppresses the cloud ice gain from Z08MH to Z08mh through the significant reduction in P_{IDW} from Z08MH to Z08mh although the collection of cloud ice by snow (P_{SACI}), P_{SFI} and conversion of cloud ice to snow (P_{SAUT}) decrease from Z08MH to Z08mh.

Z08mh shows the better rainfall simulation than H80, K95, and Z08L do because of the decreases in rain rate from H80, K95, and Z08L to Z08mh, which are mainly associated with the decrease in net condensation (Table 17.3) through the reductions in P_{CND} and vapor deposition to cloud ice (P_{DEP}) (Figs. 17.2a–c and f).

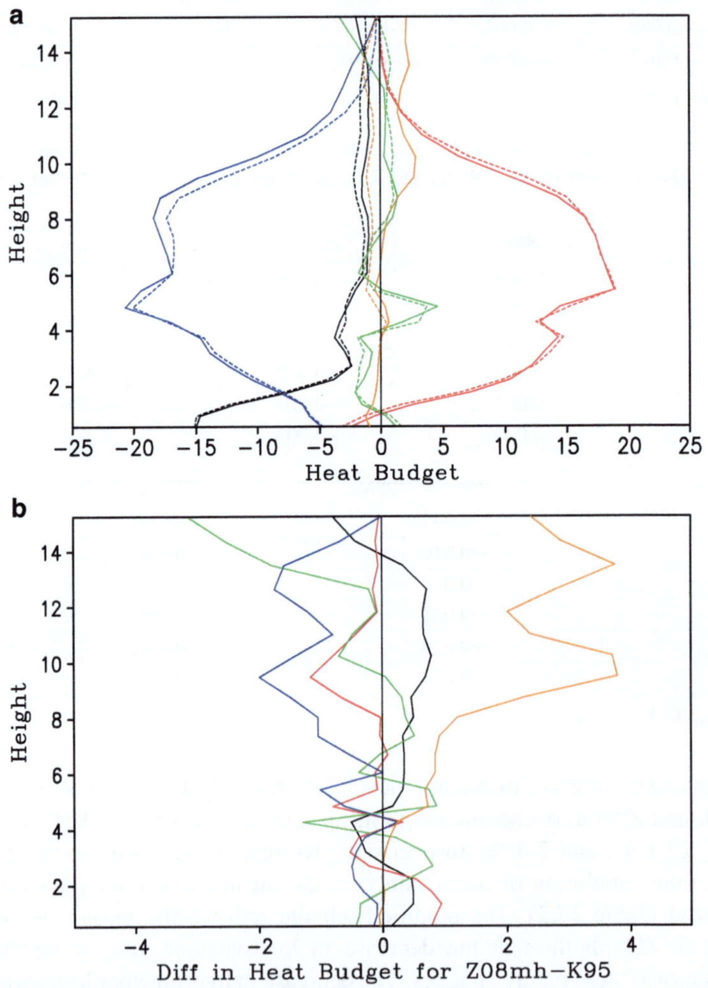

Fig. 17.3 Vertical profiles of **(a)** heat budgets [local temperature change (*black*), condensational heating (*red*), convergence of vertical heat flux (*green*), vertical temperature advection (blue), and radiation (*orange*)] averaged for 5 days over model domain in Z08mh (*solid*) and K95 (*dashed*) and **(b)** their differences for Z08mh-K95. Unit is °C d^{-1}

Table 17.5 Fractional coverage of raining stratiform (FC_{RS}), convective (FC_C), and non-raining stratiform (FC_{NRS}) regions averaged for 5 days over model domain in H80, K95K, Z08L, Z08ML, Z08MH, and Z08mh

	H80	K95	Z08L	Z08ML	Z08MH	Z08mh
FC_{RS}	23.59	24.49	25.50	24.18	50.87	26.30
FC_C	10.13	9.62	9.62	9.26	7.08	9.21
FC_{NRS}	40.56	43.90	43.01	57.04	42.06	62.52

Unit is %

The reductions in P_{CND} and P_{DEP} are associated with the suppressed release of latent heat around 8–12 km from H80, K95, and Z08L to Z08mh in responses to the change in infrared radiative tendency from cooling in H80, K95, and Z08L to warming in Z08mh in the upper troposphere (e.g., K95 versus Z08mh in Fig. 17.3). The change in infrared radiative tendency is related to the increase in cloud ice because the modified scheme in Z08mh increases cloud ice (Fig. 17.2) through the decrease in P_{SFI} and P_{SAUT} from H80 and Z08L to Z08mh and the increase in melting of cloud ice to cloud water (P_{IMLT}) from K95 to Z08mh.

Z08mh also simulates rainfall better than Z08ML does due to the decrease in rain rate from Z08ML to Z08mh, which is mainly associated with the reduction in hydrometeor loss (Table 17.3). The slowdown in hydrometeor loss from Z08ML to Z08mh is mainly related to the increase in cloud-ice gain (Table 17.4) through the decrease in P_{SFI} although P_{DEP} decreases from Z08ML to Z08mh (Fig. 17.2d, f).

The modified scheme significantly decreases and increases the fractional coverage of raining and non-raining stratiform regions, respectively from Z08MH to Z08mh (Table 17.5). The modified scheme barely changes the fractional coverage of stratiform rainfall in H08, K95, Z08L, Z08ML, and Z08mh, but it increases the fractional coverage of non-raining stratiform regions from H08, K95, Z08L, and Z08ML to Z08mh. The fractional coverage of convective rainfall (7–10 %) is much less sensitive to the parameterization scheme of depositional growth of ice crystal than that of stratiform regions (64–93 %).

References

Bergeron T (1935) On the physics of clouds and precipitation. Proc. 5th, Int Union of Geodesy Geophys 2:156–178

Findeisen W (1938) Kolloid-meteorologische Vorgänge bei Neiderschlags-bildung. Meteor Z 55:121–133

Fletcher HN (1962) The physics of rainclouds. Cambridge University Press, Cambridge, 386 pp

Hallet J, Mossop SC (1974) Production of secondary ice particles during the riming process. Nature 249:26–28

Hsie EY, Farley RD, Orville HD (1980) Numerical simulation of ice-phase convective cloud seeding. J Appl Meteor 19:950–977

Koenig LR (1971) Numerical modeling of ice deposition. J Atmos Sci 28:226–237

Krueger SK, Fu Q, Liou KN, Chin HNS (1995) Improvement of an ice-phase microphysics parameterization for use in numerical simulations of tropical convection. J Appl Meteor 34:281–287

Li X, Sui CH, Lau KM, Chou MD (1999) Large-scale forcing and cloud-radiation interaction in the tropical deep convective regime. J Atmos Sci 56:3028–3042

Shen X, Huang W, Qing T, Huang W, Li X (2014) A modified scheme that parameterizes depositional growth of ice crystal: a modeling study of pre-summer torrential rainfall case over southern China. Atmos Res 138:293–300. (c) Elsevier. Reprinted with permission

Wegener A (1911) Thermodynamik der Atmosphare. J. A. Barth, Leipzig

Zeng X, Tao WK, Lang S, Hou AY, Zhang M, Simpson J (2008) On the sensitivity of atmospheric ensembles to cloud microphysics in long-term cloud-resolving model simulations. J Meteor Soc Japan 86:45–65

Zeng X, Tao WK, Matsui T, Xie S, Lang S, Zhang M, Starr D, Li X (2011) Estimating the ice crystal enhancement factor in the tropics. J Atmos Sci 68:1424–1434

Chapter 18
Future Perspective of Cloud-Resolving Modeling

Abstract In this chapter, a set of cloud microphysical schemes are simplified based on the grid-scale data from the simulation of tropical rainfall event during TOGA COARE. The basic concept of cloud-resolving convection parameterization and its application to general circulation model simulations are discussed. Finally, cloud-resolving model is applied to the global domain for the improvement of precipitation simulations at global scale.

Keywords Cloud-resolving convection parameterization • Global cloud-resolving model • Scheme simplification

The simulations show that the performance of prognostic cloud microphysical parameterization schemes has the direct, crucial impacts on the simulations of cloud clusters in the genesis, evolution, propagation, and amplitudes. However, the computation of the full set of prognostic cloud microphysical equations is time consuming. Thus, simplified prognostic cloud microphysical parameterization schemes are discussed in this chapter. Possible replacement of cumulus parameterization in general circulation models with cloud-resoling model package and recent progresses of global cloud-resolving modeling are addressed.

18.1 Simplification of Cloud Microphysical Parameterization Schemes

Li et al. (2002) found from their analysis of cloud microphysical budgets in COARE1 that in the deep tropical convective regime, the magnitudes of 12 terms out of a total of 29 cloud microphysical processes are negligibly small. Thus, they proposed a simplified set of cloud microphysical equations (18.1), which saves 30–40 % of CPU time.

$$\begin{aligned} S_{q_c} = & -P_{SACW}\,(T < T_o) - P_{RAUT} - P_{RACW} \\ & -P_{GACW} + P_{CND}, \end{aligned} \tag{18.1a}$$

© Springer International Publishing Switzerland 2016

X. Li, S. Gao, *Cloud-Resolving Modeling of Convective Processes*,
Springer Atmospheric Sciences, DOI 10.1007/978-3-319-26360-1_18

$$S_{q_r} = P_{RAUT} + P_{RACW} + P_{GACW}(T > T_o) - P_{REVP} + P_{SMLT}(T > T_o) + P_{GMLT}(T > T_o), \quad (18.1b)$$

$$S_{q_i} = -P_{SAUT}(T < T_o) - P_{SACI}(T < T_o) - P_{SFI}(T < T_o) - P_{GACI}(T < T_o) + P_{DEP}, \quad (18.1c)$$

$$S_{q_s} = P_{SAUT}(T < T_o) + P_{SACI}(T < T_o) + P_{SFI}(T < T_o) - P_{GACS} + P_{SMLT}(T > T_o) - (1 - \delta_4) P_{WACS}(T < T_o) + P_{SACW}(T < T_o) + (1 - \delta_1) P_{SDEP}(T < T_o), \quad (18.1d)$$

$$S_{q_g} = P_{GACS} + (1 - \delta_4) P_{WACS}(T < T_o) - P_{GMLT}(T > T_o) - P_{MLTG}(T > T_o) + P_{GACI}(T < T_o) + P_{GACW}(T < T_o) - (1 - \delta_1) P_{GDEP}(T < T_o). \quad (18.1e)$$

The neglected terms in the simplified set include the accretion of cloud ice and snow by raindrops, the evaporation of melting snow, the accretion of cloud water and raindrops by snow, the accretion of raindrops and the homogeneous freezing of cloud water by cloud ice, the accretion and freezing of raindrops by graupel, the growth of cloud water by the melting of cloud ice, and the growth of cloud ice and snow by the deposition of cloud water. An experiment with the simplified set of cloud microphysical equations was conducted and compared to an experiment with the full set of cloud microphysical equations (COARE1) (see Sect. 3.1). Both experiments show similar time evolution and magnitudes of temperature and moisture profiles, surface rain rates including stratiform percentage and fractional coverage of convective, raining, and non-raining stratiform clouds (Fig. 18.1). This suggests that the original set of cloud microphysical equations could be replaced by the simplified set in simulations of tropical oceanic convection.

Li et al. (2005) proposed a prognostic cloud scheme that includes 18 cloud microphysical terms from 29 cloud microphysical terms included in the cloud-resolving model (Tao and Simpson 1993) based on the prognostic cloud scheme from Zhao and Carr (1997), which shows reasonable thermodynamic simulations. Combined with Li et al. (2002, 2005), Li and Weng (2004) proposed a new simplified prognostic cloud scheme.

$$S_{q_c} = -P_{RAUT} - P_{RACW} + P_{CND}, \quad (18.2a)$$

$$S_{q_r} = P_{RAUT} + P_{RACW} - P_{REVP} + P_{SMLT}(T > T_o) + P_{GMLT}(T > T_o) \quad (18.2b)$$

$$S_{q_i} = -P_{SAUT}(T < T_o) - P_{SACI}(T < T_o) - P_{SFI}(T < T_o) + P_{DEP}, \quad (18.2c)$$

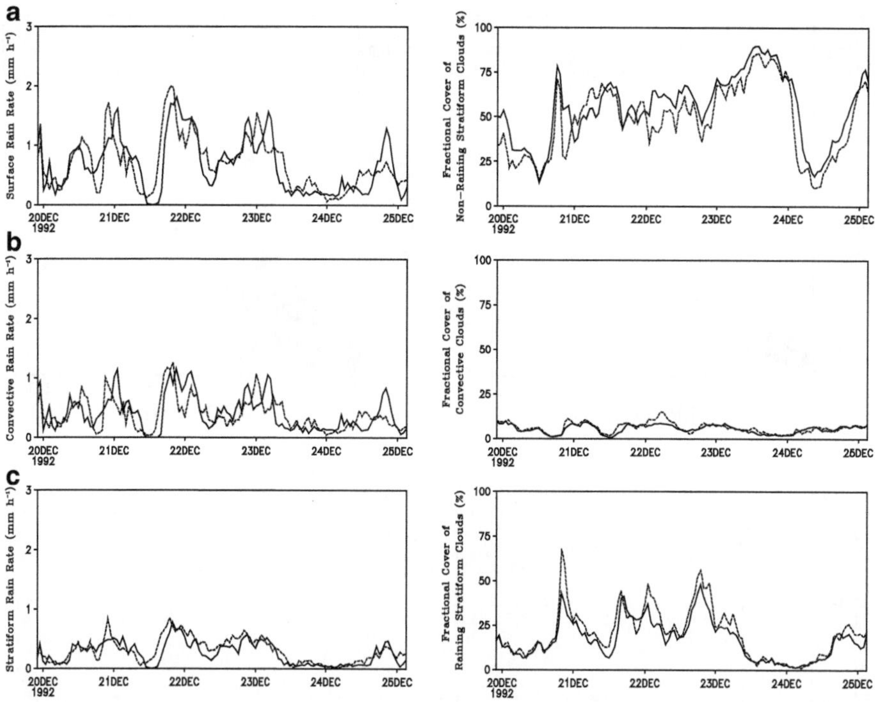

Fig. 18.1 *Left panel*: time series of (**a**) surface rain rate (mm h^{-1}), (**b**) convective rain rate, and (**c**) stratiform rain rate simulated with the simplified set of cloud microphysics equations (*solid*) and simulated in COARE1 (with the full set) (*dashed*); *Right panel*: time series of fractional coverage (%) of (**a**) non-raining stratiform regions, (**b**) convective regions, and (**c**) raining stratiform regions (After Li et al. 2002)

$$S_{q_s} = P_{SAUT}(T < T_o) + P_{SACI}(T < T_o) + P_{SFI}(T < T_o) \\ -P_{GACS} - P_{SMLT}(T > T_o) - (1 - \delta_4) P_{WACS}(T < T_o), \quad (18.2d)$$

$$S_{q_g} = P_{GACS} + (1 - \delta_4) P_{WACS}(T < T_o) - P_{GMLT}(T > T_o) \\ -P_{MLTG}(T > T_o) \quad (18.2e)$$

The experiments with both the full set of cloud microphysical equations and the simplified set of cloud equations (18.2) are carried out with the forcing calculated from NCEP/GDAS data averaged over the area of 150°–160°E and 0.5°S–0.5°N. The model is integrated from 0500 LST 18 April to 1100 LST 26 April 2003. The simulations are evaluated with the MSPPS *IWP* and *LWP* to examine the simplification of the cloud scheme. Note that roughly 6-hourly data from three satellites (NOAA-15, 16, 17) are used in the MSPPS data, whereas hourly simulation data are analyzed. The *IWP* (0.2 mm) is generally smaller than

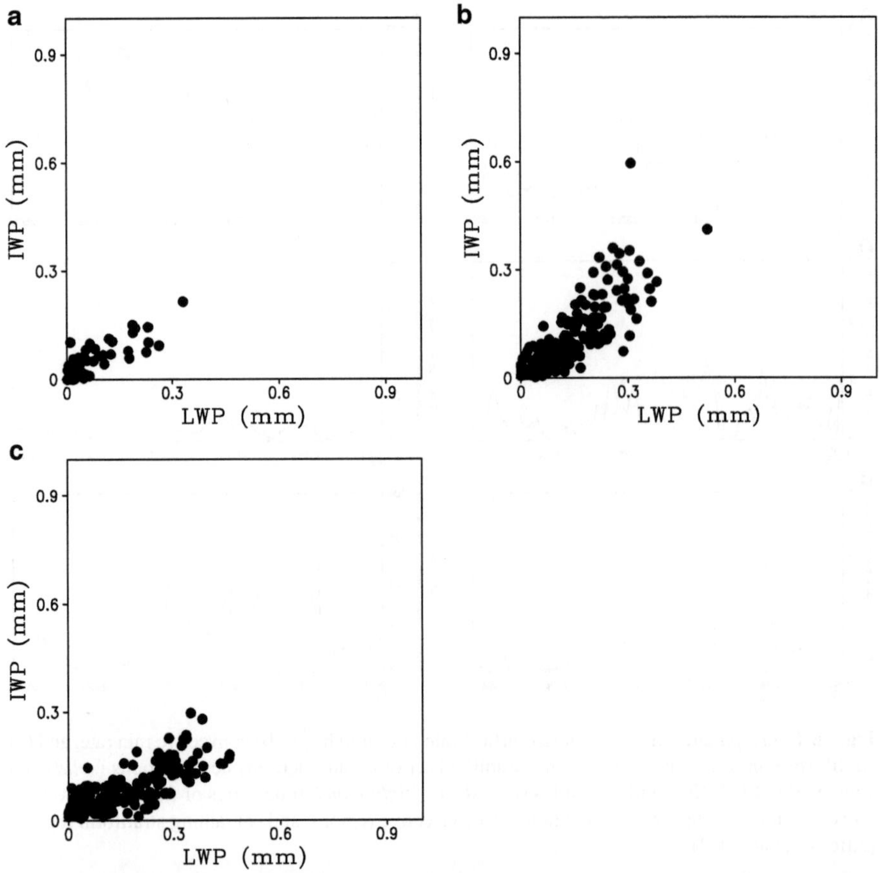

Fig. 18.2 (**a**) MSPPS *IWP* (mm) versus *LWP* (mm), (**b**) *IWP* versus *LWP* simulated by the full set of cloud equations, and (**c**) *IWP* versus *LWP* simulated by the simplified set of cloud equations. The imposed the forcing calculated from NCEP/GDAS data averaged over the area of 150°–160°E and 0.5°S–0.5°N from 0500 LST 18 April to 1100 LST 26 April 2003 (After Li and Weng 2004)

the *LWP* (0.3 mm) in the MSPPS data (Fig. 18.2a), whereas the *IWP* and *LWP* simulated with the full set of microphysical equations (Fig. 18.2b) have the same values of up to 0.4 mm. Thus, the magnitudes of *IWP* and *LWP* simulated with the full set of microphysical equations are 50 and 25 % larger than those observed in the MSPPS data, respectively, indicating that the full set of cloud microphysical equations produces relatively large cloud condensates. The *IWP* simulated with the simplified set of microphysical equations (0.3 mm Fig. 18.2c) becomes smaller than that simulated with the full set of microphysical equations (0.4 mm Fig. 18.2b) so that the *IWP* is smaller than the *LWP* in the simulation with the simplified set of microphysical equations and the ratio of the *IWP* to *LWP* is similar to that observed in the MSPPS data (Fig. 18.2a). This comparison shows that the simplified prognostic cloud scheme may produce better cloud simulations than the original one.

18.2 Cloud-Resoling Convection Parameterization

Grabowski and Smolarkiewicz (1999) and Grabowski (2001) applied a 2D cloud-resolving model into each column of a 3D large-scale model based on the assumption of scale separation between small-scale convection and large-scale circulation. The approach allows interaction between columns through the large-scale dynamics only and is referred to as Cloud-Resolving Convection Parameterization (CRCP). The comparison between the CRCP and cloud-resolving model simulation in a 2D nonrotating atmosphere with the presence of *SST* gradients shows that large-scale thermodynamic states and circulations are reasonably simulated in the CRCP, whereas the interaction between ascending and descending branches through the gravity wave mechanism and the organization of small-scale convection into mesoscale convective systems are poorly captured. The studies further show that the CRCP simulation produces the large-scale organization of convection within the equatorial waveguide. Khairoutdinov and Randall (2001) carried out a 2-month simulation with a 2D cloud-resolving model (CRM) installed in a realistic general circulation model (GCM), analyzed the results in January, and found that the cloud-resolving model (CRM) domain used in the combined GCM/CRM model produces good cloud statistics and that the model generates reasonable simulations, including January precipitation patterns in both extra-tropical storm tracks and the tropics, the Inter-Tropical Convergence Zone, the South Pacific Convergence Zone, as well as the South Atlantic Convergence Zone. Randall et al. (2003) performed an annual cycle simulation to simulate a vigorous Madden-Julian Oscillation including a slow propagation over the warm water over the Indian Ocean and western Pacific and a fast movement east of the date line. Khairoutdinov et al. (2005) conducted a 500-day simulation using the NCAR Community Atmosphere Model (CAM) with superparameterization and compared it with the simulation using the standard CAM and found that the CAM with superparameterization simulated between mean states in terms of precipitation, *PW*, top-of-atmosphere radiative fluxes, cloud radiaitve forcing, and fractional cover for high clouds in both winter and summer seasons and that the model also produced more realistic intraseasonal variability. However, the disadvantages of CRCP are arbitrary dimension configuration for cloud-resolving models and artificial scale separation between the CRM and the GCM. The Community Climate System Model (CCSM) with the CRCP is used to correctly simulate the northward propagation of convection, precipitation, and circulation as well as the eastward and westward propagations of the intraseasonal oscillations over South Asia (Khairoutdinov et al. 2014) and to realistically simulate the maximum precipitation over the continent and easterly waves associated with the West African monsoon (McMrary et al. 2014a, b).

18.3 Global Cloud-Resolving Model

Grabowski (1998) conducted a 2D cloud-resolving model simulation with a zonally varied *SST* over the large domain of 4000 km. The maximum *SST* at the center of the domain is 28 °C and the amplitude of the *SST* is 4 °C. Grabowski integrated the model into quasi-equilibrium states in 60 days. The clouds occur around the center of the domain where the *SST* is relatively high, which resembles an ascending branch of the Walker circulation. The subsidence over the cold water causes dry air, which forms a descending branch of the Walker circulation. The explicitly resolved convection shows quasi-2-day oscillations that are similar to those observed by Takayabu et al. (1996).

Satoh et al. (2005) and Tomita et al. (2005) applied the cloud-resolving model into global domain to conduct a global cloud-resolving model simulation by developing an icosahedral grid technique. The new model is referred to as Nonhydrostatic Icosahedral Atmospheric Model (NICAM). The model includes a simple two-category cloud microphysics scheme with the ice effects proposed by Grabowski (1998), a radiation parameterization scheme developed by Nakajima et al. (2000), and the Meller-Yamada level-2 subgrid-scale turbulence closure and surface flux scheme by Louis et al. (1982). Satoh et al. started a 60-day integration with a horizontal resolution of 14 km, then integrated the model for another 30 days with a grid mesh of 7 km, and finally used the data on the twentieth day of 7 km run as the initial conditions to integrate another 10 days with a horizontal resolution of 3.5 km. The analysis of 10-day simulation data in 3.5 km run shows realistic cloud simulations including multi-scale cloud structures, Intertropical Convergence Zone (ITCZ), Madden and Julian Oscillation (MJO)-like wave propagation, and diurnal rainfall variations. Miura et al. (2005) conducted an additional experiment with the increase of 2 K for *SST* in Tomita et al. (2005). Comparison between the two experiments shows that the temperature and relative humidity is 4 °C warmer but 15 % smaller in the experiment with +2 K than in Tomita et al.'s experiment in the tropics. Cloud amount and albedo are larger in the experiment with +2 K than in Tomita et al.'s experiment near the equator and in high latitudes. The results are encouraging. As the computational power increases, the global cloud-resolving modeling will bring realistic cloud simulations and be beneficial to a better understanding of cloud-radiation interaction and associated impacts on climate variability.

References

Grabowski WW (1998) Toward cloud resolving modeling of large-scale tropical circulations: a simple cloud microphysics parameterization. J Atmos Sci 55:3283–3298

Grabowski WW (2001) Coupling cloud processes with the large-scale dynamics using the cloud-resolving convection parameterization (CRCP). J Atmos Sci 58:978–997

Grabowski WW, Smolarkiewicz PK (1999) CRCP: a cloudresolving convection parameterization for modeling the tropical convecting atmosphere. Physica D 133:171–178

Khairoutdinov M, Randall DA (2001) A cloud resolving model as a cloud parameterization in the NCAR Community Climate Model: Preliminary results. Geophys Res Lett 28:3617–3620

Khairoutdinov M, Randall DA, DeMott C (2005) Simulations of the atmospheric general circulations using a cloud-resolving model as a superparameterization of physical processes. J Atmos Sci 62:2136–2154

Khairoutdinov M, Stan C, Randall DA, Shukla RP, Kinter JL III (2014) Simulation of the South Asian monsoon in a coupled model with an embedded cloud-resolving model. J Clim 27:1121–1142

Li X, Weng F (2004) An operational cloud verification system and its application to validate cloud simulations in the operational models, 13th conference on satellite meteorology and oceanography, Norfolk, Virginia, 20–24 September 2004

Li X, Sui CH, Lau KM (2002) Dominant cloud microphysical processes in a tropical oceanic convective system: a 2-D cloud resolving modeling study. Mon Weather Rev 130:2481–2491

Li X, Sui CH, Lau KM, Tao WK (2005) Tropical convective responses to microphysical and radiative processes: a 2D cloud-resolving modeling study. Meteorog Atmos Phys 90:245–259

Louis JF, Tiedke FM, Geleyn JF (1982) A short history of the PBL paramterization at ECMWF. Workshop on planetary boundary layer parameterization. ECMWF, Reading

McMrary RR, Randall DA, Stan C (2014a) Simulations of the West African monsoon with a superparameterizaed climate model. Part I: the seasonal cycle. J Clim 27:8303–8322

McMrary RR, Randall DA, Stan C (2014b) Simulations of the West African monsoon with a superparameterizaed climate model. Part II: African easterly waves. J Clim 27:8323–8341

Miura H, Tomita H, Nasuno T, Iga S, Satoh M (2005) A climate sensitivity test using a global cloud resolving model under an aqua planet condition. Geophys Res Lett 32. doi:10.1029/2005GL023672

Nakajima T, Tsukamoto M, Tsushima Y, Numaguti A, Kimura T (2000) Modeling of the radiative process in an atmospheric general circulation model. Appl Opt 39:4869–4878

Randall DA, Khairoutdinov M, Arakawa A, Grabowski WW (2003) Breaking the cloud parameterization deadlock. Bull Am Meteorol Soc 84:1547–1564

Satoh M, Tomita H, Miura H, Iga S, Nasumo T (2005) Development of a global resolving model – a multi-scale structure of tropical convections. J Earth Sim 3:1–9

Takayabu YN, Lau KM, Sui CH (1996) Observation of a quasi-2-day wave during TOGA COARE. Mon Weather Rev 124:1892–1913

Tao WK, Simpson J (1993) The Goddard cumulus ensemble model. Part I: model description. Terr Atmos Oceanic Sci 4:35–72

Tomita H, Miura H, Iga S, Nasuno T, Satoh M (2005) A global cloud-resolving simulation: Preliminary results from an aqua planetary experiment. Geophys Res Lett 32. doi:10.1029/2005GL022459

Zhao Q, Carr FH (1997) A prognostic cloud scheme for operational NWP models. Mon Weather Rev 125:1931–1953

Acronyms

1D	One-dimensional
2D	Two-dimensional
3D	Three-dimensional
AMSU	Advanced MSU
ARM	Atmospheric Radiation Measurement
CAM	Community Atmosphere Model
CAPE	Convective available potential energy
CCM2	Community Climate Model 2
CEM	Cumulus Ensemble Model
CFAD	Contoured frequency by altitude diagram
COARE	Coupled Ocean–atmosphere Response Experiment
COR	Correlation
CMPE	Cloud-microphysics precipitation efficiency
CR	Cloud ratio
CRCP	Cloud-Resolving Convection Parameterization
CRM	Cloud-resolving model
CVV	Convective vorticity vector
DVV	Dynamic vorticity vector
EQ	Equator
ERBE	Earth Radiation Budget Experiment
GARP	Global Atmospheric Research Program
GATE	GARP Atlantic Tropical Experiment
GCE	Goddard cumulus ensemble
GCM	General circulation model
GDAS	Global Data Assimilation System
GSFC	Goddard Space Flight Center
GMS	Geostationary meteorological satellite
H	Helicity
IFA	Intensive Flux Array

© Springer International Publishing Switzerland 2016

X. Li, S. Gao, *Cloud-Resolving Modeling of Convective Processes*,
Springer Atmospheric Sciences, DOI 10.1007/978-3-319-26360-1

IOP	Intensive Observing Period
IR	Infrared
ITCZ	Intertropical Convergence Zone
IWP	Ice Water Path
KWAJEX	Kwajalein Experiment
LFC	Level of free convection
LSPE	Large-scale precipitation efficiency
LST	Local standard time
LWP	Liquid Water Path
MJO	Madden-Julian Oscillation
MSPPS	Microwave Surface and Precipitation Products System
MSU	Microwave Sounding Unit
MV	Moist vorticity
MVV	Moist vorticity vector
NASA	National Aeronautics and Space Administration
NCAR	National Center for Atmospheric Research
NCEP	National Centers for Environmental Prediction
NESDIS	National Environmental Satellite, Data, and Information Service
NICAM	Nonhydrostatic Icosahedral Atmospheric Model
NOAA	National Oceanic and Atmospheric Administration
PSU	Practical salinity units
PV	Potential vorticity
PW	Precipitable water
RMS	Root-mean-square
SCS	South China Sea
SCSMEX	SCS Monsoon Experiment
SST	Sea surface temperature
TOGA	Tropical Ocean Global Atmosphere
TMI	TRMM Microwave Imager
TRMM	Tropical Rainfall Measuring Mission
UV	Ultraviolet
VIS	Visible

Index

© Springer International Publishing Switzerland 2016
X. Li, S. Gao, *Cloud-Resolving Modeling of Convective Processes*,
Springer Atmospheric Sciences, DOI 10.1007/978-3-319-26360-1

D

MIX
Papier aus verantwortungsvollen Quellen
Paper from responsible sources
FSC® C105338

If you have any concerns about our products, you can contact us on
ProductSafety@springernature.com

In case Publisher is established outside the EU, the EU authorized representative is:
Springer Nature Customer Service Center GmbH
Europaplatz 3, 69115 Heidelberg, Germany

Printed by Libri Plureos GmbH
in Hamburg, Germany